EARTH SCIENCE
A Comprehensive Study

THE PHYSICAL SETTING

Peggy Lomaga
Longwood Junior High School (retired)
Middle Island, New York

Amy Schneider
Longwood Junior High School
Middle Island, New York

EduTech, Inc., Albany, NY 12203

Editor:

William M. Marrs, Earth Science Teacher
Smithtown High School East
St. James, NY

Acknowledgments:
Special Recognition for their contributions is accorded to:

Brecka Coonradt - Jay, NY
Cover Photo: Milky Way was taken over Mount Colvin from Indian Head, in Keene, NY
Mount Colvin is the 39th highest peak in the High Peaks Region of the Adirondack Mountains, in New York State.

Kevin Lenhart - Saranac Lake
Cover Photo: Sunset was taken on the summit of Hurricane Mountain, in Keene, NY
Hurricane Mountain is a 3,694-foot mountain near Keene in the north of the High Peaks region of the Adirondacks in New York State.

©2018 by EduTech, Inc., 407 Krumkill Road, Albany, NY 12203

All Rights reserved. No part of this material protected by this copyright notice may be reproduced or utilized in any form or by any means, electronic or mechanical, including photocopying and recording, or by any information and retrieval system without written permission from the copyright owner.

Printed in the United States of America

ISBN: 978-0-692-76733-7

1 2 3 4 5 6 7 8 9 0

TABLE OF CONTENTS

Chapter 1: Scientific Inquiry and Analysis 1
Observation, Measurement, Density, Graphing

Chapter 2: Astronomy 13
The Universe, Stars

Chapter 3: The Solar System 25
Solar System, Orbits, Earth's Moon

Chapter 4: Planet Earth and Earth Motions 41
Earth's Size and Shape, Earth Motions, Latitude-Longitude, Time Zones

Chapter 5: Seasons and Insolation 57
Solar Time, Seasonal Sun Path, Insolation

Chapter 6: Properties of the Atmosphere 79
Origin and Function, Structure of the Atmosphere, Environmental Issues, Energy Transfer, Phase Changes, Atmospheric Variables

Chapter 7: Weather 103
Weather Station Models, Air Masses and Fronts, Weather Maps, Severe Weather

Chapter 8: Water Cycle and Climate 125
Water Cycle, Soil Water Movement, Factors that Control Climate, Climate Change

Chapter 9: Surface Processes 147
Weathering, Erosion, Deposition

Chapter 10: Landscapes and Mapping 171
Landscapes, Classification and Development of Landscapes, Contour Mapping

Chapter 11: Natural Resources, Minerals, and Rocks 189
Earth's Natural Resources, Minerals, Rock Cycle, Types of Rocks

Chapter 12: Earth's Interior and the Dynamic Crust 209
Earth's Interior, Crustal Movements, Plate Tectonics, Volcanoes, Earthquakes

Chapter 13: Earth's Geologic History 233
Earth's History, Geologic Time Line and Fossils, Interpreting Geologic History

Chapter 14: The Earth Science Reference Table (ESRT) 255

Chapter 15: Earth Science Skills 281
Metric Measurements, Math Skills, Eccentricity, Locating Epicenters, Classification of Rocks, Properties of Minerals, Mapping

ESRT 301

Glossary 317

Index 327

Practice Regents Exams 329

CHAPTER 1

SCIENTIFIC INQUIRY AND ANALYSIS
Observation, Measurement, Density, Graphing

OBSERVATIONS AND INFERENCES

An **observation** is a description of an object or event. We make observations by using the five senses. For example, "this rock is glassy" is an observation using the sense of sight. We extend our senses by using instruments. A triple beam balance enables us to measure the mass of an object; a telescope allows us to observe objects that are very far away.

An **inference** is a conclusion, opinion, or explanation of what was observed. A **prediction** is an inference about the future. For example, a weather forecast is an inference based on observations and measurements of atmospheric variables.

Scientists often organize their observations in a meaningful way. Earth scientists have developed many **classification systems**. Some of these include the classification of clouds, hurricanes, earthquakes, rock types, sediments, and stars.

Questions

A student is studying different mineral samples and records the following statements in her notebook. Indicate whether the statement is an observation (O) or an inference (I).

1. This mineral feels greasy. _____
2. This mineral is olivine. _____
3. Volcanic activity formed this mineral. _____
4. This mineral has a green-black streak. _____
5. The hardness of this mineral is 3.5. _____
6. This mineral has a shiny luster. _____
7. This mineral has a mass of 14.8 g. _____
8. This mineral will weather easily. _____
9. This mineral formed in water. _____
10. This mineral breaks unevenly. _____

MEASUREMENTS

A **measurement** adds accuracy and precision to observations. All measurements include a number and a unit of measurement; for example: 523.0 grams, 49.2 centimeters.

When the value of the measurement is a very large or a very small number **scientific notation** is used. For example, the average distance of Neptune from the Sun is 4,500,000,000 kilometers. This value can be written as 4.5×10^9 km.

Scientific measurements are usually rounded to the nearest tenth, unless you are instructed to round to a different place value or to the nearest significant digit. For example, if your answer is 0.0092, you should round to the nearest thousandth (0.009) not to the nearest tenth otherwise your answer would be zero.

In science, the **metric (SI) system** is used. This system uses common prefixes with each unit of measurement. Three common prefixes used in Earth Science are "*kilo*" (1000), "*centi*" (1/100), and "*milli*" (1/1000). A distance of 25.0 kilometers is equivalent to 25,000 meters. A measurement of 25.0 centimeters is equivalent to 0.25 meters.

The chart below describes some of the common measurements used in Earth Science.

Measurement	Definition	Instrument	Unit(s)
Time	interval between events	stop watch	second (s)
Temperature	average kinetic energy of molecules	thermometer	degrees (°F, °C, K)
Length	distance between two points	ruler	meter (m)
Mass	amount of matter in an object	triple beam balance	gram (g)
Volume	amount of space an object occupies	graduated cylinder or ruler	milliliter (mL) or cubic centimeters (cm³)

Questions

11. Identify the measurement indicated by each value given.

    a. 12.6 kg: _____   c. 7.3 km: _____

    b. 3000 K: _____   d. 50.5 mL: _____

12. Round each of these numbers to the nearest tenth. *(Remember when rounding to the tenth to ONLY look at the hundredth place to determine whether to round up or down).*

    a. 7.58 = _____   b. 13.845 = _____   c. 73.692 = _____   d. 0.83 = _____

13. Write each number in scientific notation.

    a. 5,600,000,000 = _____   b. 0.000075 = _____

14. Write the number represented.

    a. 4.63×10^5 = _____   b. 2.53×10^{-4} = _____

15. a. The equatorial diameter of Earth is _____ *(be sure to include the unit)*.

    b. Express this value in scientific notation, rounded to the tenth place. _____

16. a. Express the half-life of Potassium-40 in scientific notation. _____

    b. Express the half-life of Potassium-40 in standard form. _____

17. a. The Cretaceous Period began _____ million years ago.

    b. Write this number in standard form. _____ years ago

    c. Write this number in scientific notation, rounded to the tenth place. _____ years ago

Density

Density is the concentration of mass in a given volume. It is calculated by dividing the mass of an object by its volume (mass ÷ volume). **Mass** is the amount of matter present in an object. Mass is measured with a triple beam balance in grams. **Volume** is the amount of space an object occupies. Volume is calculated as l × w × h for a regular-shaped object or measured using a graduated cylinder for liquids or irregular-shaped objects.

Every material has its own unique density by which it can be identified. The density of a material will not change with size or shape. A gold coin, a gold ring, and a gold nugget each have the same density of 19.3 g/cm³.

The density of different objects can be compared by flotation. If an object is less dense than the liquid it is in, it will float; if it is more dense, it will sink. Ice floats because it is less dense than liquid water. Water is denser as a liquid than as a solid.

The density of a material will change if impurities are added, the phase (solid, liquid, gas) is changed, temperature changes, or if the pressure exerted on the object changes. Gases are less dense than liquids and solids. Solids are usually denser than liquids, except for water (ice floats). If an object is under pressure, its molecules will compress and move closer together, so the object will become denser. As temperature increases, molecules of the substance move further apart so the density decreases. Hot air rises because hot air is less dense than cold air. Cold air and cold water will sink because the molecules are closer together resulting in an increase in density. Density explains many Earth phenomena such as cloud formation and tectonic plate movements.

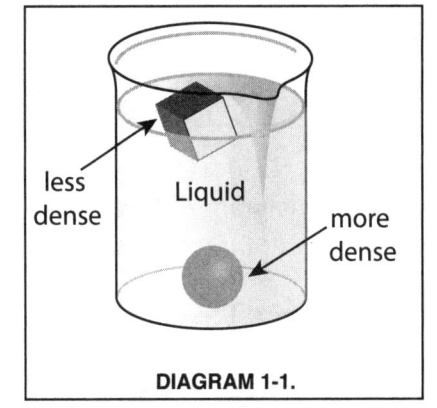

DIAGRAM 1-1.

Questions

18. For each statement indicate whether the density will *(decrease) (increase) (remain the same)*

    a. A larger amount of water is used. _____

    b. The air temperature decreases. _____

    c. The pressure in Earth's interior increases. _____

    d. Atmospheric air pressure decreases. _____

    e. The mineral you are using breaks in half. _____

    f. Liquid water freezes. _____

    g. Liquid water begins to boil. _____

    h. Limestone is metamorphosed into marble. _____

19. The diagram below illustrates convection currents in Earth's mantle layer.

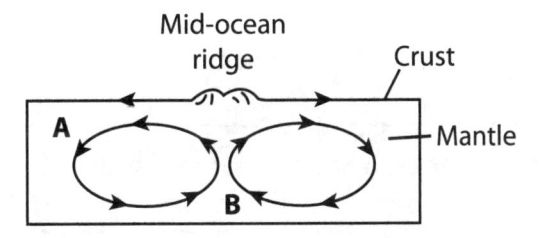

    a. Which letter indicates a location of less density? _____

    b. Which letter indicates a location of more density? _____

20. Each planet has a unique density. Use the ESRT, page 15.

    a. The densest planet is _____.

    b. Two planets that have the same density are _____ and _____.

    c. The planet that could float on water is _____.

    d. Compared to Earth, the Moon is *(less) (more)* dense.

    e. The inner terrestrial planets are *(less) (more)* dense than the outer Jovian planets.

21. In which layer of Earth would you find a density of approximately 4.8 g/cm^3? _____

22. Calculate to the nearest tenth the density of a piece of steel with a mass of 83.6 grams and a volume of 10.7 cm³. *Show solution.*

23. This diagram illustrates a mineral sample with a mass of 213.0 grams and a volume of 28.0 cm³. A chart of the densities of various minerals is also given. Refer to ESRT, page 16.

    Mineral Sample A

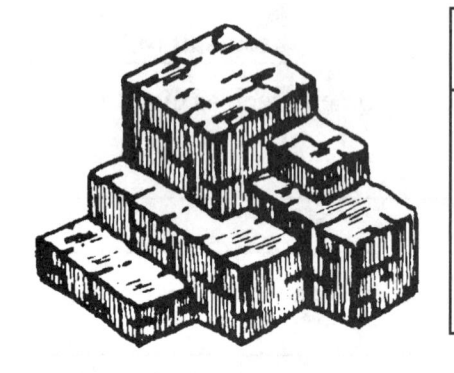

    Mineral Density Table

Mineral	Density (g/cm³)	Mineral	Density (g/cm³)
Gypsum	2.3	Hornblende	3.2
Orthoclase	2.6	Chalcopyrite	4.2
Quartz	2.7	Pyrite	5.0
Calcite	2.7	Magnetite	5.2
Dolomite	2.9	Galena	7.6
Fluorite	3.2	Copper	8.9

    a. How was this mineral's mass determined? _____

    b. How could this mineral's volume be determined? _____
    _____

    c. Identify the name of this mineral. _____
    Explain how you determined the name of this mineral. *Show solution.*

    d. What would happen to the density of this mineral if a piece was broken off and lost?
    _____

    e. The minerals quartz and calcite both have the same density. How could you tell them apart?
    _____
    _____

    f. A student determined the density of various pieces of magnetite. What conclusion should the student realize? _____

    g. A piece of copper is heated and expands. How will this change its density? _____
    _____
    Why will this occur? _____
    _____

4

GRAPHING

Graphs show the relationships between measured variables and the rate at which those variables change.

When graphing data it is important to follow these rules:

1. Label with a name and a unit the **X** (manipulated/independent) axis and the **Y** (responding/dependent) axis.
2. Number each axis. The data collected must fit along each axis of the graph. The axes must be numbered with equal intervals such as by 2 (2,4,6,8...) or by 5 (5,10,15,20......). The same interval does not need to be used for both axes.
3. Plot each data point clearly and neatly.
4. Connect only the plotted data points with a best-fit line or point-to-point line.
5. Write a title for the graph. Be sure the **X** and **Y** variables are mentioned in the title by name. Do *NOT* write: "*X vs. Y*" as a title. A better title would be:
    "*The effect of (X variable by name) on (Y variable by name)*".

Graphs illustrate **rate of change** by the slope of the line. A steep sloped line indicates a rapid rate of change; a gentle sloped line represents a slow rate of change. Rate of change can be calculated as: $\frac{\text{change in value}}{\text{time}}$

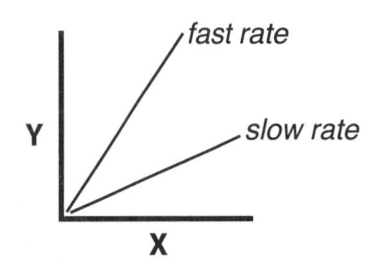

Graphs illustrate the relationship or pattern between the measured variables "X" and "Y". These graph relationships can be classified as direct, indirect, constant, or cyclic.

Type of Graph	Variable Relationship	Graph Appearance	An Example
Direct	As X increases, Y increases.		As pressure increases, density increases.
Indirect	As X increases, Y decreases or as X decreases, Y increases.		As air temperature increases, density of air decreases.
Constant	As X increases, Y stays the same. (there is no change in one variable)		As size of a material increases, density stays the same.
Cyclic	As X increases, Y increases and decreases in a pattern that is repetitive and predictable.		During the day, the height of the ocean tide increases and decreases.

Questions

24. A student measures and records the temperature of the water in a beaker for 8 minutes as shown in the data chart below.

       Start   Finish

Time	0 min	1 min	2 min	3 min	4 min	5 min	6 min	7 min	8 min
Temperature	90°C	83°C	78°C	73°C	68°C	64°C	60°C	57°C	54°C

   a. Construct a line graph of the student's data.

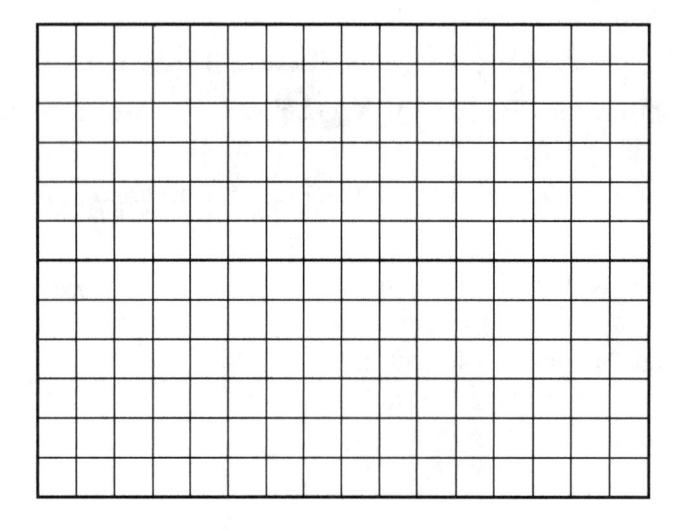

   b. State the relationship between time and temperature. _____

_____

   c. Calculate the average rate of temperature change for the water during the 8 minutes.
      Show solution, include equation used, substitution, and final answer rounded to tenth with units.

25. The diagrams show four different solid materials **A**, **B**, **C**, and **D**. Some of their physical dimensions are shown. The graph shows the relationship between the mass and volume of materials **B** and **C**.

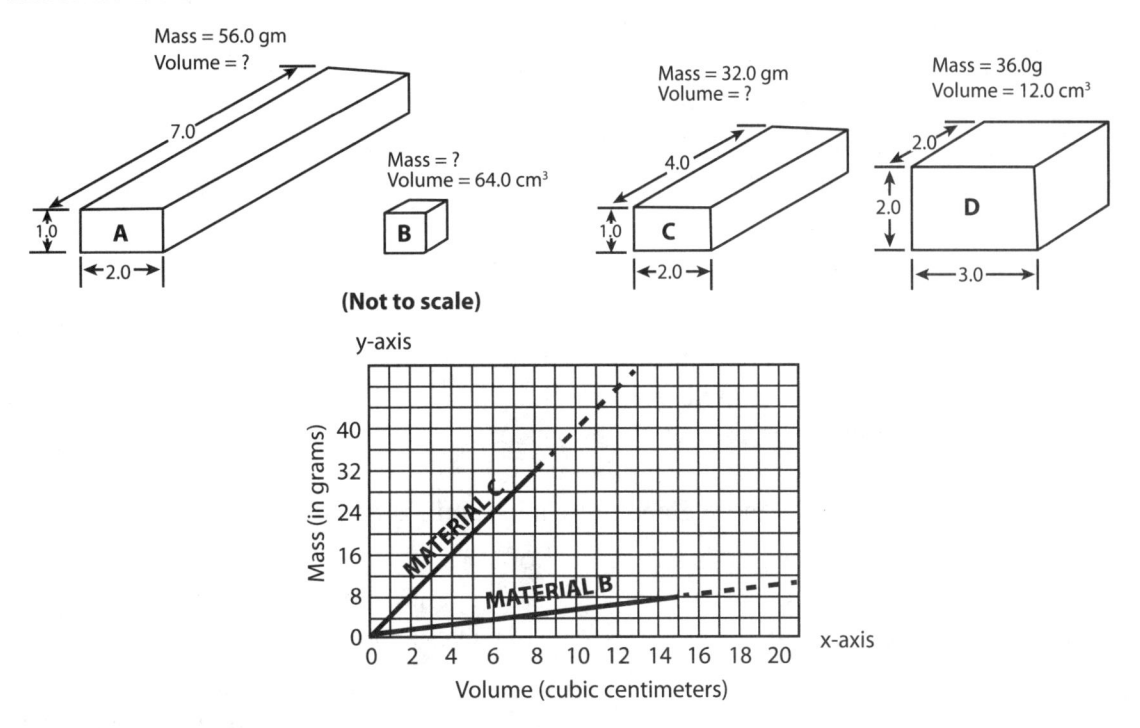

(Not to scale)

a. If material **C** were cut in half, what would happen to the slope of the line? _____

   _____

   Why? _____

b. What would be the mass of 6.0 cm³ of material **C** ? _____

c. Calculate the density of material **B**. *(show solution)*

d. On the graph, sketch the graph line for material **D**. Label the line.

   Explain why you positioned the line for **D** in this way. _____

   _____

   _____

e. If these four solids were exposed to extreme cold, how would their densities be

   affected ? _____

   Why? _____

   _____

CHAPTER 1 REVIEW

1. Which statement about a major hurricane is an inference?
   (1) The windspeed is 200 km/hr.
   (2) The central air pressure is 946.0 mb.
   (3) A rain gauge recorded three inches of rain in the past hour.
   (4) Damage from this storm is expected to be extensive.

2. A student classifies several rock samples into three categories: igneous, sedimentary, and metamorphic. This student's classification is based on
   (1) inferences (2) interpretations (3) observations (4) predictions

3. Which statement illustrates a classification system?
   (1) The rate at which a glacier melts each year.
   (2) The depth of the ocean across the Atlantic Ocean.
   (3) The snowfall predictions for this coming winter.
   (4) The grouping of stars by their temperature and luminosity.

4. The rising and setting of the Sun is an example of a
   (1) noncyclic, predictable event   (3) cyclic, unpredictable event
   (2) noncyclic, unpredictable event   (4) cyclic, predictable event

5. An inference is a
   (1) description (2) measurement (3) fact (4) possible explanation

6. A centimeter is 0.01 meters. This measurement can be expressed as
   (1) 1.0×10^{-1} m (2) 1.0×10^{0} m (3) 1.0×10^{-2} m (4) 1.0×10^{2} m

7. The equatorial diameter of Venus is approximately 12,000 kilometers. This distance in scientific notation is written as:
   (1) 1.2×10^{2} km (2) 1.2×10^{3} km (3) 1.2×10^{4} km (4) 1.2×10^{-4} km

8-9. The diagrams below show several measuring tools used in Earth Science.

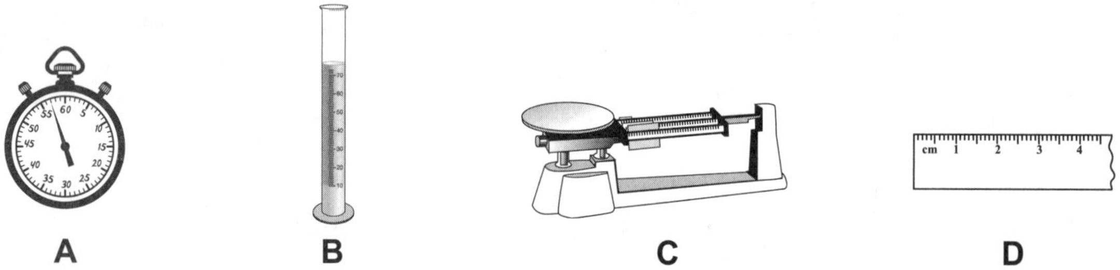

A   B   C   D

8. Which would be used to measure the volume of a pebble?
   (1) A (2) B (3) C (4) D

9. Which tools would be used to find the density of a block of wood?
   (1) A and D (2) B and D (3) C and B (4) C and D

10. The diagram below represents a solid object with a mass of 120.0 grams. What is the density of the object?
    (1) 0.5 g/cm³
    (2) 2.0 g/cm³
    (3) 6.0 g/cm³
    (4) 10.0 g/cm³

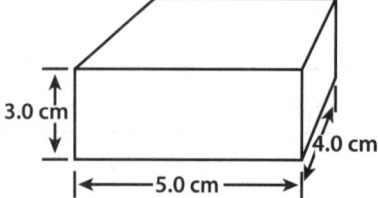

11. The graph to the right shows the relationship between the mass and volume of a mineral. What is its density?

    (1) 3.0 g/cm³
    (2) 4.5 g/cm³
    (3) 6.0 g/cm³
    (4) 9.0 g/cm³

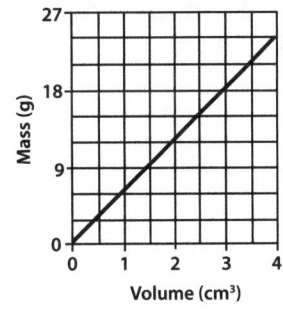

12. Substances **A**, **B**, **C**, and **D** are placed in a container of water. Which substance is the least dense?

    (1) A
    (2) B
    (3) C
    (4) D

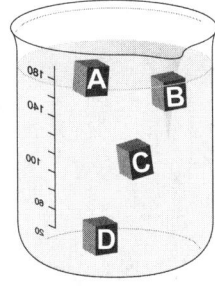

13-14. The graph shows the heating of four different materials.

13. Which material heated at the slowest rate?

    (1) iron
    (2) basalt
    (3) dry air
    (4) liquid water

14. The approximate rate at which the basalt heated was:

    (1) 5.0°C/min
    (2) 15.0°C/min
    (3) 20.0°C/min
    (4) 30.0°C/min

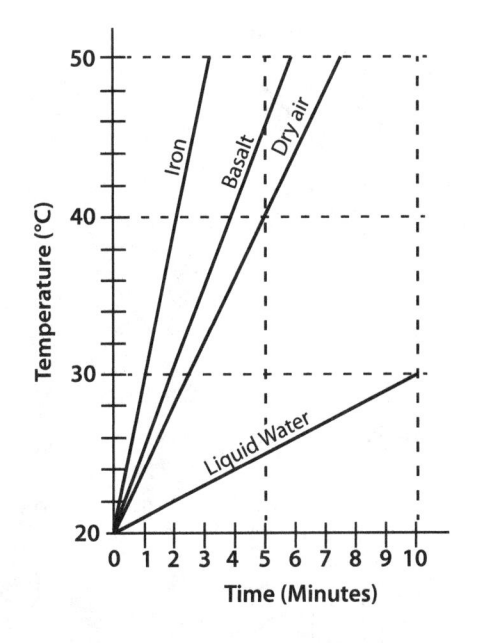

15. The diagram to the right shows an empty 1,000 mL container with a mass of 250.0 grams.
    When a liquid is placed in the container, the container and the liquid have a combined mass of 1300.0 grams.

    What is the density of the liquid?

    (1) 1.00 g/mL
    (2) 1.05 g/mL
    (3) 1.30 g/mL
    (4) 0.95 g/mL

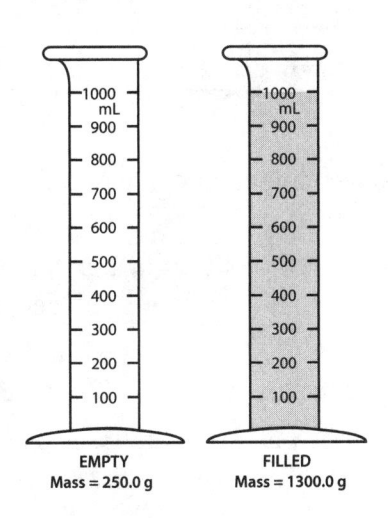

16. The diagram shows a container with four different liquids **W, X, Y,** and **Z**. A piece of solid quartz with a density of 2.7 g/cm³ is placed on the surface of liquid **W** and released. The quartz will pass through

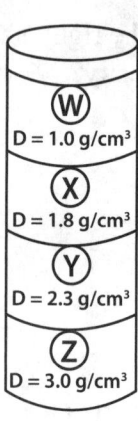

    (1) W but not X, Y, or Z
    (2) W and X, but not Y or Z
    (3) W, X, and Y, but not Z
    (4) W, X, Y and Z

    W D = 1.0 g/cm³
    X D = 1.8 g/cm³
    Y D = 2.3 g/cm³
    Z D = 3.0 g/cm³

17. Which bar graph best represents the average densities of the Sun, Moon, and Earth?

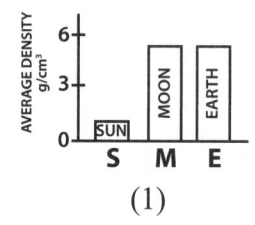

    (1) 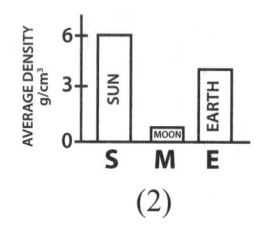 (2) 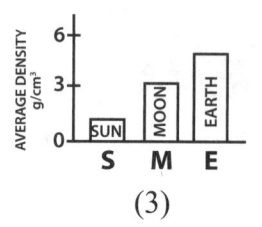 (3) 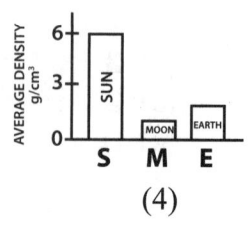 (4)

18. Which graph best represents the relationship between the density of five different pieces of iron each having different volumes?

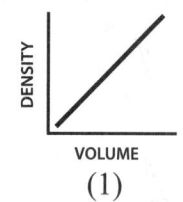

    (1) 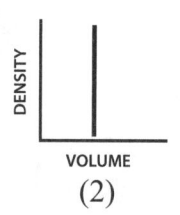 (2) 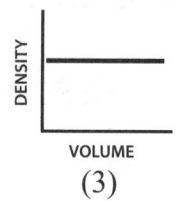 (3) 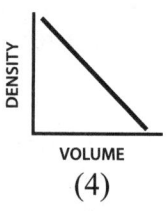 (4)

19. As water cools from 4°C to 0°C, its density will
    (1) decrease   (2) increase   (3) remain the same

20. Which bar graph best represents the relationship between the different phases of matter excluding water?

    [Key: S = solid, L = liquid, G = gas]

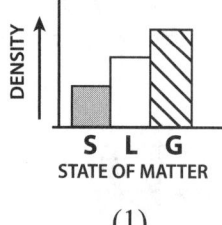

    (1) 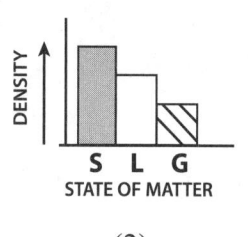 (2) 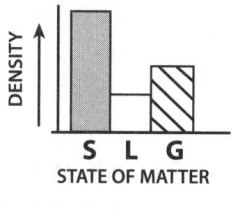 (3) 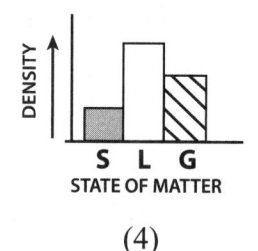 (4)

21. As the temperature of a parcel of air decreases, its density will
    (1) decrease   (2) increase   (3) remain the same

22. Compared to the air in a **cP** air mass, the air in a **mT** air mass is generally
    (1) less dense   (2) denser   (3) same density

23. As pressure on a parcel of air decreases, the air density will
    (1) decrease   (2) increase   (3) remain the same

24. As a rock is metamorphosed by extreme pressure, the rock's density will
    (1) decrease   (2) increase   (3) remain the same

25. Which correctly lists the planets in increasing density?
    (1) Mars, Neptune, Uranus, Saturn
    (2) Uranus, Mars, Mercury, Neptune
    (3) Jupiter, Neptune, Venus, Mercury
    (4) Earth, Neptune, Mars, Mercury

26-29. The diagrams below show four solid objects made of the same material. Mass and volume measurements are given for all except the sphere and the bar.

(Not drawn to scale)

Cube
Mass 81.0 g
Volume 27.0 cm³

Sphere
Mass 75.0 g
Volume ?

Bar
5.0 cm
2.0 cm
3.0 cm
Mass 90.0 g
Volume ?

Cylinder
Mass 60.0 g
Volume 20.0 cm³

26. Calculate the density of the **cube**. *Show solution.* _____

27. State how the densities of these four materials compare and explain why. _____
_____
_____

28. When the sphere is heated, its mass will _____, its volume will _____, and its density will _____.

29. Calculate the volume of the **bar**. *Show solution and explain.*

11

30-35. The graph below illustrates the volumes and masses of four different solids: **A**, **B**, **C**, and **D**.

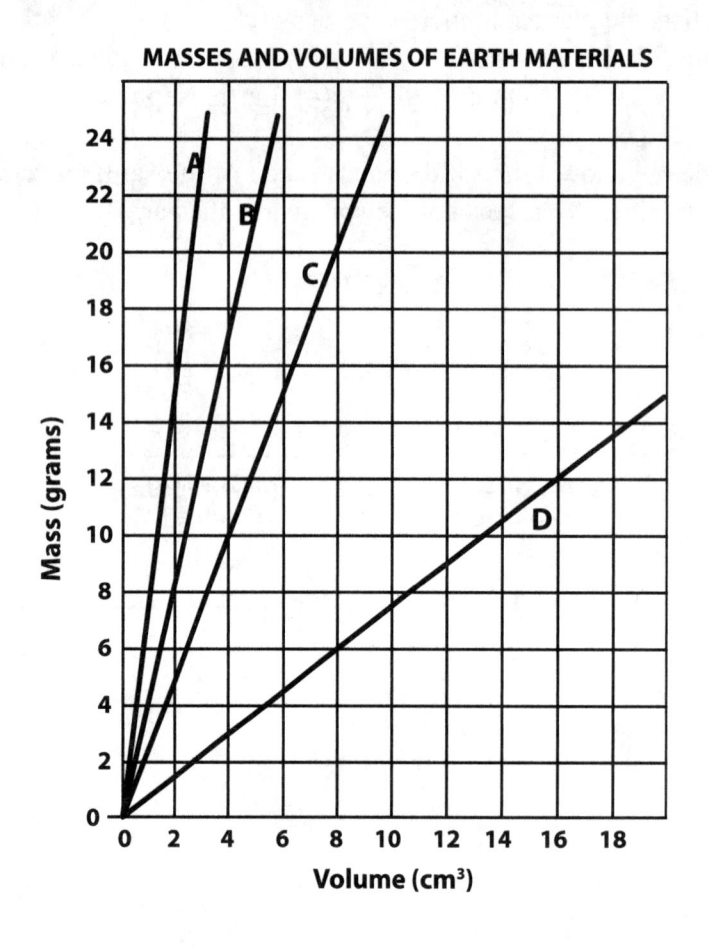

MASSES AND VOLUMES OF EARTH MATERIALS

Mass (grams) vs *Volume (cm³)*

30. A 12.0 g piece of **B** has a volume of _____ cm³.

31. As the volume of each material increased, the mass *(decreased)* *(increased)* *(remained the same)*.

32. Which material has the greatest density? _____

    How do you know? _____

    _____

33. Calculate the density of material **D** *(show solution in the space below)*

34. If material **A** was cut in half, its density would *(decrease)* *(increase)* *(remain the same)*.

35. If these four solids were exposed to extreme pressure, the graph lines would become *(less steep)* *(steeper)* *(unchanged)*.

12

CHAPTER 2

ASTRONOMY
The Universe, Stars

THE UNIVERSE

The **universe** is composed of empty space, matter, and energy. Most of the universe is empty space where no matter exists. The matter in space includes subatomic particles, gases, dust, rock debris, solid bodies, and stars. The most common element in the universe is hydrogen. The **electromagnetic spectrum** classifies all the energies of the universe by wavelength. These energies travel at the speed of light through the vacuum of empty space.

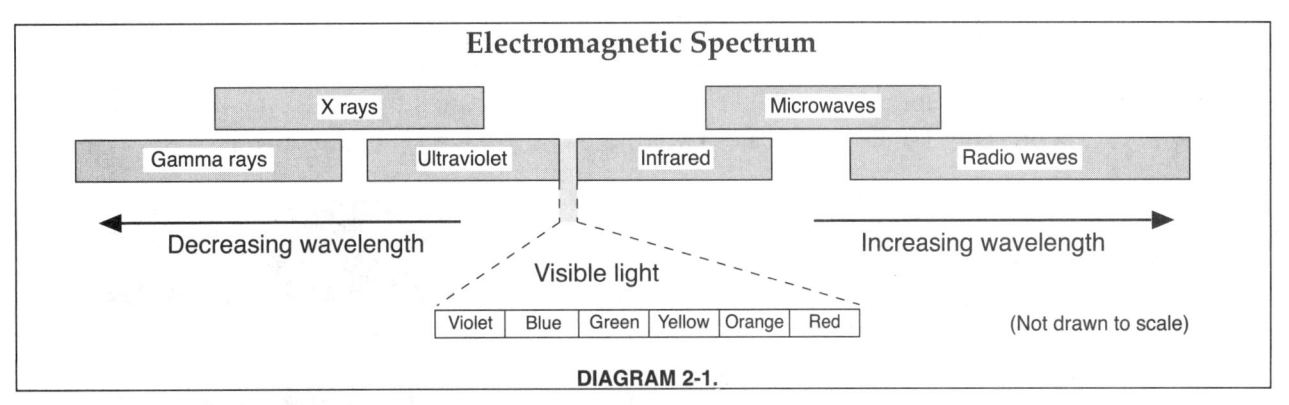

Electromagnetic Spectrum

Gamma rays | X rays | Ultraviolet | Visible light | Infrared | Microwaves | Radio waves

← Decreasing wavelength   Increasing wavelength →

Visible light: Violet | Blue | Green | Yellow | Orange | Red

(Not drawn to scale)

DIAGRAM 2-1.

The universe is estimated to be 10 to 14 billion years old. The **Big Bang Theory** explains that initially all matter and energy were concentrated as a hot, dense mass. A sudden "big bang" expansion occurred that resulted in matter moving away from each other. Since that time the universe continues to expand.

There are two evidences for the Big Bang Theory. One evidence for the "big bang" event is the detection in space of cosmic background radiation caused by this massive expansion. A second evidence is that the light from all distant stars is shifted towards the red end of the visible light spectrum. This is known as the **doppler red shift**. Light waves spread apart and shift towards the red end of the spectrum as stars move away from each other in the expanding universe.

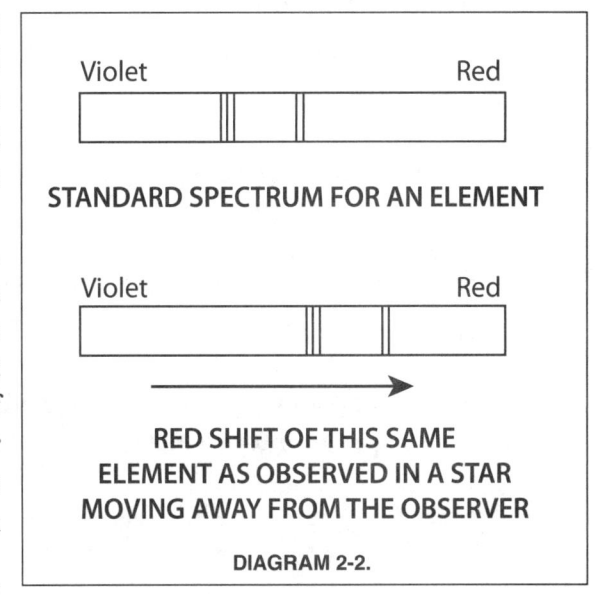

STANDARD SPECTRUM FOR AN ELEMENT

RED SHIFT OF THIS SAME ELEMENT AS OBSERVED IN A STAR MOVING AWAY FROM THE OBSERVER

DIAGRAM 2-2.

After the "big bang" event, the universe cooled and matter began to organize into subatomic particles (protons, electrons, and neutrons). These particles formed the elements of hydrogen, helium, and lithium. The most common element formed was hydrogen. These elements form stars where the heavier elements are made.

Questions

1. Use the diagram of the "*Electromagnetic Spectrum*" on page 14 of the ESRT

    a. The energy with the longest wavelength is _____.

    b. The energy with the shortest wavelength is _____.

    c. Heat energy is also known as _____.

    d. Some wavelengths of microwaves are similar to _____ and _____.

    e. List these energies in order from longest to shortest wavelength:

    visible light, x-ray, ultraviolet, radio

    _____ , _____ , _____ , _____

2. The diagram below represents the development of our universe from the time of the "big bang" event to the present. Letter **A** indicates two of the many galaxies in the universe.

    a. The total "time" represented by this diagram is estimated to be between _____ and _____ billion years.

    b. This diagram shows that the universe is (*contracting*) (*expanding*) in size.

    c. Name the element that is most abundant in the galaxies at letter **A**. _____

    d. State the *two* evidences for the "Big Bang" expansion.

    _____ and

    _____.

3. The diagram below represent the standard dark-line spectrum for an element.

    Violet   Red

    a. Sketch the spectrum as it would appear for a star moving away from Earth. Be sure to include all the spectral lines with the same spacing as shown above.

    Violet   Red

    b. The sketch you drew above is known as the doppler _____ and is evidence that the universe is _____.

STARS

Stars are spherical, luminous masses of gases, mainly hydrogen. They produce energy by **nuclear fusion**. Fusion occurs in a star's very hot core where hydrogen protons combine to form helium and energy.

Stars vary in color and luminosity. The color of a star is caused by its temperature. Cool stars (2000 K to 3000 K) are red; the hottest stars (~20,000 K) are blue. A star's **luminosity** is the amount of energy given off compared to the Sun. Generally the larger the star, the more luminous it is. Dwarf stars are dim; giant stars are bright.

Stars are classified based on temperature and luminosity. The "*Characteristics of Stars*" diagram illustrates the classification system for stars. Ninety percent of stars are **main sequence** stars. Main sequence stars increase in luminosity as their size and temperature increases. Red dwarfs are small, cool, and dim; blue giants are large, hot, and bright. The cool **red supergiants** and **giants** are bright because of their huge size. The very hot **white dwarfs** are dim because of their very small size.

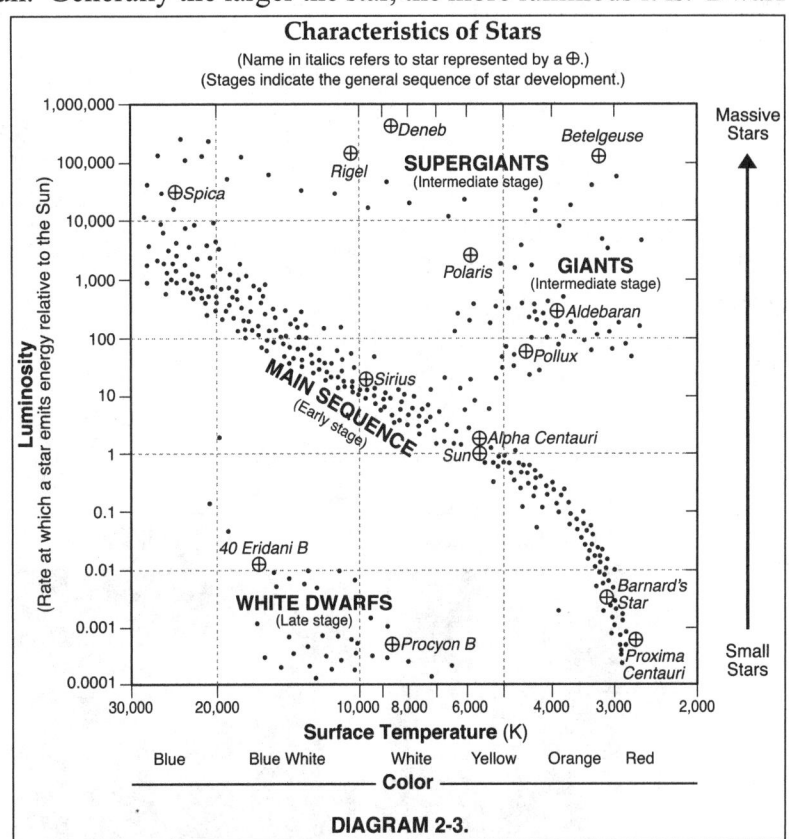

A star forms within a cloud of dust and gas called a **nebula**. The mass of dust and gas in the nebula begins to contract due to the force of gravity. This makes the nebula denser and hotter. The nebula will first become a **protostar**. When the center of the protostar becomes dense and hot enough for fusion of hydrogen to begin, a star is "born." The star will spend most of its life as a **main sequence** star. After millions to billions of years the star will expand to become a **red giant** or **supergiant**. A very massive giant star will eventually **supernova** (explode) and then become an object so dense that no light can escape called a **black hole**. A low mass red dwarf star will collapse to form a **white dwarf** which is extremely hot and dense. Over time the white dwarf will cool to become a black dwarf. Hotter stars will go through this life cycle more quickly because they use their hydrogen "fuel" faster.

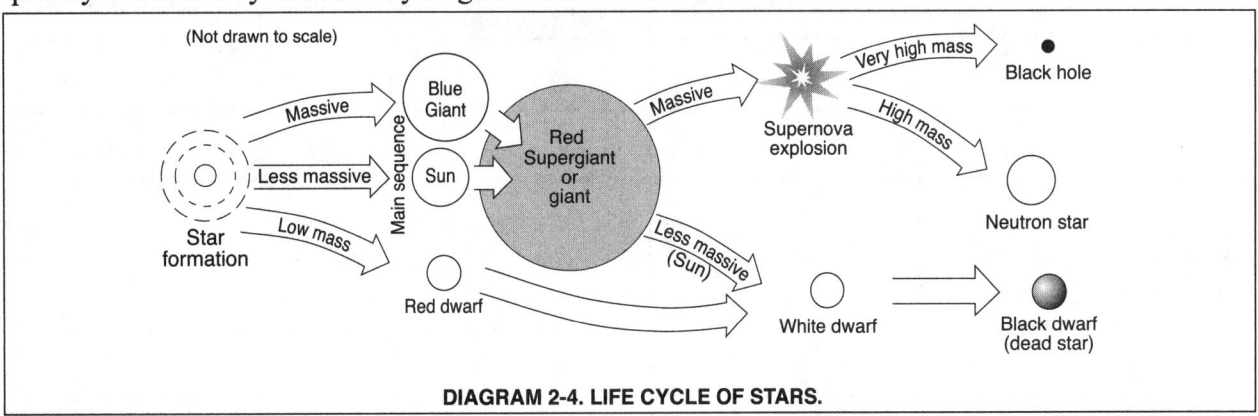

DIAGRAM 2-3.

DIAGRAM 2-4. LIFE CYCLE OF STARS.

The **Sun** is an average main sequence star. It is the star nearest to Earth. The Sun formed approximately 4.6 billion years ago and is near the center of our solar system. The Sun's surface temperature is about 6,000 K. Some of the features of the Sun include solar prominences and flares which can generate a **solar wind** of charged particles which travel through the solar system. In our upper atmosphere these charged particles will cause brilliant **auroras** and can disrupt communications. **Sunspots** are dark, cooler areas on the Sun's surface. The number of sunspots vary in a cyclic pattern which repeats every 10 to 11 years.

Galaxies are huge systems of billions of stars. Galaxies are classified according to their shape. Irregular-shaped galaxies are rare and contains the youngest stars. Spiral galaxies are disc-shaped with extending "arms." The Sun and solar system are located along an arm of the spiral-shaped **Milky Way Galaxy**. Elliptical-shaped galaxies are the most common and contain the oldest stars.

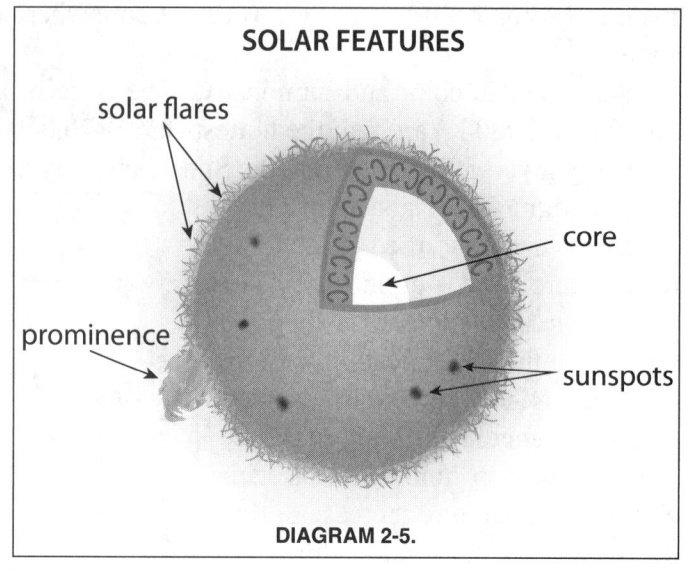

SOLAR FEATURES

solar flares, core, prominence, sunspots

DIAGRAM 2-5.

Irregular galaxy | Spiral galaxy | Elliptical galaxy

DIAGRAM 2-6.

Questions

4. Complete this chart based on the *"Characteristics of Stars"* diagram on page 15 of the ESRT.

star name	temperature (K)	luminosity	stage in life cycle	color	star type
Sun		1			
Rigel					supergiant
40 Eridani B			late stage		
Proxima Centauri	2750 K				
Spica				blue	
Sirius					

5. Place an "X" in the correct box to indicate the temperature and luminosity for each star as compared to the Sun.

Stars	Temperature Hotter	Temperature Cooler	Luminosity Brighter	Luminosity Dimmer
Procyon B				
Barnard's Star				
Rigel				

6. The diagram below is a simple version of the "*Characteristics of Stars*" diagram in the ESRT. Write the letter(s) from the diagram that apply to the statement.

   a. The hottest stars. _____
   b. Red Giants. _____
   c. White Dwarfs. _____
   d. Blue Giants. _____
   e. Main Sequence. _____
   f. "Late stage" stars. _____
   g. Next life stage for Sun. _____
   h. Cold, small stars. _____
   i. Hot, small stars. _____
   j. *Aldebaran*. _____
   k. *Barnard's Star*. _____
   l. *Sirius*. _____

7. The star *Betelgeuse* is much further from Earth than *Aldebaran*. Why does *Betelgeuse* appear more luminous than *Aldebaran*? _____

8. The stars *Betelgeuse* and *Barnard's Star* are both the same temperature. Why is *Barnard's Star* so much dimmer than *Betelgeuse*? _____

9. Write a statement that describes the relationship between the age of a star and the amount of hydrogen present in the star. _____

10. Arrange these terms in order of size from largest to smallest.

    a. *Polaris, Procyon B, Sirius*. _____
    b. galaxy, star, universe. _____

11. List these five stars in order of decreasing luminosity.

    Aldebaran, Alpha Centauri, Betelgeuse, Polaris, Sirius

    _____, _____, _____, _____, _____

12. The diagram to the right represents the shape of our *Milky Way Galaxy*.

    a. This shape is described as _____.

    b. Place an "**X**" where the Sun would be located.

13. The graph below shows the changes in the Sun's magnetic activity and number of sunspots for a period of 100 years.

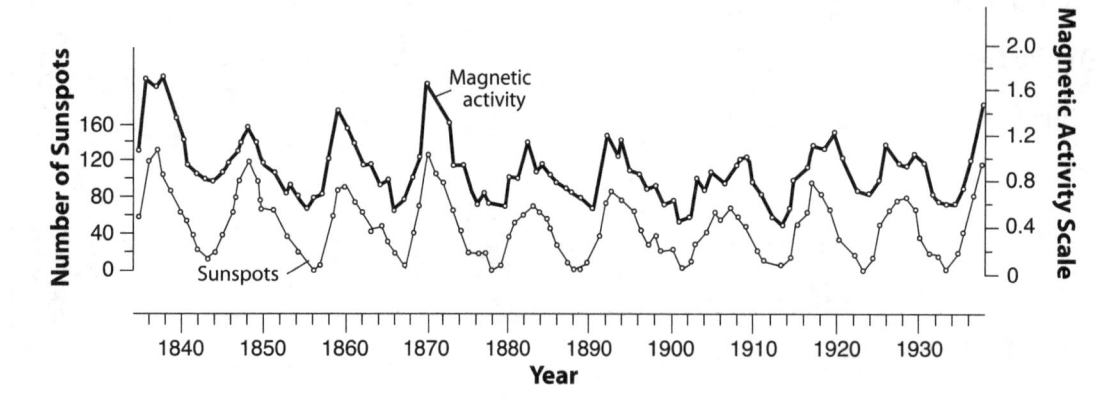

    a. This graph can be described as *(cyclic) (non-cyclic)*.

    b. Write a statement that describes the relationship between magnetic activity and sunspots.
    _____

    c. In 1874 there was approximately _____ sunspots.

    d. According to the graph, the number of sunspots peaks about every *(5) (10) (20)* years.

    e. Other than sunspots, name *two* other solar features. _____ and _____

14. The diagram below shows two possible sequences in the life cycle of stars beginning with their formation from a nebula.

The Life Cycles of Stars

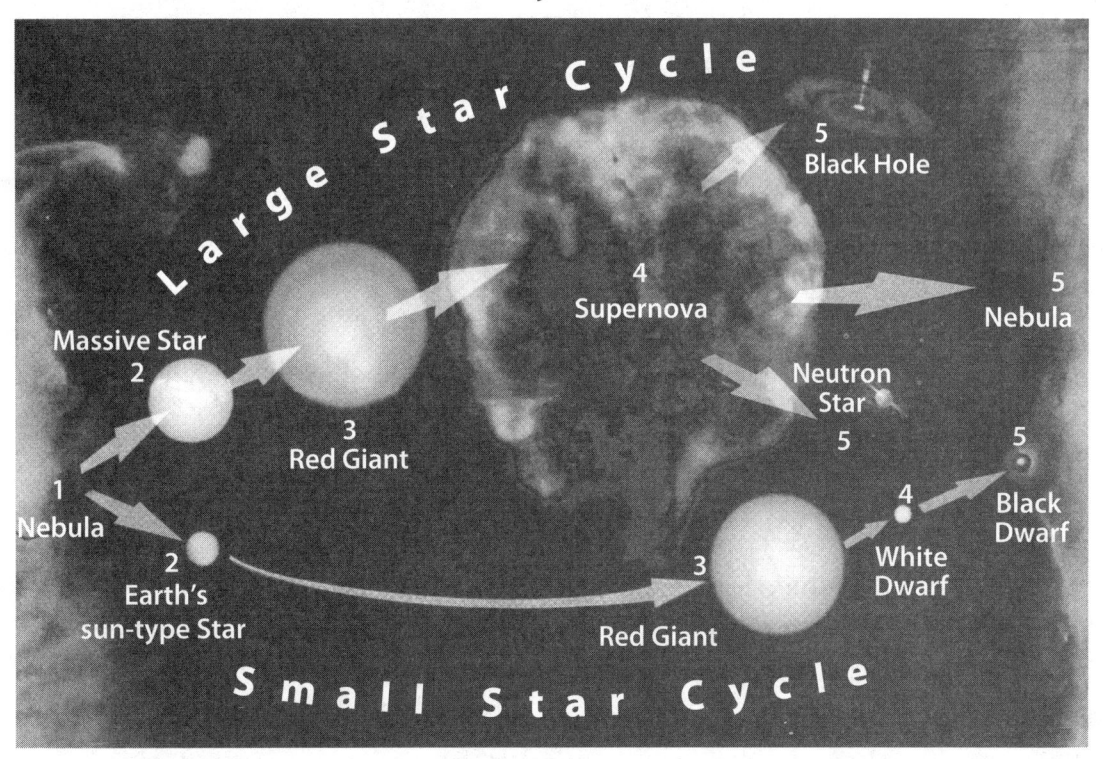

a. Name the most common element in the nebula at **#1**. _____

b. Name the force responsible for organizing the nebula into a star at **#2**. _____

c. Name the process that must begin in order for a star to form at **#2**. _____

d. The star *Aldebaran* would be found at # _____ in the life cycle diagram.

e. At the end of its life cycle our Sun will become a _____.

f. As life stage **#3** begins, the star's luminosity will *(decrease)* *(increase)* *(remain the same)*.

g. Stars which supernova are generally *(large)* *(small)* *(size has no effect)*.

h. Many astronomers consider our Sun a second generation star because of the heavier elements found in the Sun. Based on this diagram, how is it possible for a star to have formed from another star?

_____

_____

_____

CHAPTER 2 REVIEW

1. The universe is primarily composed of
   (1) energy (2) stars (3) dust (4) empty space

2. The most common element in the universe is
   (1) helium (2) hydrogen (3) nitrogen (4) oxygen

3. As a star ages it will be composed of
   (1) more hydrogen and less helium
   (2) more helium and less hydrogen
   (3) more oxygen and less carbon
   (4) more nitrogen and less oxygen

4. Which information best supports the inference that the universe began with a rapid expansion?
   (1) measurements of the rate of decay using carbon-14
   (2) observations of cosmic background radiation
   (3) calculations of the distance from the Sun to each asteroid in the asteroid belt
   (4) calculations of the temperature and luminosity of stars

5. An immense cloud of gas and dust in the universe is called a
   (1) pulsar (2) nova (3) supernova (4) nebula

6. The Big Bang Theory, describing the formation of the universe, is best supported by the
   (1) blueshift of light from distant galaxies
   (2) redshift of light from distant galaxies
   (3) the number of stars in the universe
   (4) presence of craters on the Moon

7. Most astronomers agree that at the present time the universe is
   (1) contracting (2) expanding (3) wobbling (4) staying the same size

8. Which statement best describes the movement of galaxies in the universe?
   (1) galaxies are moving towards each other
   (2) galaxies are moving away from each other
   (3) galaxies are moving randomly
   (4) galaxies do not move

9. Compared to the universe, the Sun, Earth, and planets are:
   (1) younger (2) older (3) the same age

10. When a star ages it will leave the main sequence and become a
    (1) blue giant (2) red giant (3) white dwarf (4) red dwarf

11. Stars which form from a very small nebula will initially become a
    (1) blue giant (2) red giant (3) white dwarf (4) red dwarf

12. During its life cycle when a star's size increases, its luminosity will
    (1) decrease (2) increase (3) remain the same

13. By which process do stars convert mass into large amounts of energy?
    (1) nuclear fission (2) radioactive decay (3) convection (4) nuclear fusion

14. The approximate temperature and luminosity of the star *Deneb* is
    (1) 12,000 K and 750,000
    (2) 12,000 K and 500,000
    (3) 8500 K and 750,000
    (4) 8500 K and 500,000

15. Compared to the star *Polaris*, the star *Sirius* is
    (1) hotter and more luminous
    (2) hotter and less luminous
    (3) cooler and less luminous
    (4) cooler and more luminous

16. Compared to the star *Proxima Centauri*, the star *Aldebaran* is
    (1) smaller and less luminous
    (2) smaller and more luminous
    (3) larger and more luminous
    (4) larger and less luminous

17. Which is a blue giant star?
    (1) *Sirius* (2) *Spica* (3) *Betelgeuse* (4) *Proxima Centauri*
18. Which star is the oldest?
    (1) *40 Eridani B* (2) *Alpha Centauri* (3) *Spica* (4) *Deneb*
19. The star *Elnath* has a surface temperature of 13,700 K and a luminosity of 700. This star would be classified as a
    (1) main sequence (2) white dwarf (3) supergiant (4) giant
20. The stars *Polaris* and *Alpha Centauri* are most similar in
    (1) size (2) life stage (3) luminosity (4) temperature
21. During the beginning stages of star formation, as a giant nebula collapses its density will
    (1) decrease (2) increase (3) remain the same
22. Compared to early stage stars, intermediate stage stars are generally
    (1) smaller (2) larger (3) the same size
23. Which star is in its final stage of development?
    (1) *Spica* (2) *Polaris* (3) *Procyon B* (4) *Barnard's Star*
24. Galaxies are classified based on their
    (1) color (2) temperature (3) luminosity (4) shape
25. The diagram below represents the present position of our solar system in a side view of the Milky Way Galaxy. The distances across are measured in light-years (ly).
    The approximate distance in light-years from the center of the galaxy to our solar system is
    (1) 20,000 ly
    (2) 30,000 ly
    (3) 50,000 ly
    (4) 110,000 ly

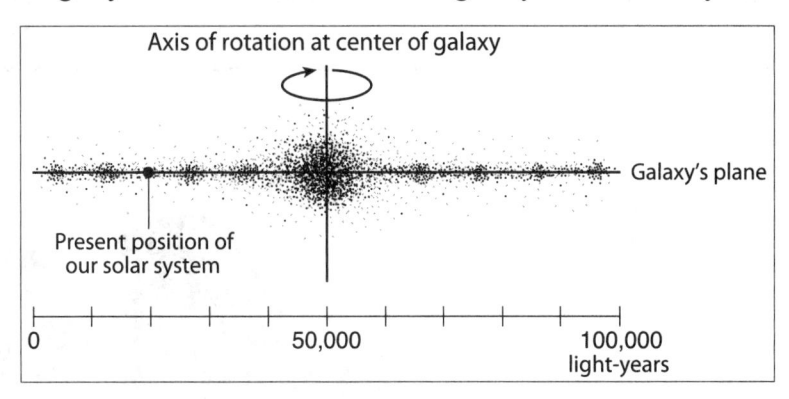

26-29. Base your answers to questions **26** through **29** on the reading below and your knowledge of Earth Science.

> Hydrogen gas is the main source of fuel that powers the nuclear reactions that occur in the Sun. But just like many sources of fuel, the hydrogen is in limited supply. As the hydrogen gas is used up, scientists predict that the helium created as an end product of earlier nuclear reactions will begin to fuel new nuclear reactions. When this happens, the Sun is expected to become a red giant star with a radius that would extend out past the orbit of Venus and possibly out as far as Earth's orbit. Earth will probably not survive this change in the Sun's size. But no need to worry at this time. The Sun is not expected to expand to this size for a few billion years.

26. Name the "nuclear reaction" referred to in this passage that combines hydrogen gas to form helium and produce the Sun's energy. _____

27. What will cause the Sun to expand to a red giant star. _____
_____
_____

28. As the Sun becomes a red giant, its luminosity will _____.

29. On the diagram below of the planets and the Sun's surface draw a vertical line to represent the inferred location of the Sun's surface when it becomes a red giant star.

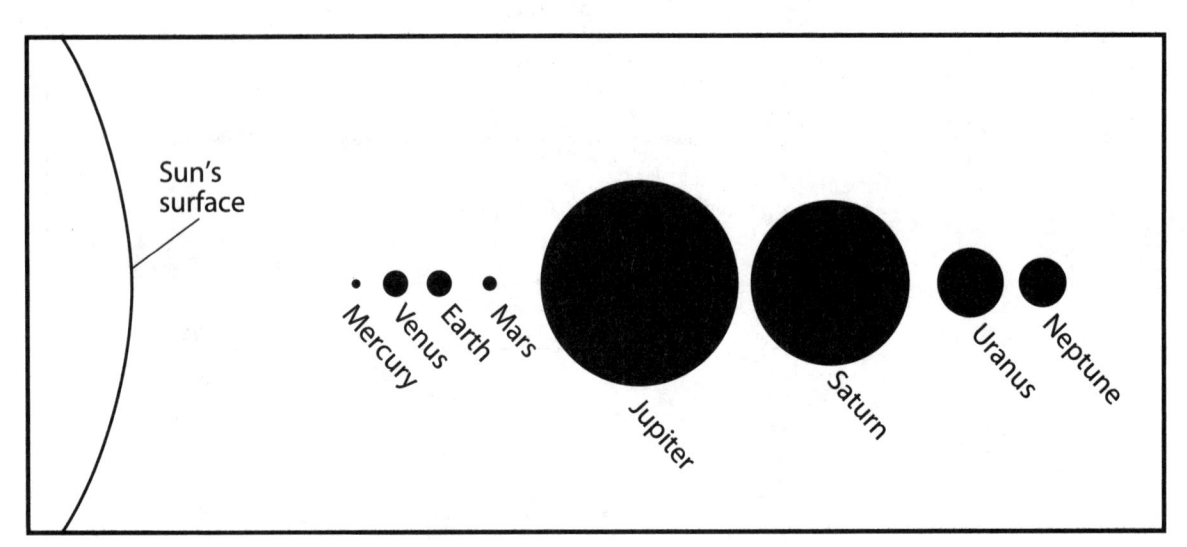

(Distances are not drawn to scale)

30-35. The graph below shows the early formation of main sequence stars with different masses (**M**). The arrows represent the temperature and luminosity changes as each star becomes part of the main sequence. The time needed for this to occur is shown.

Formation of Main Sequence Stars

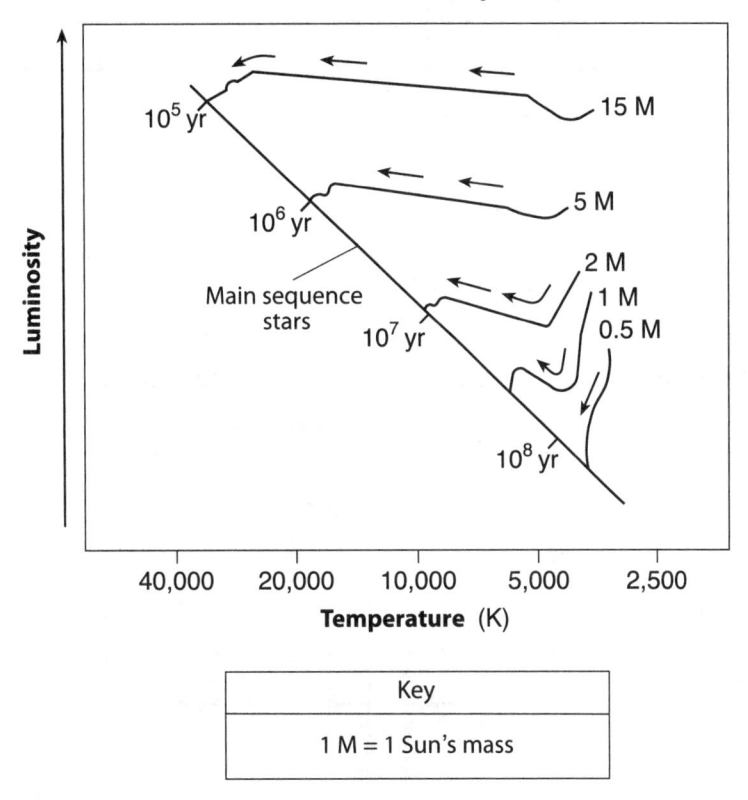

[Graph: Luminosity (y-axis) vs Temperature (K) (x-axis: 40,000 — 20,000 — 10,000 — 5,000 — 2,500). Curves labeled 15 M, 5 M, 2 M, 1 M, 0.5 M with formation times 10^5 yr, 10^6 yr, 10^7 yr, 10^8 yr along the main sequence line.]

Key: 1 M = 1 Sun's mass

30. How many years does it take for a 2 M star to become a main sequence star?

   a. Express in scientific notation. _____ yrs.

   b. Write in standard form. _____ yrs.

31. Describe the relationship between the original mass of the star and the length of time needed for it to become a main sequence star. _____

_____

32. Describe the change in luminosity of a star with an original mass of 0.5 M as it progresses to a main sequence star.

_____

_____

33. Place an "**X**" for the present location of our Sun on this diagram.

34. Older main sequence stars tend to have (*less*) (*more*) mass.

35. Hot, bright main sequence stars tend to be (*younger*) (*older*) than other main sequence stars.

23

36-40. The data table below lists some galaxies, their distance from Earth, and the velocities at which they are moving away from Earth.

Name of Galaxy	Distance (million light-years)	Velocity (thousand km/s)
Virgo	70	1.2
Ursa Major 1	900	15
Leo	1100	19
Bootes	2300	40
Hydra	3600	61

One light-year = distance light travels in one year

36. Construct a line graph which shows the relationship between each galaxy's distance and the velocity at which it is moving away. Use an "**X**" to plot each data point.

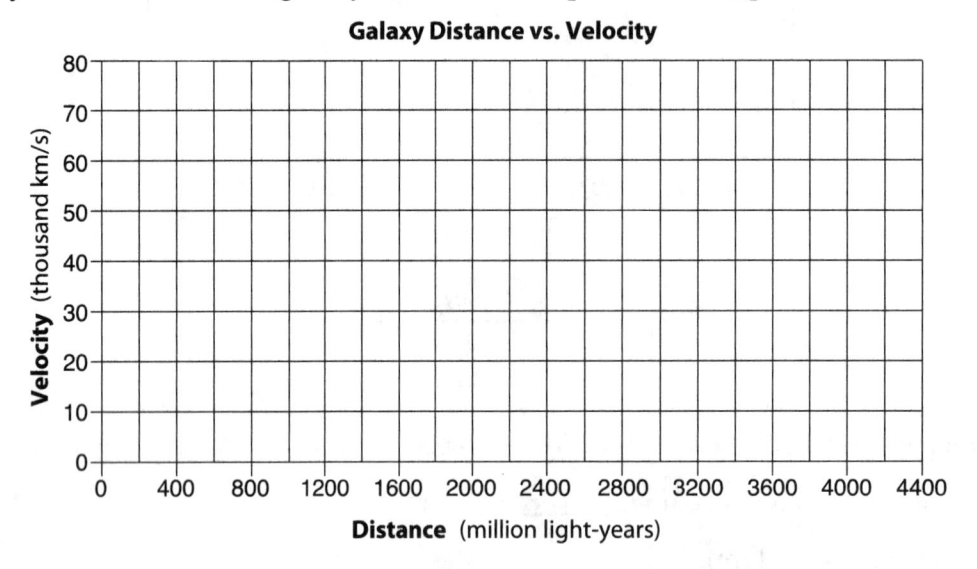

Galaxy Distance vs. Velocity

37. State the relationship between a galaxy's distance from Earth and the velocity at which the galaxy is moving away. _____

38. Another galaxy is moving at a velocity of 30 thousand kilometers per second. Estimate the galaxy's distance from Earth. _____ million light-years.

39. What event occurred which continues to cause matter in the universe to move away from each other? _____

40. What phenomena observed in the spectral light from stars led to the inference that the galaxies are moving away from Earth? _____

CHAPTER 3

THE SOLAR SYSTEM

Solar System, Orbits, Earth's Moon

THE SOLAR SYSTEM

The **solar system** formed as a by-product of star formation. This occurred about 4.6 to 5.0 billion years ago when a giant **nebula** of gas and dust collapsed. The center of this collapsed mass became the Sun, while the outer parts cooled and coalesced to become the solid objects of the solar system.

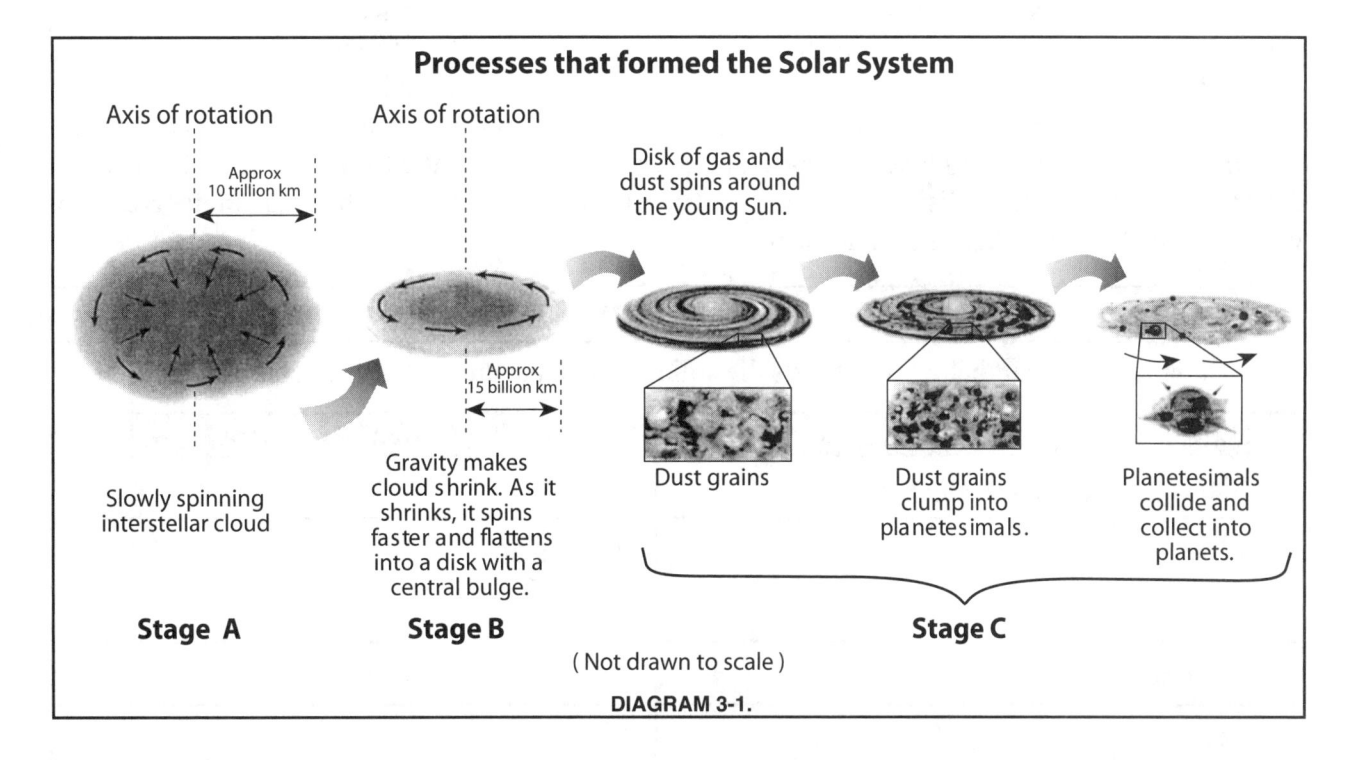

Processes that formed the Solar System

Stage A: Slowly spinning interstellar cloud (Axis of rotation, Approx 10 trillion km)

Stage B: Gravity makes cloud shrink. As it shrinks, it spins faster and flattens into a disk with a central bulge. (Axis of rotation, Approx 15 billion km)

Stage C: Disk of gas and dust spins around the young Sun. Dust grains. Dust grains clump into planetesimals. Planetesimals collide and collect into planets.

(Not drawn to scale)

DIAGRAM 3-1.

Our solar system is heliocentric. In a **heliocentric** system, the objects of the solar system **revolve** around **(orbit)** the Sun. The period of revolution defines a "year" on a planet. Most of the solar system objects also **rotate** on an axis. The period of rotation defines a "day" on a planet.

Eight solid, spherical **planets** orbit the Sun. The inner **terrestrial planets** (Mercury, Venus, Earth, and Mars) are small, dense, rocky, have few moons, and no rings. The outer **Jovian planets** (Jupiter, Saturn, Uranus, and Neptune) are large, low density, have many moons, and rings. **Moons** orbit planets. There are hundreds if not thousands of **dwarf planets** orbiting the Sun. Most are found beyond the orbit of Neptune. These dwarf planets are solid spheres, smaller than Mercury whose orbits cross the paths of other objects. Pluto and Eris are dwarf planets.

Also orbiting the Sun are **asteroids** which are irregular-shaped rocks and **comets** which are masses of ice and rock. Comets orbit the Sun in very elliptical paths. When the comet comes close to the Sun it will vaporize and form a tail of gases and rock debris which reflects sunlight. **Meteoroids** are smaller pieces of rock and dust scattered throughout the solar system. When they impact planets, moons, and asteroids they are called **meteorites** and may form craters. Meteoroids often disintegrate in Earth's atmosphere and cause a brilliant **meteor** display which are mistakenly called "shooting stars."

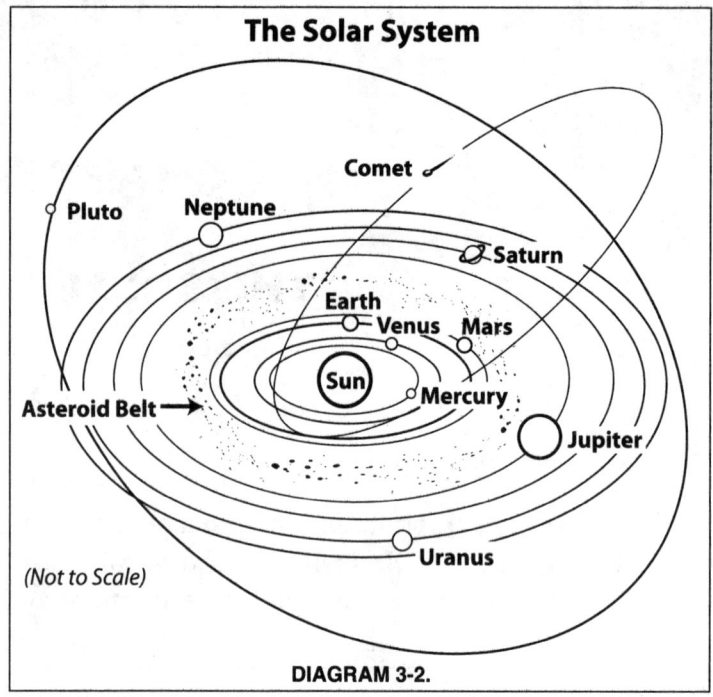

The Solar System

DIAGRAM 3-2.

Questions

1. Complete the chart below which lists the solid objects that are part of the solar system.

Object	Description	Location/Movement
Planet	solid, spherical shape	
Moon		orbits a planet
Asteroid		
Meteoroid		
Comet		
Dwarf Planet		

2. The line below shows the average distance between the Sun and Earth.
   Place a dot "O" where Jupiter would be and label the dot "Jupiter."

   Sun   Earth

   EXPLAIN how you determined this position for Jupiter. _____
   _____
   _____

3. **a.** A "*day*" on Saturn equals _____.

   b. A "*year*" on Mercury equals _____.

4. The diagram to the right shows the relative diameters of the planets compared to the Sun.

   a. Circle the terrestrial planets.

   b. Place an "**X**" on the planet with the lowest density.

   c. How many times larger is the diameter of the Sun than the diameter of Jupiter. _____
   Show solution.

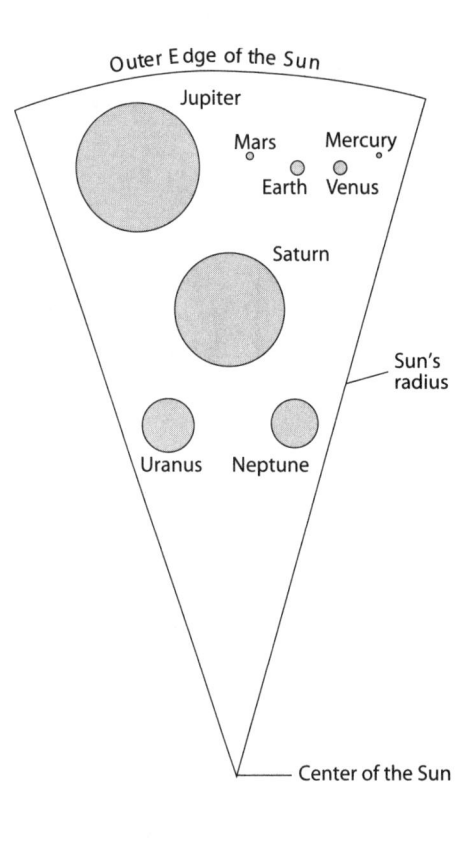

5. List the Jovian planets from least mass to greatest mass.

   _____, _____, _____, _____

6. Use the "*Solar System Data*" chart in the ESRT to name the planet described.

   a. longest "day" _____
   b. longest "year" _____
   c. least density _____
   d. greatest density _____
   e. greatest mass _____
   f. two planets with similar shaped orbits are _____ and _____
   g. two planets with similar periods of rotation are _____ and _____
   h. slowest rotation _____
   i. slowest revolution _____
   j. least eccentric orbit _____
   k. least mass _____
   l. furthest from Sun _____

7. The bar graph below shows the equatorial diameter of Earth.

   a. Construct bars to represent the equatorial diameters of Mars, Venus, and Earth's Moon.

   b. Calculate how many times bigger Earth is than its Moon. _____
   Show solution.

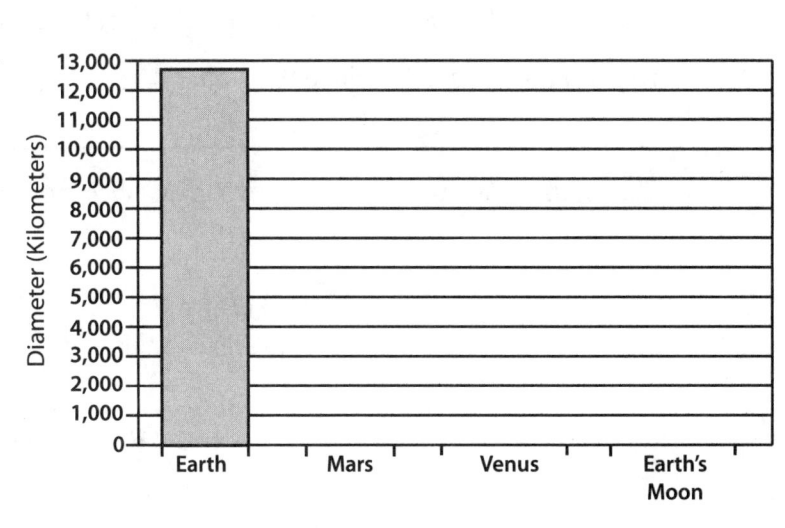

27

8. The diagram below illustrates some of the objects in our solar system.

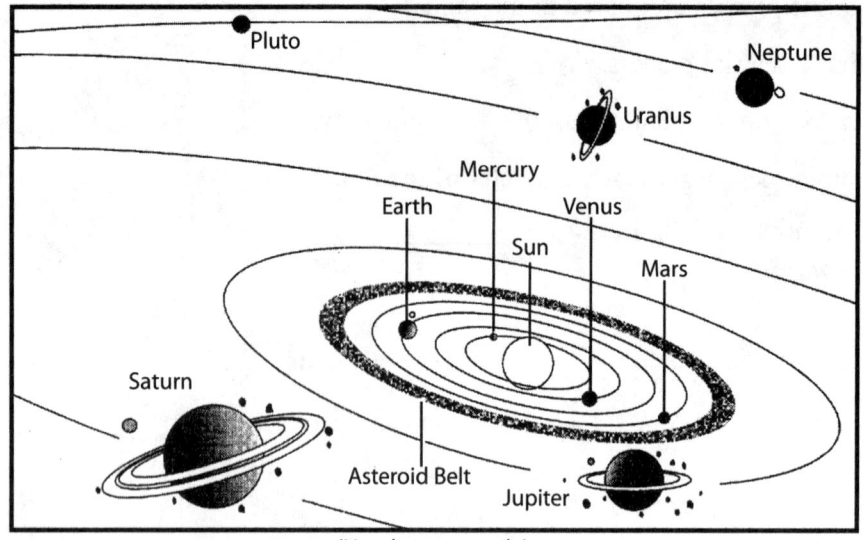

(Not drawn to scale)

a. Name *two* solar system objects that are missing from this diagram.
_____ and _____

b. Based on this diagram explain why Pluto is no longer classified as a planet. _____
_____

c. Based on this diagram what distinguishes the terrestrial planets from the Jovian planets?
_____
_____

d. Describe *two* other ways that the terrestrial planets are different from the Jovian planets.
_____ and _____

e. Name the force that holds the solar system together. _____

f. Planets that are further from the Sun have longer periods of revolution. Why? _____
_____
_____

g. The average distance of the asteroid belt from the Sun is _____ million kilometers.
Explain how you determined this value. _____
_____

h. Sketch the graph relationship between distance from Sun and period of revolution.

period of revolution

distance from Sun

CHARACTERISTICS OF ORBITS

The orbits of all objects in the solar system are **ellipses**. A circle is drawn relative to one center or focal point and all the points on the circle are equidistant from the center. An ellipse is drawn relative to two **focus** points (foci). There is no center and all the diameters are not the same in an ellipse.

Eccentricity describes the shape of an ellipse. Eccentricity measures how far from circular the orbit is. The eccentricity of a circle is zero. Therefore the closer the eccentricity is to zero the more circular; the larger the eccentricity the more elliptical the orbit is. Comets have very elliptical orbits. Eccentricity is calculated as:

<u>distance between the foci (d)</u>
length of major axis (L)

Eccentricity is ALWAYS rounded to the thousandth place and does NOT have a unit!

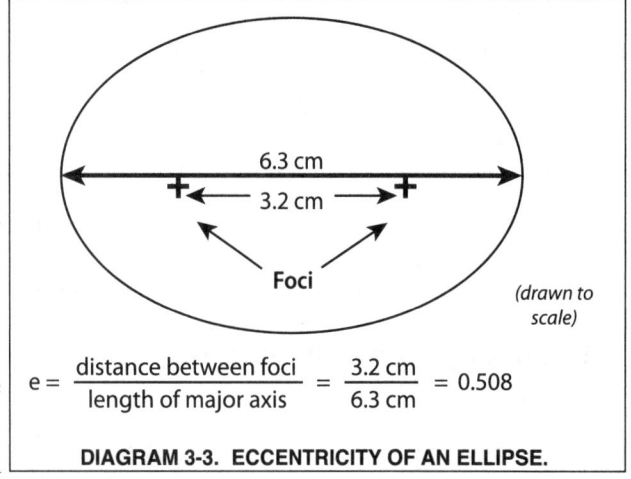

$$e = \frac{\text{distance between foci}}{\text{length of major axis}} = \frac{3.2 \text{ cm}}{6.3 \text{ cm}} = 0.508$$

DIAGRAM 3-3. ECCENTRICITY OF AN ELLIPSE.

All planets, comets, and asteroids orbit the Sun in an elliptical path with the Sun at one of the focal points. The distance from a planet to the Sun varies in a cyclic pattern. This is why the apparent diameter of the Sun as viewed from Earth changes during the year. Moons have elliptical orbits with a planet at one of the focal points.

As a planet orbits closer to the Sun (**perihelion**), its orbital velocity increases (it revolves faster). This occurs because the Sun is exerting a greater gravitational pull at perihelion. When a planet, asteroid, or comet is at perihelion, it has more orbital velocity and more kinetic energy. At **aphelion** (furthest from the Sun), there is less gravitational pull so the orbital velocity is less with more potential energy. The rate of rotation is not affected by the changing distance to the Sun. Due to the changing speed of revolution the area of space "swept out" by an orbiting object is the same for equal time intervals. In diagram 3-4, the area of "Section X" equals the area of "Section Y" since both are "swept out" in an equal time of ten days.

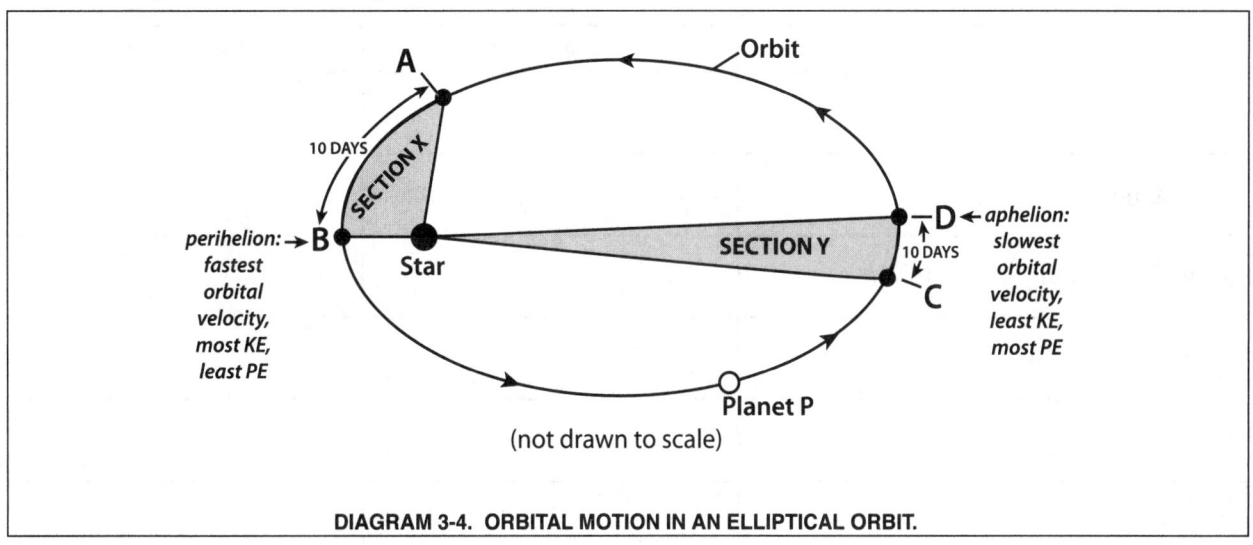

DIAGRAM 3-4. ORBITAL MOTION IN AN ELLIPTICAL ORBIT.

Questions

9. Name the planet described.

   a. has the most circular orbit. _____

   b. has the most eccentric orbit. _____

   c. eccentricity is 0.093 _____

   d. has eccentricity similar to Earth's Moon. _____

10. List these planets from most to least eccentric : *Earth, Neptune, Saturn, Venus*

    _____ , _____ , _____ , _____

11. The diagram below shows the orbit of an asteroid around the Sun. **A** and **B** are two position in its orbit.

    a. The Sun is at one of the focus points. Locate the other focus point of this asteroid's orbit with a "o" and label it "**F**"

    b. In the chart below indicate which position (**A** or **B**) is *aphelion* and which is *perihelion* by writing the word in the chart.

    c-g. For parts **c** through **g** write the word *least*, *greatest*, or *the same* to compare Position **A** with Position **B**.

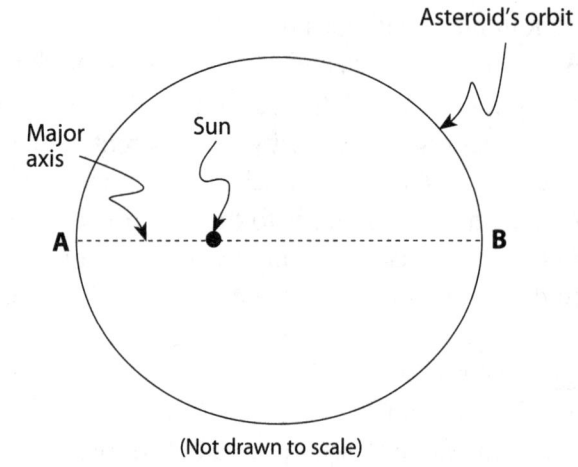

	Position A	Position B
b. aphelion **or** perihelion?		
c. distance from Sun		
d. gravitational pull from Sun		
e. orbital velocity		
f. kinetic energy		*least*
g. potential energy		

12. The diagram to the right shows an elliptical orbit around the Sun.
    a. Calculate the eccentricity of this orbit.
       Show solution.

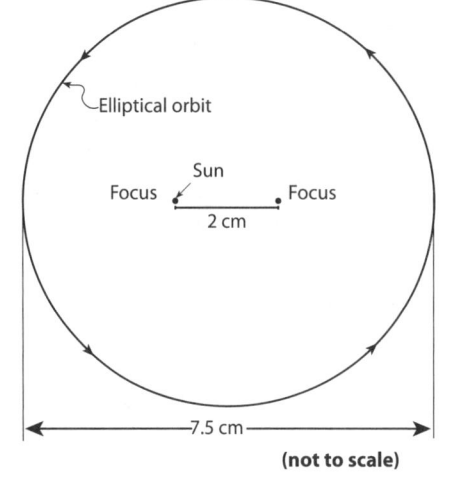

    b. Name the planet whose orbit is most similar
       to this diagram. _____
       Explain why you selected this planet. _____
       _____
       _____

    c. On the diagram label *perihelion* and *aphelion*
    d. Place an "**X**" on the orbit where the orbital velocity is the greatest.
    e. Place a "**P**" on the orbit where there is the greatest potential energy.

13-14. This diagram represents a satellite in an elliptical orbit around Earth. The foci of the orbit are F_1 and **Earth**. Daily positions in the satellite's orbit are indicated as **1** through **12**.

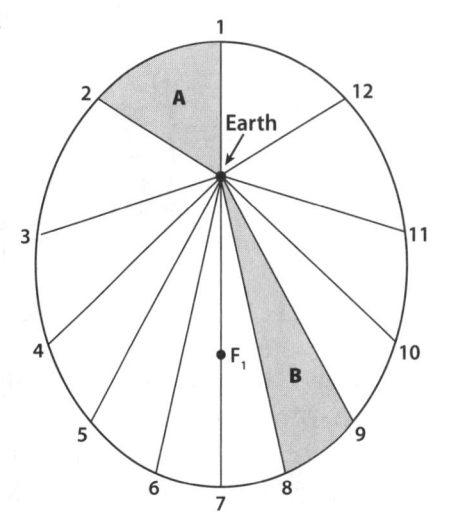

13. *Complete using the diagram.*
    a. How does the area of sections **A** and **B** compare? _____
    b. Shade in another section that has the same area as section **A**.
    c. This satellite orbits Earth once every _____ days.
    d. Explain why this is a geocentric orbit. _____
       _____

14. Use the above diagram of the satellite's orbit around Earth. Select the graph which represents the following relationships.

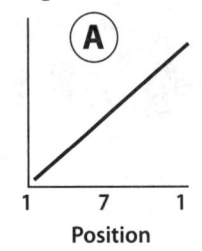

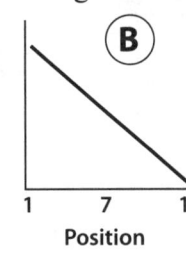

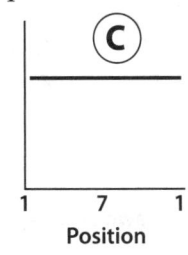

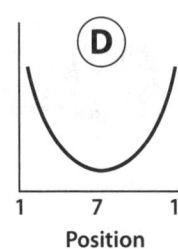

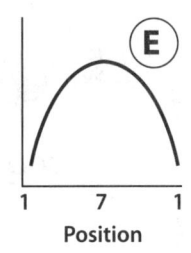

    a. Position in orbit vs. orbital velocity: _____
    b. Position vs. kinetic energy: _____
    c. Position vs. gravitational attraction: _____
    d. Orbital velocity if orbit was a perfect circle: _____
    e. Position vs. area swept out: _____
    f. Positon vs. potential energy: _____

31

Earth's Moon

Earth's Moon is visible because it reflects sunlight. The Moon is a natural satellite of Earth. The Earth-Moon system is a **geocentric** system in that the Moon is orbiting Earth. The Moon rotates counterclockwise on its axis. It revolves counterclockwise around Earth in an elliptical path. The Moon's period of rotation is $27\frac{1}{3}$ days and its period of revolution is $27\frac{1}{3}$ days. Due to the synchronous (equal) rates of rotation and revolution, the same side of the Moon always faces Earth.

The **phases of the Moon** are the different portions of the Moon that we see from Earth. As the Moon revolves around Earth the appearance or phase of the Moon changes. One complete cycle of phases occurs every 29.5 days.

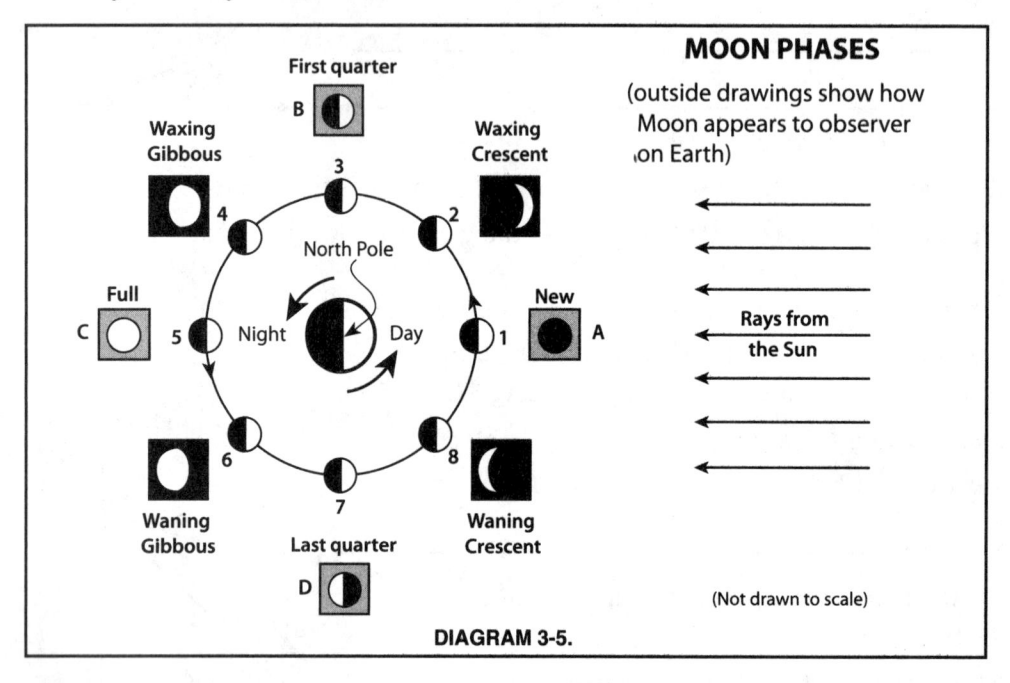

DIAGRAM 3-5.

The movement of Earth and Moon relative to the Sun causes **eclipses**. Eclipses only occur when the Earth, Moon, and Sun are lined up in space. During a **solar eclipse** the Moon is between Earth and Sun so that the Moon's shadow falls on Earth blocking the view of the Sun. During a **lunar eclipse** Earth is between the Sun and the Moon so that Earth's shadow darkens the full Moon. Eclipses are predictable because of the known motions of Earth and Moon.

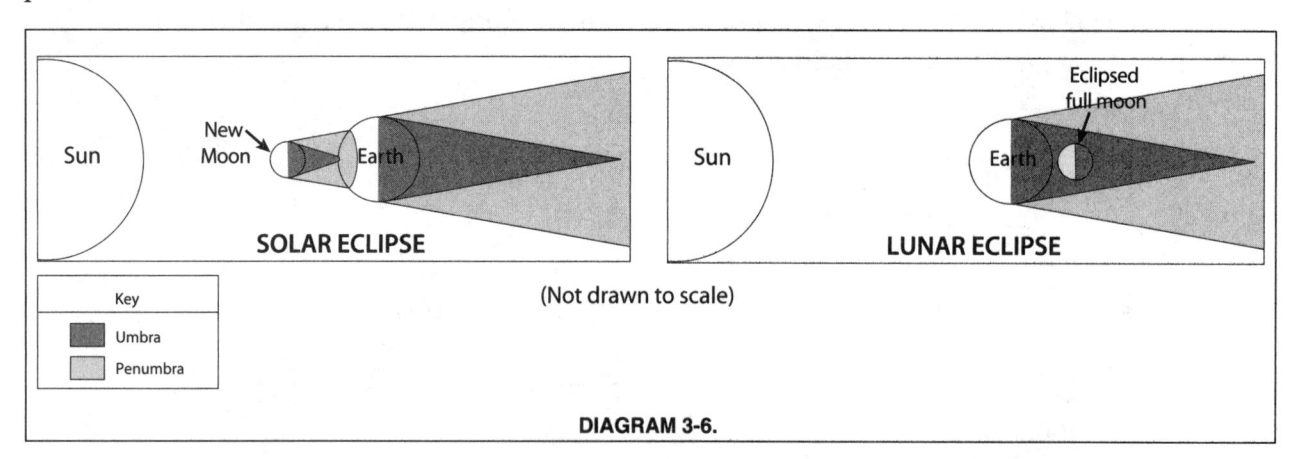

DIAGRAM 3-6.

Earth's gravitational force keeps the Moon in orbit. The Moon also exerts a gravitational pull on Earth. This "pull" results in the twice daily change in ocean **tides** as the seawater rises and falls along the shoreline. The Sun has a slight effect on tides.

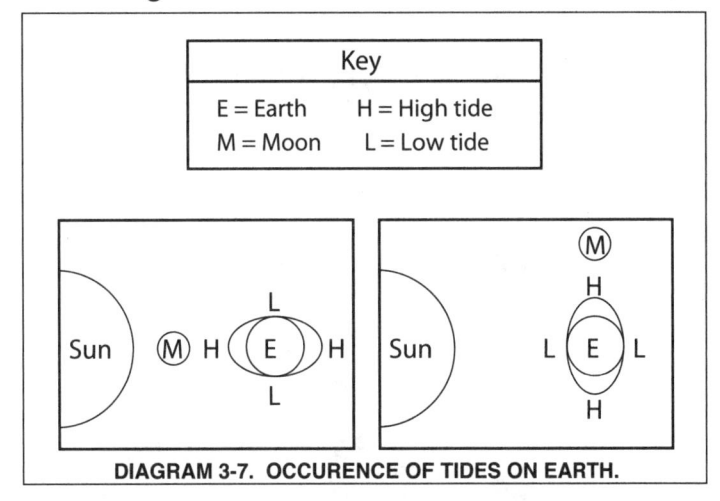

Key
E = Earth   H = High tide
M = Moon   L = Low tide

DIAGRAM 3-7. OCCURENCE OF TIDES ON EARTH.

Questions

15. Use the diagram which shows eight positions of the Moon around Earth. The shading on Earth and the Moon indicate nighttime.

    a. Draw an arrow between each Moon position to show direction of the Moon's revolution.

    b. Draw in four horizontal lines to show the direction of the Sun's rays. Label these as "Sun."

    c. The time it takes for the Moon to orbit from position **A** to position **B** is about_____ days.

    d. The Moon orbits Earth in a(n) _____ path with Earth at a(n) _____.

    e. Compared to Jupiter, the orbit of the Moon is more *(circular) (elliptical)*.

    f. Name the phase at **A** _____ **D** _____

16. Why does the Moon have a greater effect than the Sun on Earth's ocean tides?
    _____

17. Using the terms "rotation" and "revolution" explain why the same side of the Moon always faces Earth._____
    _____
    _____

33

18. The diagram shows the Moon in eight different positions during a month relative to Earth and Sun. Location "X" is on Earth.

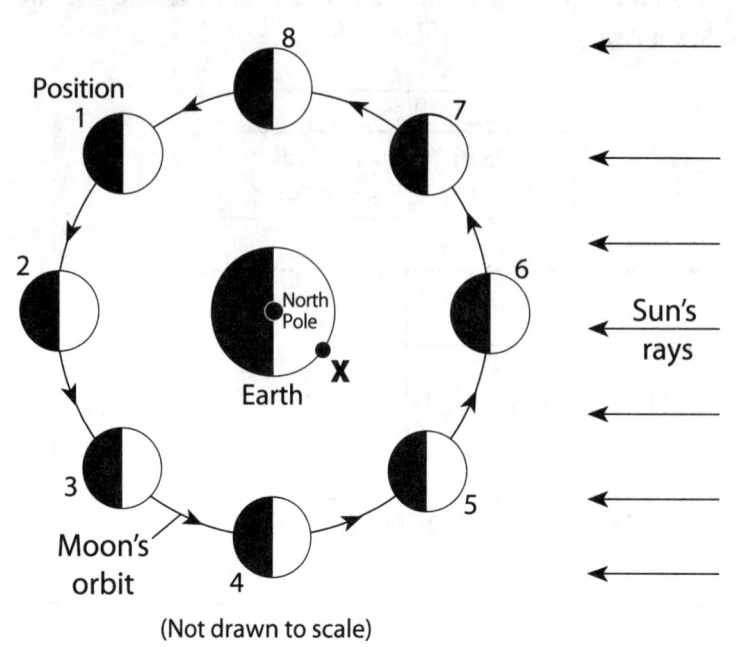

(Not drawn to scale)

a. The Moon goes through a cycle of phases once every _____ days.

b. The Moon goes through a cycle of phases because _____.

c. Location "X" would be experiencing a low tide if the Moon is in position(s) _____.

d. A solar eclipse could occur when the Moon is in position _____ only if _____
_____.

e. Shade in the circle to show the Moon's appearance as seen from Earth at **position 3**.

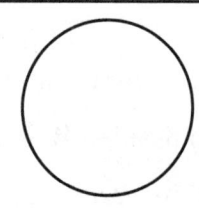

f. Give the number of the Moon position for the occurrence of these phases:

    Full Moon: _____   First Quarter: _____

    Waning Crescent: _____   Waxing Gibbous: _____

CHAPTER 3 REVIEW

1. The solar system most likely formed from a
   (1) nebula   (2) red giant   (3) black hole   (4) comet

2. Which sequence is correct for planets with increasing periods of rotation?
   (1) Venus, Mars, Jupiter   (3) Mars, Jupiter, Neptune
   (2) Neptune, Mars, Mercury   (4) Jupiter, Venus, Earth

3. Which planet has completed less than one orbit of the Sun in the last 100 years?
   (1) Mercury   (2) Neptune   (3) Mars   (4) Uranus

4. Compared to Saturn and Neptune, Mercury and Mars have greater
   (1) periods of revolution   (3) distances from the Sun
   (2) orbital velocities   (4) equatorial diameters

5. The further a planet is located from the Sun, the
   (1) shorter its periods of rotation   (3) longer its period of revolution
   (2) shorter its period of revolution   (4) longer its period of rotation

6. On which planet would there be the longest time between sunrise and sunset?
   (1) Earth   (2) Mars   (3) Mercury   (4) Venus

7. The orbits of the planets are
   (1) elliptical with Earth at one of the foci   (3) circular with the Sun at one of the foci
   (2) elliptical with the Sun at one of the foci   (4) circular with Earth at one of the foci

8. Which diagram best shows the relative size of Earth and Mars?

   Earth   Mars   Earth   Mars   Earth   Mars   Earth   Mars
      (1)   (2)   (3)   (4)

9. Venus is considered Earth's "twin" because of similarities in
   (1) time of revolution   (3) eccentricity of orbit
   (2) time of rotation   (4) size

10. Which does not describe the inner terrestrial planets?
    (1) low mass and large densities
    (2) short periods of rotation and long periods of revolution
    (3) long periods of rotation and short periods of revolution
    (4) closest to Sun and fastest orbital velocities

11. One of Saturn's moons, Titan, has a density of 1.881 g/cm^3. What planet has a similar density?
    (1) Jupiter   (2) Mercury   (3) Neptune   (4) Saturn

12. A comet which crosses Earth's orbit will leave rock debris and dust in Earth's path. When these particles enter Earth's atmosphere they are called
    (1) meteoroids   (2) meteorites   (3) meteors   (4) supernovae

13. The elliptical shape of Earth's orbit around the Sun results in
    (1) changes in the orbital velocity of Earth   (3) oblate spheroid shape of Earth
    (2) tilting of Earth's axis   (4) phases of the Moon

14. If the distance between the Moon and Earth were double its present distance, the Moon's cycle of phases would occur
    (1) in reverse order and more slowly   (3) in the same order but more slowly
    (2) in reverse order but more quickly   (4) in the same order but more quickly

15. The diagram below shows the relative position of Earth and Mars in their orbits on a day in winter of 2007.

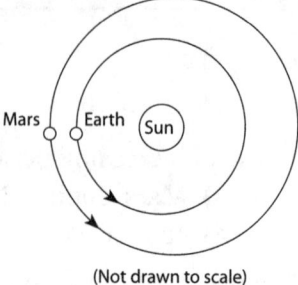

(Not drawn to scale)

Which diagram shows the location of Earth and Mars on the same date in winter of 2008?

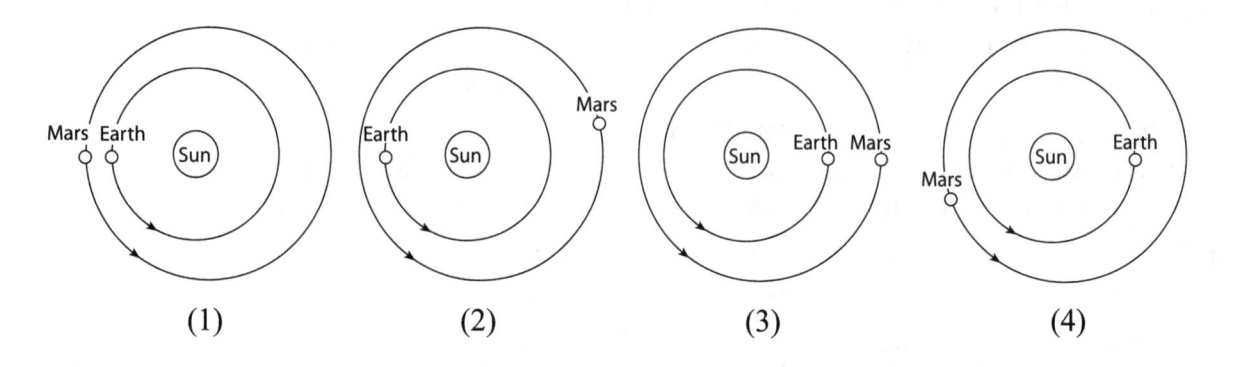

(1)   (2)   (3)   (4)

16-19. The diagram below shows the orbital path of a planet around a star. Locations P_1 through P_6 are various points in the planet's orbit.

16. The greatest gravitational attraction occurs at position:
    (1) P_1 (3) P_3
    (2) P_2 (4) P_4

17. The greatest angular diameter of the star will appear when the planet is at position:
    (1) P_1 (3) P_3
    (2) P_2 (4) P_4

18. The planet is orbiting at its slowest rate at position:
    (1) P_1 (3) P_3
    (2) P_2 (4) P_4

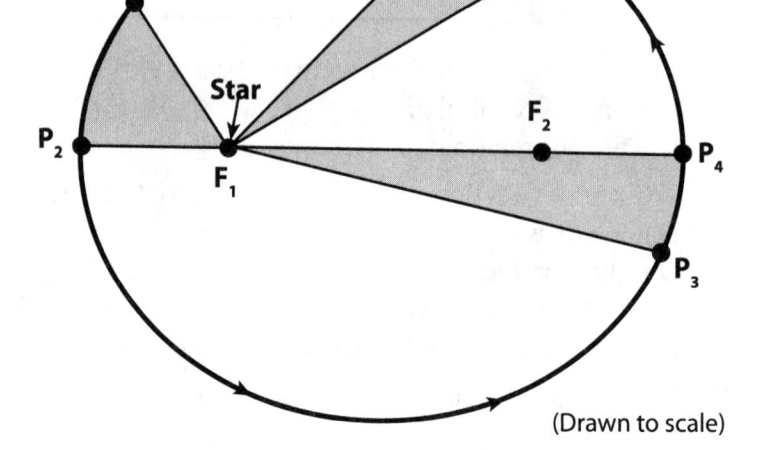

(Drawn to scale)

19. If the shaded areas are equal then the time period from P_1 to P_2 is equal to the time period from
    (1) P_2 and P_3 (2) P_3 and P_4 (3) P_4 and P_5 (4) P_6 and P_1

20. The diagram shows two photographs of the Moon, **A** and **B**, taken at full moon phase several months apart. The photographs were taken with the same magnifications. Each photograph was cut in half and the halves placed next to each other.

What most likely caused the difference in the apparent size of the Moon in photographs **A** and **B**?
(1) The Moon phase changed.
(2) The Moon expanded.
(3) The distance from Earth to Moon changed.
(4) The Moon rotated.

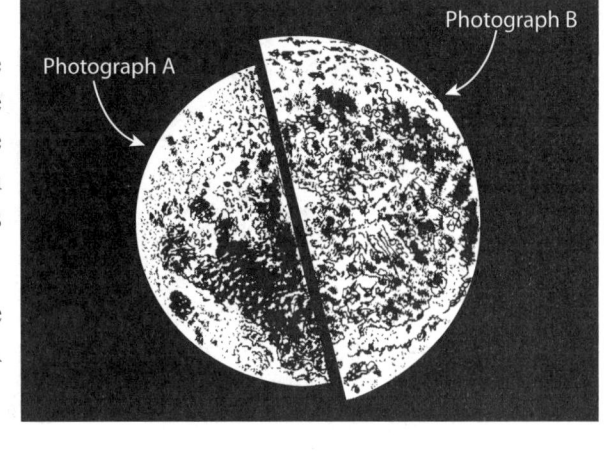

21. The photographs show the Moon as seen from Earth during an 80-minute interval in a single night. Which motion caused this changing appearance?

(1) The Moon moved into Earth's shadow.
(2) The Moon moved into the shadow of the Sun.
(3) The Sun moved into Earth's shadow.
(4) The Sun moved into the shadow of the Moon.

22. What motion causes the Moon to go through a cycle of phases every 29.5 days?

(1) Moon's rotation on its axis
(2) Moon's revolution around Earth
(3) Earth's revolution around the Sun
(4) Earth's rotation on its axis

23-25. The diagram shows eight positions of Moon around Earth. The direction of Sun rays are shown.

23. At which numbered position could a lunar eclipse occur?
(1) 1   (3) 3
(2) 7   (4) 5

24. A high tide will occur at location **X** on Earth when the Moon is in position:
(1) 1   (3) 5
(2) 2   (4) 7

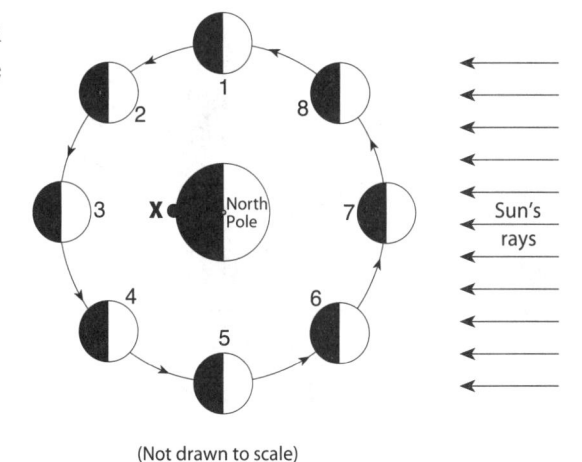

25. Which Moon phase would be seen by an observer in New York State when the Moon is at position 2?

(1)   (2)   (3)   (4)

37

26-27. The diagram below shows the position of Earth, Moon, and Sun for one day.

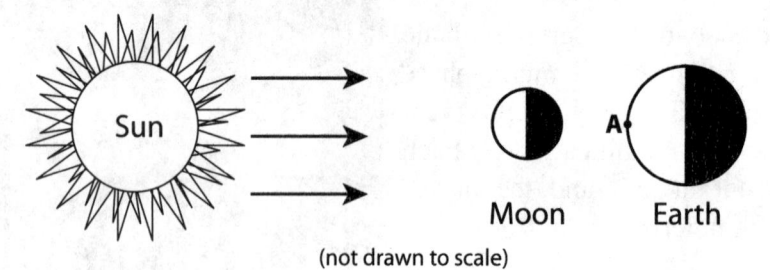

(not drawn to scale)

26. What phase would Earth appear in if viewed from the Moon? _____

27. Location **A** is experiencing a higher than normal high tide. What is causing this?
_____
_____

28-30. The diagram below shows the Moon, Earth, and Sun positions for one day.

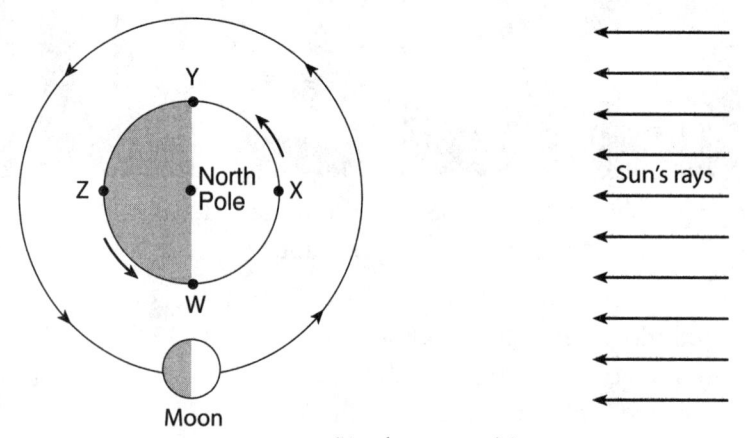

(Not drawn to scale)

28. Shade in the circle to indicate how the Moon would appear for an observer at **W**.

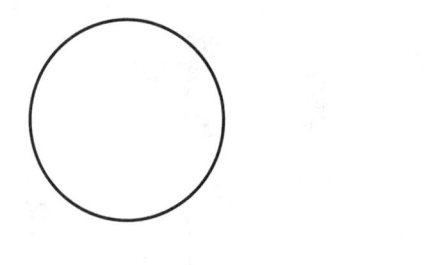

29. What phase does the Moon appear in for this position? _____

30. Which position(s) would be experiencing low tide? _____

38

31-35. The diagram shows the orbits of Earth and a comet.

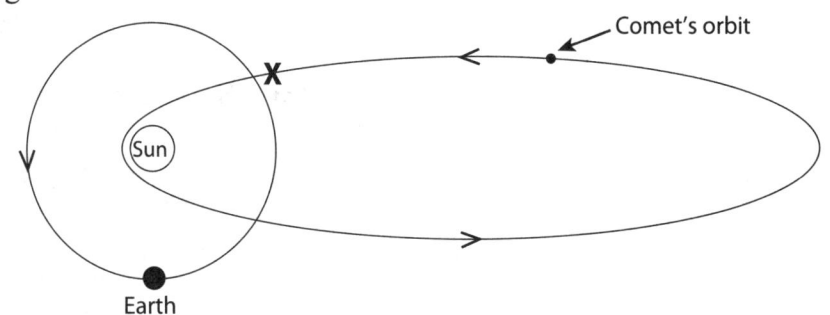

31. How does this comet's orbit illustrate the heliocentric model of our solar system.
_____

32. Why is the time required for one revolution of the comet more than the time for one Earth revolution. _____
_____

33. Place an "S" in the comet's orbit where it would be orbiting the slowest.

34. The "X" is one position on the comet's orbit where a "tail" will most likely occur. At this position, draw the comet's tail as it would form relative to the Sun.

35. Place an "F" at location of the other foci in the comet's orbit.

36-37. Refer to the reading below.

> **The Moon Is Moving Away While Earth's Rotation Slows**
>
> Tides on Earth are primarily caused by the gravitational force of the Moon acting on Earth's surface. The Moon causes two tidal bulges to occur on Earth: the direct tidal bulge occurs on the side facing the Moon, and the indirect tidal bulge occurs on the opposite side of Earth. Since Earth rotates, the bulges are swept forward along Earth's surface. This advancing bulge helps pull the Moon forward in its orbit, resulting in a larger orbital radius. The Moon is actually getting farther away from Earth, at a rate of approximately 3.8 centimeters per year.
>
> The Moon's gravity is also pulling on the direct tidal bulge. This pulling on the bulge causes friction of ocean water against the ocean floor, slowing the rotation of Earth at a rate of 0.002 second per 100 years.

36. Explain why the force of gravity between the Moon and Earth will decrease over time. _____
_____
_____

37. In 100,000 years, the rotation of Earth will be slower by how many seconds? _____
(Show solution)

38-40. Table 1 shows average planetary distance from Sun in Astronomical Units (AU) and average orbital speed. Neptune's orbital speed has been left blank. Pluto is a dwarf planet.

Table 1

Planet	Average Distance from Sun (AU)	Average Orbital Speed (km/sec)
Mercury	0.4	48.0
Venus	0.7	35.0
Earth	1.0	30.0
Mars	1.5	24.0
Jupiter	5.2	13.0
Saturn	9.6	10.0
Uranus	19.0	7.0
Neptune	30.0	
Pluto	39.0	4.7

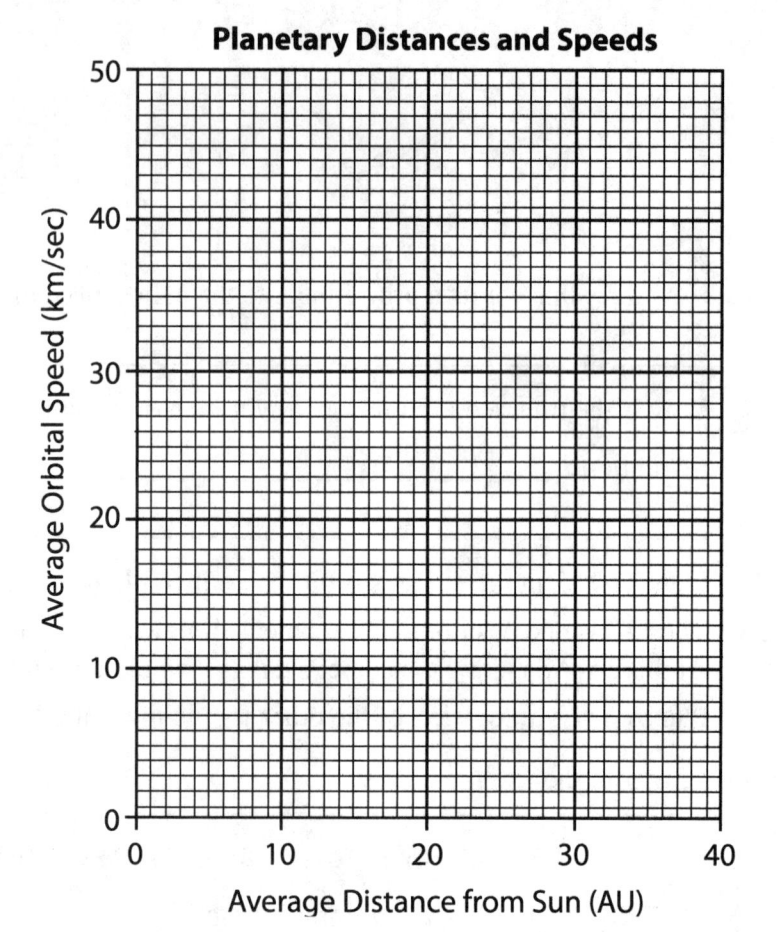

Planetary Distances and Speeds

38. Construct a line graph which shows the relationship between planetary distance and orbital speed.

39. Based on your graph, estimate to the nearest whole number the average orbital speed of Neptune. _____

40. 1 Astronomical Unit (AU) is approximately equal to _____ million km.

CHAPTER 4

PLANET EARTH & EARTH MOTIONS
Earth's Size & Shape, Earth Motions, Latitude-Longitude & Time Zones

EARTH'S SIZE AND SHAPE

Earth has a spherical shape. Many celestial and terrestrial observations led people to understand that Earth was a sphere. Sailing ships appear to "sink" below the horizon. Celestial objects, such as the Sun, Moon, and planets, were observed to be round. Travelers noticed that as they changed their location, the stars, especially the Sun and *Polaris* (the North Star) were often higher or lower in the sky. Today pictures of Earth taken from space prove Earth's spherical shape.

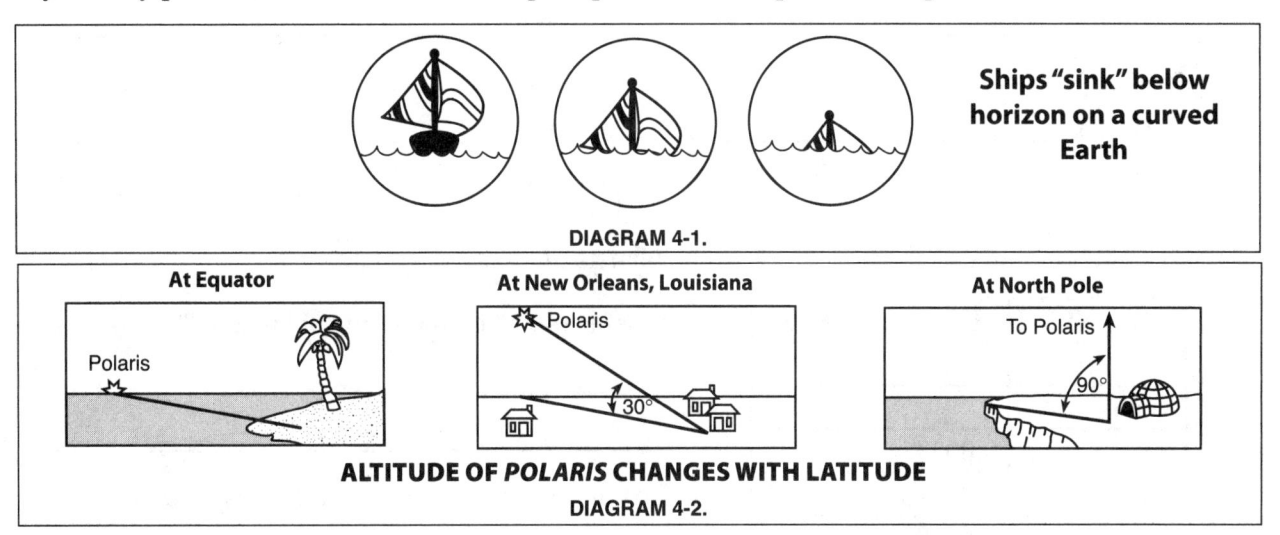

Ships "sink" below horizon on a curved Earth

DIAGRAM 4-1.

ALTITUDE OF *POLARIS* CHANGES WITH LATITUDE
DIAGRAM 4-2.

Earth is not a perfect sphere, it is an **oblate spheroid**. Its equatorial diameter is slightly larger than its polar diameter. At the Equator you are further from the center of Earth so the gravitational force (F_g) on you is less. Therefore you weigh slightly less at the Equator than at the Poles. Models of Earth, such as classroom globes, show a perfect sphere because Earth is <u>slightly</u> oblate. This oblateness is only noticed with careful, exact measurements.

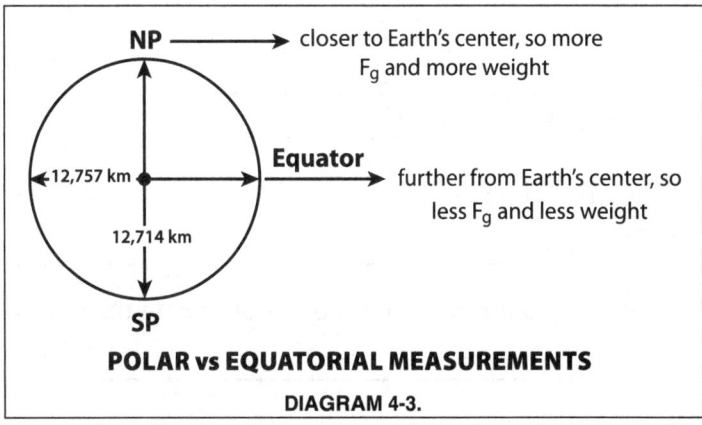

NP → closer to Earth's center, so more F_g and more weight

Equator → further from Earth's center, so less F_g and less weight

12,757 km
12,714 km
SP

POLAR vs EQUATORIAL MEASUREMENTS
DIAGRAM 4-3.

The oblate shape of Earth is caused by its rotation. Jupiter has a very oblate shape because of its fast rate of rotation. Saturn not only rotates quickly but its rings exert a gravitational pull that causes its obvious oblate shape.

There are three specific parts of Earth called spheres. The **atmosphere** is a very thin layer of gases that surround Earth. The lower part of the atmosphere is called the **troposphere**. Most of the gases of the atmosphere are found within the troposphere which extends only 11 km above Earth's surface. The **hydrosphere** refers to the waters of Earth. The average depth of the oceans is 4 km. In comparison to Earth's average diameter of 12,756 km, the hydrosphere is very thin. The **lithosphere** is the thin layer of rock that surrounds Earth. The lithosphere includes the continental and oceanic crust and upper rigid mantle and is about 75 to 100 kilometers in thickness. Compared to Earth's size, these spheres are all very thin.

Questions

1. Explain why your weight will be slightly more at the North and South Poles than at the Equator.

   _____
   _____
   _____

2. To what extent does Earth's shape differ from a perfect sphere? _____

3. Select the Earth sphere referred to: (*atmosphere*) (*hydrosphere*) (*lithosphere*).

   a. beach. _____   e. sea floor. _____
   b. mountains. _____   f. river. _____
   c. ocean. _____   g. air. _____
   d. icecap. _____   h. crust. _____

4. The chart below lists the oblateness of the planets of the solar system.

PLANET	OBLATENESS
Mercury	0
Venus	0
Earth	0.00335
Mars	0.00648
Jupiter	0.06487
Saturn	0.09796
Uranus	0.02293
Neptune	0.01708

   a. What is meant by the term "oblateness"?

      _____
      _____

   b. List these planets in order of increasing oblateness:

      Jupiter, Mercury, Neptune, Uranus

      _____, _____, _____, _____

   c. Based on the ESRT, why is Uranus more oblate than

      Mars? _____

   d. Explain *two* reasons for Saturn being the most oblate of the planets.

      (1)_____
      _____
      (2)_____
      _____

Earth Motions

Earth is constantly moving. It **rotates** (spins) on a tilted axis in a counterclockwise direction (west to east). One rotation takes approximately 24 hours.

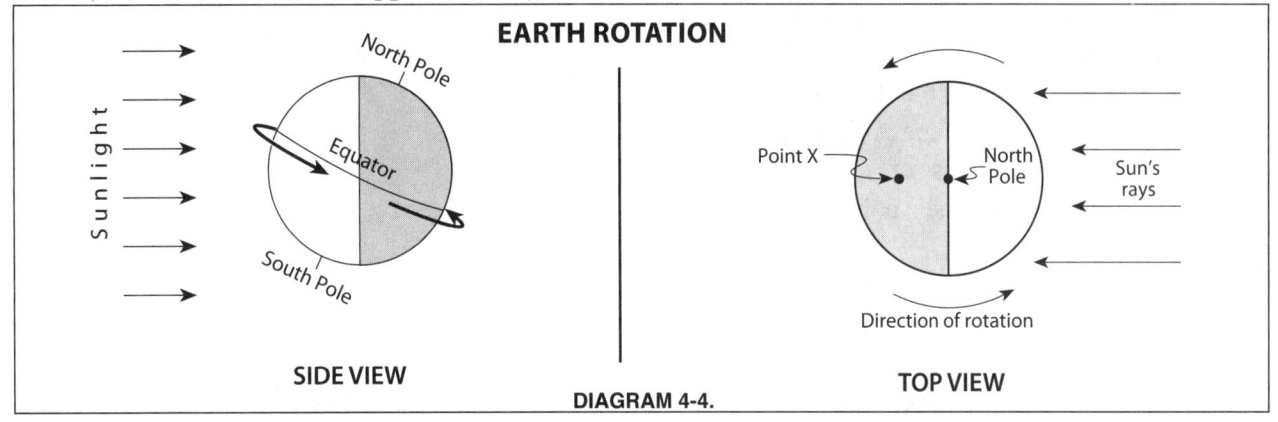

EARTH ROTATION

SIDE VIEW | TOP VIEW

DIAGRAM 4-4.

Evidence for rotation is based on terrestrial (Earth) observations. The **Foucault pendulum** appears to change its direction of swing as Earth rotates beneath it. The **Coriolis effect** causes winds and ocean currents to curve to the right in the Northern Hemisphere and to the left in the Southern Hemisphere.

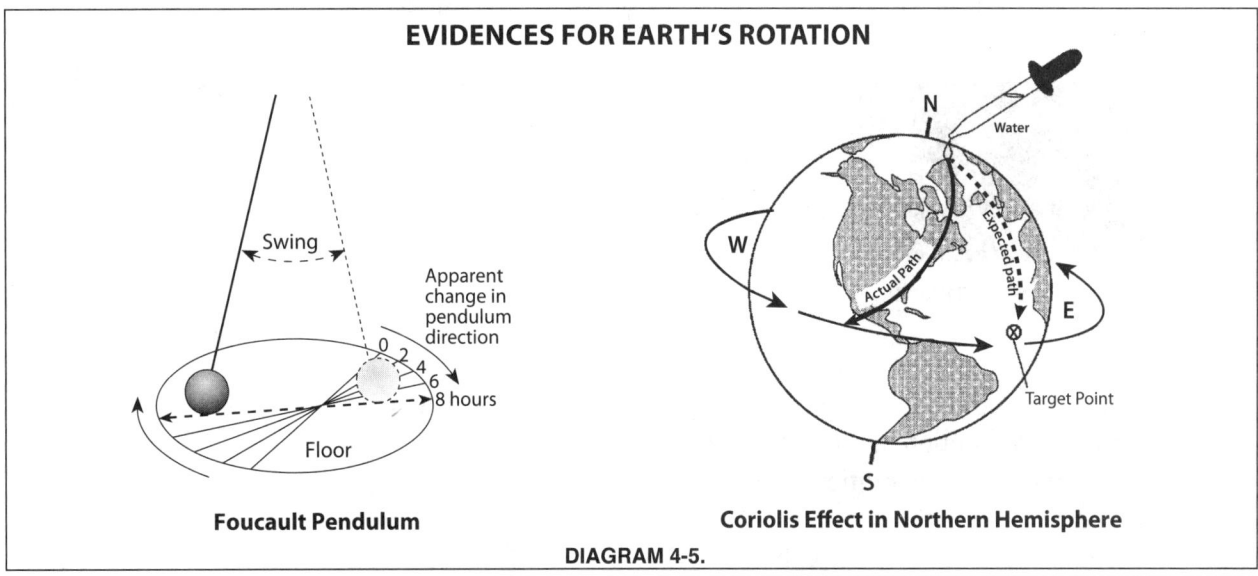

EVIDENCES FOR EARTH'S ROTATION

Foucault Pendulum | Coriolis Effect in Northern Hemisphere

DIAGRAM 4-5.

Some of the effects caused by Earth's rotation are day and night, and sunrise and sunset. Celestial objects such as the Sun, stars, and Moon, appear to move westward across the sky at a rate of 15° per hour, the same rate as Earth's rotation. The time of day changes as Earth rotates.

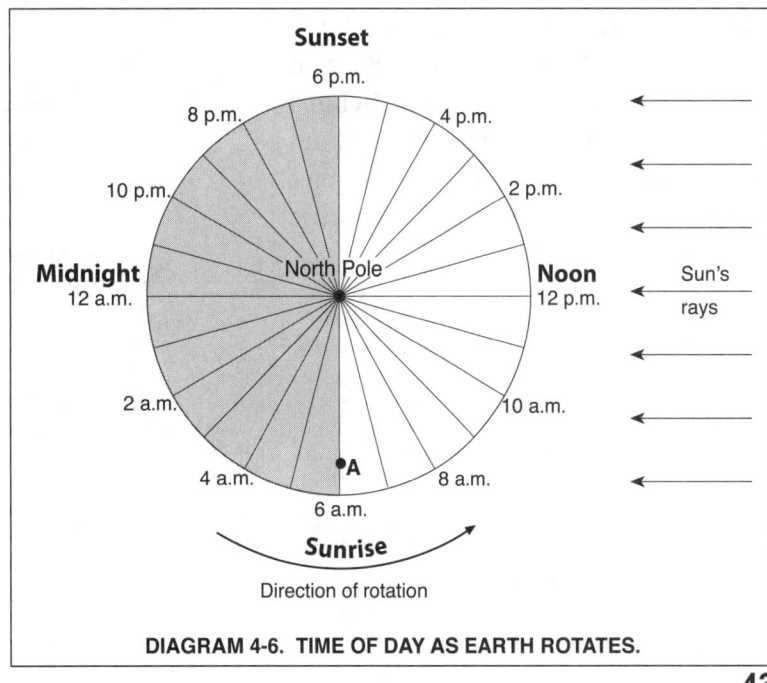

DIAGRAM 4-6. TIME OF DAY AS EARTH ROTATES.

43

The axis of rotation is tilted 23.5° and points towards the "north pole star." Earth wobbles on its axis like a toy top that is slowing down. This motion is called **precession**. One cycle of precession occurs every 26,000 years. The tilt of Earth's axis changes during precession. Since the axis of rotation "wobbles" (precesses), the pole star changes over time.

Today the north pole star is *Polaris*, also known as the North Star. The stars that are near *Polaris* do not appear to rise or set. Instead they make circular star trails around *Polaris* each day. These stars are referred to as **circumpolar**. Other stars, such as the Sun, appear to rise and set each day.

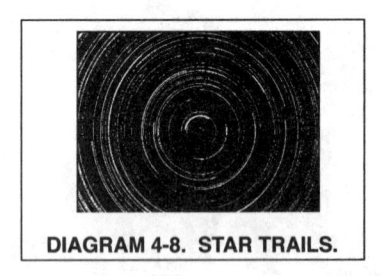

DIAGRAM 4-7. PRECESSION OF AXIS.

DIAGRAM 4-8. STAR TRAILS.

Earth **revolves** around the Sun. Earth's period of revolution is 365 ¼ days. The direction of revolution is counterclockwise. Earth revolves approximately 1° each day. Our path around the Sun is not a perfect circle, it is an ellipse. We are closest to the Sun on or about January 3rd (**perihelion**) and furthest from the Sun on or about July 3rd (**aphelion**). Earth's changing distance from the Sun does NOT cause seasons!!

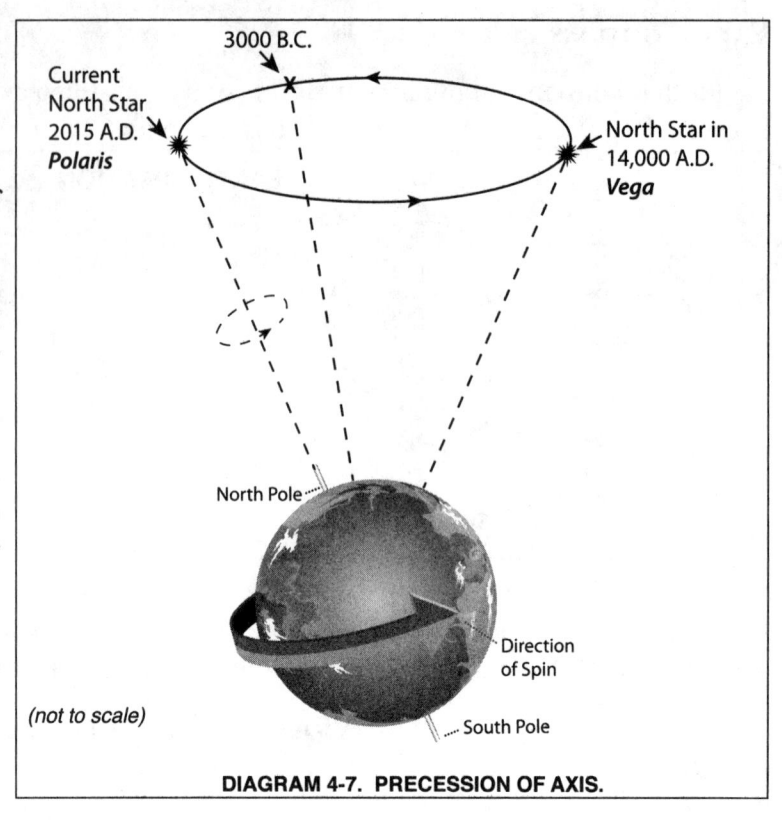

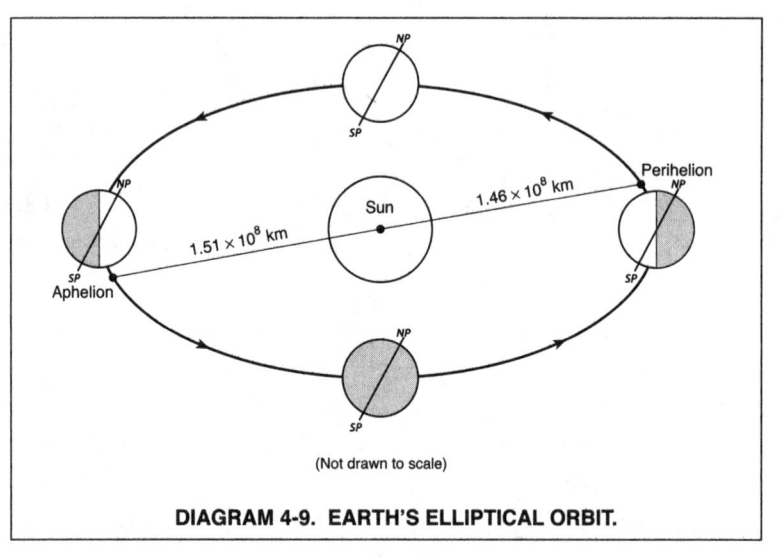

DIAGRAM 4-9. EARTH'S ELLIPTICAL ORBIT.

There are two theoretical models of the solar system. The first one is the **geocentric** model. In this model, all objects in the solar system move around Earth and Earth does not move. The Coriolis effect and Foucault pendulum prove that Earth moves and that the geocentric theory is incorrect. The Sun is the central body of the **heliocentric** model in which Earth and the planets, asteroids, comets, and meteoroids orbit the Sun. The appearance of different constellations during each month, such as the zodiac constellations, prove a heliocentric solar system.

DIAGRAM 4-10. CHANGE IN ZODIAC CONSTELLATION AS EARTH REVOLVES.

Questions

5. The diagram below shows Earth's orbit around the Sun and the zodiac constellations visible at different times of years. The diagram is NOT to scale.

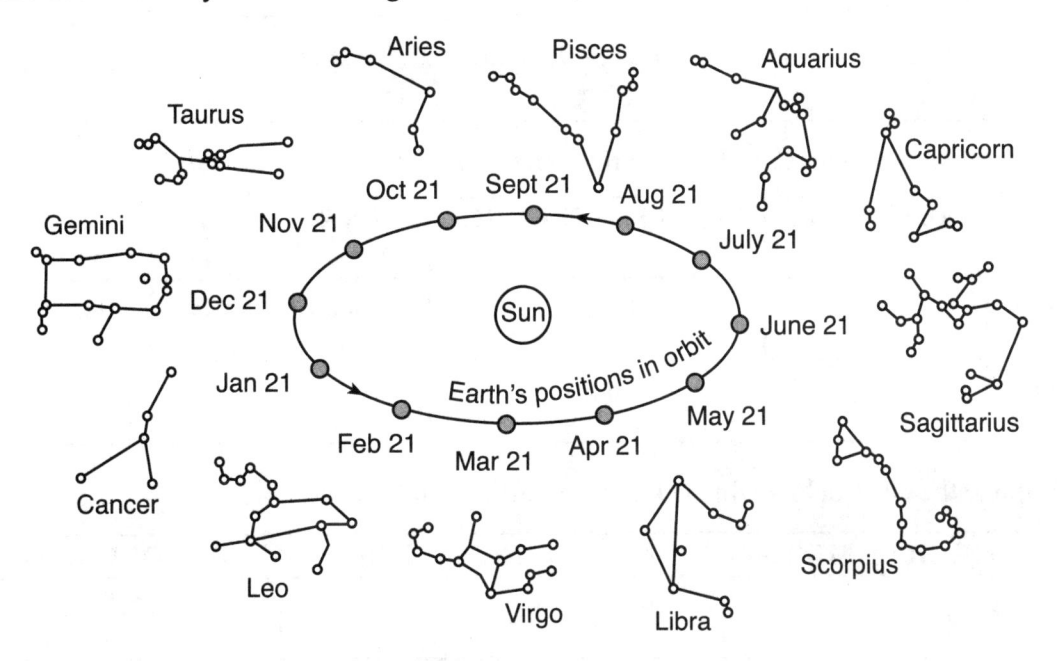

   a. Which motion of Earth is indicated by the arrows ? _____
   b. On February 21 which zodiac constellation is visible at midnight?_____
   c. On September 21 the noon Sun is blocking from our view which constellation? _____
   d. Circle the position which marks the first day of winter.
   e. Write the word "aphelion" nearest that position on the orbit.
   f. Define the term "aphelion." _____
   g. At aphelion the orbital velocity of Earth will be the *(greatest) (least)*.
   h. This diagram illustrates the *(geocentric) (heliocentric)* model of the solar system.

45

6. The diagram below shows several visible constellations in the night sky. These constellations appear to move counterclockwise around the star in the center. Straight lines are at 15-degree intervals. *Merak* and *Dubhe* are two stars in the Big Dipper.

   a. Identify the star located in the center of this star field. _____

   b. What motion causes the counterclockwise motion of these constellations? _____

   c. How many degrees will each star appear to move in 3 hours? _____

   d. Explain your answer to **c**. _____

   e. Place an "**X**" where the star *Dubhe* will be 2 hours later.

   f. The stars *Merak* and *Dubhe* are in the same galaxy as our Sun. Name this galaxy. _____

   g. These constellations are classified as "circumpolar." Explain. _____

   h. Complete this chart for the classification of stars. *One has been done for you.*

STAR	LUMINOSITY	TEMPERATURE (K)	CLASSIFICATION	COLOR
Merak	50	10,000	main sequence	blue-white
Dubhe	230	4800		

7. Complete the chart below which compares Earth's rotation and revolution.

EARTH MOTION	ROTATION	REVOLUTION
a. definition		
b. direction of…		
c. period of …		
d. rate of . . (express as degrees per time)		
e. evidence(s) for . . .		

46

8. The diagrams below show constellations observed at 9 p.m. and 11 p.m. on the same night.

Diagram 1 — 9:00 p.m.

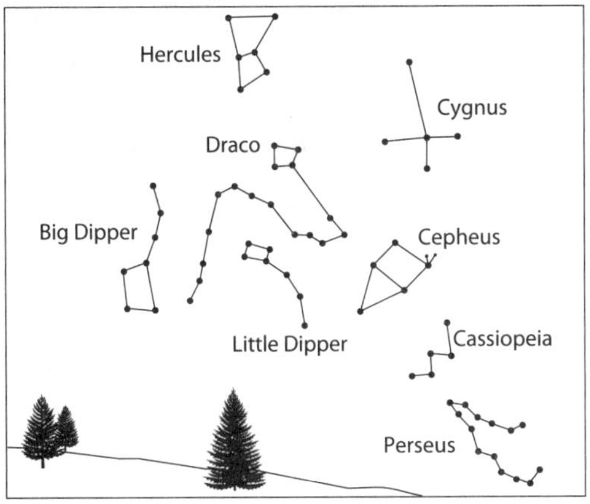

Diagram 2 — 11:00 p.m.

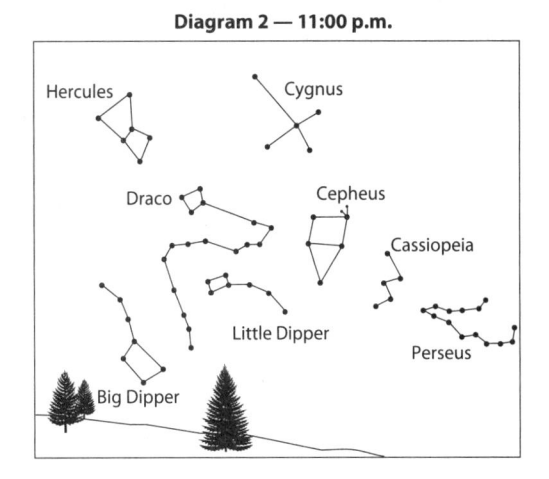

a. Circle and label *Polaris* in both diagrams.

b. Name the direction the observer is facing. _____

c. This observer is in the *(Northern) (Southern)* Hemisphere.

   How do you know? _____

d. What change has occurred from 9 p.m. to 11 p.m.? _____
   _____

e. Why has this change occurred? _____

f. Hercules is not a circumpolar constellation. During the night what will happen to this constellation? _____

g. Cepheus is a circumpolar constellation. On Diagram **2** use an arrow to draw the path that Cepheus will follow during the night.

LOCATING POSITIONS AND TIME ZONES ON EARTH

To locate positions on Earth, a coordinate system is used where every position has two quantities: latitude and longitude. Since Earth is spherical, degrees are used as the unit of measurement. A circle is 360 degrees. Each degree is divided into 60 arc minutes, and each minute is divided into 60 arc seconds. Every latitude and longitude position is written with a measurement and a compass direction; for example: 42°40' N, 66°00' W.

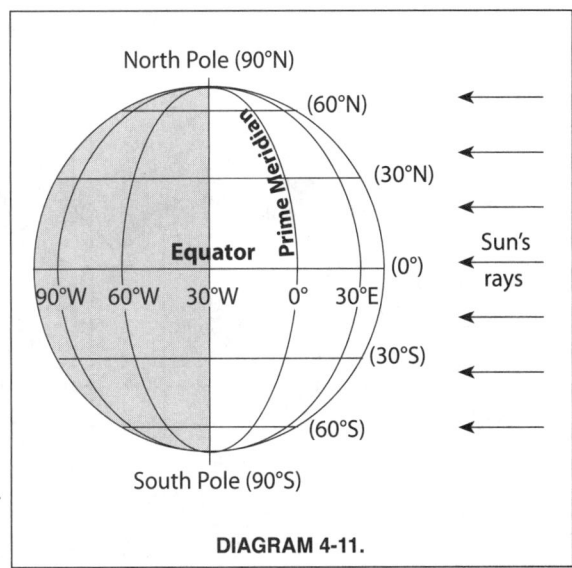

Latitude uses the **Equator** as its reference line. The Equator divides Earth into two hemispheres: Northern Hemisphere and Southern Hemisphere. Parallels of latitude "run" east and west parallel to the Equator and each other. **Latitude** is the angular distance north and south of the Equator. It ranges from 0° to 90° North and South. Latitude will remain the same if you move east or west. As you move north or south from the Equator your latitude will increase.

DIAGRAM 4-11.

47

Latitude is equal to the altitude of *Polaris* (North Star) in the Northern Hemisphere. Earth's axis of rotation points directly towards *Polaris*. During the night, *Polaris* does not rise or set but remains stationary while other stars circle around it. In the Southern Hemisphere, the pole star is a very dim star named *Sigma Octantis*. A **sextant** is the instrument used to measure the altitude of a star.

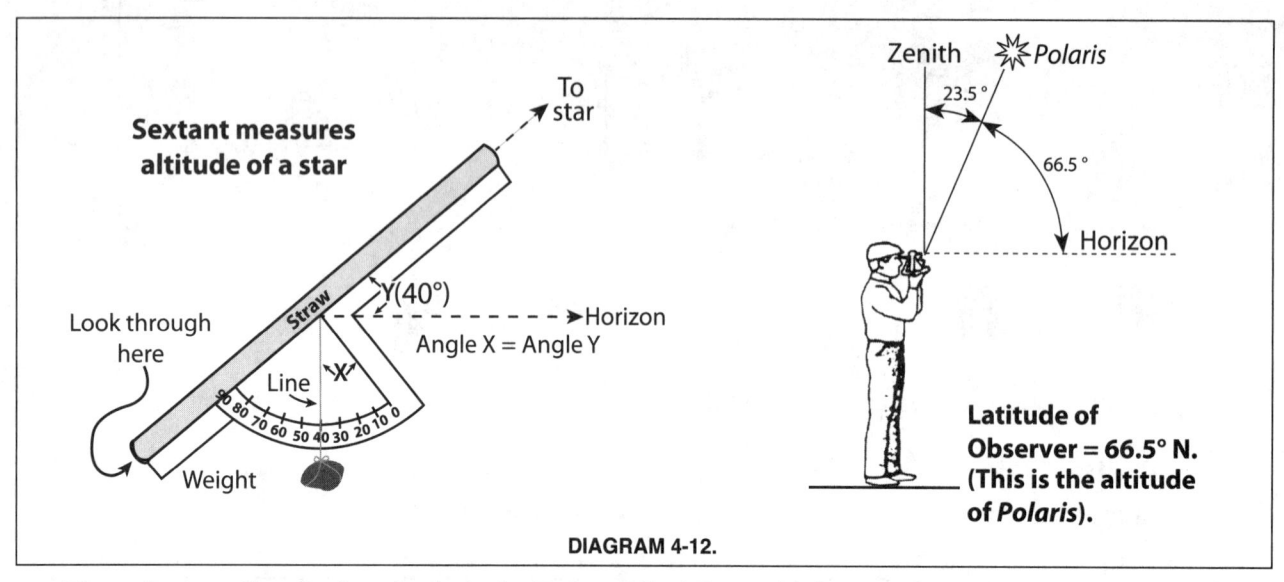

DIAGRAM 4-12.

The reference line for longitude is the **Prime Meridian** which goes through Greenwich, England. Meridians of longitude "run" north to south and meet at the poles. **Longitude** is the angular distance east and west of the Prime Meridian. It ranges from 0° to 180° East and West. Longitude only changes if you move east and west. It remains the same if you move north and south.

A **chronometer** indicates the time on the Prime Meridian which is known as **GMT** (Greenwich Mean Time). Longitude is based on the time difference between a known location and an unknown location. Every one-hour time difference equals 15° longitude. Locations to the east are ahead in time; locations to the west are behind in time. There are 24 time zones in the world, each separated by approximately 15° of longitude.

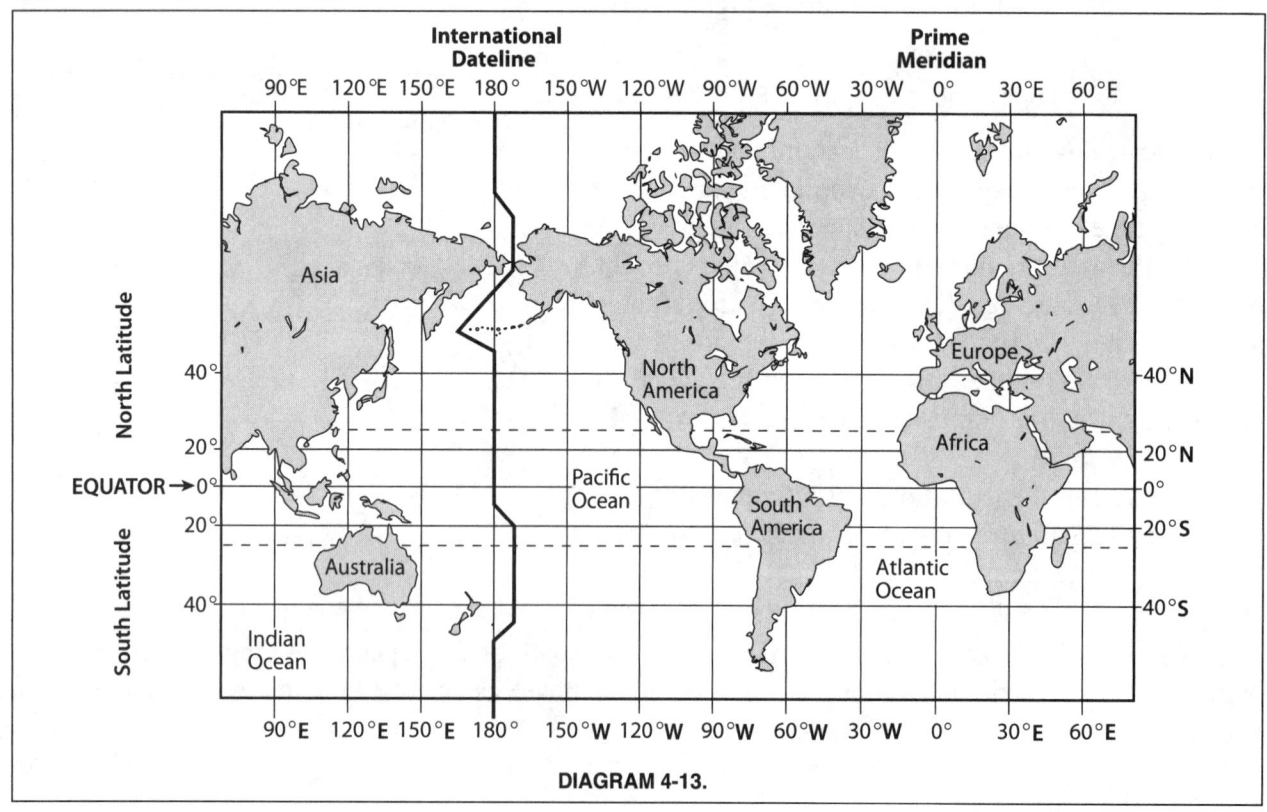

DIAGRAM 4-13.

DIAGRAM 4-14. TIME ZONES IN THE UNITED STATES.

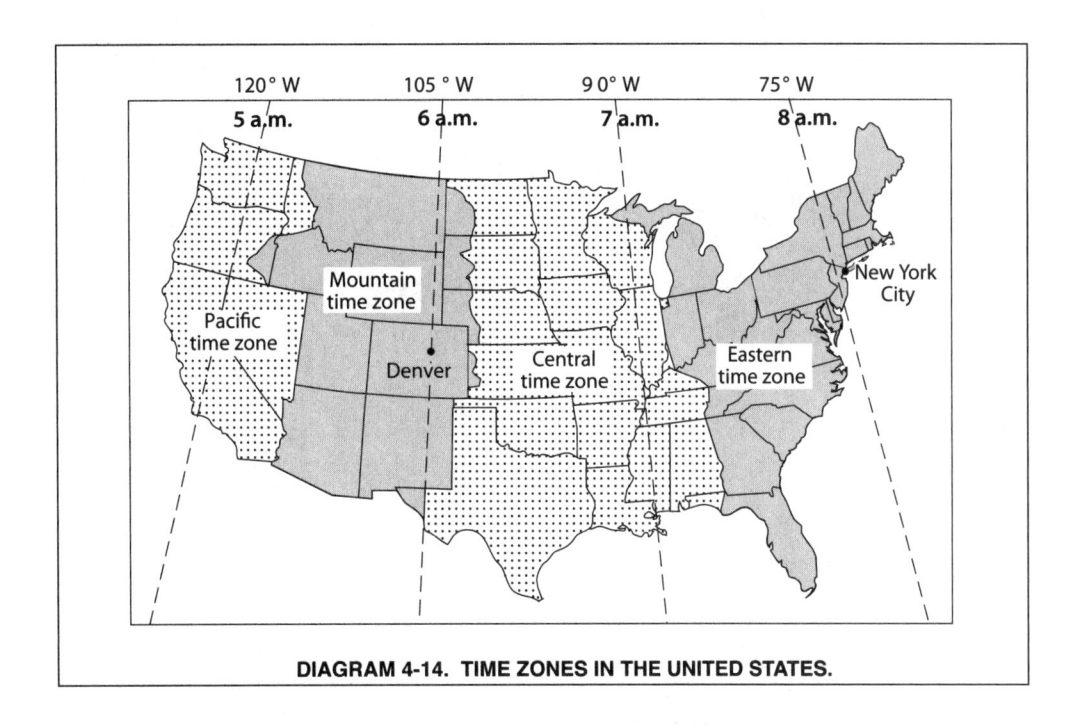

Questions

9. Complete the chart below to compare latitude and longitude.

	Latitude	Longitude
a. definition		
b. reference line (0°)		
c. range		
d. determined by		
e. will change if you move		

10. The diagrams below show observations of *Polaris* (North Star) or *Sigma Octantis* (South Star) at different locations on Earth's surface. Give the latitude for each observer.

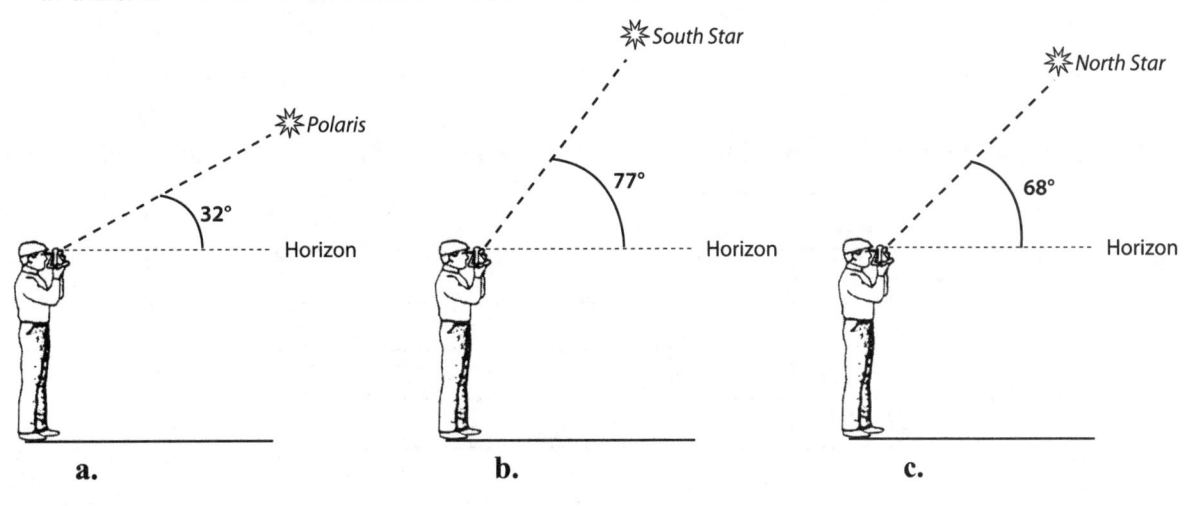

a.   b.   c.

49

11. Use the world map below.

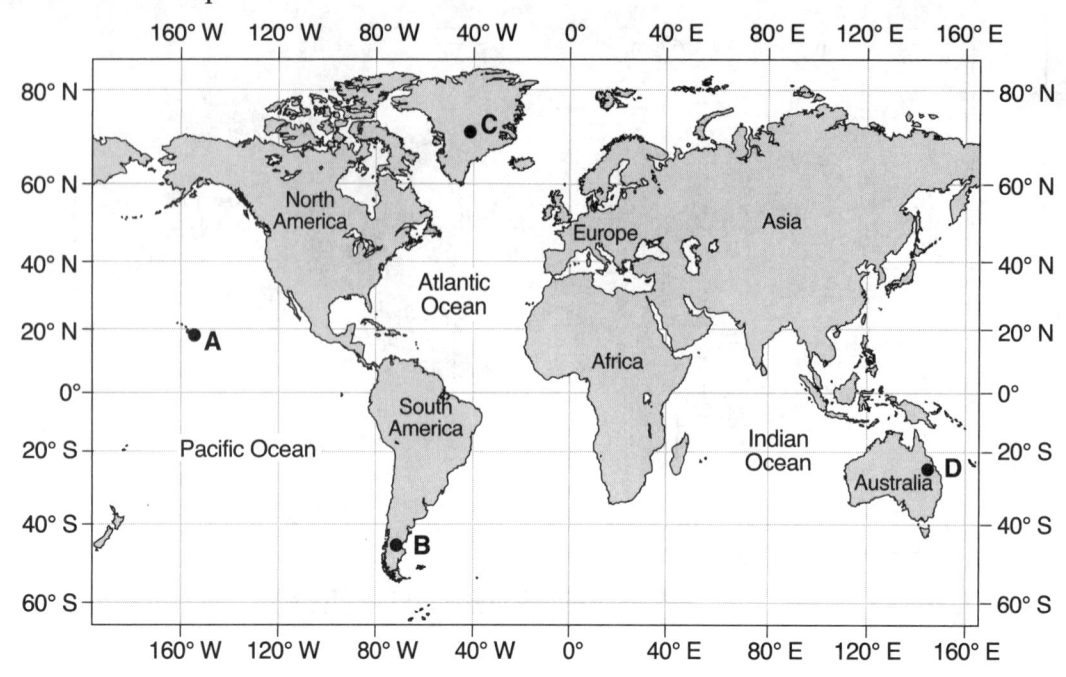

a. Draw in the *Tropic of Capricorn* and label by name and latitude.
b. Draw in the *Artic Circle* and label by name and latitude.
c. Label the Prime Meridian.
d. Place an "**X**" at the location that has the same value for latitude and longitude.

e. At which lettered position(s) will *Polaris* be highest in the sky? _____

f. At which lettered position(s) will *Polaris* <u>not</u> be visible? _____

Why not? _____

_____

g. Give the latitude and longitude to the nearest degree for the following positions.

position	latitude	longitude
A		
B		
C		
D		

h. Label these lettered points on the world map based on the given latitude-longitude.

	latitude, longitude	located in
E	50° N, 110° E	
F	5° S, 130° W	Pacific Ocean
G	25° S, 25° E	
H	35° N, 90° W	

50

12. A chronometer is a clock which indicates the time at Greenwich, England (GMT).

    a. What is the importance of Greenwich, England? _____

    b. If at your solar noon, the chronometer reads 10 a.m. What is your longitude? _____
       (show solution)

    c. It is 2 p.m. at your location when the chronometer reads 7 p.m.
       What is your longitude? _____   *(show solution)*

13. The diagram shows parallels of latitude that are 10 degrees apart and meridians of longitude that are 15 degrees apart.

    a. What is the latitude of the North Pole?
       _____ Equator? _____

    b. Name a city located on the Prime Meridian.
       _____

    c. What is the latitude and longitude for point "X"? _____

    d. A person at city **A** observes Polaris at 30 degrees above the horizon and is 2 hours ahead of "X". Locate city **A** on the diagram.

    e. Place a letter "S" at a location that has the same time as "X".

    f. Draw a dashed line to show the *Tropic of Cancer*. Label the line.

14. The diagram shows a part of Earth's surface and its latitude-longitude coordinates.

    a. The latitude of point **A** is _____

    b. The longitude of point **D** is _____

    c. Name *two* points with the same longitude?
       _____ & _____

    d. Name *two* points that have the same solar time.
       _____ & _____

    e. Which point(s) are ahead of **D** in solar time?
       _____

    f. What is the compass direction from point **A** to point **D**? _____

51

15. Complete the following chart for cities in New York State. Latitude and longitude should be written to the nearest degrees and arc minutes. (*One latitude is completed for you.*)

city	latitude	longitude
Oswego		
Ithaca		
Old Forge	43°45' N	
Niagara Falls		

16. The diagram below shows the angular altitude of *Polaris* above the horizon at a certain location.

   a. The latitude of the observer is _____

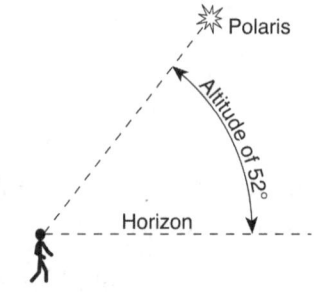

   b. Draw an arrow from the observer to zenith.
      Label the point "zenith".

   c. Place an "X" where Polaris would be for an observer at the Arctic Circle.

   d. How would this diagram change if the observer was in the Southern Hemisphere?
   _____

17. The diagram below shows a North Pole view of Earth, direction of Earth motion, and the shaded nighttime side of Earth. Some latitude and longitude lines have been labeled.

   a. Draw in three Sun rays to show the direction of the Sun for this diagram.

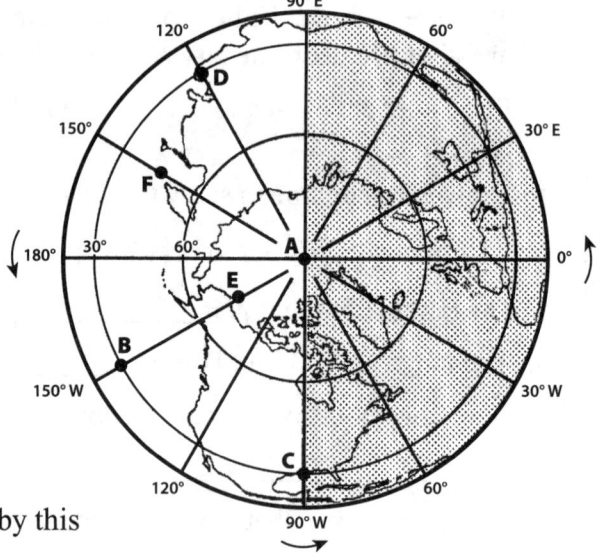

   b. What Earth motion is indicated by the arrows? _____

   c. The time at point **C** is approximately _____

   d. Which points have the same latitude? _____

   e. The latitude and longitude for location **D** is _____, _____

   f. The latitude and longitude for location **E** is _____, _____

   g. What day(s) of the year are represented by this diagram? _____

   h. The altitude of Polaris for observer at **C** would be _____

   i. If it is 8 a.m at location **D**, what time is it at location **F**? _____

52

CHAPTER 4 REVIEW

1. The bedrock of the continents and the sea floor are part of the
   (1) troposphere   (2) atmosphere   (3) hydrosphere   (4) lithosphere

2. In which group are the spheres of Earth listed in order of increasing density?
   (1) atmosphere, hydrosphere, lithosphere   (3) atmosphere, lithosphere, hydrosphere
   (2) lithosphere, hydrosphere, atmosphere   (4) hydrosphere, lithosphere, atmosphere

3. In the diagrams below the dark zone at the surface of each wedge of Earth represents the average depth of the oceans. Which segment is drawn most to scale?

4. Which diagram most accurately represents the shape of Earth?

5. Accurate measurements of Earth show Earth to be
   (1) greatest in diameter from Pole to Pole
   (2) greatest in diameter at the Equator
   (3) a perfect sphere

6. The polar circumference of Earth is 40,008 km. What is the equatorial circumference?
   (1) 12,740 km   (2) 25,000 km   (3) 40,008 km   (4) 40,076 km

7. A gravity meter is used to measure the gravitational pull at Earth's North Pole and Earth's Equator. How would these readings of gravitational pull compare?
   (1) The reading would be lower at the North Pole than the Equator.
   (2) The reading would be higher at the North Pole than at the Equator.
   (3) The readings would be the same at the North Pole and at the Equator.

8. Compared to the weight of a person at the North Pole, the weight of the same person at the Equator would be
   (1) slightly less, because the person is farther from the center of Earth
   (2) slightly less, because the person is closer to the center of Earth
   (3) slightly more, because the person is closer to the center of Earth
   (4) slightly more, because the person is farther from the center of Earth

9. The curving of planetary winds (Coriolis effect) is evidence for
   (1) Earth's rotation   (3) Earth's revolution
   (2) high and low pressure belts   (4) the tilt of Earth's axis

10. The time required for one Earth rotation is about
    (1) one hour   (2) one day   (3) one month   (4) one year

11. The diagram below shows a large pendulum in motion over an eight-hour time period. What is the main reason the pendulum appears to change its direction of swing over time?
    (1) tilt of Earth on its axis
    (2) rotation of Earth on its axis
    (3) revolution of Earth in its orbit
    (4) speed of Earth in its orbit

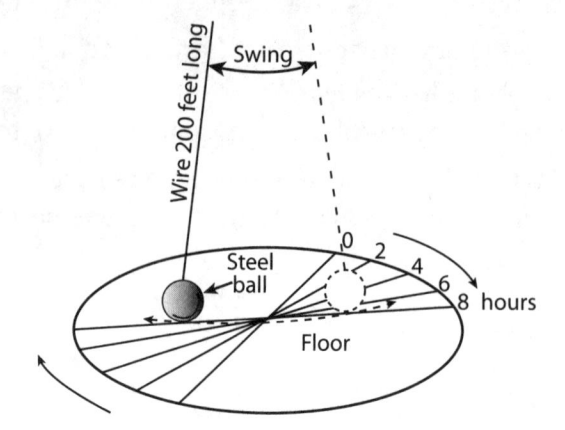

12. Which diagram correctly shows how surface winds are deflected (curved) in the Northern and Southern Hemispheres due to Earth's rotation?

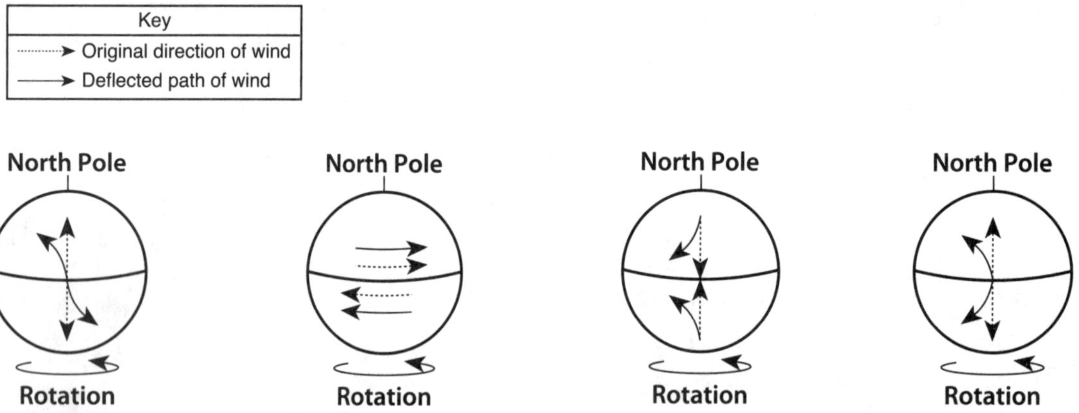

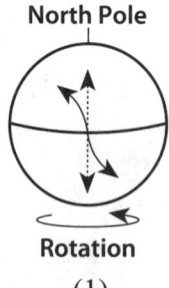

13. How would a three-hour time exposure photograph of stars in the northern sky appear if Earth did *not* rotate?

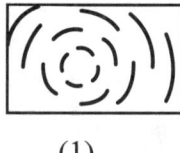

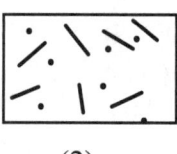

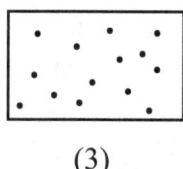

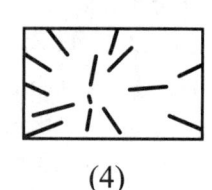

14. Which statement provides evidence that Earth revolves around the Sun?
    (1) Winds at different latitudes are deflected.
    (2) The Sun follows an apparent arc across the sky each day.
    (3) Stars appear to follow circular paths around *Polaris*.
    (4) Different star constellations are visible from Earth at different times of the year.

15. The day and the year, as units of time, are based on the motions of the
    (1) Earth   (2) Moon   (3) Sun   (4) distant stars

16. What is the total number of degrees Earth rotates on its axis in 12 hours?
    (1) 1°   (2) 15°   (3) 180°   (4) 360°

17. What change would occur if Earth's rate of rotation increased?
    (1) The day would be shorter.
    (2) The day would be longer.
    (3) The year would be longer.
    (4) The year would be shorter.

18. Which change would occur if Earth's rate of revolution were to decrease?
    (1) The day would be shorter.
    (2) The day would be longer.
    (3) The year would be longer.
    (4) The year would be shorter.

19. The time required for one Earth orbit is about
    (1) one hour
    (2) one day
    (3) one month
    (4) one year

20. Which statement best explains the apparent daily motion of the stars around *Polaris*?
    (1) Earth rotates on its axis.
    (2) Earth revolves around the Sun.
    (3) Earth's orbit is an ellipse.
    (4) Earth is an oblate spheroid.

21. To an observer in New York, stars appear to rise toward the
    (1) north
    (2) south
    (3) east
    (4) west

22. A person observes a star that is due east and just above the horizon. During the next two hours the distance between the star and the horizon will
    (1) decrease
    (2) increase
    (3) remain the same

23. Which motion can be classified as geocentric?
    (1) Earth rotates on its axis.
    (2) The Moon revolves around Earth.
    (3) Earth revolves around the Sun.
    (4) A meteoroid falls towards Earth's surface.

24. Why is *Polaris* used as a celestial reference point for Earth's latitude system?
    (1) *Polaris* always rises at sunset and sets at sunrise.
    (2) *Polaris* is located over Earth's axis of rotation.
    (3) *Polaris* can only be seen in the Northern Hemisphere.
    (4) *Polaris* is a very bright star.

25. As a ship crosses the Prime Meridian, the altitude of *Polaris* is 65°. What is the ship's location?
    (1) 0° latitude, 65° W longitude
    (2) 0° latitude, 65° E longitude
    (3) 65° N latitude, 0° longitude
    (4) 65° S latitude, 0° longitude

26. What is the difference in mean solar time between 30° N, 75° W and 30° N, 120° W?
    (1) 1 hour
    (2) 2 hours
    (3) 3 hours
    (4) 4 hours

27. The location of Slide Mountain is:
    (1) 74° 25'N, 42° W
    (2) 42° N, 74° 25' W
    (3) 42° N, 74° W
    (4) 42° 25' N, 74° 45' W

28. Which altitude of *Polaris* could be observed in Hawaii?
    (1) 10°
    (2) 22°
    (3) 43°
    (4) 150°

29. As a person travels south from Canada towards the Equator, the altitude of *Polaris* will
    (1) decrease
    (2) increase
    (3) remain the same

30. As a person travels due west across the United States, the altitude of *Polaris* will
    (1) decrease
    (2) increase
    (3) remain the same

31-33. The map below shows a portion of Earth's system of latitude and longitude and five locations.

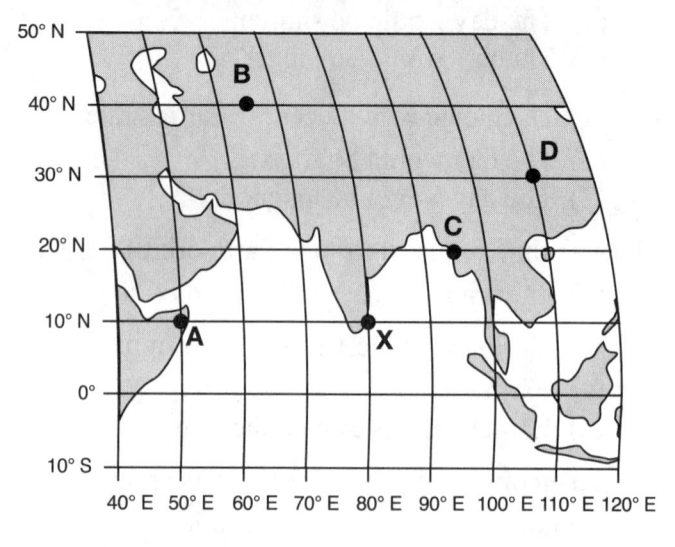

31. If it is solar noon at location **"X"**, at which location will solar noon occur next?
    (1) A (3) C
    (2) B (4) D

32. If it is 4 a.m. at location **A**, what time is it at location **B**?
    (1) 2 a.m. (3) 4 a.m.
    (2) 3 a.m. (4) 5 a.m.

33. At which location would Polaris be closest to the horizon?
    (1) A (3) C
    (2) B (4) D

34-35. At a location in North America a camera was placed outside and pointed towards the northern sky. The lens was kept open to produce a time exposure photograph. *Polaris* and six star trails are shown.

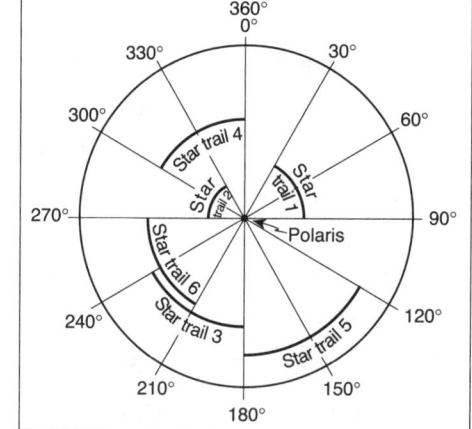

34. Why is *Polaris* stationary? _____
    _____
    _____

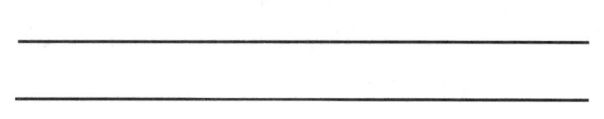

35. Calculate how many hours this film was exposed for. _____
    (*show solution*)

36-40. The diagram below shows the motions of Earth and Moon by lettered arrows **A, B, C,** and **D**.

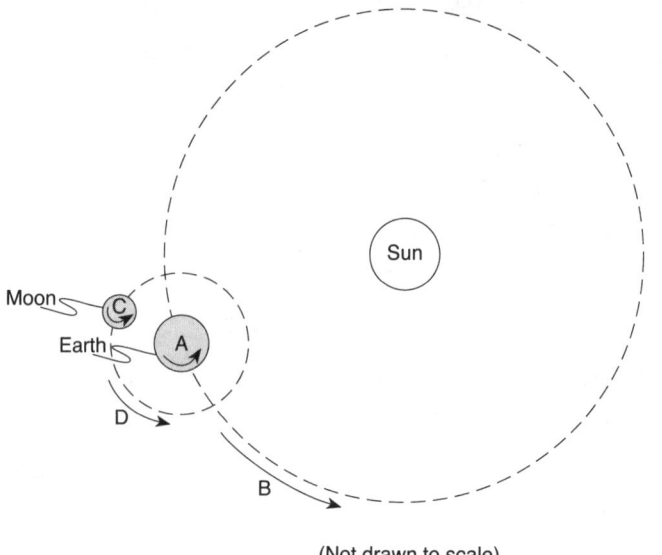

36. Which letter represents Earth's revolution? _____

37. Which letter represents the Moon's rotation? _____

38. As viewed from Earth the Moon would be in _____ _____ phase.

39. Place an **"X"** on Earth where solar noon would occur

40. Based on this diagram and the position of Earth as drawn, draw a circle on Earth's orbit to represents its position three months from now.

56

CHAPTER 5

SEASONS AND INSOLATION
Solar Time, Seasonal Sun Path, Insolation

SOLAR TIME

Every day the Sun appears to move across the sky because Earth rotates. The path of the Sun across our sky appears arc-shaped. The apparent path of the Sun changes slightly each day as the tilted Earth revolves around the Sun.

Time of day is determined by the position of the Sun in the observer's sky. The Sun rises towards the east and set towards the west. At solar noon the Sun will be at its highest position in the sky for the day. In the morning (a.m.) hours the Sun is in the eastern part of the sky; in the afternoon (p.m.) hours the Sun is in the western part of the sky. The Sun appears to move 15° per hour towards the west.

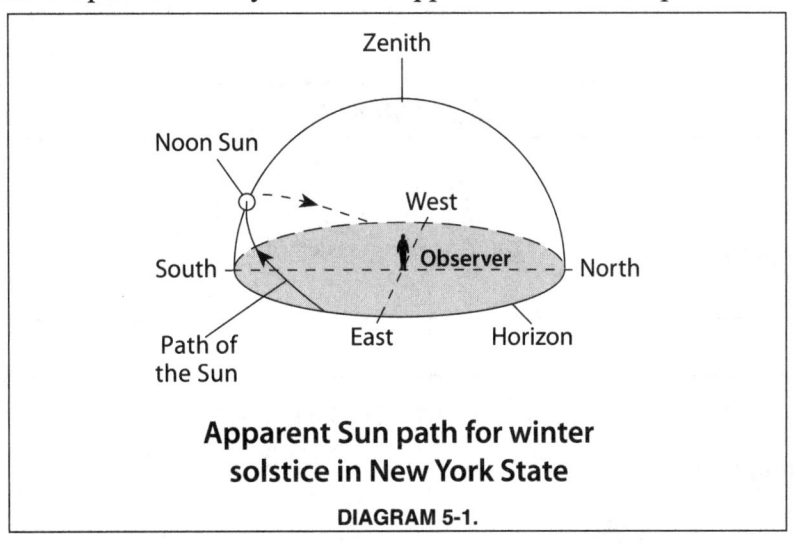

Apparent Sun path for winter solstice in New York State

DIAGRAM 5-1.

Only in the tropics, between the Tropic of Cancer (23½° N) and the Tropic of Capricorn (23½° S), will the noon Sun appear at zenith. Locations south of the tropics look to the north to see the noon Sun; locations north of the tropics look south to see the noon Sun.

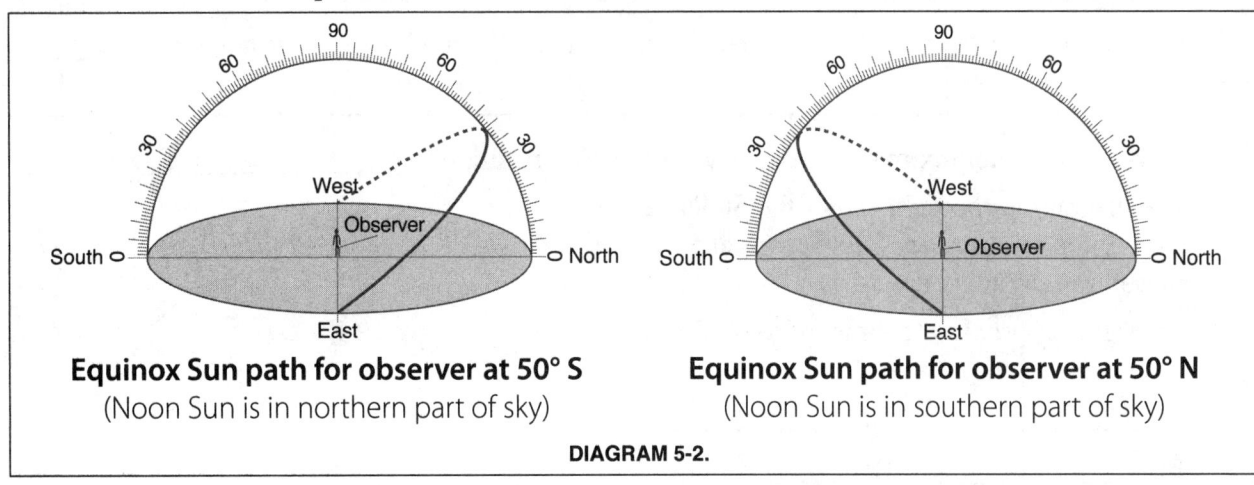

Equinox Sun path for observer at 50° S
(Noon Sun is in northern part of sky)

Equinox Sun path for observer at 50° N
(Noon Sun is in southern part of sky)

DIAGRAM 5-2.

As Earth rotates, the Sun's position will change at a rate of 15° per hour. The diagrams below show solar times for one moment on Earth for the equinoxes as seen from a polar view and a side view.

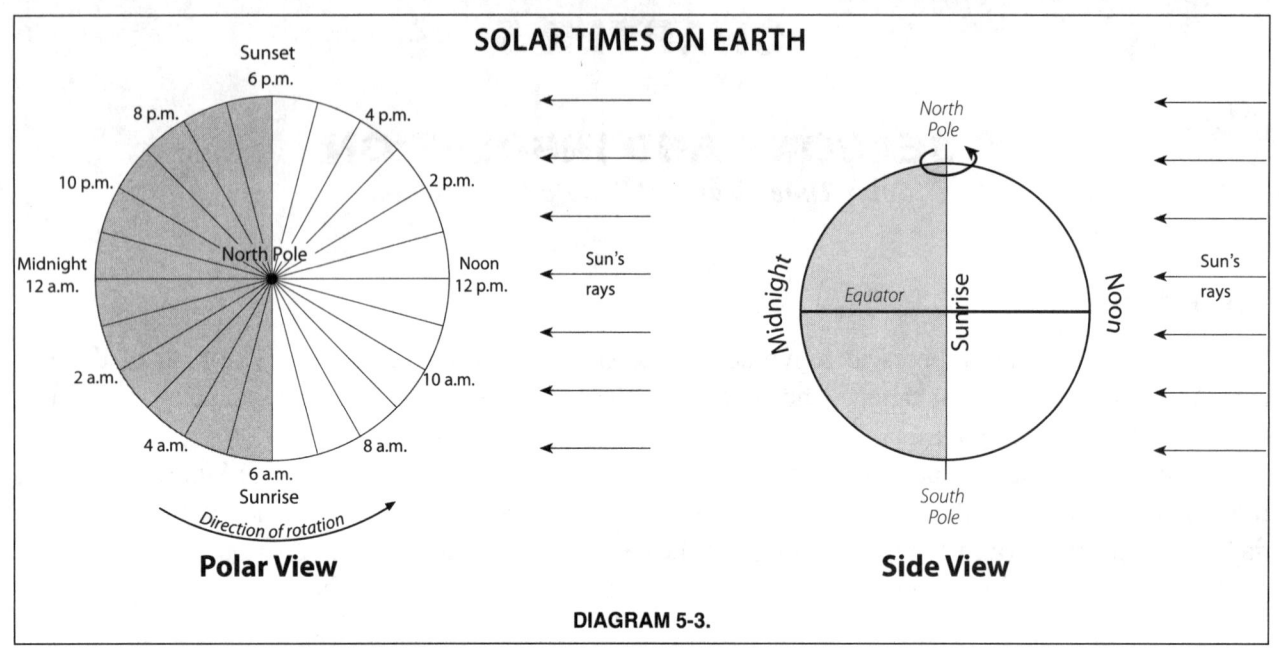

SOLAR TIMES ON EARTH

Polar View | **Side View**

DIAGRAM 5-3.

Questions

1. The time of day is based on the position of the _____ in the observer's sky.

2. Describe the position of the Sun in the observer's sky at:

   a. solar noon: _____

   b. sunset: _____

3. The diagram to the right shows the Sun's path for an observer on March 21, an equinox.

   a. Draw the noon Sun and label.

   b. Draw arrows to show the apparent direction the Sun moves during the day.

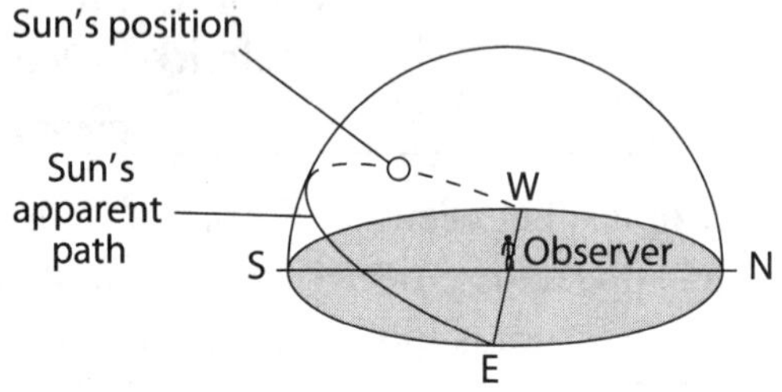

   c. Label the *zenith* point.

   d. This observer is in the Northern Hemisphere. How do you know this is correct? _____

   _____

   e. What is the approximate time of day for this Sun position? _____

4. The diagram to the right shows the Sun's path across the sky for an observer. Sunrise was at 5:00 a.m. and sunset was at 7:00 p.m.

   Give the approximate time of day for each Sun position.

   A: _____   C: _____

   B: _____   D: _____

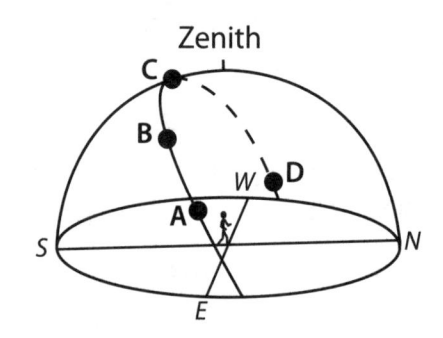

58

5. The diagram below shows a side view of Earth on March 21. The directions of the Sun's rays and Earth rotation are shown.

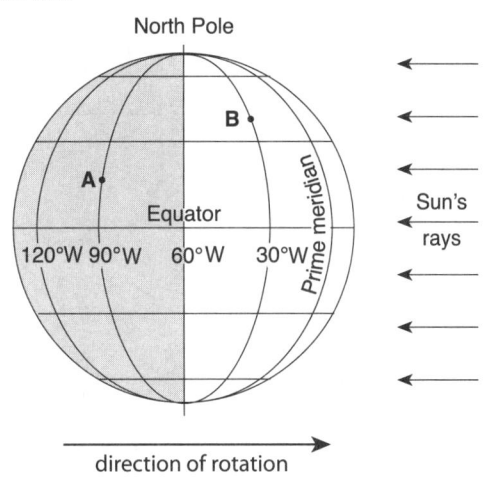

   a. Label with an **N** a position experiencing solar noon (12 p.m.)
   b. Label with an **M** a position experiencing midnight (12 a.m.)
   c. Positions along the 60°W meridian are experiencing (*sunrise*) (*sunset*).
   d. It is 4 a.m. at position **A**. What is the time for position **B**? _____

6. The diagram below shows a polar view of Earth on the Equinox when sunrise is at 6 a.m. and sunset is at 6 p.m. The meridians of longitude are labeled. Time zones are the same along each meridian of longitude. The directions of the Sun's rays and Earth's rotation are shown.

   Give the time of day for each lettered position.

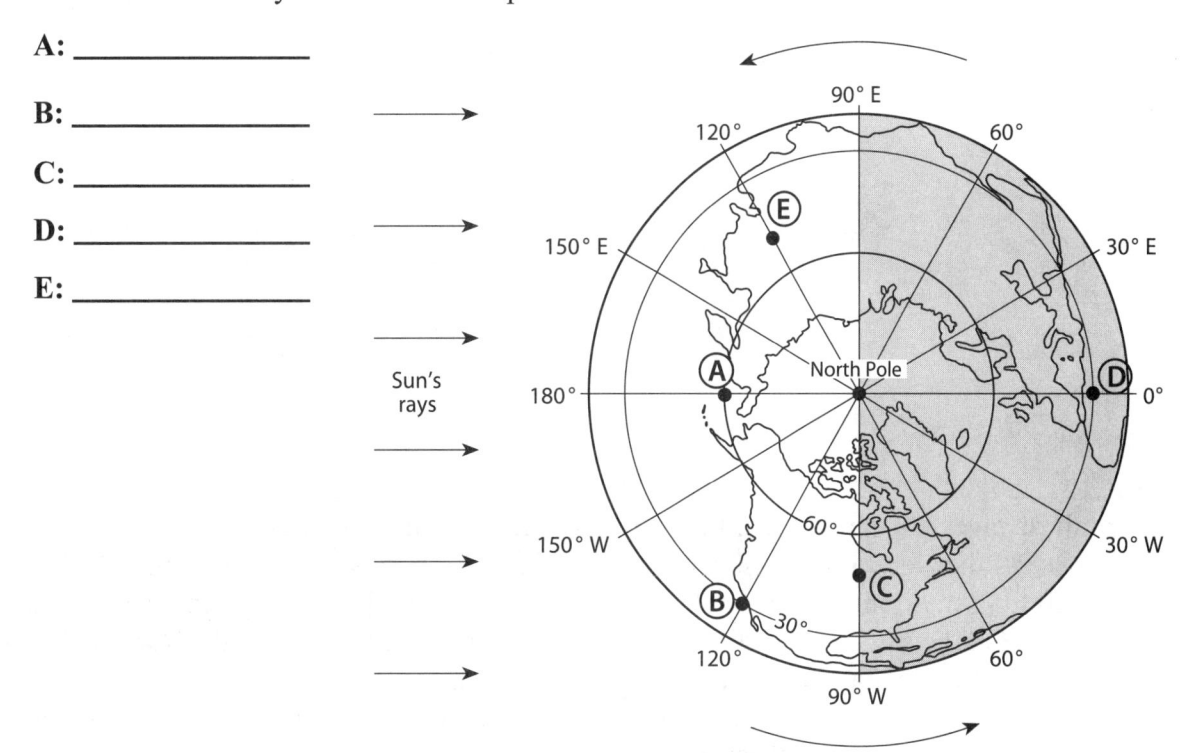

   A: _____
   B: _____
   C: _____
   D: _____
   E: _____

59

Seasonal Sun Path

The Sun's apparent path changes during the year in a cyclic pattern. The altitude of the noon Sun, direction of sunrise/sunset, and the length of day and night changes. As the path of the Sun varies, the seasonal weather changes.

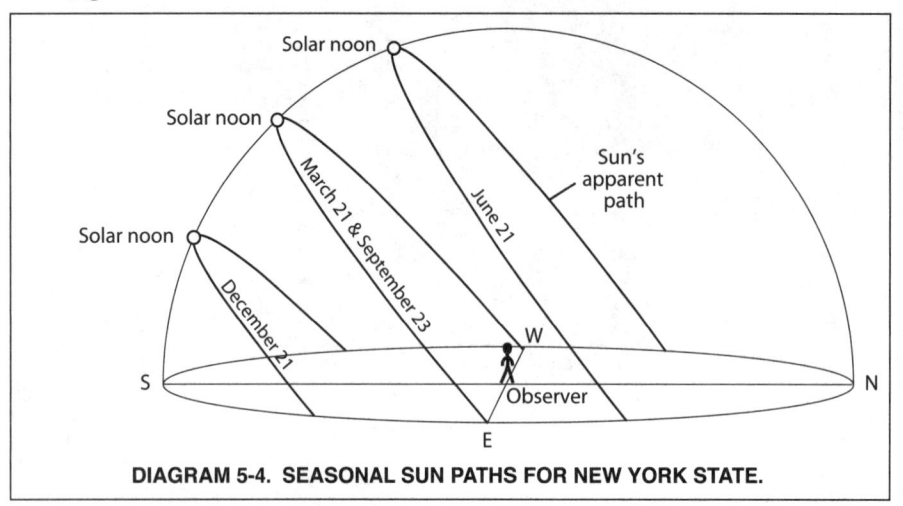

DIAGRAM 5-4. SEASONAL SUN PATHS FOR NEW YORK STATE.

There are three factors that cause the cyclic change in the seasonal Sun paths. The first cause is Earth's revolution around the Sun. The second is that Earth's axis is tilted at an angle of 23½° so different locations on Earth receive different intensities of sunlight. The third is that Earth's axis always points towards *Polaris*. This is called parallelism of the axis. These three factors cause the apparent Sun path to change throughout the year for everyone on Earth.

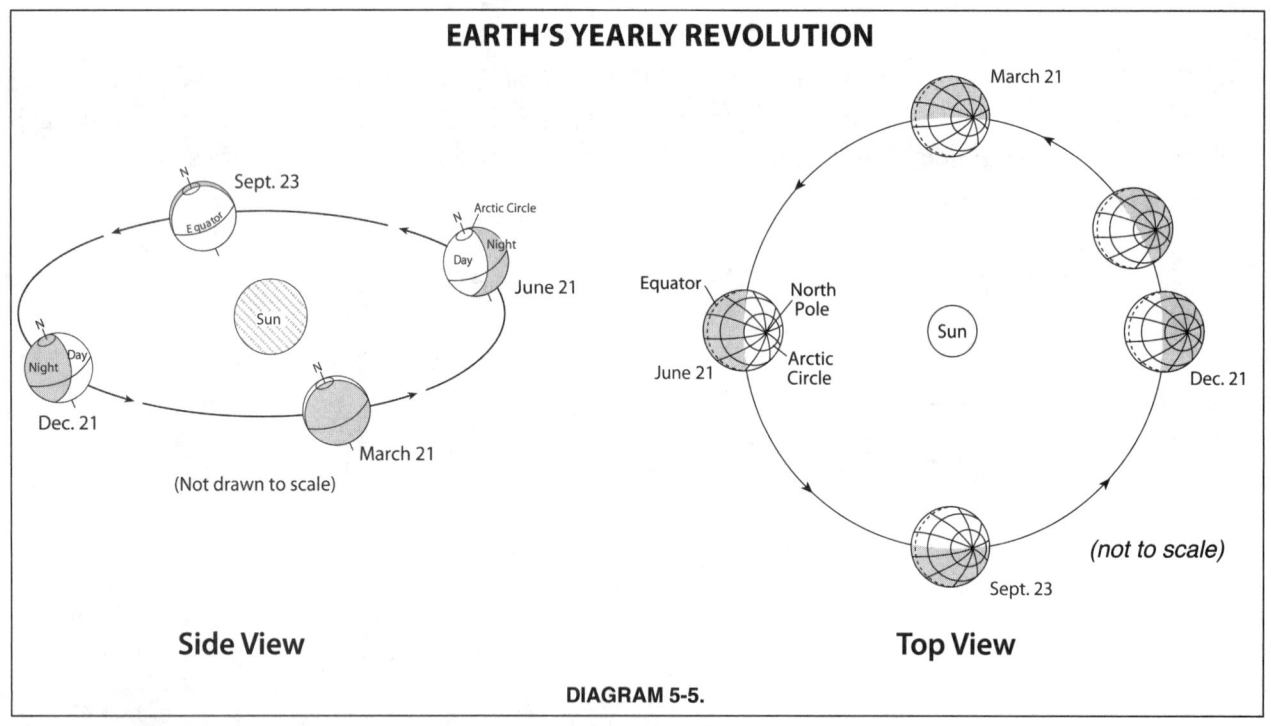

EARTH'S YEARLY REVOLUTION

Side View | Top View

DIAGRAM 5-5.

As Earth revolves around the Sun, the North Pole tilts towards the Sun and then away from the Sun. On June 21, the North Pole is tilted towards the Sun. From June 21 to December 21, the North Pole begins to tilt away from the Sun so that the Sun gets lower in the sky, daylight becomes shorter, and it gets colder in the Northern Hemisphere. From December 21 to June 21, the North Pole begins to tilt towards the Sun so that the Sun gets higher in the sky, daylight gets longer, and temperatures begin to rise in the Northern Hemisphere. In the Southern Hemisphere the seasonal Sun paths and seasons are opposite to that in the Northern Hemisphere.

TILT OF AXIS RELATIVE TO THE SUN AND SEASONAL EVENTS

December 21 Winter Solstice	March 21 & September 23 Spring & Fall Equinox	June 21 Summer Solstice
North Pole tilts away from Sun	the tilted Earth does not lean towards or away from Sun	North Pole tilts towards the Sun
near perihelion (closest to Sun)	average distance to Sun	near aphelion (furthest from Sun)
direct Sun ray at Tropic of Capricorn (23½°S)	direct Sun ray at Equator (0°)	direct Sun ray at Tropic of Cancer (23½°N)
area north of Arctic Circle in continual darkness	12 hours of darkness world wide	area south of Antarctic Circle in continual darkness
area south of Antarctic Circle in continual daylight	12 hours of daylight world wide	area north of Arctic Circle in continual daylight
Sun rises south of east, Sun sets south of west	Sun rises due east, Sun sets due west	Sun rises north of east, Sun sets north of west
Northern Hemisphere has low noon Sun	Northern Hemisphere has average noon Sun altitude	Northern Hemisphere has high noon Sun
Northern Hemisphere has shortest daylight	12 hours daylight world wide	Northern Hemisphere has longest daylight

Questions

7. The diagram below shows Earth at the four seasonal positions relative to the Sun.

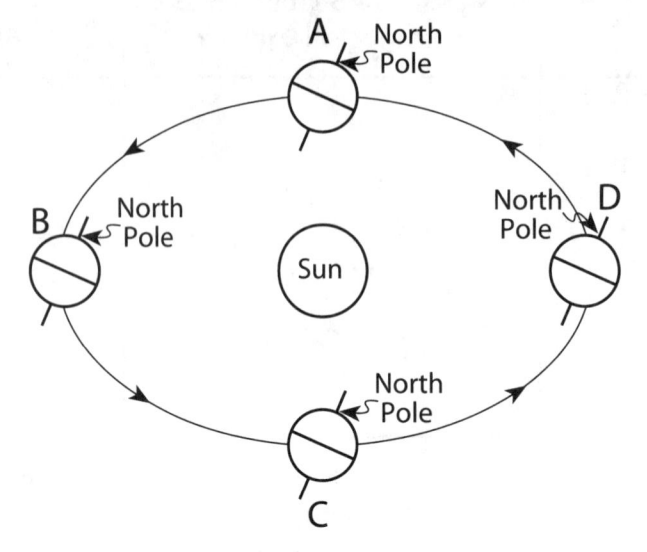

(Not drawn to scale)

   a. The arrows in the diagram represent the Earth motion of _____
   b. Label each position with seasonal date: *March 21, June 21, Sep 23, Dec 21*
   c. On position **B** label the *Equator*.
   d. On position **D** shade in areas experiencing nighttime.
   e. The time it takes for Earth to move from **A** to **C** is about _____ months.

8. The diagram to the right shows Earth relative to the Sun's rays for one day.

   a. Name the season. _____
   b. Draw the latitude line that receives the direct ray of the Sun. **Label** it's latitude.
   c. Carefully shade in the part of Earth in darkness.
   d. Which pole is having 24 hours of daylight? _____

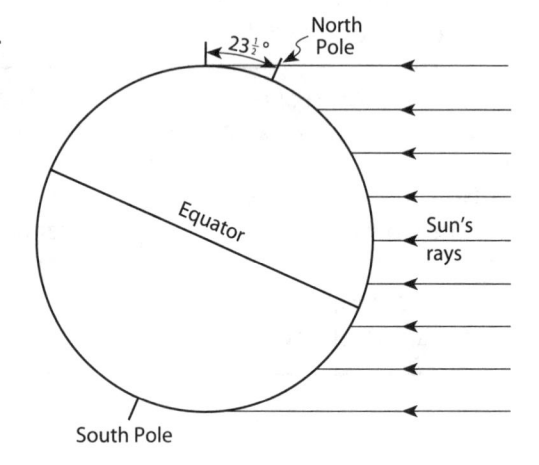

9. What would be the effect on summer and winter in the mid-latitudes if Earth was tilted:
   a. more than 23.5°? _____
   b. less than 23.5°? _____
   c. 0° (no tilt)? _____

62

10. For each statement select the seasonal date or dates when this event occurs.

    (March 21) (June 21) (September 23) (December 21)

    a. Spring Equinox. _____
    b. Winter Solstice. _____
    c. Earth is near perihelion. _____
    d. Sunrise is due east. _____
    e. Sunset is southwest. _____
    f. 12 hrs daylight worldwide. _____
    g. More direct sunlight for Northern Hemisphere. _____
    h. North of Arctic Circle has 24 hours of darkness. _____
    i. Longest daylight for NYS. _____
    j. Longest daylight for Australia. _____
    k. Noon Sun at zenith for Equator. _____
    l. Noon Sun is vertical at 23½°N. _____
    m. Antarctica has 24 hours of daylight. _____
    n. Sun is at lowest altitude for Florida. _____

11. Earth is closer to the Sun on January 3 than it is on July 3. Explain why it is colder in the Northern Hemisphere on January 3.

    _____

    _____

12. The diagram shows the apparent paths of the Sun in New York State for the first day of each season.

    a. Label on the diagram the directions: *East* and *West*

    b. Name the season for each Sun path:

       DAG: _____

       EBH: _____

       FCI: _____

    c. Zenith is how many degrees above the horizon? _____
    d. When the Sun is on the horizon it will have an altitude equal to _____.
    e. Which letter(s) represent a position of the noon Sun? _____
    f. Which letter is the sunrise point for the equinox? _____
    g. Which letter is the sunset point for the winter solstice? _____
    h. Which letter is the noon Sun for summer solstice? _____
    i. At which letter will the noon Sun cast the shortest shadow? _____
    j. When the Sun follows path **EBH**, how many hours of daylight are there? _____
    k. How will the Sun paths appear for next year? _____
    l. Why? _____

13. Use the diagram below of the celestial sphere which shows a Sun path for June 21. Daylight is approximately 16 hours. This observer is located at 50°N latitude.

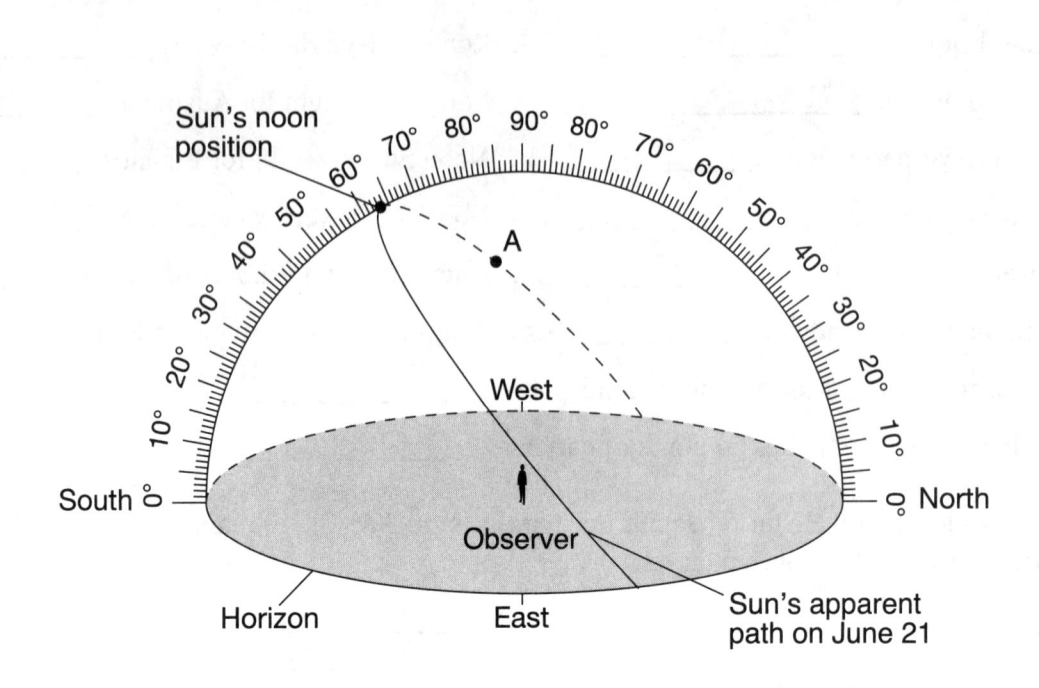

a. The altitude of noon Sun for June 21 is _____ above the *(north) (south)* horizon.

b. On the June 21 path, draw a circle to show where the Sun would be at 4 p.m.

c. Draw arrows on the path to show the direction of the Sun's apparent movement.

d. Why does the Sun appear to move across the sky each day? _____

e. Place a "*" where *Polaris* would appear in the sky. Label the * *Polaris*.

f. At this location the altitude of the noon Sun for March 21 is 40°. Draw in the Sun path for March 21 and label. What other date has this same path? _____

g. Draw December 21 Sun path and label. (*hint: altitude of noon Sun changes by 23.5°*)

h. The Sun path changes during the year because of _____
_____

i. How would these Sun paths appear for a location at 50° S latitude. _____
_____
_____

14. Describe the change in direction of sunrise and sunset from:

a. December to June: _____

b. June to December: _____

64

INSOLATION

Insolation (**in**coming **so**lar radi**ation**) is the energy Earth receives from the Sun. Insolation includes ultraviolet radiation, visible light, and infrared (heat) energy. Visible light is the most intense. The atmosphere absorbs most of the ultraviolet and infrared radiations, so very little of these energies reach Earth's surface.

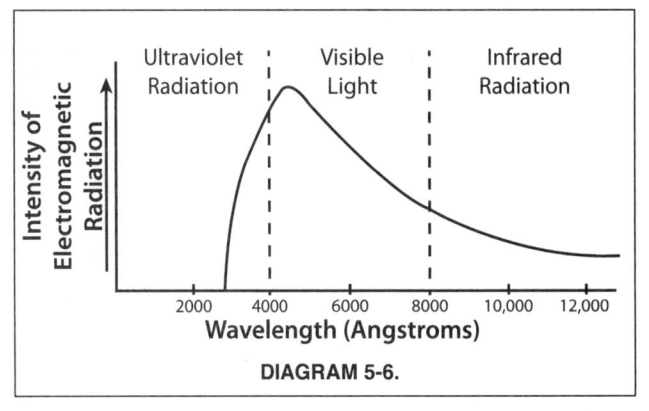

The **intensity of insolation** refers to the strength and amount of solar energy received. The intensity of insolation received by Earth's surface is affected by the atmosphere, the type of Earth surface, the angle (altitude) of the Sun, and the duration (length of daylight hours) of insolation.

DIAGRAM 5-6.

EFFECTS OF ATMOSPHERE AND EARTH'S SURFACE ON INSOLATION

Insolation must first pass through Earth's atmosphere. It can be **reflected** (returned unchanged) by clouds of ice crystals and/or raindrops. Aerosols (suspended solids and liquids in the air such as water droplets, ice crystals, dust, pollen, and soot) will also cause insolation to be reflected. Insolation can be **refracted** (bent) and **scattered** in different directions by the atmospheric aerosols and clouds. Reflection, refraction, and scattering reduce the intensity of insolation and cause lower temperatures. Insolation can be **absorbed** (taken in) by atmospheric gases. Ultraviolet radiation from the Sun is absorbed by ozone in the stratosphere. Infrared (heat) from the Sun is absorbed by greenhouse gases such as water vapor and carbon dioxide.

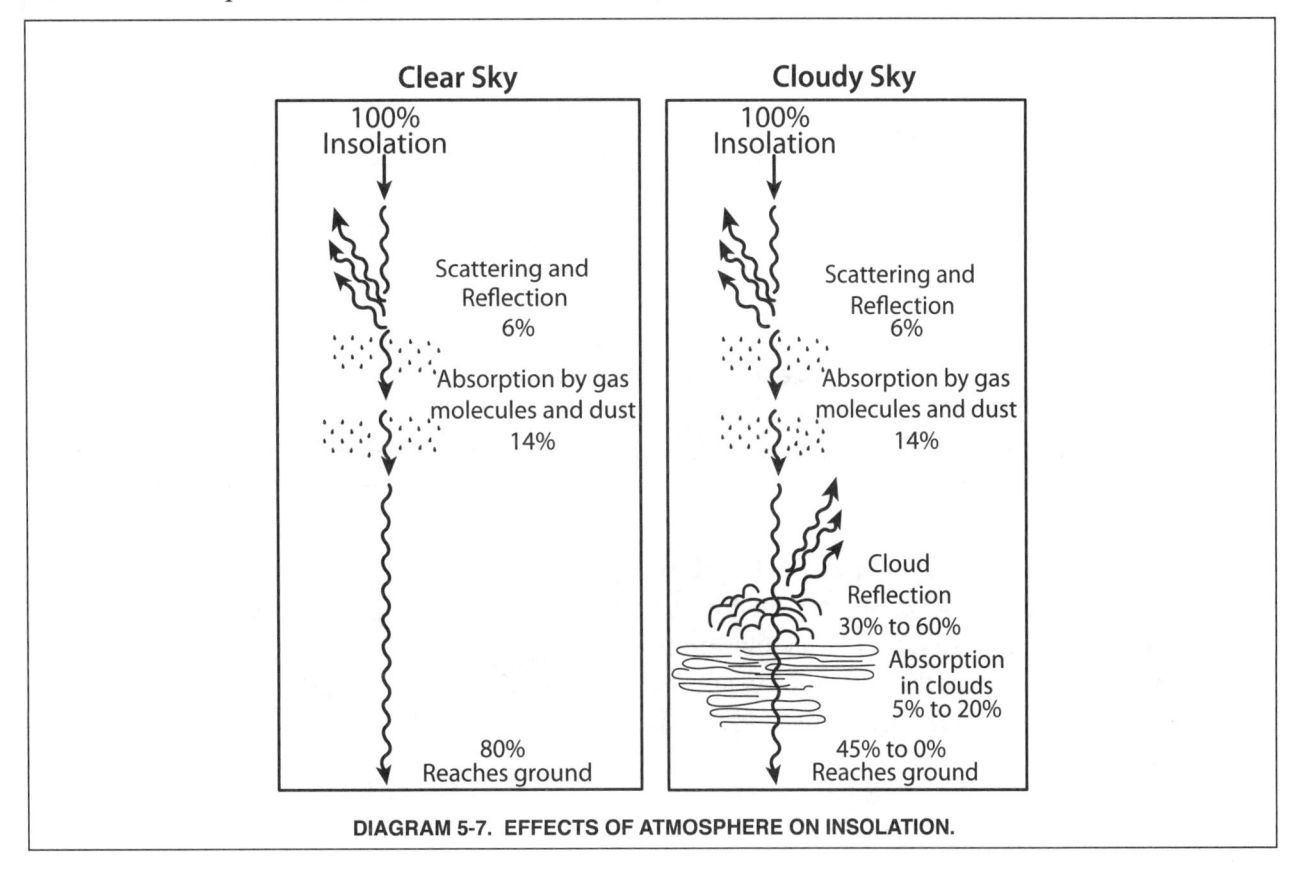

DIAGRAM 5-7. EFFECTS OF ATMOSPHERE ON INSOLATION.

Insolation that reaches Earth's surface will be absorbed and reflected in varying degrees depending on the surface properties. Dark, rough surfaces absorb more insolation and become warmer. Light, smooth surfaces reflect more insolation than they absorb. Vegetation and dark soil absorb insolation. Smooth water surfaces and snowfields will reflect insolation.

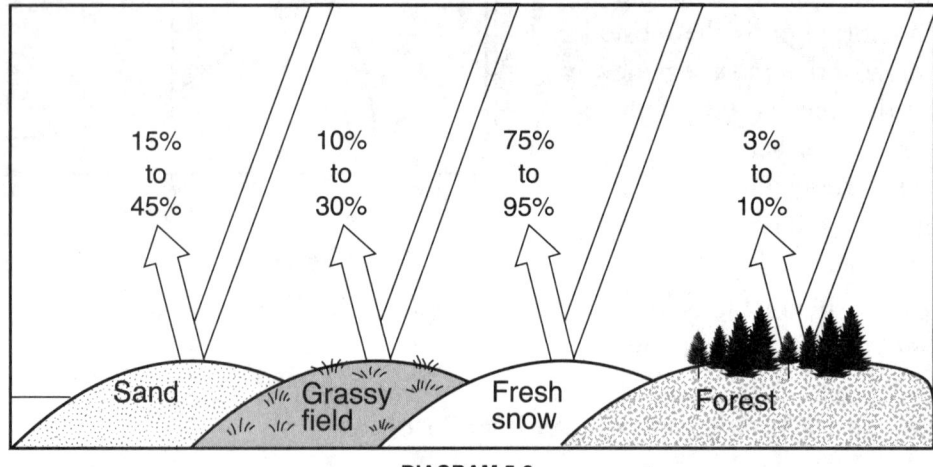

DIAGRAM 5-8.

Percent of insolation reflected off different surfaces:
- Sand: 15% to 45%
- Grassy field: 10% to 30%
- Fresh snow: 75% to 95%
- Forest: 3% to 10%

ANGLE OF INSOLATION

The **angle of insolation** is the angle at which the Sun's rays strike Earth's surface. It is measured as the altitude of the Sun above the horizon. The angle of insolation affects the amount of sunlight received by Earth. The intensity of insolation is greatest when the Sun is at zenith causing temperatures to increase. At low angles, there is more reflection, sunlight is spread out and less intense, and therefore it is cooler.

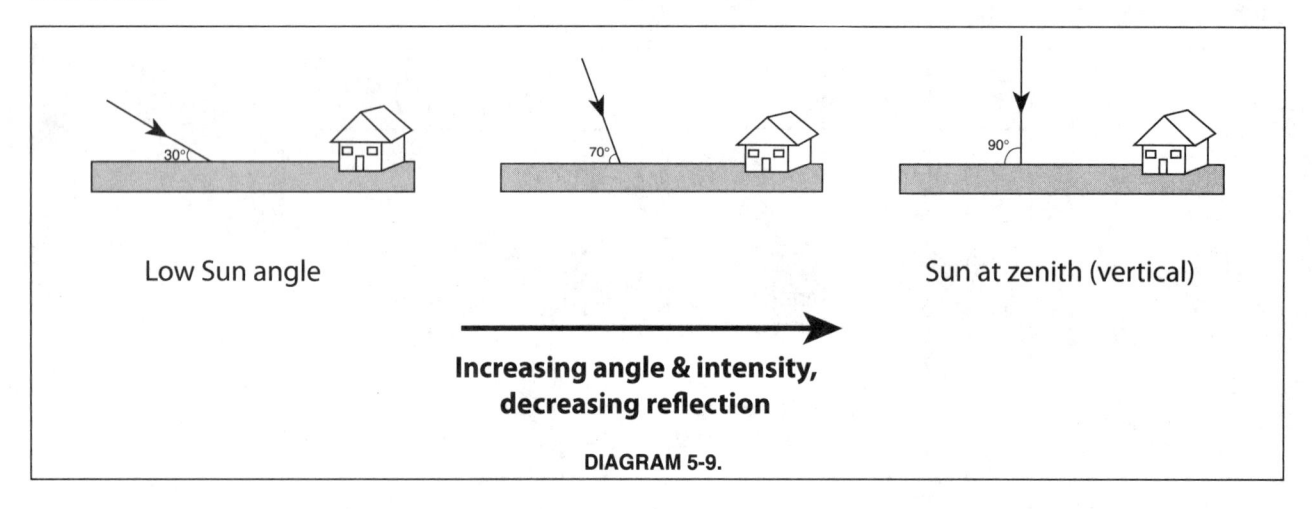

Low Sun angle (30°) → (70°) → Sun at zenith (vertical) (90°)

Increasing angle & intensity, decreasing reflection

DIAGRAM 5-9.

The angle of the Sun determines the length of shadows. A high Sun produces shorter shadows; a low Sun produces longer shadows. Shadows point in the opposite direction of the Sun. For example, if the Sun is in the southern sky, the shadow will point north.

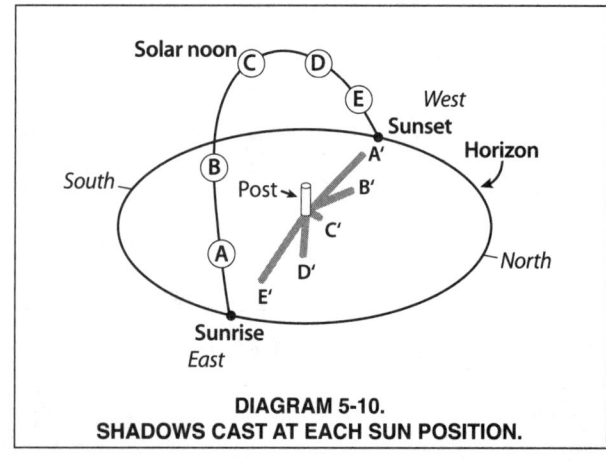

DIAGRAM 5-10.
SHADOWS CAST AT EACH SUN POSITION.

There are many factors that affect the angle of insolation. The angle is least at sunrise and sunset, and greatest at noon. Therefore the Sun is most intense at solar noon, but it is hottest at about 2 p.m. in the afternoon. This is known as a **heating lag**. Earth takes time to absorb insolation and reradiate the heat to the atmosphere. So although the Sun's intensity is greatest at solar noon, temperatures will continue to rise for the next few hours.

The altitude of the Sun also changes with the season due to Earth's tilt and revolution around the Sun. In the Northern Hemisphere, the maximum angle is on June 21, and the lowest angle is on December 21. However, June 21 is not the hottest day of the year, nor is December 21 the coldest day of the year. This is due to Earth's heating and cooling lags. It takes time for Earth to heat up and cool down, so that the hottest month is usually July and the coldest month is usually January in the Northern Hemisphere.

The Sun's altitude changes with latitude because Earth is a sphere. The Sun is highest in the sky in the tropics and lowest at the Poles. The direct, perpendicular ray of the noon Sun is only observed between the Tropic of Cancer ($23\frac{1}{2}°N$) and the Tropic of Capricorn ($23\frac{1}{2}°S$).

Season	Location of vertical Sun ray
Spring, Mar 21	Equator, 0°
Summer, Jun 21	Tropic of Cancer, 23½°N
Fall, Sept. 23	Equator, 0°
Winter, Dec 21	Tropic of Capricorn, 23½°S

DIAGRAM 5-11.

DURATION OF INSOLATION

Duration of insolation refers to the length of daylight. As duration of insolation increases, temperature increases. Duration varies with time of year and latitude. The Equator always has 12 hours of daylight. All locations on Earth have 12 hours of daylight on the Equinoxes. The North and South Poles have the greatest variation in daylight, it ranges from 0 to 24 hours.

Location	Duration on June 21	Duration on Dec 21	Duration on Mar 21 & Sept 23
North Pole	24 hours	0 hours	12 hours
42° N	15 hours	9 hours	12 hours
Equator	12 hours	12 hours	12 hours
42° S	9 hours	15 hours	12 hours
South Pole	0 hours	24 hours	12 hours

INSOLATION AND LATITUDE

The diagram to the right illustrates how insolation varies with time of year and latitude. The Equator receives nearly constant high insolation which peaks on the equinoxes when the Sun reaches zenith. The North and South Poles receive the least insolation due to low Sun angles, reflection off snow and ice, and 24 hours of darkness in the winter. The maximum insolation in the Northern Hemisphere is received on June 21.

The Equator receives the most constant insolation because of high Sun and a constant 12 hours duration of insolation every day. The Polar regions have low Sun and the duration of insolation varies from 0 to 24 hours.

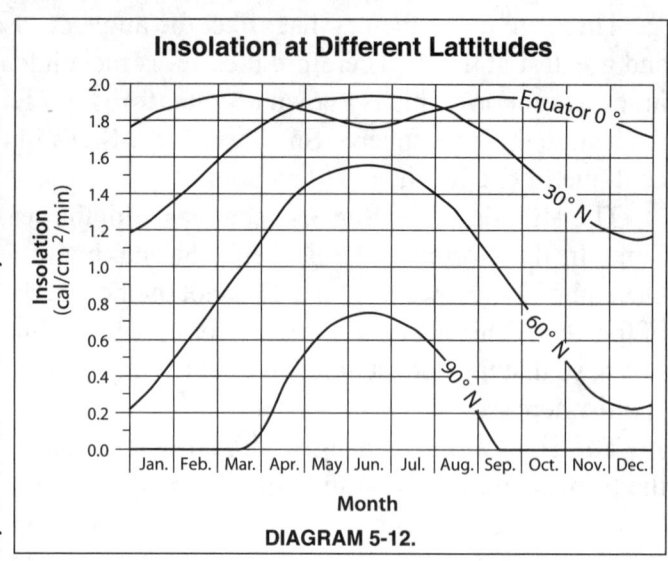

Insolation at Different Lattitudes

DIAGRAM 5-12.

Polar zone: low angles of insolation

Tropics: high angles of insolation

Polar zone: greatest seasonal variation in duration of insolation, 0 to 24 hours

Equator: 12 hours duration of insolation every day

DIAGRAM 5-13. THE ANGLE AND DURATION OF INSOLATION VARY WITH LATITUDE.

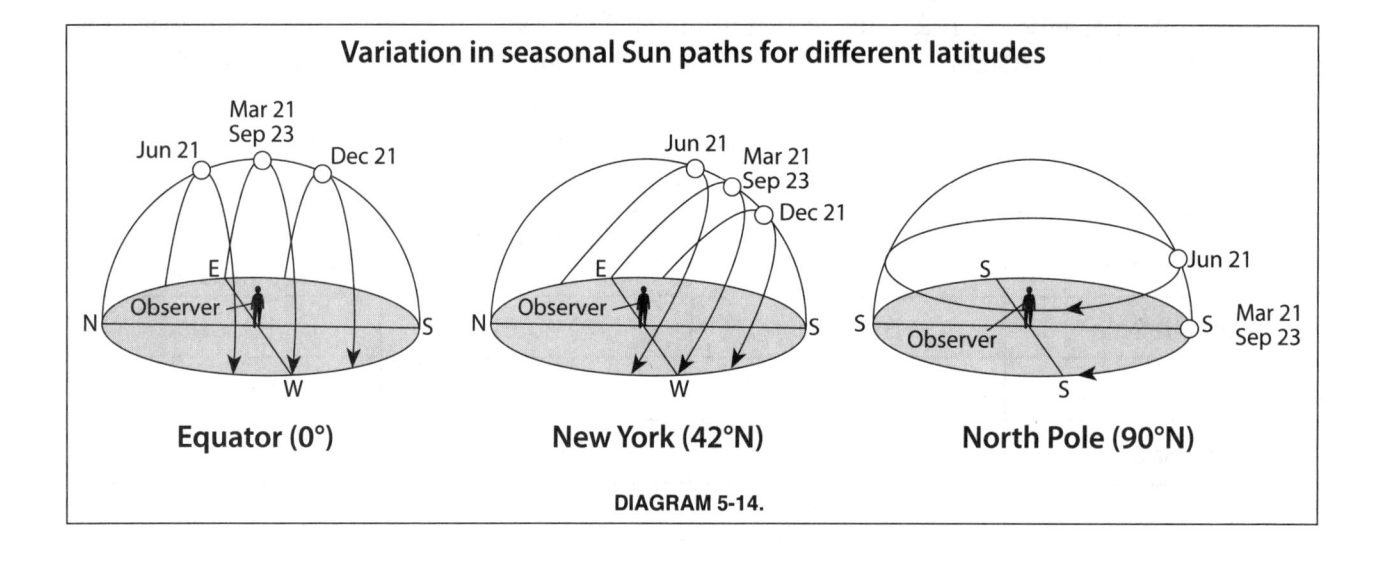

Variation in seasonal Sun paths for different latitudes

Equator (0°) New York (42°N) North Pole (90°N)

DIAGRAM 5-14.

Questions

15. State how the Sun's altitude and the length of Sun path are related to the season.

_____

_____

16. What is the range of latitude in which the noon Sun is directly overhead (at zenith) at some time during the year? _____

17. Describe the relationship between latitude and annual intensity of insolation. _____

_____

18. How does the time of maximum insolation differ for the Northern Hemisphere and the Southern Hemisphere? _____

19. Describe the change in angle of insolation from sunrise to sunset. _____

_____

20. Describe the change in angle of insolation for the United States from:

    a. June 21 to December 21: _____

    b. December 21 to June 21: _____

21. Describe the change in duration of insolation for the United States from:

    a. summer to winter: _____

    b. winter to summer: _____

22. Describe the change in duration of insolation from the Equator to the North Pole on:

    a. March 21: _____

    b. June 21: _____

    c. September 23: _____

    d. December 21: _____

23. On June 21, the Arctic Circle to the North Pole receives 24 hours of daylight.

    a. What causes this? _____

    b. What factors account for it not getting very warm despite continual sunlight? _____

_____

_____

24. The diagram shows Earth's tilt relative to the Sun for one day in the year.

   a. The season in the Southern Hemisphere would be _____

   b. The date when Earth is tilted this way relative to the Sun is _____

   c. The number of degrees between the Equator and the Tropic of Cancer is _____

   d. The number of hours of daylight for the South Pole is _____

   e. The number of hours of daylight for the Equator is _____

   f. The number of degrees between the Arctic Circle and the North Pole is _____

   g. Location which receives the Sun's most intense vertical ray is _____

   h. The time of day at **A** and **B** is _____

25. The diagram to the right shows the noontime shadow cast by a pole in New Jersey on March 21.

   a. On the diagram write the compass direction for each lettered position.

   b. Draw the shadow cast by the pole as the Sun sets for this day.

   c. Describe how the noontime shadow will be different for

   June 21: _____

   Dec 21: _____

   d. What other day of the year will have the same noontime shadow as shown? _____

   e. Describe the noon time shadow for this day at the Equator: _____

   f. For this location the shadows cast during the day will never point towards **A**. Why not ?

   _____

26. The diagram below shows lettered positions on Earth's surface on a certain date. The altitude of the noon Sun is shown for locations **W**, **X**, and **Z**.

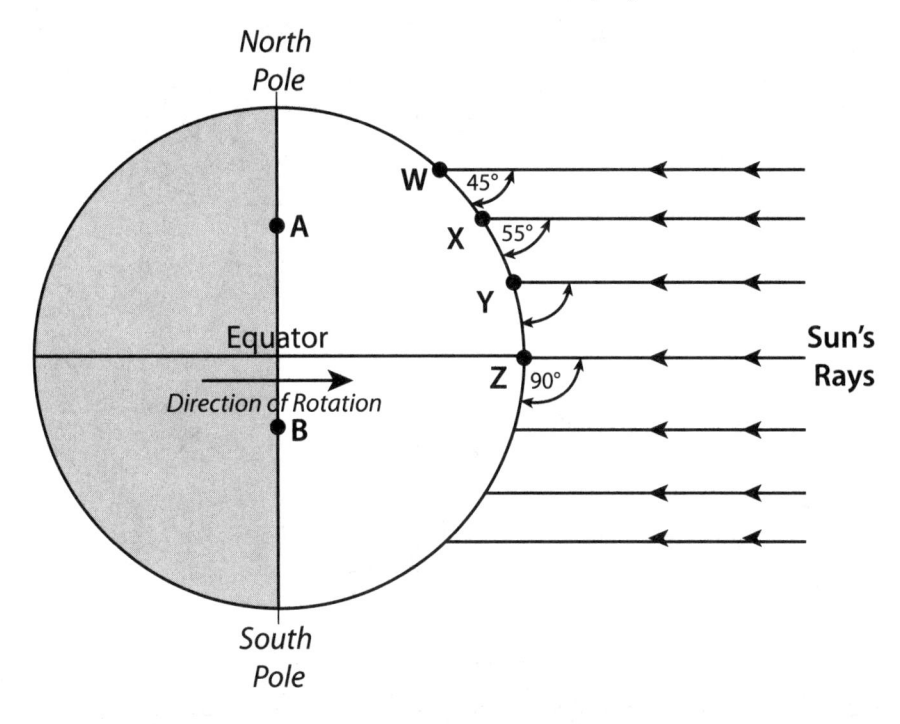

   a. What date(s) could Earth be in this position? _____

   b. The approximate altitude of the noontime Sun at position **Y** would be _____

   c. At which lettered position is the noontime Sun most direct? _____

   d. The time of day at location **A** is closest to _____

   e. To see the sunrise, an observer at **X** would look towards the _____

   f. Positions **W**, **X**, **Y** and **Z** will have the same _____

   g. On the diagram below draw in Earth's axis of rotation as it would appear on December 21.

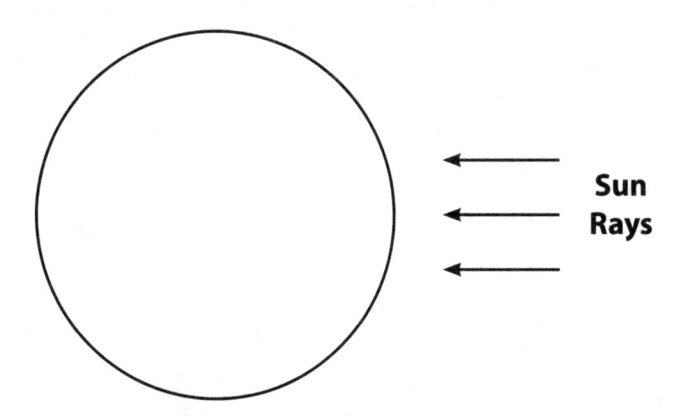

71

27. The diagrams below show the Sun paths for four different locations on **March 21**.

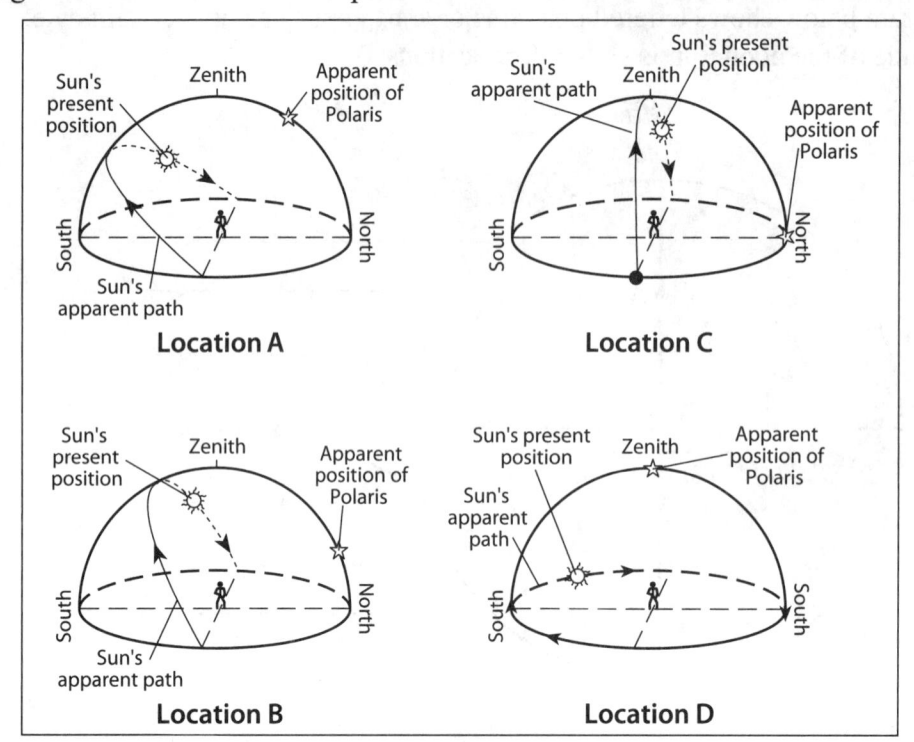

Location A

Location C

Location B

Location D

a. Based on the Sun's present position, the observer at location **A** casts a shadow towards what compass direction? _____

b. As the Sun sets for observer at location **A**, the shadows will become *(longer) (shorter)*.

c. Based on the present Sun position, the approximate time of day for location **B** is _____.

d. The intensity of insolation is greatest for location **C** than the other three locations. Why? _____

e. Location **C** is at the Equator. How do you know? _____

f. The observer at **C** will not cast a shadow at noon for this day. Why not? _____
_____

g. Location **D** is at a higher latitude than the other locations. State one way that this conclusion can be made from the diagram. _____
_____

h. What is the latitude for location **D**? _____

i. State the other day of the year when the Sun's apparent paths will be the same as these four paths. _____

j. What is the approximate duration of insolation for locations **A**, **B**, and **C**? _____

k. For locations **A**, **B**, and **C** the Sun will set in the _____

l. What observation about the Sun paths confirm that these are for the Spring Equinox?
_____

m. For location **A**, the Sun path for June 21 will be
(higher and longer) (higher and shorter) (lower and shorter) (lower and longer)

CHAPTER 5 REVIEW

1. What time of day is occurring at point **X**?

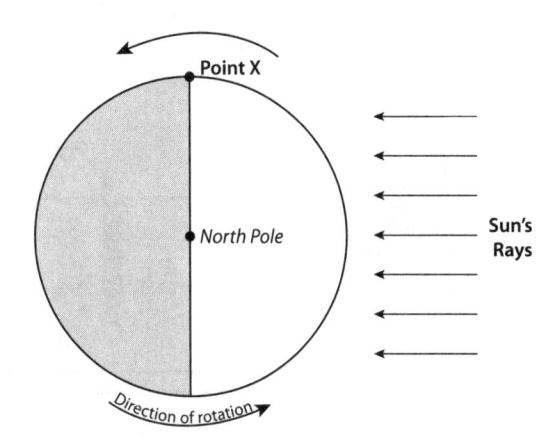

   (1) 12 a.m.   (2) 6 a.m.   (3) 12 p.m.   (4) 6 p.m.

2. Seasonal changes on Earth are caused by
   (1) parallelism of the Sun's axis as the Sun revolves around Earth
   (2) changes in the distance between Earth and the Sun
   (3) elliptical shape of Earth's orbit around the Sun
   (4) tilt of Earth relative to the Sun as Earth orbits the Sun

3. During which season is Earth closest to the Sun?
   (1) spring   (2) summer   (3) fall   (4) winter

4. This diagram represents Earth at four different positions **A**, **B**, **C**, and **D** in its orbit around the Sun. Between which positions would New York State be experiencing summer season?
   (1) A and B
   (2) B and C
   (3) C and D
   (4) D and A

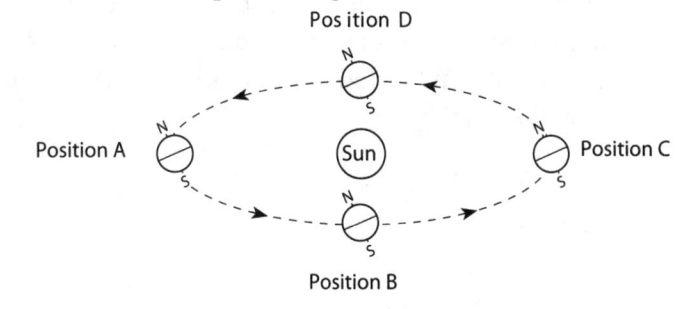

   (Not drawn to scale)

5. Which of these cities has the most daylight on December 21?
   (1) Sydney, Australia   (3) Reykjavik, Iceland
   (2) Miami, Florida   (4) London, England

6. On which date is the vertical ray of the noon Sun south of the Equator?
   (1) January 1   (2) April 1   (3) July 1   (4) September 1

7. The diagram shows the noontime shadow cast by a flagpole in New York State. Which letter indicates a location west of the pole?

    (1) A
    (2) B
    (3) C
    (4) D

8. The diagram shows the shadows cast by a vertical pole at various times during the day. Which shadow was cast when the greatest intensity of insolation was being received?

    (1) A
    (2) B
    (3) C
    (4) D

9. On a given day, which factors have the most effect on the amount of insolation received at any location on Earth's surface?
    (1) longitude and elevation
    (2) longitude and time of day
    (3) latitude and time of day
    (4) latitude and elevation

10. The lowest surface air temperatures in the Southern Hemisphere usually occur during the month of
    (1) January
    (2) April
    (3) July
    (4) October

11. From March 21 to September 23, the altitude of the noon Sun as observed at the Equator will
    (1) decrease
    (2) increase
    (3) increase then decease
    (4) decrease then increase

12. On September 23, as latitude increases, the angle of insolation will
    (1) decrease
    (2) increase
    (3) remain the same

13. The graph shows the relationship between the altitude of the noon Sun at various times during the year for a given location. What latitude would have this data shown in the graph?

    (1) 23.5° N
    (2) 66.5° N
    (3) 23.5° S
    (4) 66.5° S

14. In New York State, the length of a shadow cast by a tree at noon from January to May will
    (1) continuously decrease
    (2) continuously increase
    (3) remain the same
    (4) increase then decrease

15. At which latitude will the Sun never appear directly overhead at noon?
    (1) 42° S
    (2) 20° N
    (3) 0°
    (4) 5° S

16. During which month will the Sun rise north of east in New York State?
    (1) February
    (2) July
    (3) October
    (4) December

17. On June 21, some Earth locations have 24 hours of daylight. These locations are between the latitudes of
    (1) 0° and 23½° N
    (2) 23½° N and 23½° S
    (3) 30° N and 66½° N
    (4) 66½° N and 90° N

18. The diagram shows the apparent paths of the Sun in NYS for June 21 and December 21. Which statement best explains the cause of this apparent change in the Sun's path?
    (1) The Sun's orbital velocity changes as it revolves around Earth
    (2) Earth's orbital velocity changes as it revolves around the Sun
    (3) Earth axis is tilted 23.5°
    (4) The Sun's axis is tilted 23.5°

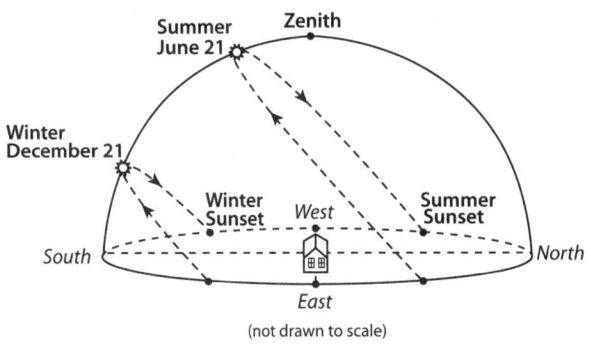

19. Which graph shows the relationship between angle of insolation and intensity of insolation?

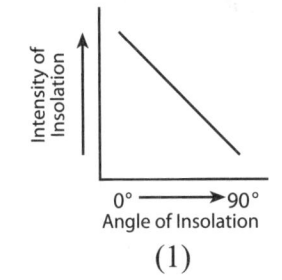

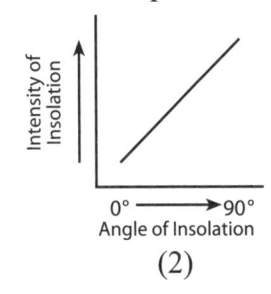

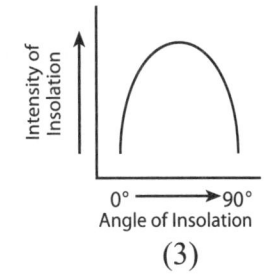

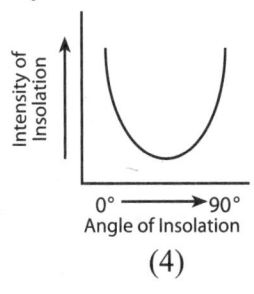

20. Which diagram best represents an apparent path of the Sun for December 21?

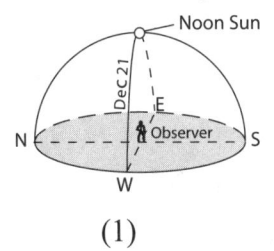

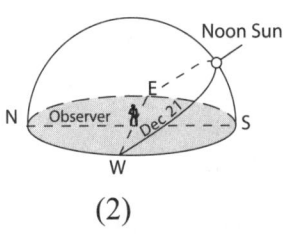

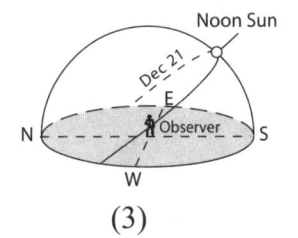

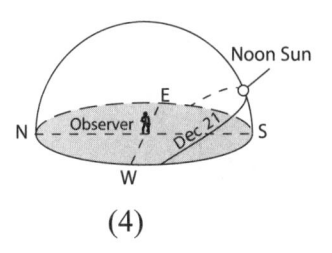

21. Which latitude and date does Earth receive the most intense insolation?
    (1) Tropic of Cancer on June 21
    (2) Equator on December 21
    (3) Antarctic Circle on June 21
    (4) South Pole on March 21

22. Which latitude and date receives the greatest duration of insolation?
    (1) Tropic of Cancer on June 21
    (2) Equator on March 21
    (3) Antarctic Circle on December 21
    (4) South Pole on June 21

23. In New York State, the number of hours of daylight continuously increases from
    (1) March 1 to May 1
    (2) June 1 to August 1
    (3) September 1 to November 1
    (4) December 1 to February 1

24. This diagram shows the apparent paths of the Sun for the first day of each season in New York.

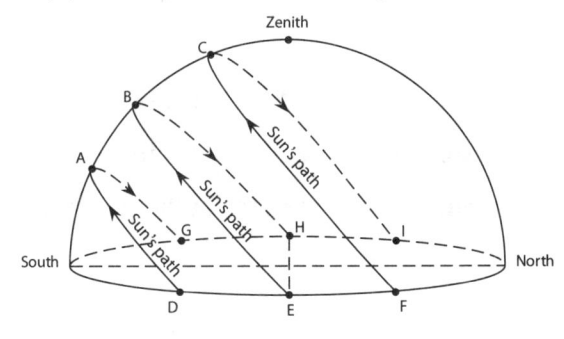

    Which point represents sunrise on the first day of winter?
    (1) A   (2) G   (3) F   (4) D

25. A student read in a newspaper that the maximum length of daylight for the year had just occurred for Boston, Massachusetts. What was the date of the newspaper?
    (1) March 21   (2) June 21   (3) September 23   (4) December 21

26. Which type of surface absorbs the greatest amount of insolation on a sunny day?
    (1) smooth, shiny, and dark in color
    (2) rough, dull, and dark in color
    (3) smooth, shiny, and light in color
    (4) rough, dull, and light in color

27. What is the cause of the usual decrease in temperature from sunset to sunrise?
    (1) cloud cover   (2) ground radiation   (3) strong winds   (4) heavy precipitation

28. Which form of radiation given off by Earth causes the heating of the atmosphere?
    (1) X-ray
    (2) visible light
    (3) ultraviolet radiation
    (4) infrared radiation

29. A student in New York drew the following diagram to show the positions of sunrise, **A**, **B**, and **C**, at three different times during the year.

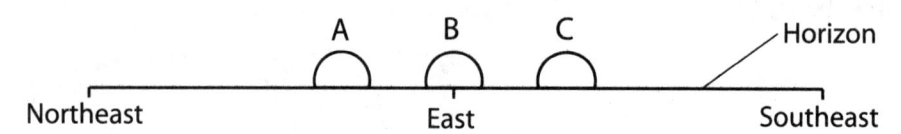

    Which list correctly pairs the location of sunrise to the time of year?
    (1) A - June 21   (2) A - December 21   (3) A - March 21   (4) A - June 21
        B - March 21   B - March 21   B - June 21   B - March 21
        C - September 23   C - June 21   C - December 21   C - December 21

30. Which graph best represents the relationship between the time of day and the length of a shadow cast by a tree on March 21?

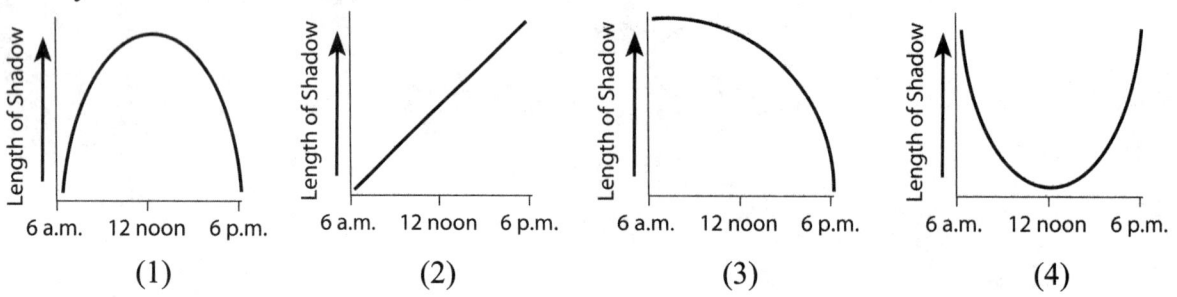

31-33. The diagrams show the Sun path at four different locations at different times during the year.

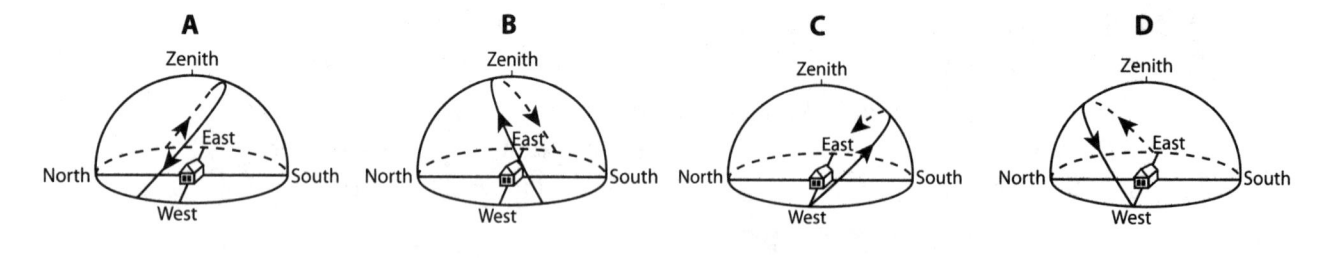

31. Which Sun path(s) are for December 21? _____

32. Which Sun path(s) are for the Equinoxes? _____

33. Which Sun paths(s) are for Southern Hemisphere? _____

34-39. The diagram shows the observations of the Sun and *Polaris* made by a sailor who fell off a ship and floated to a deserted island.

34. Draw an arrow on the Sun path to show the direction of apparent movement of the Sun during the day.

35. Place an "X" where the Sun is at solar noon.

36. Draw the Sun path for March 21.

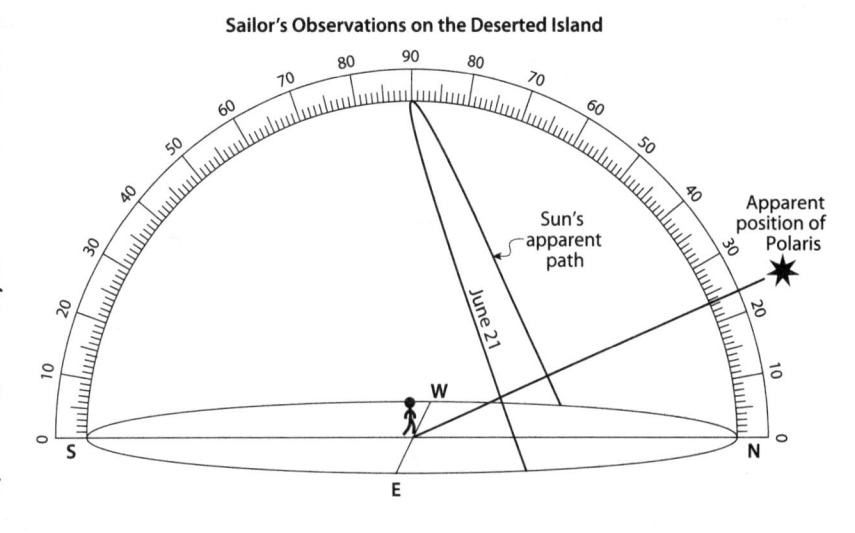

37. This labeled path is for June 21. How do you know this is correct? _____
_____

38. What is the latitude of this deserted island? _____
How do you know? _____
_____

39. The sailor noted that at solar noon on the island, his chronometer read 8 a.m. What is the longitude of this island? _____ (*show solution*)

40-43. Base your answers on the diagram of a house located in Pennsylvania.

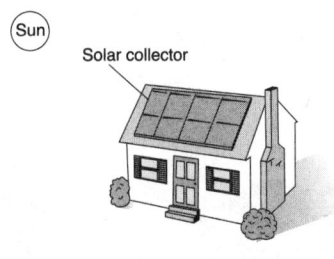

40. Explain why solar energy can still be collected on a cloudy day. _____
_____

41. State one advantage of using solar energy instead of burning fossil fuels to produce electrical energy. _____
_____

42. Based on this house's geographic location, the solar panel should be placed on the roof facing _____ for best results.

43. The best color for the solar collector is (*black*) (*white*) because _____
_____

77

44-47. The diagram below shows four apparent paths of the Sun, labeled **A**, **B**, **C**, and **D**, observed in Jamestown, New York. The June 21 and December 21 paths are labeled. Compass directions are shown along the horizon.

44. The greatest duration of insolation is along path _____

45. At what time of day is the Sun at position **S**?

_____

46. Give an approximate date when the Sun will appear to travel path **B**.

_____

47. Place an "X" on the December 21 path where the Sun would be at 3:00 p.m.

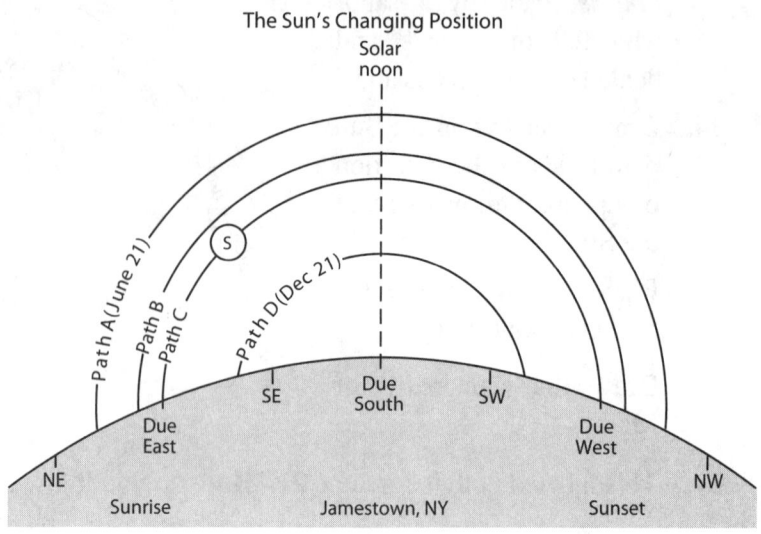

48-50. The diagram below shows Earth's orbit around the Sun as viewed from space. Earth is at eight different positions, labeled **A** through **H**. Earth's North Pole, Arctic Circle and Equator have been labeled at position **C**.

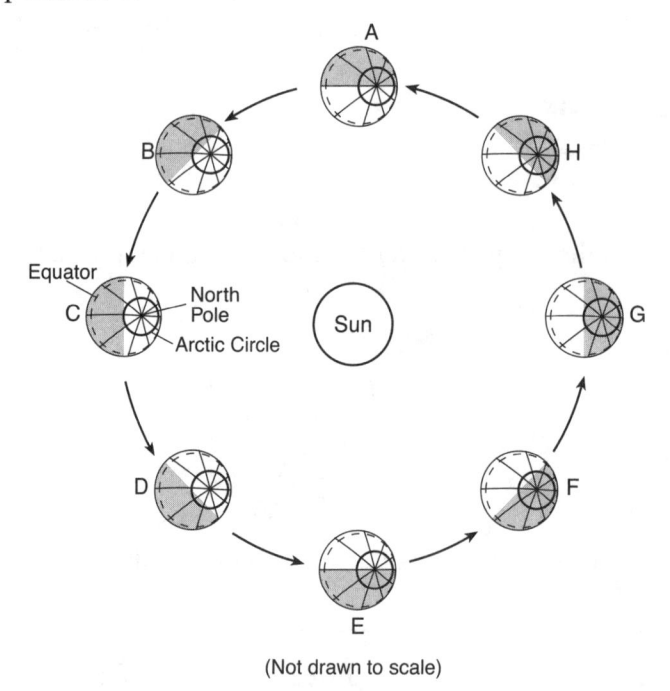

48. Approximately how many days does it take Earth to go from **A** to **D**? _____

49. Approximate date when Earth will be at position **H**. _____

50. At which position(s) will the Antarctic Circle have continual daylight? _____

78

CHAPTER 6

PROPERTIES OF THE ATMOSPHERE
*Origin and Functions, Structure of the Atmosphere, Environmental Issues,
Energy Transfer, Phase Changes, Atmospheric Variables*

ORIGIN AND FUNCTIONS OF THE ATMOSPHERE

The **atmosphere** is a mixture of gases and aerosols surrounding Earth. Most of the atmospheric gases are within the first 5 km of the atmosphere. Atmospheric gases include: nitrogen (78%), oxygen (21%), and other gases (1%) such as water vapor, ozone, and carbon dioxide. Atmospheric **aerosols** are small solids and liquids suspended in the air such as water droplets, ice crystals, soot, dust, ash, and pollen.

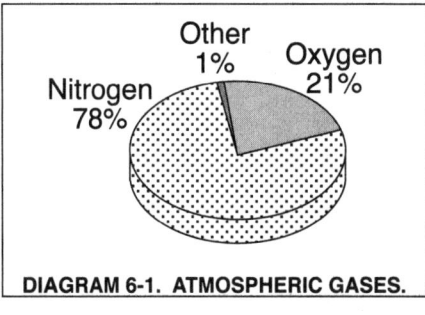

DIAGRAM 6-1. ATMOSPHERIC GASES.

The early (primordial) atmosphere was composed of carbon dioxide, ammonia, and methane. There was no oxygen, no ozone, and no water vapor. Outgassing from volcanic activities released water vapor, nitrogen, and other gases that were trapped in Earth's interior. As Earth cooled, water vapor condensed and clouds formed. As it began to rain the oceans and surface waters formed. Ocean waters absorbed carbon dioxide which led to deposition of carbonate rocks such as limestone. **Photosynthesis** by cyanobacteria in the ocean absorbed carbon dioxide and released oxygen. Once oxygen was available, the protective ozone layer formed. Our present atmosphere is young; it is about 800 million years old.

The atmosphere has many functions. It supports life by providing moisture and breathing gases. The atmosphere protects life on Earth from meteoroid impacts and dangerous ultraviolet radiation which is absorbed by ozone in the stratosphere. **Greenhouse gases** in the atmosphere such as water vapor, carbon dioxide, and methane absorb terrestrial radiation from Earth and infrared radiation from the Sun. This helps to maintain a constant Earth temperature.

Questions

1. How does Earth's first atmosphere compare to the present atmosphere? (*be specific*)

   _____

   _____

2.a. Name the process responsible for the development of oxygen in the atmosphere?

   _____

   b. Name the first group of oxygen-producing organisms._____

3.a. According to the ESRT, how many million of years ago did oxygen begin to enter the

   atmosphere?_____

   b. What eon was this? _____

Structure of the Atmosphere

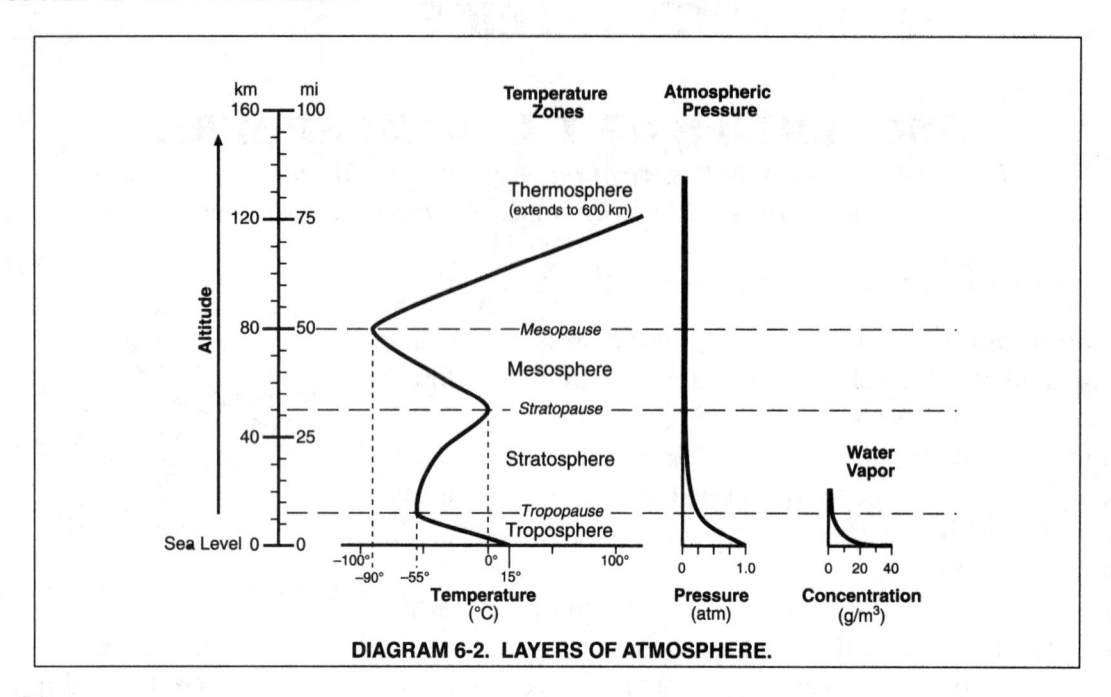

DIAGRAM 6-2. LAYERS OF ATMOSPHERE.

The atmosphere is divided (stratified) into four major layers based on temperature changes. The lowest layer is the **troposphere**. This layer has the smallest vertical extent. In this first 11 kilometers of the atmosphere most of the atmospheric gases are found. The troposphere contains most of the moisture in the atmosphere, so weather occurs here. Air pressure and temperature decrease with height in this layer as the air molecules expand. The troposphere ends at the **tropopause** when temperature stops decreasing.

Above the troposphere is the **stratosphere**. It extends from 11 to 50 km in altitude. The air is dry and clouds are rare. Air pressure continues to decrease because there are less atmospheric gases above. The temperature in the stratosphere rises because the **ozone** layer is absorbing ultraviolet radiation from the Sun and converting it to heat. The stratosphere ends at the **stratopause** when temperature stops increasing.

In the **mesosphere** there is no water vapor or aerosols. It is in this layer that meteoroids disintegrate due to friction with the atmosphere. Air pressure and temperature decrease until the **mesopause**.

The top layer of the atmosphere is the **thermosphere**. It extends from 80 km and beyond. Absorption of solar radiation causes high molecular motion and therefore high molecular temperature. The gases in the thermosphere are separated by density. A wind of solar particles can excite the gases in this layer causing **auroras**.

Questions

4. State the physical property of the atmosphere that is used to divide it into four layers.

5. For each statement, name the atmospheric layer(s) described.
     (troposphere)   (stratosphere)   (mesosphere)   (thermosphere)

   a. ozone is found here. _____
   b. meteoroids disintegrate. _____
   c. found at 60 km altitude. _____
   d. air pressure of 0.1 atm. _____
   e. weather takes place here. _____
   f. temperature increases. _____
   g. temperature of -80°C. _____
   h. highest amount of water vapor. _____

6. Calculate the ratio of nitrogen to oxygen in the atmosphere to the nearest whole number. *(show solution)*

7. The approximate temperature at 30 km: _____; 30 miles: _____ ; 65 miles: _____

8. In the mesosphere, as altitude increases, the temperature _____

9. In the troposphere, as altitude increases, the temperature _____

10. As altitude above sea level increases, air pressure _____

ENVIRONMENTAL ISSUES

The future of life on Earth depends on how wisely we treat our environment. There are many environmental concerns regarding the atmosphere.

An increase in the **greenhouse effect** has led to a concern about **global warming**. Like a greenhouse, the atmosphere allows short-wave visible light to pass through. The light is absorbed and changed to heat (infrared), which has a longer wavelength and is trapped inside. The atmosphere is like the glass of the greenhouse. Water vapor, methane, and carbon dioxide in the atmosphere trap in heat and keep Earth's atmosphere warm.

Human activities add carbon dioxide to the air by the burning of fossil fuels (coal, oil, and natural gas). As carbon dioxide increases in the atmosphere, more heat is trapped and the temperatures on Earth increase.

Global warming can have serious effects. An increase in heat waves will affect human health. As ocean temperatures increase, the sea water expands causing a rise in sea level. Higher polar temperatures cause the icecaps and glaciers to melt also adding to sea level rise. As a result floods have affected coastal locations. An increase in tropospheric temperature results in more water vapor in the air, an increase in atmospheric convection, and more severe storms. The distribution of rainfall can be altered as Earth's temperature changes. We need to decrease the burning of fossil fuels, have more efficient cars, machines, and factories, and use alternative energy sources. Life on Earth exists because of Earth's temperature. A change in temperature on Earth will change life on Earth.

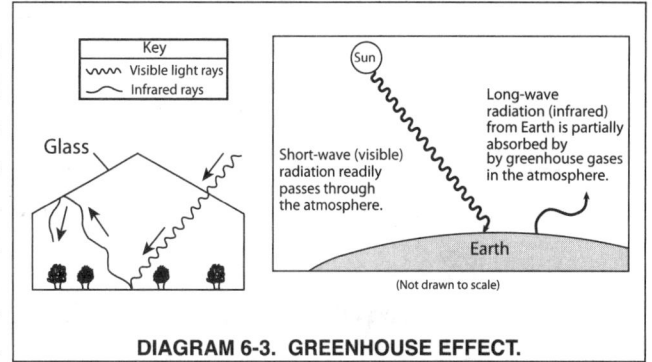

DIAGRAM 6-3. GREENHOUSE EFFECT.

The **ozone** layer in the stratosphere has been depleted by the use of CFCs (chlorofluorocarbons). These chemicals were used in the production of the refrigerant freon and styrofoam. This decrease in the ozone layer was discovered in the 1980s. There was concern that this could result in an increase in skin cancer, damage to eyes, and damage to land and ocean life due to more ultraviolet radiation reaching the Earth's surface. Since 2000, freon in air conditioners and refrigerators no longer contained CFCs. Hopefully we will see a natural "repair" to the ozone layer.

Another atmospheric concern is the increase in **smog** in some locations. Smog occurs when air pollutants, especially those from the burning of fossil fuels, chemically react in the presence of sunlight.

Acid precipitation is another concern. It forms when air pollutants from the burning of fossil fuels combine with atmospheric moisture during cloud formation. This results in precipitation having a lower pH than normal. Acidic water is harmful to aquatic life on Earth, damages leaves of trees, and causes metals to be leached from soil and enter the surface waters.

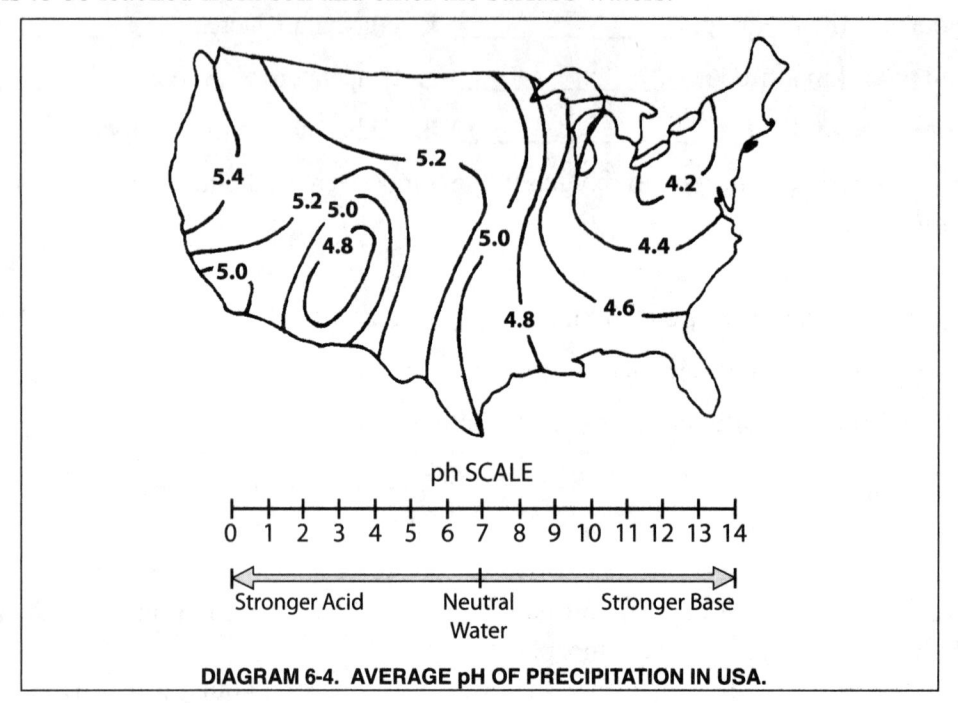

DIAGRAM 6-4. AVERAGE pH OF PRECIPITATION IN USA.

Questions

11. Explain the role of ozone in the atmosphere. _____
_____

12. What human activity is contributing to an increase in smog, acid precipitation, and global warming? _____

13. Name *two* gases which contribute to an increase in the "greenhouse effect" (global warming).
_____ and _____

14. List *two* negative effects of global warming.

    a. _____
    _____

    b. _____
    _____

15. The graph shows the changes in carbon dioxide concentration in Earth's atmosphere over a 140-year period. Concentrations of carbon dioxide are in parts per million (ppm).

   a. Name the human activity responsible for the increase in carbon dioxide. _____

   b. In this 140-year period the rate of change in atmospheric carbon dioxide has

   (*decreased*) (*increased*) (*stayed the same*)

   c. What action can we take to slow down this increase in atmospheric carbon dioxide.

   _____

   d. Calculate the rate of change of carbon dioxide in ppm/year from 1900 to 2000.

   _____ (*show solution*)

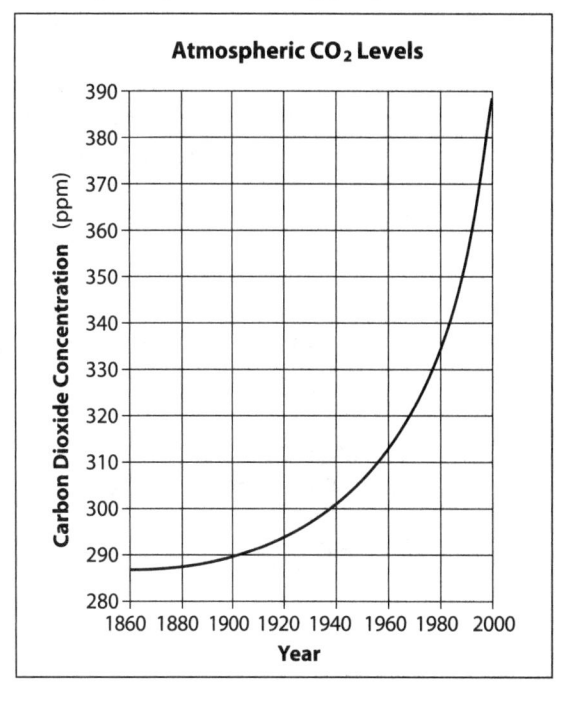

ENERGY TRANSFER METHODS

The energies in the universe and in Earth's systems are transferred from place to place or within a material by one of three methods. Energy always moves from a higher energy state to a lower energy state, such as hot to cold. Materials with a low **specific heat** will absorb and radiate energy faster than those with high specific heats.

Method of Transfer	Description	Examples
Conduction	energy is transferred within a solid from molecule to molecule; energy transferred between two objects that touch each other	• Hot magma intrudes into rock.
Convection	energy flows and moves within a fluid (liquid or gas) due to density differences as hot fluids rise and cool fluids sink; energy circulates by convection currents	• Earth's interior heat rises at hot spots. • Heat energy flows from the tropics to the poles through the atmosphere and through the oceans.
Radiation	energy is transferrd from one location to another without any medium in between or direct contact; energy is given off by a radiating object; can occur in empty space	• Starlight travels through space. • Lava cools on Earth's surface. • Earth radiates heat to the atmosphere.

Questions

16. The diagram shows a beaker of water with a dye pellet in it. As the water is heated by the flame, the pellet will dissolve.

  a. Draw arrows to show the direction the dye will move as the water is heated.

  b. Name the type of energy transfer that will occur within the water.

  _____

  c. What causes this type of energy movement? _____

  _____
  _____

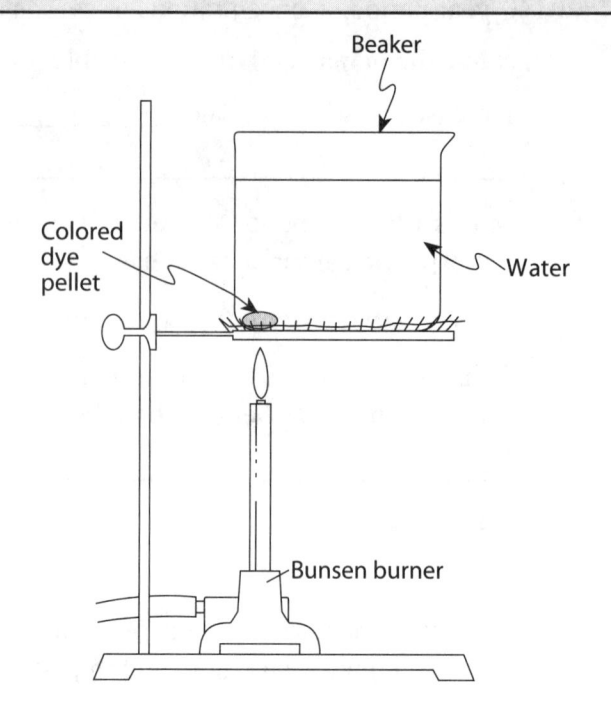

17. Soil and water both with a starting temperature of 20°C were heated for ten minutes by a heat lamp. The diagram shows their final temperatures.

  a. State the type of energy transfer shown.

  _____

  b. Explain why the soil's temperature increased more.

  _____
  _____

  c. Calculate the rate of heating for the soil in °C/min. _____
  (*show solution*)

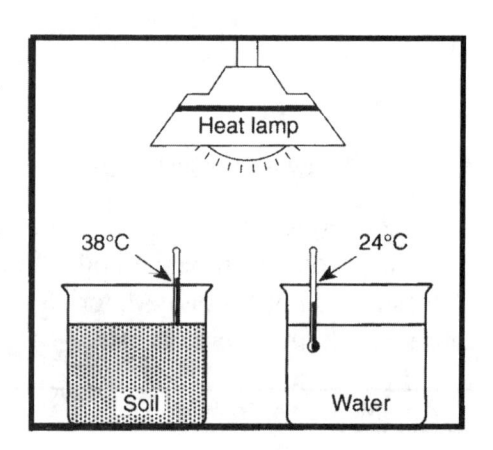

84

PHASE CHANGES

Water changes phase (state) on Earth's surface and in the atmosphere. The addition or removal of heat energy causes water to go through phase changes. These phase changes store and release heat energy to the atmosphere. When liquid water evaporates, the vapor will hold a tremendous amount of heat energy. This heat energy is later released when the vapor condenses back to liquid during cloud formation. This released heat energy "powers" a storm.

Phase Change	Term	Is Energy Released or Gained?	Example
solid to liquid	melt	gained	snow melts
liquid to gas	evaporation	gained	surface water enters air as vapor
solid to gas	sublimation	gained	surface snow vaporizes
gas to solid	deposition	released	frost forms on surfaces
gas to liquid	condensation	released	water vapor forms cloud droplets
liquid to solid	freeze	released	cloud droplets freeze to form hail

Phase changes can be illustrated by a graph. This graph shows the heating of ice starting at –100° C. When the ice reaches a temperature of 0° C it will melt. During this time of melting (**B** to **C**) the heat energy added is used to separate the molecules as the ice changes to liquid. Once it is all a liquid, the heat energy will cause the temperature to rise again (**C** to **D**). At 100° C the liquid water will evaporate to become a gas (**D** to **E**). If this is a closed system the gas at **E** will continue to increase in temperature as heat is added.

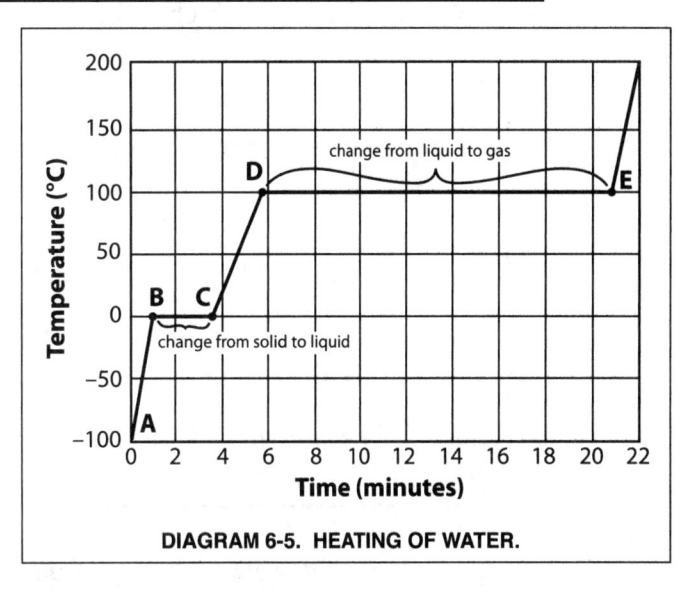

DIAGRAM 6-5. HEATING OF WATER.

Questions

18. State the phase change described:

    (*melting*) (*evaporation*) (*condensation*) (*freezing*) (*sublimation*) (*deposition*)

    a. Rain changes to sleet as it falls. _____

    b. Dew forms on the lawn during the night. _____

    c. Snow turns to rain as it falls. _____

    d. Frost forms on a cold night. _____

    e. Fog forms. _____

    f. Water droplets accumulate above 1000 feet to form a cloud. _____

    g. The soil dries out. _____

19-23. The graph below shows the heating of a sample of water from –50°C to 100°C.

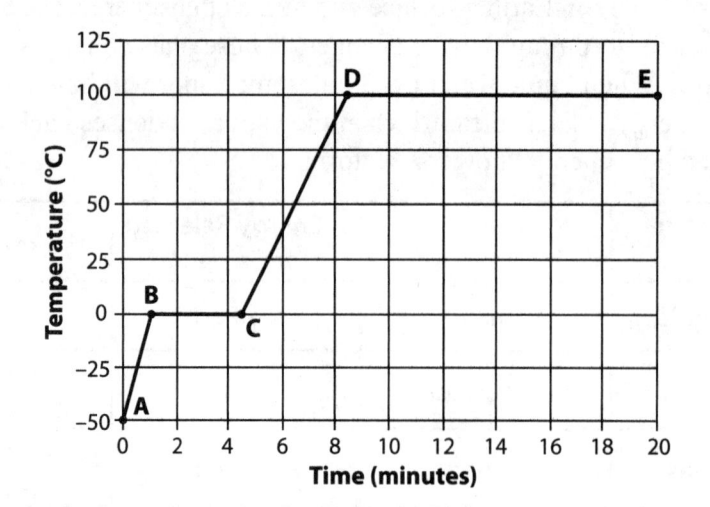

19. Name the phase or phases present from:

   a. A to B: _____   c. C to D: _____

   b. B to C: _____   d. D to E: _____

20. In the graph above between which points is the following occurring:

   a. melting: _____   b. evaporation: _____

21. Calculate the rate of temperature change from **C** to **D**. _____
    (*show solution*)

22. Compared to **C** to **D**, the rate of temperature change from **A** to **B** is

    (*less*) (*greater*) (*the same*). How do you know? _____

    _____

23. How long did it take the water to change from –50°C to 75°C? _____

24. The greatest amount of heat is released when water (*freezes*) (*evaporates*) (*condenses*).

25. Heat energy is gained when water (*freezes*) (*melts*) (*condenses*).

86

ATMOSPHERIC VARIABLES

The current conditions of the atmosphere can be described by its **atmospheric variables**. The variables include temperature, air (barometric) pressure, wind speed and direction, relative humidity, sky cover, visibility, precipitation, and dewpoint. These are monitored, measured, and recorded at weather stations.

ATMOSPHERIC VARIABLE: TEMPERATURE

The atmosphere warms due to the absorption of solar and terrestrial radiation. Heat is also released into the atmosphere by the processes of condensation (vapor to liquid droplets) and deposition (vapor to ice crystals) during cloud formation.

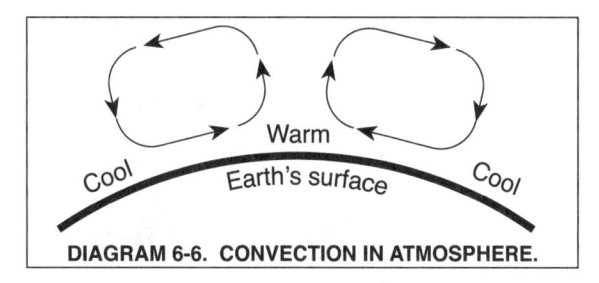

DIAGRAM 6-6. CONVECTION IN ATMOSPHERE.

Convection currents transfer heat in the atmosphere. These are air movements caused by density differences. Hot air is less dense than cool air, so hot air rises and cool air sinks.

In the atmosphere air temperature changes **adiabatically** due to the expansion and compression of air molecules. As warm air rises it expands. The expanding air molecules move further apart so they collide less. As a result, air temperature decreases. This is why it gets colder as altitude increases. Conversely, when cooler air sinks, it compresses, the molecules are closer and collide more often, so temperature increases due to friction by collisions.

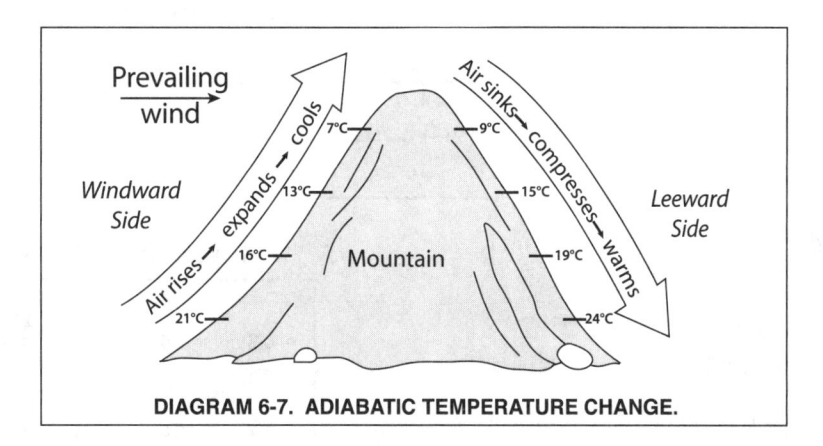

DIAGRAM 6-7. ADIABATIC TEMPERATURE CHANGE.

Air temperature on Earth varies with time of day, latitude, elevation, season, and location on a land mass (inland or coastal). Temperatures are warmer after solar noon when Earth has absorbed sunlight and reradiated heat to the air. Higher temperatures occur in the tropics and during summer because angles of insolation are greater. Coastal areas have a smaller range in seasonal temperature than inland locations. This occurs because water has a high specific heat causing water to heat and cool more slowly than land.

Temperature is measured with a **thermometer** in Celsius or Fahrenheit. On a weather map **isotherms** can be drawn which connect points of equal air temperature. These isotherms show the temperature variations and patterns on Earth at a given time.

When drawing isolines be sure to follow these guidelines:
1. Isolines connect adjacent points of the same value with a smooth line.
2. Isolines do not touch or cross each other.
3. Isolines continue to the end of the map or form closed circles.
4. Isolines tend to be parallel to each other.
5. Isolines do not have sharp corners, they are smooth curved lines.

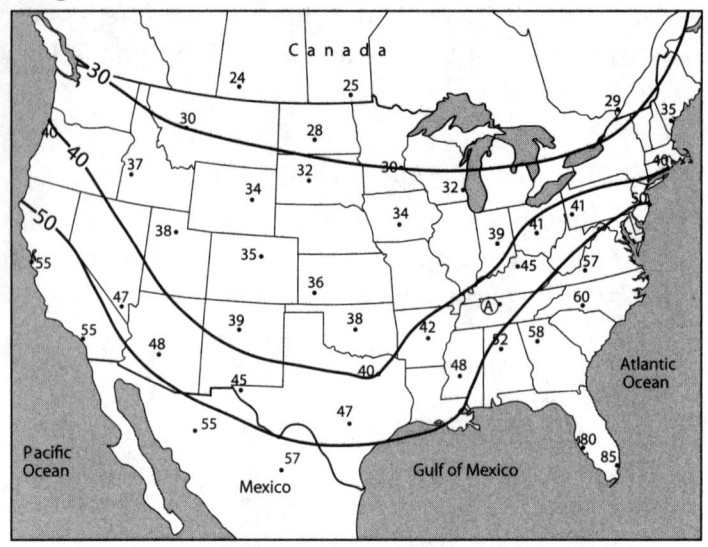

DIAGRAM 6-8. EXAMPLE OF ISOTHERMS FOR EVERY 10°F.

Questions

26. Convert these temperatures: 25° C = _____ °F; 48° F = _____ °C; –10° C = _____ °F

27. Explain what causes colder air temperatures at the top of mountains. _____
_____

28. Sinking air will (*compress*) (*expand*) and (*cool*) (*warm*) because _____
_____

29. Explain why it is colder at polar locations. _____
_____

30. These temperatures were taken in a classroom.

    a. Draw isotherms for every 2° C starting at 20° C.

    b. Heat energy will move from point (*A* to *B*) (*B* to *A*).

    c. Calculate the temperature gradient between **A** and **B**. _____
    (*show solution*)

Temperature Field Map (°C)

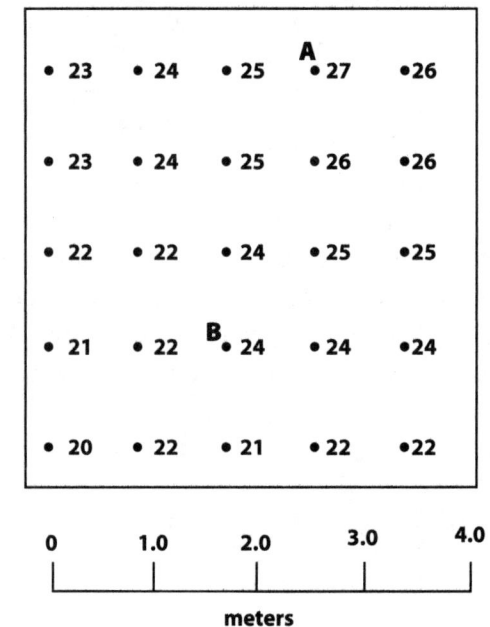

• 23	• 24	• 25	**A** • 27	• 26
• 23	• 24	• 25	• 26	• 26
• 22	• 22	• 24	• 25	• 25
• 21	• 22	**B** • 24	• 24	• 24
• 20	• 22	• 21	• 22	• 22

0   1.0   2.0   3.0   4.0
meters

31. The map below records the temperature readings taken at 1 p.m. on a winter day.

Map 1—Temperatures (°F)

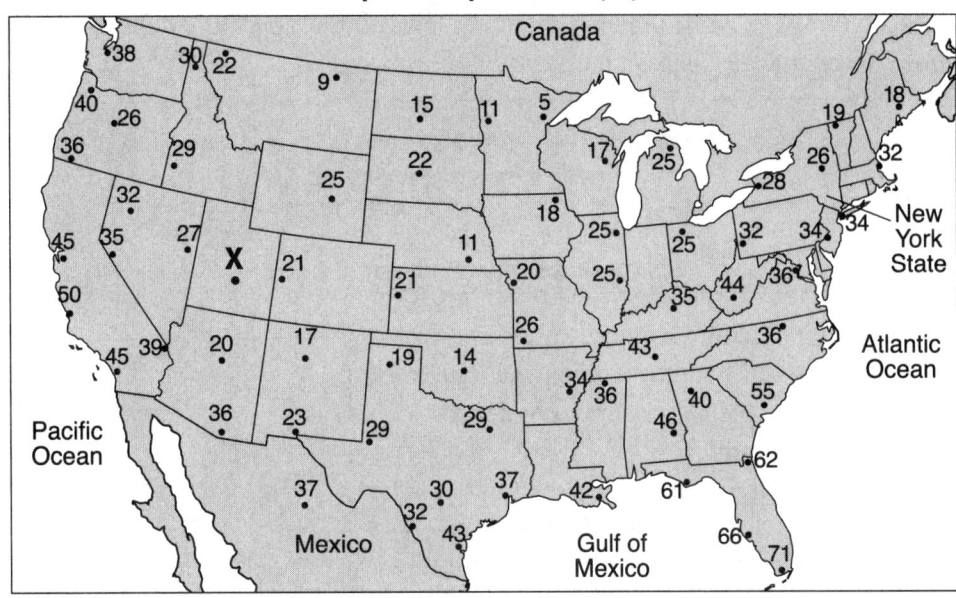

a. Draw isotherms for every 10° F starting at 10° F.
b. Compared to coastal temperatures, inland temperatures are (*lower*) (*higher*).
c. Based on your isotherms, estimate the temperature of location **X** to the nearest whole number. _____

ATMOSPHERIC VARIABLES: MOISTURE AND PRECIPITATION

Atmospheric moisture exists in all three states (solid, liquid, and gas). Ice, snow, hail, sleet, and frost are solids. Water droplets, rain, dew, and fog are all liquids. Invisible water vapor (humidity) is the gaseous form of water. Clouds are made of microscopic ice crystals and water droplets. When the cloud droplets join together and become bigger, gravity will pull them to Earth as precipitation.

Moisture in the atmosphere comes from **evaporation** of surface waters and **transpiration** from plants. The two processes together are called **evapotranspiration**. Several factors affect the rate at which water will evaporate. As temperature increases, evaporation increases. As the wind increases, so does evaporation. Evaporation will be greater if there is a large exposed surface area. However, if there is a lot of moisture in the air (high humidity) evaporation will decrease. A wet towel will dry much quicker on a hot, dry, windy day compared to a cool, humid, calm day.

Evaporation occurs when liquid water molecules are heated and energized. This process takes a large amount of energy (2260 joules/gram). When water molecules gain enough energy they will escape into the air as water vapor. As water vapor enters the air, humidity increases. The air becomes saturated when it can not hold any more moisture. Equilibrium is reached when the number of molecules which are evaporating equals the number of molecules which are condensing (vapor changing to liquid).

Relative humidity is the ratio of water vapor in the air to the amount of water the air can hold. Warm air can hold more water vapor than cold air. Relative humidity is expressed as a percent. If there is 0% relative humidity, there is no moisture in the air. Air that has 100% humidity is saturated. Relative humidity will increase if evaporation increases or if temperature decreases.

Dewpoint is the temperature at which condensation (vapor to liquid) or deposition (vapor to solid) occurs. At dewpoint temperature the air is saturated. As air temperature gets closer to dewpoint temperature, the chance of precipitation increases.

A **psychrometer** is used to determine dewpoint and relative humidity. The dry-bulb temperature is the air temperature. The wet-bulb temperature is affected by evaporation. When the air is dry, more evaporation occurs lowering the wet-bulb temperature. Dewpoint temperature depends on the moisture in the air. Dewpoint temperature rises as humidity increases.

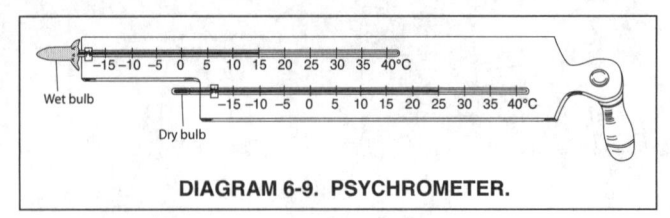

DIAGRAM 6-9. PSYCHROMETER.

Clouds are made of microscopic water droplets and/or ice crystals suspended in the air. They form when air becomes saturated with water vapor and the water vapor condenses onto aerosols called **condensation nuclei**.

Warm, moist air rises and cools to dewpoint. The temperature at the base of a cloud is equal to dewpoint. For clouds to form the air temperature must be at dewpoint, air must be saturated, and there must be condensation nuclei. If the dewpoint in the cloud is below freezing then ice crystals will form in the cloud.

Precipitation is the falling of liquid and solid water from clouds. It occurs when cloud droplets are large enough so that gravity pulls them to Earth. This is why there are clouds in the sky even when it is not raining, the cloud droplets are not large enough to fall.

Precipitation has a cleaning effect on the air because inside each raindrop or snowflake there is a solid particle, an aerosol. Acid precipitation forms when pollutants in the air mix with water vapor and become part of the precipitation. Acidic rain, snow, and fog are harmful to life on Earth.

DIAGRAM 6-10. CLOUD FORMATION

(Not drawn to scale)

DIAGRAM 6-11. WATER VAPOR CONDENSING ON TO CONDENSATION NUCLEI (SALT, DUST).

The processes that occur prior to precipitation are:
1. The Sun warms Earth and evaporation of surface waters occurs.
2. Warm, moist air which is less dense rises.
3. Rising air expands and adiabatically cools.
4. Air cools to dewpoint.
5. Water vapor in the air condenses on to aerosols, clouds begins to form.
6. In the cloud, tiny cloud droplets collect together to form raindrops large enough to fall as precipitation. If dewpoint is below freezing, ice crystals form in the cloud.

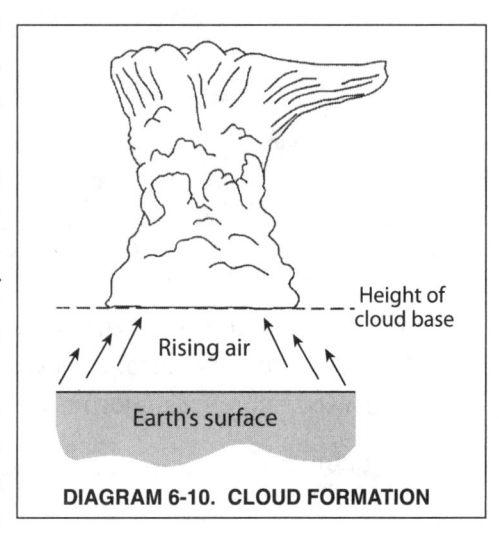

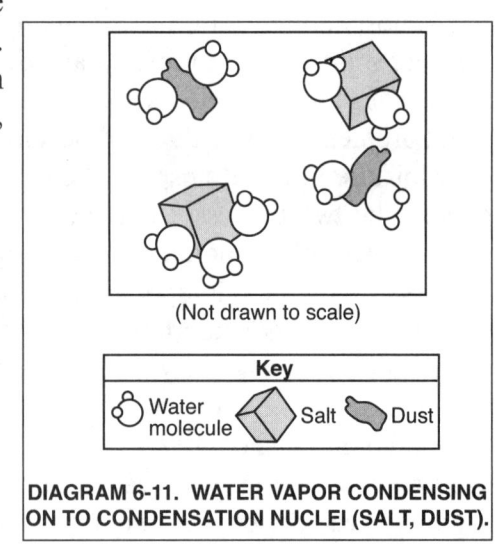

Questions

32. Complete the statements about the relationship between the two variables by using the term (*decrease*), (*increase*), or (*remains the same*).

    a. As air temperature increases, rate of evaporation will _____

    b. As wind speed increases, rate of evaporation will _____

    c. As humidity in the air increases, rate of evaporation will _____

    d. As evaporation occurs, relative humidity will _____

    e. As relative humidity increases, dewpoint temperature will _____

    f. As evaporation increases, wet-bulb temperature will _____

    g. As air temperature approaches dewpoint, chance of precipitation will _____

33. Use the ESRT to complete the chart below.

	Dry-bulb Temperature	Wet-bulb Temperature	Dewpoint Temperature	Relative Humidity
a	18° C	10° C		
b	21° C	17° C		
c	6° C		4° C	
d	12° C			48%
e	2° C		–11° C	

34. The graph shows air temperature and relative humidity at a location for one day.

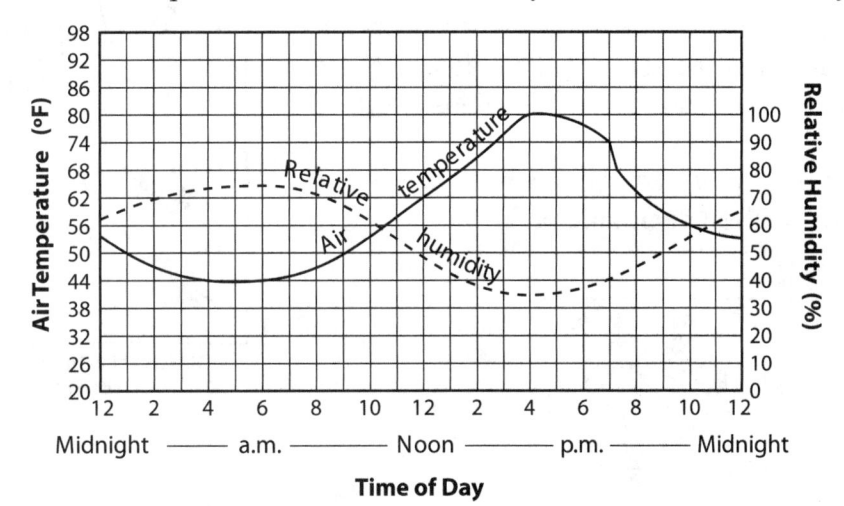

    a. The change in relative humidity from 12 noon to 4 p.m. was _____%.

    b. The greatest rate of evaporation had most likely occurred at _____

    because _____

    c. At 4 a.m. the air temperature was _____ and the relative humidity was _____.

91

ATMOSPHERIC VARIABLES: AIR PRESSURE AND WINDS

Air molecules have weight. The weight of the atmospheric gases exerts a downward force on Earth. This is known as **barometric air pressure**. It is measured with a **barometer** and the units are millibars or inches of mercury. **Isobars** are lines that connect points of equal air pressure on a weather map.

Dense air has more weight. If density increases, air pressure increases. Cold air is denser than warm air so it exerts more pressure. As temperature increases, air pressure decreases.

Air pressure changes with altitude. At high elevations there is less atmosphere above us and therefore less air pressure. As altitude increases, air pressure decreases.

Air pressure also changes as the moisture content of the air changes. Water vapor has less mass than the air molecules of nitrogen and oxygen. So as humidity increases, air pressure decreases. Low pressure systems have more humidity.

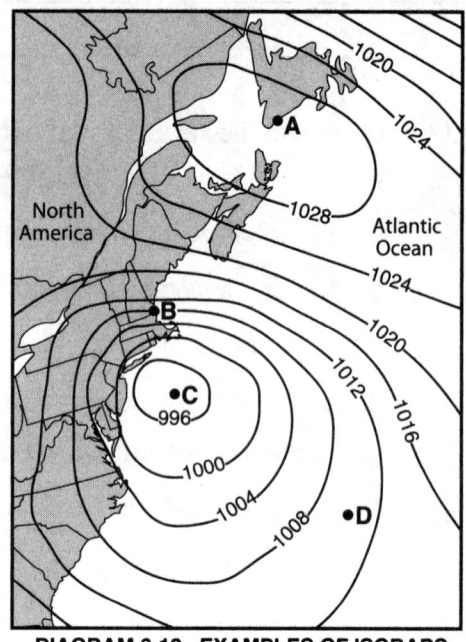

DIAGRAM 6-12. EXAMPLES OF ISOBARS FOR EVERY 4 mbs.

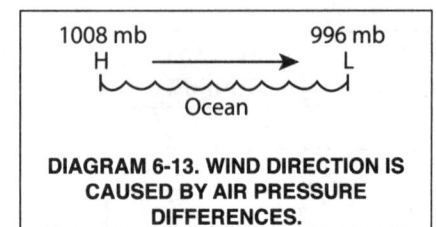

DIAGRAM 6-13. WIND DIRECTION IS CAUSED BY AIR PRESSURE DIFFERENCES.

Winds are horizontal air movements which move parallel to the ground. They are caused by differences in air pressure. Winds move from areas of high pressure (H) to areas of low pressure (L).

Wind speed is caused by the rate at which the pressure changes (pressure gradient). The diagram to the right shows isobars for a region at one time. When isobars are close together (**B**), the wind speed will be high. Isobars that are spread apart (**C**) indicate gentle winds due to a low pressure gradient.

Wind direction is measured with a **wind vane**. Wind speed is measured with an **anemometer**.

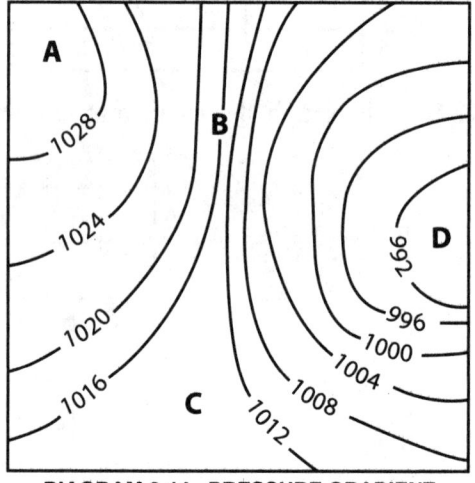

DIAGRAM 6-14. PRESSURE GRADIENT DETERMINES WIND SPEED.

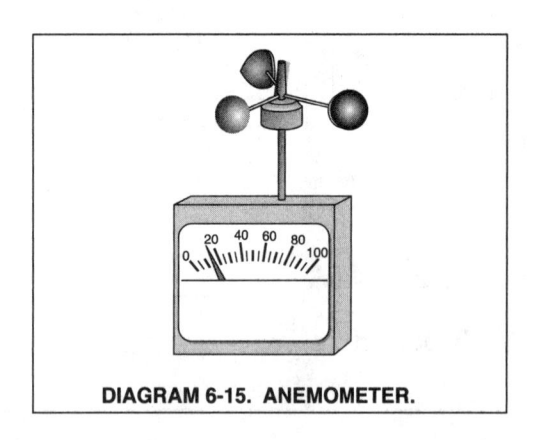

DIAGRAM 6-15. ANEMOMETER.

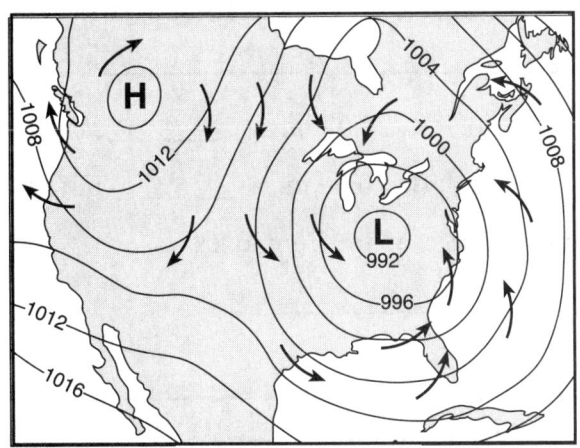

DIAGRAM 6-16. AIR MOVEMENT AROUND A HIGH AND A LOW IN THE NORTHERN HEMISPHERE.

The Coriolis effect due to Earth's rotation causes winds in the Northern Hemisphere to curve to the right and in the Southern Hemisphere to curve to the left. Winds in a low pressure system (**cyclone**) move inward and counterclockwise. Winds in a high pressure system (**anticyclone**) move outward and clockwise.

Weather systems are classified as "High" or "Low." These systems result from differences in temperature and moisture content.

High Pressure (anticyclone)	Low Pressure (cyclone)
Dry air	Moist air
Cooler temperatures	Warmer temperatures
Clear weather	Stormy weather
Winds blow out and CW	Winds blow in and CCW

Planetary (global) winds are worldwide air movements that affect the entire Earth. They are caused by uneven heating of the air at different latitudes and differences in air pressure. Planetary winds shift north or south with the Sun's vertical ray as the seasons change. These planetary winds control the movement of weather systems.

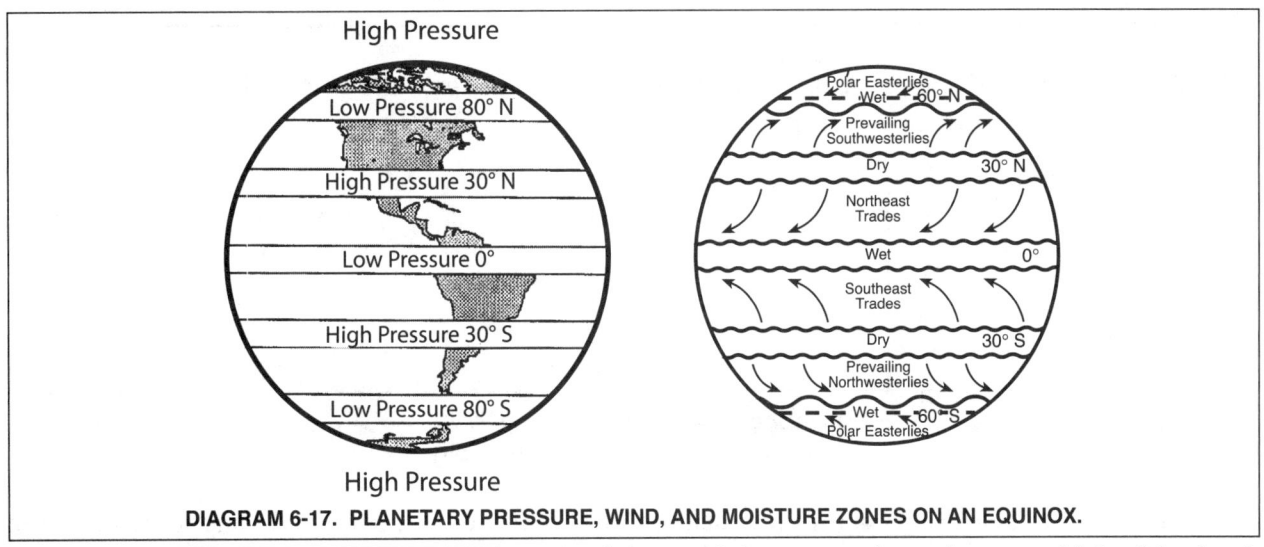

DIAGRAM 6-17. PLANETARY PRESSURE, WIND, AND MOISTURE ZONES ON AN EQUINOX.

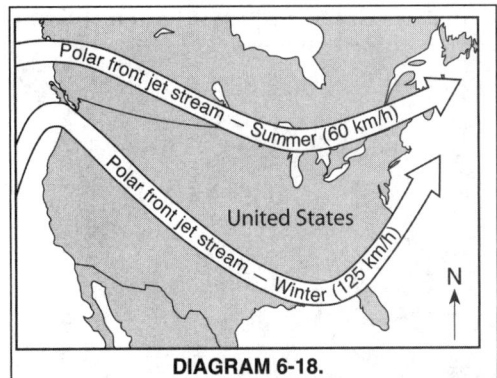

At the top of the troposphere there are high altitude air movements that encircle the globe called **jet streams**. They move quickly from west to east. The jet stream moves air masses and storms (lows). In winter it can move cold air from the arctic southward causing severe cold weather in the United States that can extend down to Florida.

DIAGRAM 6-18.

Ocean currents are affected by the planetary winds and tend to follow the same pattern. Planetary winds move heat energy in the atmosphere between the tropics and the poles. The ocean currents move heat energy in the oceans.

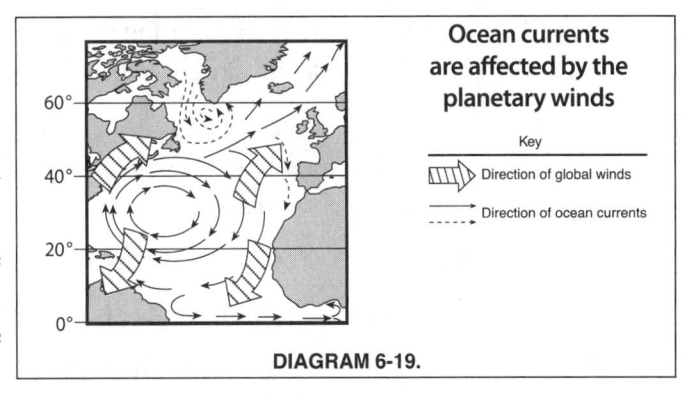

Ocean currents are affected by the planetary winds

Key
- Direction of global winds
- Direction of ocean currents

DIAGRAM 6-19.

93

Questions

35. Convert these air pressures: **a.** 1010.0 mb = _____ in. **c.** 29.08 in. = _____ mb

b. 1030.5 mb = _____ in. **d.** 30.06 in. = _____ mb

36. Name the planetary winds that occur at the following latitudes on the Equinoxes.

a. 10°N: _____ **b.** 55°S: _____ **c.** 75°N: _____

37. Complete the chart below for surface ocean current.

Location	Name of Ocean Current	Warm or cool?	Moving from what direction?
west coast of South America			
east coast of Greenland			
north of Antarctica			
15°N, 160°W			
50°N, 30°W			

38. Air pressure is affected by many atmospheric conditions. Draw the graph relationships between the following variables and air pressure.

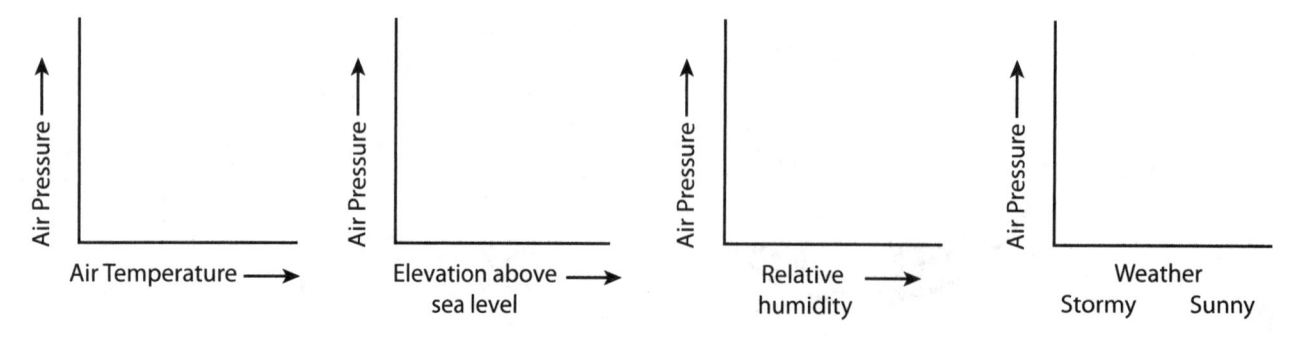

(Graph 1: Air Pressure vs Air Temperature)
(Graph 2: Air Pressure vs Elevation above sea level)
(Graph 3: Air Pressure vs Relative humidity)
(Graph 4: Air Pressure vs Weather — Stormy, Sunny)

39. The weather map below shows the location of a high-pressure center (**H**) and a low-pressure center (**L**) over a portion of North America. The isolines indicate surface air pressure in millibars.

a. What is the maximum surface air pressure that could occur in the center of the High (**H**) to the nearest tenth? _____ millibars

b. Place an "**X**" on the map for a location that could have an air pressure of of 1017 mb.

c. Place a "**W**" at a location that would be experiencing high wind speeds.

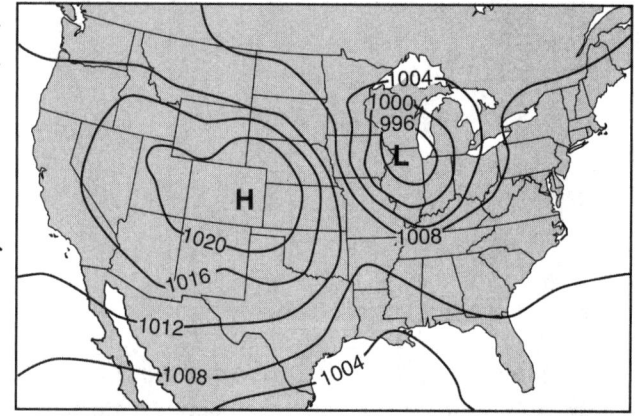

d. Draw four arrows around the **H** to show the air movements.

e. Draw four arrows around the **L** to show the air movements.

f. Describe the sky conditions for the high pressure system. _____

94

CHAPTER 6 REVIEW

1. In the early history of Earth, as the number of green plants increased, what changes occurred in the atmosphere?
   (1) both oxygen and carbon dioxide decreased
   (2) both oxygen and carbon dioxide increased
   (3) oxygen decreased and carbon dioxide increased
   (4) oxygen increased and carbon dioxide decreased

2. Which gas was absent from Earth's original atmosphere?
   (1) ammonia   (2) carbon dioxide   (3) methane   (4) oxygen

3. Water vapor and carbon dioxide cause warming of Earth's atmosphere because they
   (1) have high specific heats   (3) absorb infrared radiation
   (2) reflect insolation   (4) absorb ultraviolet radiation

4. Which gas in Earth's upper atmosphere is beneficial to humans and life on Earth because it absorbs large amounts of ultraviolet radiation from the Sun?
   (1) water vapor   (2) methane   (3) nitrogen   (4) ozone

5. Deforestation increases the greenhouse effect on Earth because deforestation causes the atmosphere to contain
   (1) more carbon dioxide which absorbs infrared radiation
   (2) less carbon dioxide which absorbs short-wave radiation
   (3) more oxygen which absorbs infrared radiation
   (4) less oxygen which absorbs short-wave radiation

6. Pollutants in the air are most likely removed by
   (1) evaporation   (2) volcanic activity   (3) precipitation   (4) transpiration

7. As altitude increases from sea level to 50 km, the air temperature will
   (1) decrease only   (3) increase, then decrease
   (2) increase only   (4) decrease, then increase

8. At what altitude is the atmospheric temperature -75° C ?
   (1) 14 km   (2) 40 km   (3) 70 km   (4) 80 km

9. What form of energy given off by Earth causes the heating of the atmosphere?
   (1) x-ray   (2) ultraviolet   (3) visible light (4) infrared

10-11. The cross section below shows the general air movement within a portion of Earth's atmosphere between 30° N and 30° S latitude.

10. Which layer of Earth's atmosphere is shown in this cross section?
    (1) troposphere   (3) stratosphere
    (2) mesosphere   (4) thermosphere

11. The air movement shown in the cross section is due to the process of
    (1) conduction   (3) convection
    (2) radiation   (4) condensation

12. Which process requires the addition of the most energy to water?
    (1) cooling of water   (3) condensation of water
    (2) freezing of water   (4) vaporization of water

13. In which phase does water have the highest specific heat?
    (1) solid   (2) liquid   (3) gas

14. Liquid water will continue to evaporate from Earth's surface until
    (1) transpiration occurs
    (2) the relative humidity decreases to 50%
    (3) the air becomes saturated
    (4) temperature of the air becomes greater than dew point

15. Which gas in the atmosphere has the most influence on day-to-day weather changes?
    (1) ozone   (2) oxygen   (3) water vapor   (4) carbon dioxide

16. Under which set of conditions does water evaporate the fastest?
    (1) warm, calm winds, high humidity   (3) cold, calm winds, low humidity
    (2) warm, high winds, low humidity   (4) cold, high winds, high humidity

17. In which air sample will condensation most likely to occur?

Air Temperature = −2° C Dewpoint = −4° C Clean Filtered Air	Air Temperature = 5° C Dewpoint = 5° C Air Containing Tiny Particles	Air Temperature = 10° C Dewpoint = 7° C Air Containing Tiny Particles	Air Temperature = 20° C Dewpoint = 20° C Clean Filtered Air
(1)	(2)	(3)	(4)

18. By which process are clouds, dew, and fog formed?
    (1) melting   (2) precipitation   (3) evaporation   (4) condensation

19. The incomplete flowchart below shows some of the changes that occur as warm air rises to form a cloud.

    Warm air rises. ⇨ The air expands due to lower pressure. ⇨ ⬚ ⇨ Condensation occurs and a cloud forms.

    Which statement should be placed in the empty box?
    (1) The air warms as its expands.
    (2) The air's relative humidity decreases to zero.
    (3) The air cools to dewpoint temperature.
    (4) The air enters the mesosphere.

20. Why is it possible for no rain to be falling from a cloud?
    (1) The water droplets are too small to fall.
    (2) The cloud is entirely water vapor.
    (3) The dewpoint has not been reached in the cloud.
    (4) There are no condensation nuclei in the cloud.

21. What is the dewpoint when air temperature is 26° C and relative humidity is 77%?
    (1) 3° C   (2) 20° C   (3) 22° C   (4) 23° C

22. A student used a psychrometer to measure the humidity of the air. If the relative humidity was 54% and the dry-bulb temperature was 10° C, what was the wet-bulb temperature?
    (1) 4° C   (2) 5° C   (3) 6° C   (4) 10° C

23. In New York State, what weather conditions are most likely to exist when the air pressure is much greater than 30 inches?
    (1) cold, dry air with clear skies   (3) strong south winds with hail warnings
    (2) warm, moist air with cloudy skies   (4) approaching thunderstorms

24. The diagram shows a cross section of a cumulus cloud. Line **AB** is the base of the cloud.

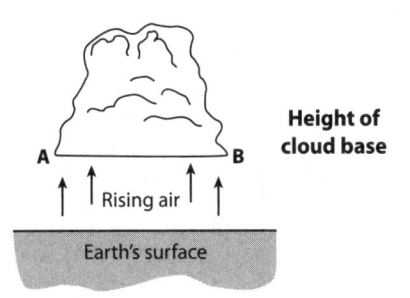

Which graph best represents the temperature measured along line **AB**?

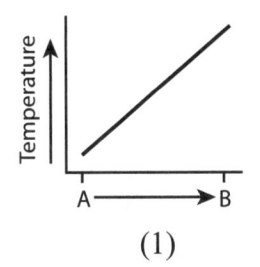

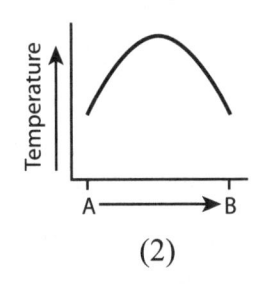

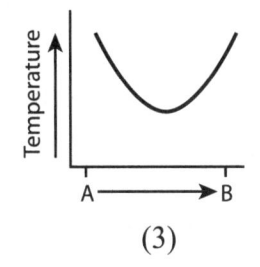

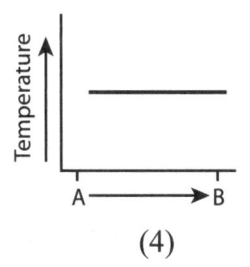

(1)   (2)   (3)   (4)

25. Air pressure is usually lowest when air is
    (1) warm and humid   (3) cold and humid
    (2) warm and dry   (4) cold and dry

26. The graph below shows the changes in air pressure and dewpoint temperature over a 24-hour period at a location. At what time was the relative humidity the lowest?

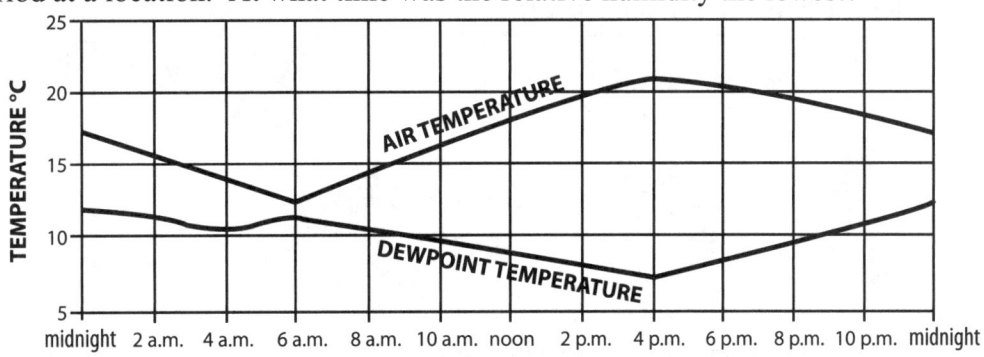

    (1) midnight   (2) 6 a.m.   (3) 10 a.m.   (4) 4 p.m.

27. The cross section shows a sea breeze blowing from the ocean to the land. The air pressure at the land surface is 1013 millibars. The air pressure over the ocean surface a few miles from shore is most likely
    (1) 994 mb   (3) 1013 mb
    (2) 1005 mb   (4) 1017 mb

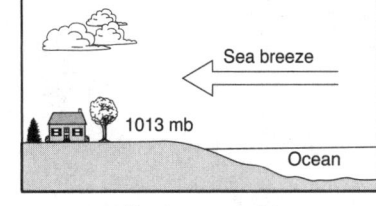

28. Earth's surface winds generally blow from regions of higher
    (1) air temperature to lower air temperature   (3) latitudes to lower latitudes
    (2) air pressure to lower air pressure   (4) elevations to lower elevations

29. Which weather variable would most likely decrease as a storm approaches?
    (1) wind speed   (2) air pressure   (3) cloud cover   (4) relative humidity

97

30. Which map best represents the surface wind patterns associated with high-pressure and low-pressure systems in the Northern Hemisphere?

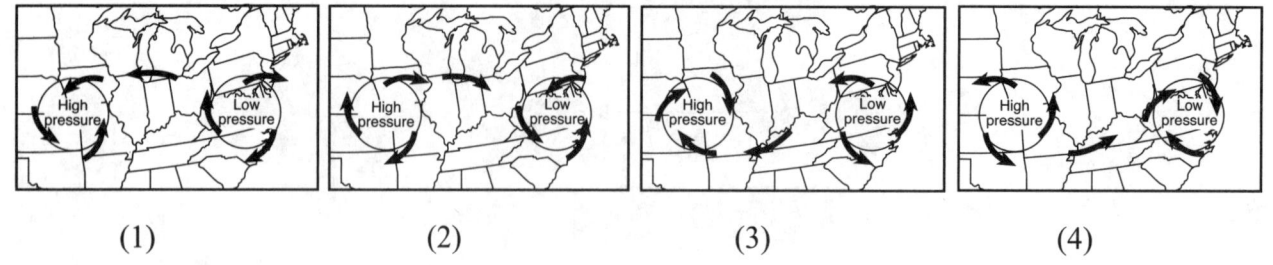

(1)   (2)   (3)   (4)

31. Which New York State location is most likely to experience the heaviest winter snowfall when the surface winds are blowing from the west or northwest?
    (1) New York City   (2) Binghamton   (3) Oswego   (4) Plattsburgh

32. At what latitude are high altitude jet streams?
    (1) 0° and 90° N and S
    (2) 0° and 30° N and S
    (3) 30° and 60° N and S
    (4) 60° and 90° N and S

33. What is the general direction of planetary winds at 20° S latitude?
    (1) northeast   (2) northwest   (3) southeast   (4) southwest

34. Which ocean current transports warm water away from the Earth's equatorial region?
    (1) Brazil Current
    (2) Guinea Current
    (3) Falkland Current
    (4) California Current

35. The arrows labeled **A** through **D** on the map below show the general paths of abandoned boats that have floated across the Atlantic ocean.

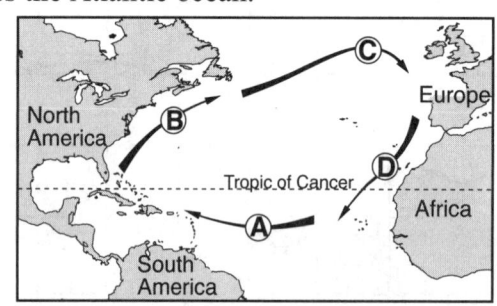

    Which sequence of ocean currents was responsible for the movement of the boats?

    (1) South Equatorial → Gulf Stream → Labrador → Benguela
    (2) South Equatorial → Australia → West Wind Drift → Peru
    (3) North Equatorial → Koroshio → North Pacific → California
    (4) North Equatorial → Gulf Stream → North Atlantic → Canary

36-40. The data table below shows the air temperatures and air pressures recorded by a weather balloon rising over Washington, DC.

36. On the grid, construct a graph of altitude above sea level and temperature .... use a "**dot**" for each data point, connect with a solid line, and label the line.

37. On the same grid, construct a graph of altitude above sea level and pressure.......... use an "**X**" for each data point, connect with a dashed line, and label the line.

38. At 2000 meters, what was the approximate air pressure? _____

39. State the relationship between altitude above sea level and air pressure._____

40. If the dewpoint at 1500 meters was 8° C, what was the relative humidity at 1500 meters? _____

Altitude Above Sea Level (m)	Air Temperature (°C)	Air Pressure (mb)
300	16.0	973
600	16.5	937
900	15.5	904
1,200	13.0	871
1,500	12.0	842
1,800	10.0	809
2,100	7.5	778
2,400	5.0	750
2,700	2.5	721

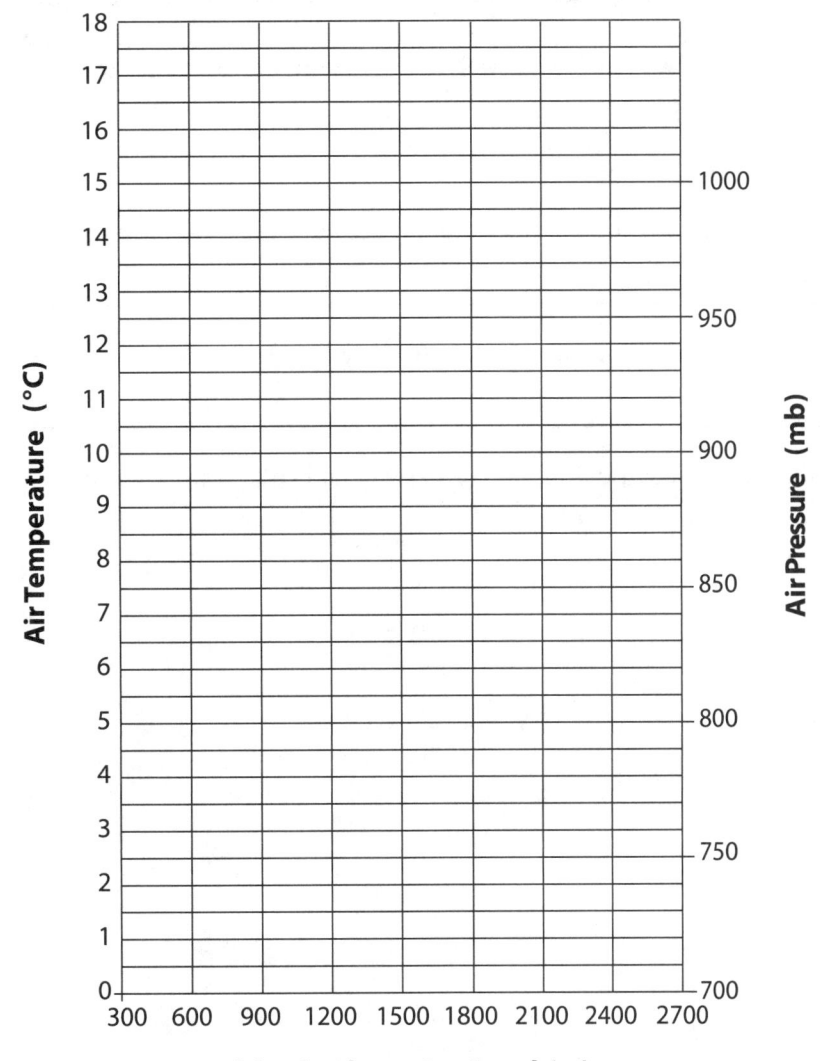

41-44. The graphs below show the air pressure, air temperature, and dewpoint temperature at a mid-latitude city on two consecutive days during the summer.

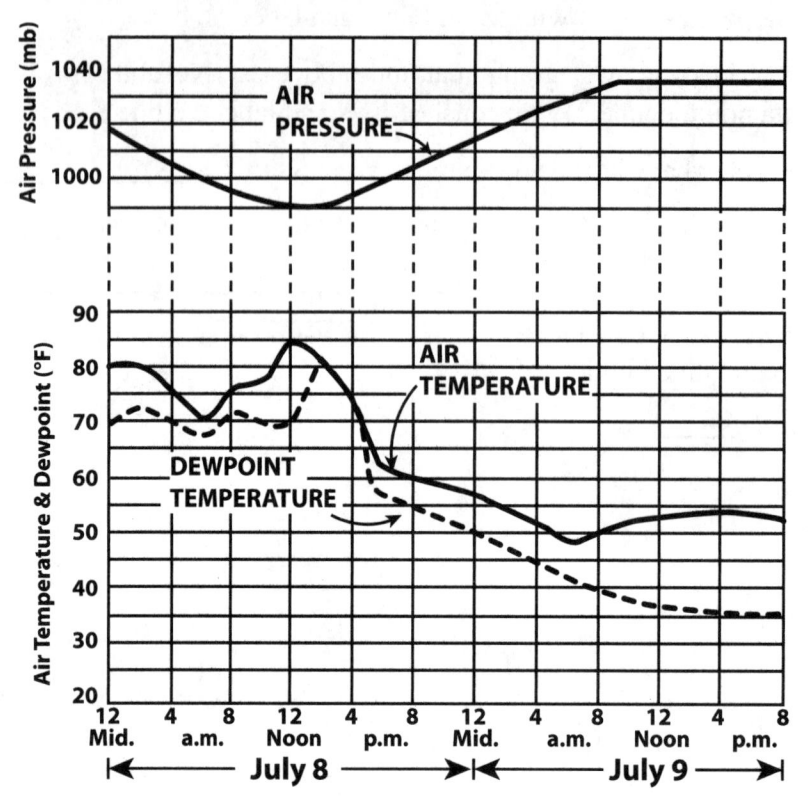

41. Name the instrument that was used to determine the dewpoint. _____

42. Based on this data, describe the relationship between air temperature and air pressure.
_____

43. The chance of precipitation was greatest on July 8 at 3 p.m. Why? _____
_____

44. On July 9th, the air became (*cooler*) (*warmer*) and (*less*) (*more*) humid.

45-48. The data table contains weather data that was recorded at 9 a.m. during a four-day period for a location in Massachusetts.

DATA TABLE

	Temperature (to nearest degree)	Air Pressure (mb)	Dewpoint (to nearest degree)	Wind Direction and Speed (knots)
9 a.m. Monday	24° C (75°F)	996.4	20° C (68°F)	NW 10
9 a.m. Tuesday	20° C (68°F)	962.4	19° C (66°F)	SSE 25
9 a.m. Wednesday	17° C (63°F)	1013.8	12° C (54°F)	W 15
9 a.m. Thursday	7° C (45°F)	1020.2	−2° C (28°F)	N 10

45. Determine the relative humidity on Monday. _____

46. The air pressure gradient was the greatest on _____. How do you know?_____
_____

47. Precipitation was most likely to occur on _____. How do you know?_____
_____

48. From Tuesday to Thursday the air pressure increased. What caused this to occur?
_____
_____

101

49-52. The diagram below shows an atmospheric cross section of a winter storm system and changes in precipitation. Zones **A**, **B**, **C**, and **D** are located on a west to east line across New York State. The storm is moving west to east.

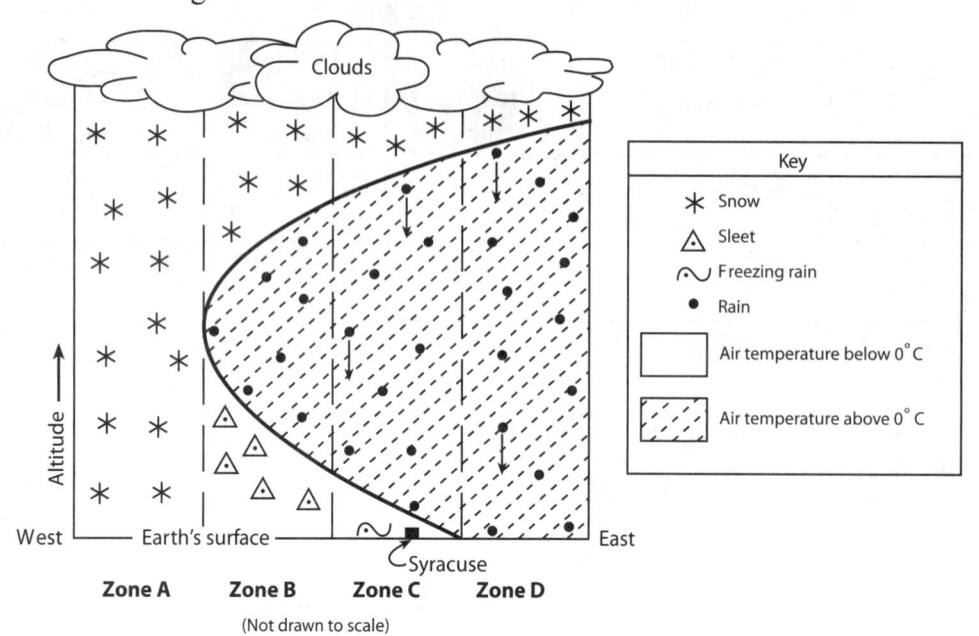

49. Describe the change in precipitation from the cloud to the ground in Zone **C**.

_____

50. In Zone **B**, why does sleet form at the Earth's surface? _____

_____

51. In Zone **A** why does the type of precipitation falling from the cloud *NOT* change as it falls to Earth's surface? _____

52. As the storm moves east, Syracuse will experience a change in precipitation. State the type(s) of precipitation that will follow the freezing rain. _____

53-55. On April 5, 1982 the eruption of El Chichon in Mexico occurred. The maps below show the spread of the volcanic ash as seen from weather satellites.

53. State the direction toward which the ash cloud spread from April 5 to April 25. _____

54. What caused the ash cloud to spread in this pattern from April 5 to April 25? _____
_____
_____

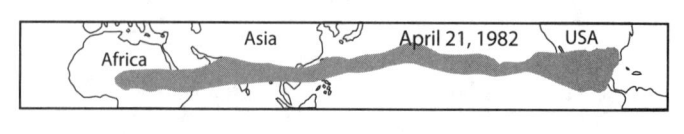

55. State the most likely effect the ash cloud would have on the temperatures of the areas under the cloud on April 25. _____
_____

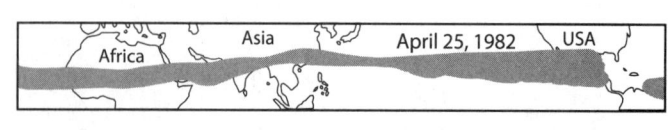

CHAPTER 7

WEATHER
Weather Station Models, Air Masses and Fronts, Weather Maps, Severe Weather

WEATHER AND STATION MODELS

Weather is the current conditions of the atmosphere that constantly change. One hour it might be sunny and the next hour it is raining. The weather variables that describe the atmospheric conditions include temperature, air pressure, wind speed and direction, humidity, precipitation, sky cover, visibility, and dewpoint.

Meteorology is the study of the atmosphere and weather. Weather is difficult to predict because of the many variables. Weather predictions are based on observed patterns. The atmospheric conditions are constantly monitored and measured at over six hundred weather stations in the USA. Technologies such as satellites and radar have greatly improved weather predictions.

The driving force of weather is the Sun. The atmosphere is heated unevenly because different latitudes receive different intensities of insolation. Since the atmosphere is "fluid," it moves and distributes heat energy by the process of convection. Cold polar air sinks and moves toward the tropics, and warm tropical air rises and moves towards the poles. The constant movement of the atmosphere and interactions between cold and warm air cause the changes in weather.

A weather station measures and records the atmospheric variables at the surface for that location. The station will launch weather balloons to record weather conditions in the upper troposphere. A **weather station model** is used to illustrate the weather conditions at a location. The data from many weather stations is placed on a map for interpretation and prediction.

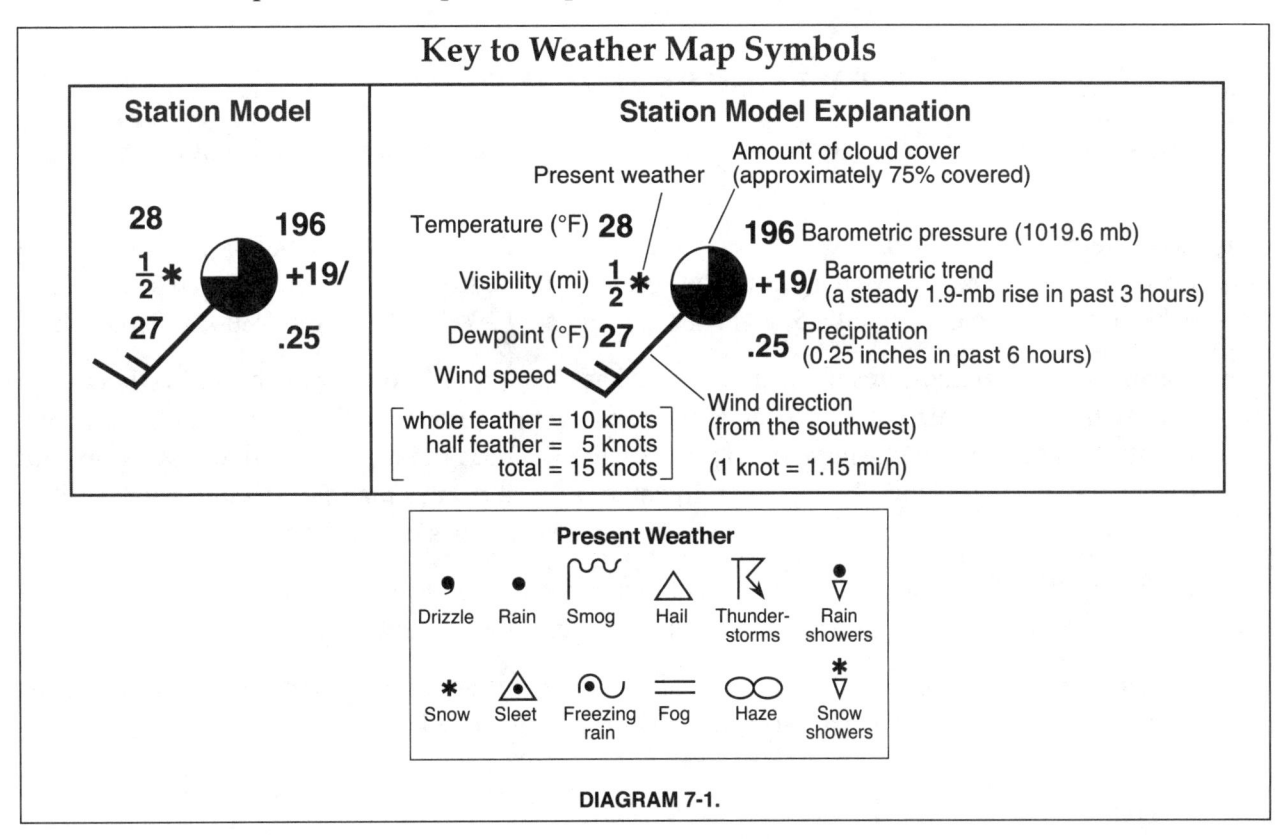

Key to Weather Map Symbols

Station Model

```
28         196
½ *       +19/
27         .25
  \
```

Station Model Explanation

- Temperature (°F) **28**
- Visibility (mi) **½**
- Present weather
- Dewpoint (°F) **27**
- Wind speed
  - whole feather = 10 knots
  - half feather = 5 knots
  - total = 15 knots
- Amount of cloud cover (approximately 75% covered)
- **196** Barometric pressure (1019.6 mb)
- **+19/** Barometric trend (a steady 1.9-mb rise in past 3 hours)
- **.25** Precipitation (0.25 inches in past 6 hours)
- Wind direction (from the southwest)
- (1 knot = 1.15 mi/h)

Present Weather

| Drizzle | Rain | Smog | Hail | Thunder-storms | Rain showers |
| Snow | Sleet | Freezing rain | Fog | Haze | Snow showers |

DIAGRAM 7-1.

The interpretation of a sample station model is shown on page 13 of the ESRT. The weather variables are always put in the same place, except for wind direction.

- **Air temperature** and **dewpoint** are recorded in °F. Only the number is written.

- **Visibility** is the distance in miles that one can clearly see at that location. It is written as a whole number or as a fraction. No unit is written.

- **Present weather** describes any weather event that is occurring at that location. The symbol is placed to the right of the visibility. A key for the symbols used is given on page 13 of the ESRT.

- Winds are named for the direction they are coming from. A north wind is blowing from the north. **Wind direction** is shown by a line pointing in that compass direction.

- **Wind speed** is measured in knots (1 knot = 1.15 mile/hour). The wind speed is indicated by "feathers" (lines) extending off the wind direction line. Each whole "feather" (line) equals 10 knots; a half "feather" (half line) equals 5 knots. An indented half-line equals 5 knots. A darkened triangular "flag" represents 50 knots. Add these "feathers" together to get the wind speed. If there is no wind direction line then it is calm.

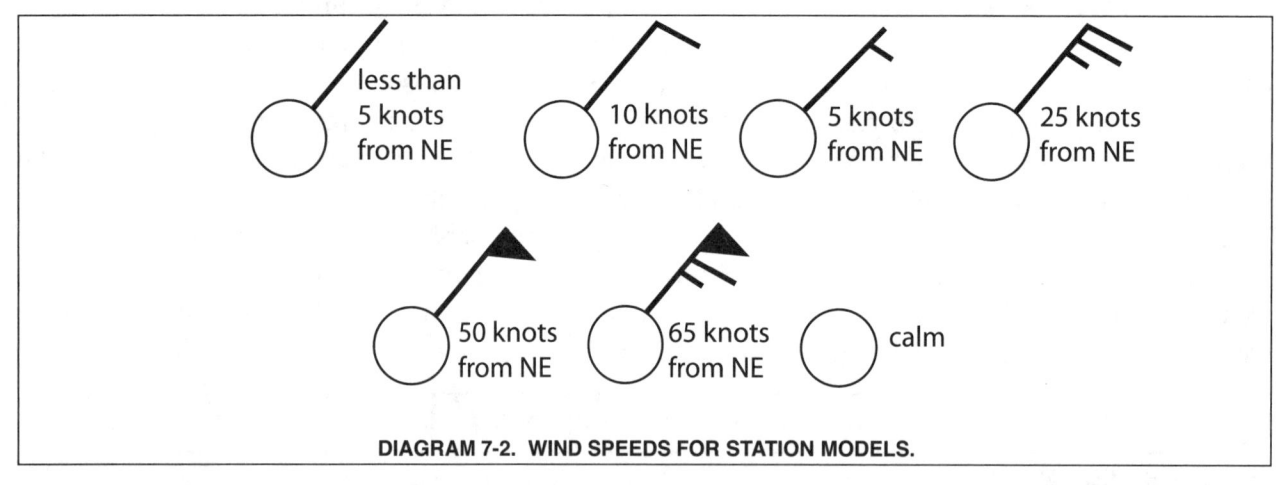

DIAGRAM 7-2. WIND SPEEDS FOR STATION MODELS.

- The station model circle is filled in to show the amount of **cloud cover**. If it is not filled in, skies are clear.

- **Barometric air pressure** is written as a code in three digits. When decoding pressure place a decimal between the last two numbers. Then place a "10" in front if the first digit is a 0 to a 4. Place a "9" in front if the first digit is a 5 or greater. *Examples*: "299" = 1029.9 mb; "804" = 980.4 mb

- **Barometric trend** indicates if the air pressure has risen or fallen during the past three hours. To decode barometric trend place a decimal between the two numbers. If only one number is given, the decimal goes before the number. If the pressure has been rising, a "+" or line sloping upward (/) is drawn. A "– " or a line sloping downward (\ ) indicates that the pressure has fallen. A straight line (–) means that the pressure has been steady, it has not changed.

  Example: "- 45\" means air pressure has fallen 4.5 mb in past three hours.
  " +7/" means air pressure has risen 0.7 mb in past three hours.

- **Precipitation** is written in decimal form. This is the amount of precipitation in inches that has fallen at that location in the past six hours. No unit is included.

Questions

1. Based your answers on the weather station model shown to the right.

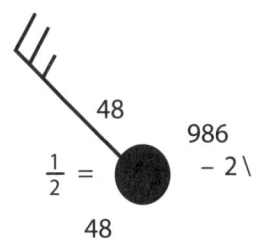

    a. air temperature:_____
    b. dewpoint:_____
    c. visibility:_____
    d. wind speed:_____
    e. wind direction:_____
    f. cloud cover: _____
    g. air pressure: _____
    h. barometric trend: _____
    i. air pressure 3 hrs ago: _____
    j. precipitation: _____
    k. current weather: _____
    l. relative humidity: _____

2. Complete the station model using the following data.

    a. air temp = 25°F
    b. dewpoint = 24°F
    c. present weather = snow
    d. visibility = ¼ mile
    e. winds from the southeast at 30 knots
    f. air pressure = 987.2 and falling
    g. cloud cover = 100%

3. The weather station models show weather conditions in a city in western New York at 3 p.m. on four consecutive days.

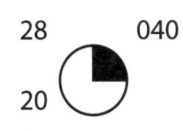

    Tuesday   **Wednesday**   **Thursday**   **Friday**

    a. The air pressure on Wednesday was: _____; Thursday: _____

    b. Describe the change in air pressure from Tuesday to Friday. _____

    c. On which day was the chance of precipitation the greatest? _____

    What type of precipitation was likely on this day? _____

    Why? _____

    d. On Tuesday winds were from the west at 5 knots. Show this on the station model for Tuesday.

    e. Determine the relative humidity on Wednesday. _____%
    Hint: weather station model uses Fahrenheit (°F)

105

4. The diagrams below show weather data from four U.S. weather stations recorded at the same time on the same day.

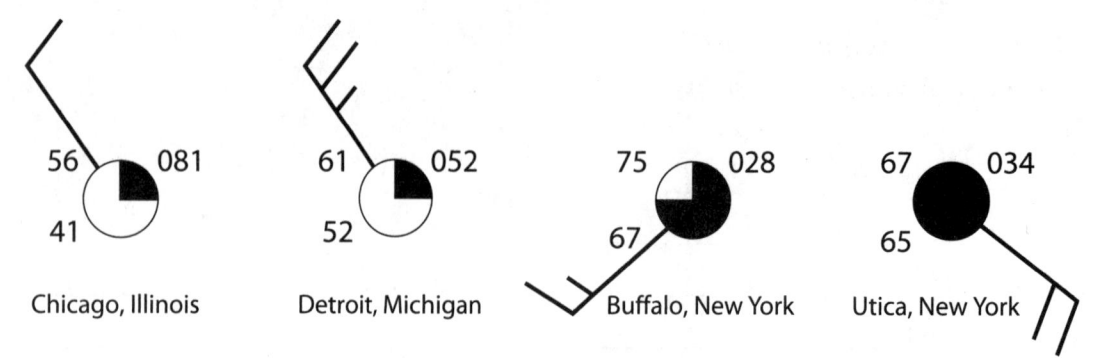

Chicago, Illinois Detroit, Michigan Buffalo, New York Utica, New York

a. Place each weather station in order from the lowest to highest wind speed.

_____ , _____ , _____ , _____

b. Which city has the greatest probability of precipitation? _____

Explain _____

_____

c. Which city is in an area of the highest pressure gradient? _____

How do you know? _____

_____

d. Which city has a steady wind from the southwest? _____

e. Air pressure for Chicago is _____ mb = _____ in.

f. Which city had the lowest relative humidity? _____

How do you know? _____

_____

5. What was the air pressure at this weather station 3 hours ago? _____

Show solution

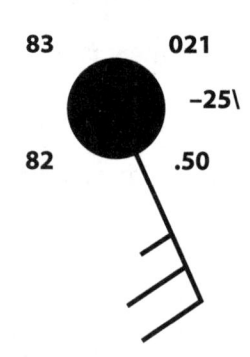

106

AIR MASSES AND FRONTS

Air masses are large bodies of air in the troposphere that have similar weather conditions throughout. They form when air stagnates or stops over a region. When this occurs, the air mass acquires the temperature and moisture characteristics of this **source region**. For example, an air mass that forms over the North Pacific will be moist and cold. Air masses control weather for a few days until another air mass moves in. Air masses are named for their source region.

Name	Abbreviation	Moist or dry?	Cold, warm, or hot?
Maritime polar	mP	Moist	Cool
Maritime tropical	mT	Moist	Warm
Continental tropical	cT	Dry	Hot
Continental polar	cP	Dry	Cold
Continental arctic	cA	Dry	Very Cold

Note: **The first letter is lower case and the second is capitalized. Do not reverse the letters. For example, maritime polar is mP not Pm.**

Air masses in the USA are moved by the prevailing westerlies. Their general movement is from SW to NE. Some of the technologies used to monitor and track air mass movement include ground based instruments, weather balloons, airplanes, satellites, and radar.

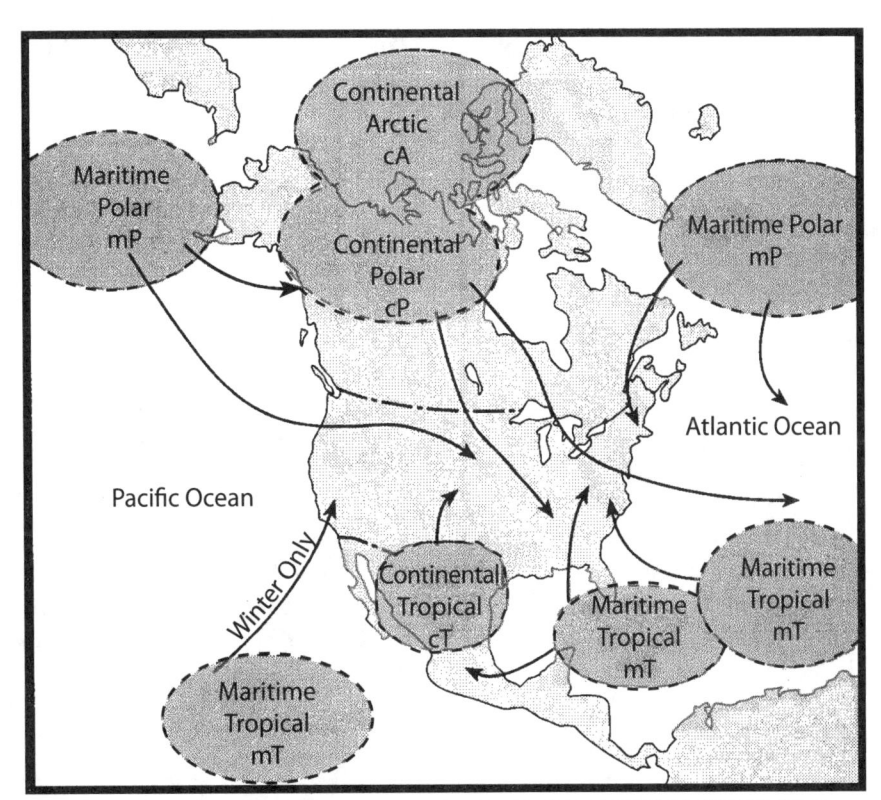

DIAGRAM 7-3. SOURCE REGIONS FOR NORTH AMERICAN AIR MASSES.

107

A **front** is the boundary or interface between two different air masses. At a front, the air is unstable because warm, less dense air rises above the cooler air. As the warm air rises clouds form and precipitation may occur. It may become windy and changes in air pressure and temperature occur. Frontal weather is short-lived.

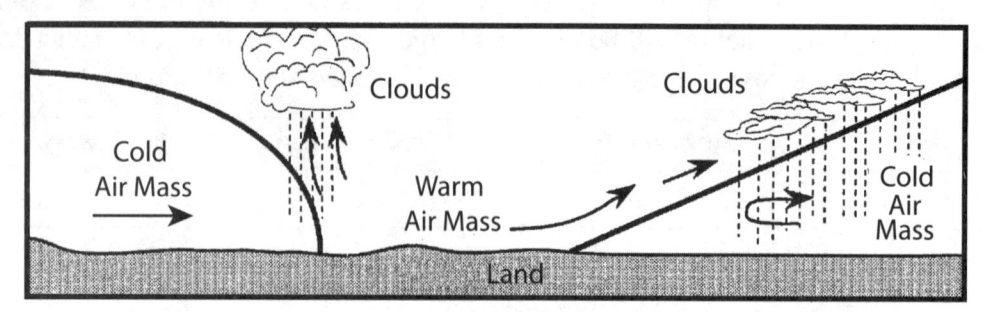

DIAGRAM 7-4. FRONTAL BOUNDARIES BETWEEN DIFFERENT AIR MASSES.

There are four types of fronts. For each front symbol used on a weather map, the symbol (solid half-circles or solid triangles) points in the direction that the front and the air mass behind the front are moving towards.

type of front and map symbol	cross-section	description	weather that occurs
cold front		cold air moves in and pushes the warm air up	cumulus clouds form, heavy precipitation of short duration, thunderstorms and hail possible
warm front		warm air moves in and slowly rises over the colder air	stratus clouds, light precipitation for six or more hours
stationary front		warm air and cold air meet and both stop moving	clouds and precipitation can last for a few days
occluded front		cold front overtakes a warm front, warm air is trapped between two cold air masses	intense rain or snow storms can occur

108

Questions

6. A **mP** air mass is _____ and _____.

7. A continental air mass is dry because _____.

8. Which type of air mass is associated with hot, dry weather? _____

9. An air mass forming over the North Atlantic Ocean would be labeled _____.

10. Air masses generally move from _____ to _____ across the United States; Air masses generally move from _____ to _____ across Australia.

11. Compared to a **mT** air mass, a **cP** air mass is *(cooler) (warmer)* and *(drier) (moister)*.

12. An **mT** air mass forms over *(water) (land)* and is *(cool) (warm)*.

13. Explain why a **cT** air mass is hotter than a **mT** air mass. _____
    _____

14. Classify the type of air mass that would form at each location. Use ESRT page 4.
    a. 25° North, 105° West _____
    b. 55° North, 50° East _____
    c. 65° North, 125° West _____
    d. 20° North, 60° West _____

15. Identify the type of front associated with each weather pattern.
    a. light precipitation, temperature rises: _____
    b. steady rain for three days: _____
    c. winds shift from the SE to the NW as hail falls in Albany: _____
    d. forward edge of cold air: _____

16. This map shows frontal systems over New York State.
    a. What type of front will pass **A** in the next few hours? _____
    b. Shade in the areas where precipitation is most likely taking place.
    c. Name the front that is north of Lake Ontario? _____
    d. Label the area *"warm air"* that is most likely a warm air mass.
    e. Use the thunderstorm symbol to indicate a location that could be experiencing thunderstorms.

109

17. The map below shows a low-pressure system with two fronts extending from its center (**L**). Points **A**, **B**, **C**, and **D** are locations on Earth's surface.

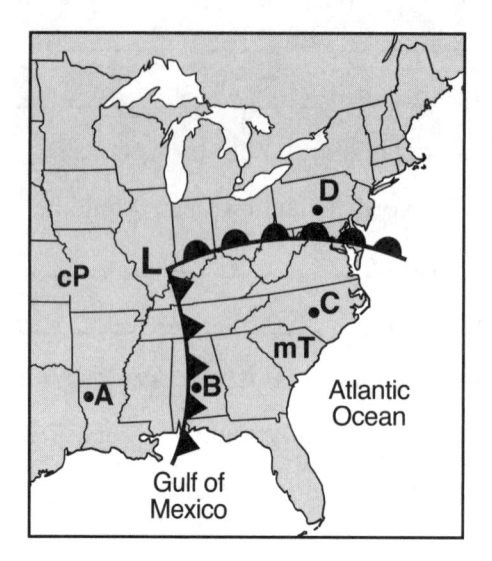

   a. In what direction is the warm front moving? _____

   b. What change in temperature is location **B** experiencing? _____

   c. What is the most probable source of the **mT** air mass? _____

   d. Describe the air conditions in the **cP** air mass. _____

   e. The center of the low-pressure system (**L**) will move towards *(A) (B) (C) (D)*.

   f. Which lettered location(s) are most likely experiencing precipitation? _____

   g. Which lettered location(s) could be experiencing thunderstorms? _____

18. The map below shows surface air temperatures, in degrees Fahrenheit, reported by weather stations in north central United States. Letter **X** represents an air mass moving in the direction shown by the arrow. A solid line marks the frontal boundary advancing in a southeasterly direction.

   a. The air mass at letter **X** is most likely classified as _____.

   b. Draw in the correct front symbol for the frontal boundary.

   c. Within the next two days, the weather stations to the southeast of the front can expect the temperatures to:

   (decrease) (increase) (remain the same).

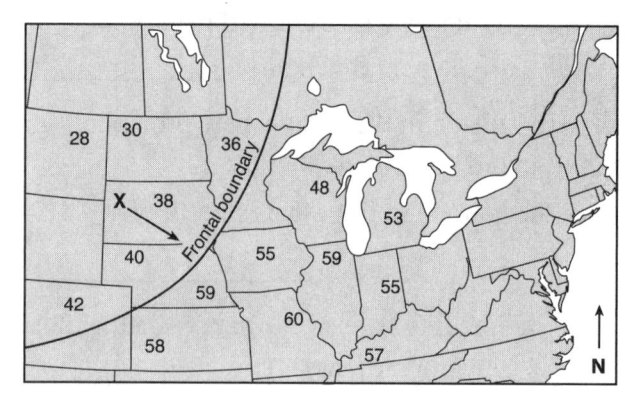

110

19. The diagram below shows the movement of air along a front. **A** and **B** are two locations at the surface.

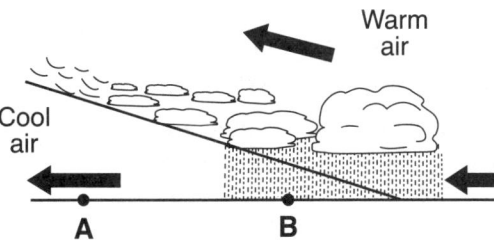

a. Name the type of front shown. _____

b. Why did you choose this frontal type? _____
_____

c. Write a weather forecast for location **A**. _____
_____
_____

d. Explain why the warm air rises at the frontal boundary. _____
_____

e. Explain the processes that cause clouds to form along the front. _____
_____
_____
_____

INTERPRETING A WEATHER MAP

The National Weather Service gathers weather data from hundreds of weather stations in the country. The data from each station is placed on a map. From this data, meteorologists will draw isobars, connecting points of the same air pressure. They will locate the centers of high pressure (**H**) and low pressure (**L**). Air masses and fronts will be located on the map. Based on this information weather forecasts will be made.

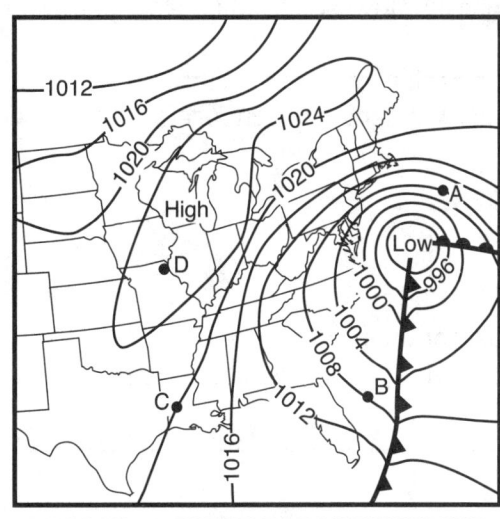

DIAGRAM 7-5. EXAMPLE OF A WEATHER MAP.

111

Questions

20. The diagram below shows a U.S. weather map for one day in the spring. Isobars are drawn for every 4 millibars. The centers of the high (**H**) and low (**L**) pressure systems are shown. Frontal boundaries are shown. Data for some weather stations is indicated.

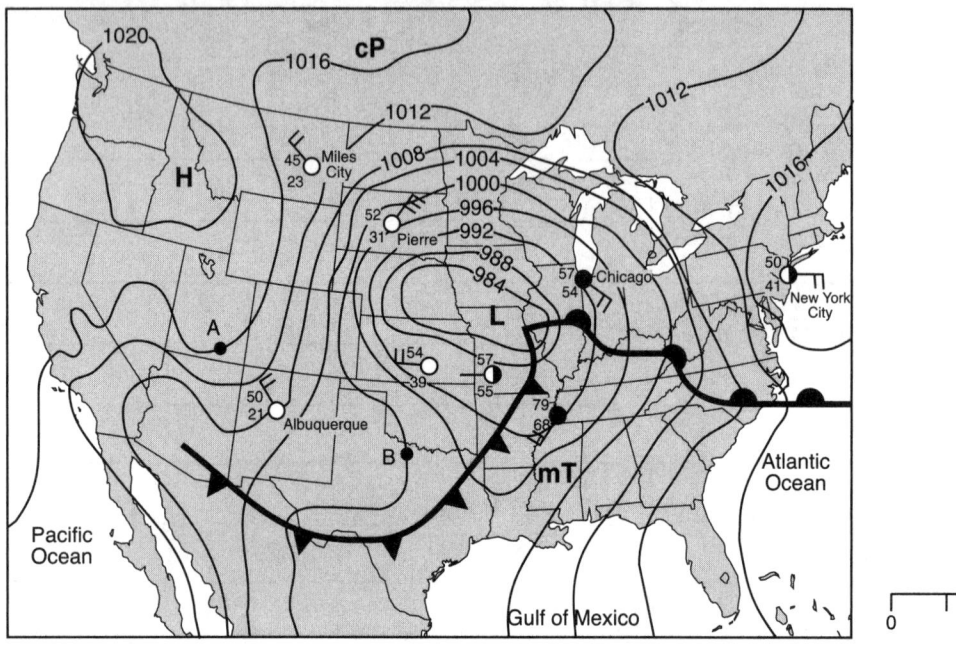

    a. Name the front that has passed location **B**? _____

    b. For Chicago the air temperature is _____ and air pressure is _____.

    c. What is the most likely source of the **cP** air mass? _____

    d. Draw four large arrows to show wind patterns around the **H** and the **L**.

    e. State the possible range for air pressure within the center of the **H**._____

    f. Describe the sky conditions for the center of the **H**. _____

    g. Why is the wind speed greatest for Pierre, South Dakota? _____
        _____

    h. List the cities from highest to lowest humidity: *Albuquerque, Chicago, New York City*
        _____, _____, _____

    i. Calculate the air pressure gradient between locations **A** and **B** to the nearest thousandth. *Show solution.*

    j. Write a weather forecast for the next day for New York City. Include the changes expected in temperature, pressure, wind, cloud cover, and precipitation.

        _____

        _____

        _____

21. The weather map below shows the center of a low pressure system off the east coast of the United States. The shaded portion represents the area of precipitation.

   a. Name the front extending to the east of the Low. _____

   b. Describe the type of weather associated with this front. _____
   _____
   _____

   c. Draw an arrow to show direction that the **Low** will move towards.

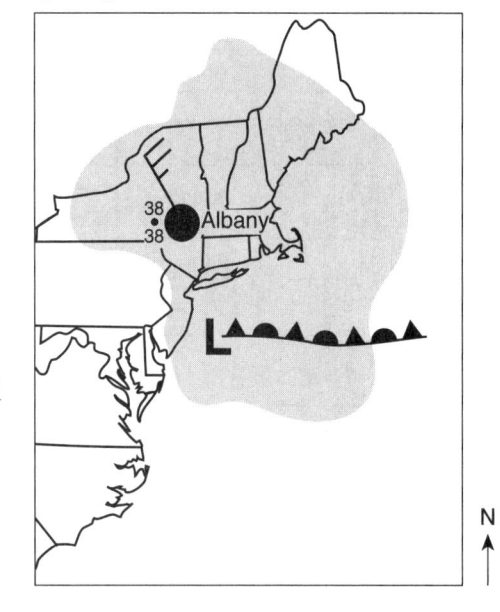

SEVERE WEATHER

Severe weather refers to any dangerous weather event with the potential to cause damage to property, serious disruption to human society, injury, and loss of life. In the event of severe weather you should be prepared with emergency supplies. This should include nonperishable canned food and bottled water to last each person in your family for three days. Also needed are medications, batteries, flashlights, and a battery powered radio. You should have in place a family communication plan so that your family will know where to find one another if you are separated during an emergency.

As the National Weather Service monitors and tracks storms they may issue an advisory telling people to keep alert to weather changes. A **storm watch** is issued when there is a possibility that a storm may occur and people should begin to make preparations. A **storm warning** is issued when there is certainty that the storm will impact the region. When a storm warning is issued, emergency preparations and possible evacuations must occur immediately.

A **tornado** is a rapidly rotating, extremely low-pressure air funnel that extends down from a thunderstorm cloud. It forms at the interface or front between two air masses that have a large difference in temperatures, for example cP vs. mT. Tornadoes are common in the midwest plains of the United States where cold, dry air from Canada (cP) meets warm, moist air from the Gulf of Mexico (mT).

Tornadoes may be on the ground for a few minutes or for hours. They can devastate an area in just seconds. Their paths are usually erratic and difficult to predict. Radar monitors their formation and warning signals are sent out. One of the most dangerous hazards of tornadoes is flying debris. If you are in an area where there are frequent tornados, you should know what precautions to take before the tornado strikes and stay alert to changing weather conditions and storm possibilities. If a tornado warning is issued go to a basement or storm cellar. If there is no basement, go to an interior room that does not have windows. If you are outside, lie down in the lowest area and try to cover your head. Protect yourself from flying debris.

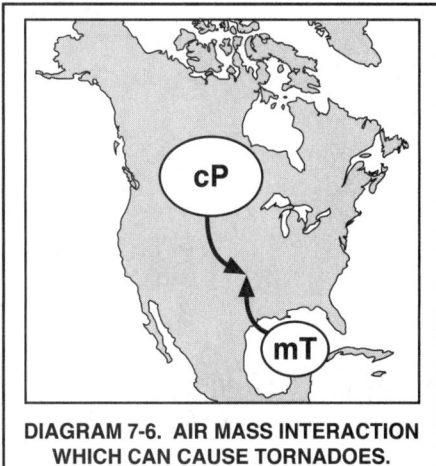

DIAGRAM 7-6. AIR MASS INTERACTION WHICH CAN CAUSE TORNADOES.

Fujita Scale for Tornadoes

F-Scale Number	Wind Speed (mph)	Type of Damage Done
F–0	40–72	some damage to chimneys; breaks branches off trees; pushes over shallow-rooted trees; damages sign boards
F–1	73–112	peels surface off roofs; mobile homes pushed off foundations or overturned; moving autos pushed off the roads; attached garages may be destroyed
F–2	113–157	considerable damage; roofs torn off frame houses; mobile homes demolished; boxcars pushed over; large trees snapped or uprooted; light-object missiles generated
F–3	158–206	roof and some walls torn off well-constructed homes; trains overturned; most trees in forest uprooted
F–4	207–260	well-constructed houses leveled; structures with weak foundations blown off some distance; cars thrown and large missiles generated
F–5	261–318	strong frame houses lifted off foundations and carried considerable distances to disintegrate; automobile-sized missiles fly through the air in excess of 100 meters; trees debarked; steel-reinforced concrete structures badly damaged

Thunderstorms usually form along a cold front where warm air is rapidly rising. Large cumulonimbus clouds form. Within the cloud, convection currents cause static electricity to form due to friction between the moving air molecules.

Thunderstorms can cause flash floods, hail, strong winds, and dangerous lightning. If a thunderstorm watch is posted stay alert. When a thunderstorm starts, seek shelter inside, do not use water, and unplug electrical appliances and devices. If you are outside, go to a low area and stay away from tall objects such as trees.

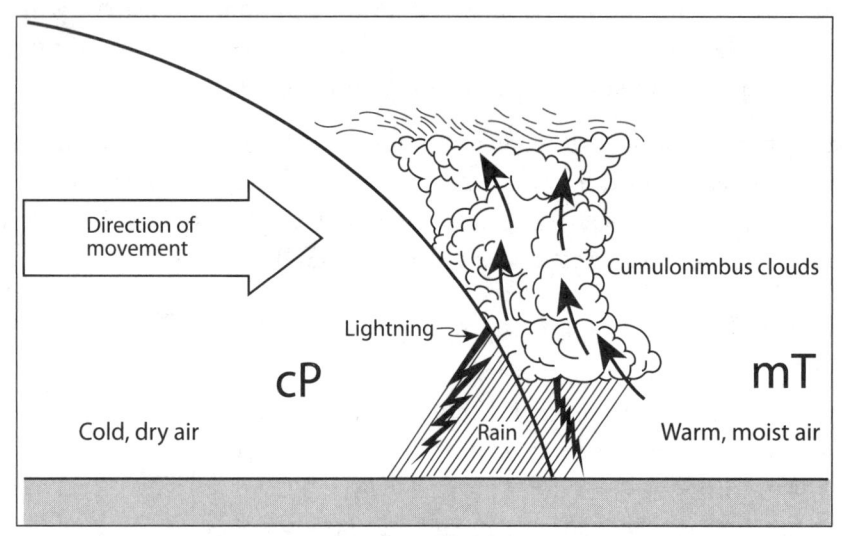

DIAGRAM 7-7. THUNDERSTORM FORMING ALONG A COLD FRONT.

Blizzards (winter storms) have heavy snowfall and winds over 35 mph. Before a blizzard strikes, make sure your emergency plans and supplies are in place. Be sure you have sufficient heating fuel, shovels, and warm clothing. Once the blizzard begins, stay inside and do not drive. Winter storm *Nemo* paralyzed most of Long Island in February of 2013 trapping hundreds of drivers inside their cars.

Hurricanes are low pressure systems that form over warm, tropical water such as the South Atlantic Ocean or the Gulf of Mexico. The water must be at least 80°F in order for a hurricane to form. The storm strengthens from the energy released during the condensation of water vapor. The **Safir-Simpson scale** rates the strength of hurricanes based on air pressure and wind speed.

Safir-Simpson Hurricane Scale

Hurricane Category	Central Air Pressure (mb)	Windspeed (km/hr)	Expected Storm Surge Height (m)	Expected Damage
1	over 979	119-153	1.2–1.5	Minimal
2	965-979	154-177	1.6–2.4	Moderate
3	945-964	178-209	2.5-3.6	Extensive
4	920-944	210-250	3.7-5.4	Extreme
5	below 920	over 250	over 5.4	Catastrophic

Hurricane season is from June to November. Hurricanes can cause catastrophic damage to coastlines as well as inland areas. Strong hurricane winds along the coast cause large waves and a surge of sea water which may flood coastal areas. Before hurricane season begins your family should have emergency supplies. If you live in an evacuation area, you should be aware of evacuation routes, emergency shelters, and how to get to higher ground. Plan to secure your home and property. During a hurricane you should listen to a radio, stay inside away from windows, and go into a small interior room. Bring in anything from the outdoors which may cause damage. Windows should be covered with storm shutters or plywood.

Questions

22. What is the source region for the an air mass that forms a hurricane? _____
_____

23. A hurricane with a central air pressure recorded at 28.70 inches has an expected storm surge of _____ meters.

24. A hurricane with wind speeds of 186 km/hr is classified as a category _____.

25. Describe the damage caused by a **F-4** tornado. _____
_____

26. Give the wind speed for a **F-3** tornado: _____; **F-0** tornado: _____

27. Explain why the midwest area of the USA experiences the most tornados. _____
_____
_____
_____

28. Complete the chart below.

	Dangers (list at least two)	**Preparedness** (what you should do before the storm)	**What do you do during the storm event?**
Blizzard			
Hurricane			
Thunderstorm			
Tornado			

29. The map below shows snowfall amounts from a December snowstorm at various locations in New York, New Jersey, and Pennsylvania. On the map draw the 30.0, 20.0, and 10.0-inch snowfall isoline. Assume that the decimal point for each snowfall depth marks the exact location where the snowfall was measured.

December Snowfall Amounts (inches)

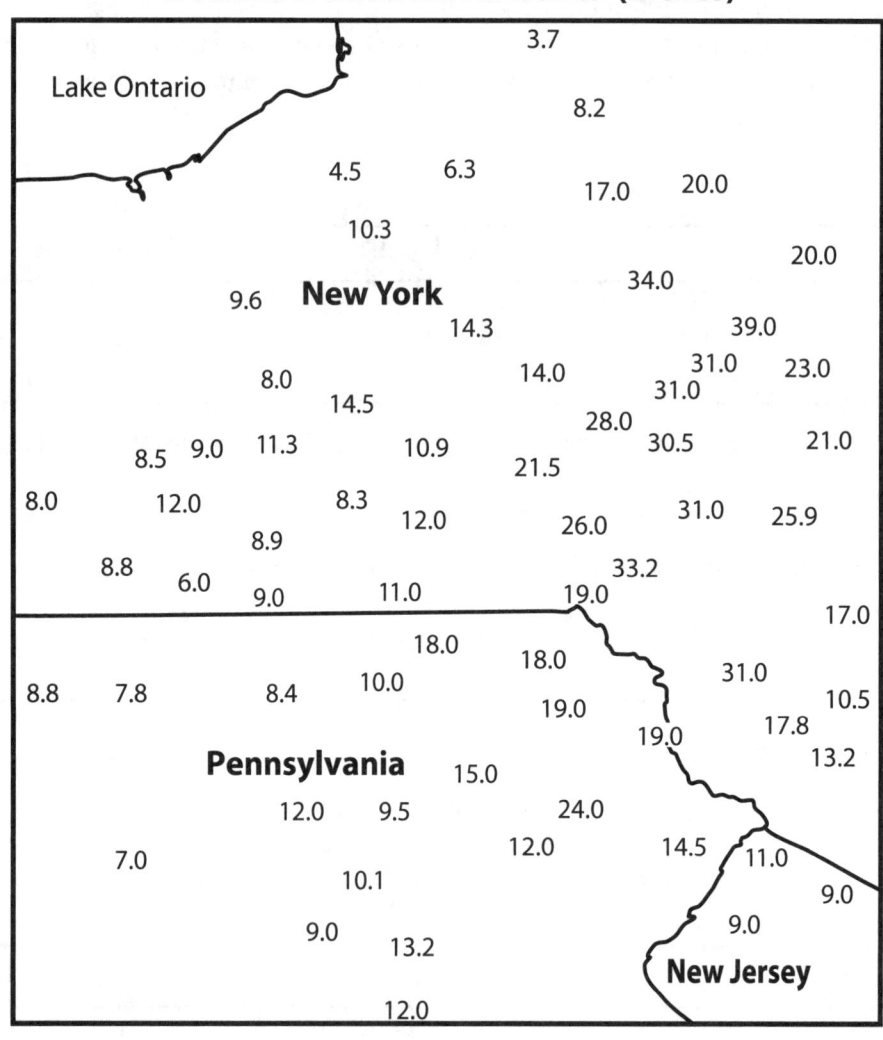

116

30. The data table shows the average number of days with thunderstorms that occur over land areas at different latitudes.

Data Table

Latitude	Average Number of Days a Thunderstorm Occurs Over Land
60° N	5
45° N	14
30° N	19
15° N	30
0° (equator)	56
15° S	44
30° S	21
45° S	8
60° S	0

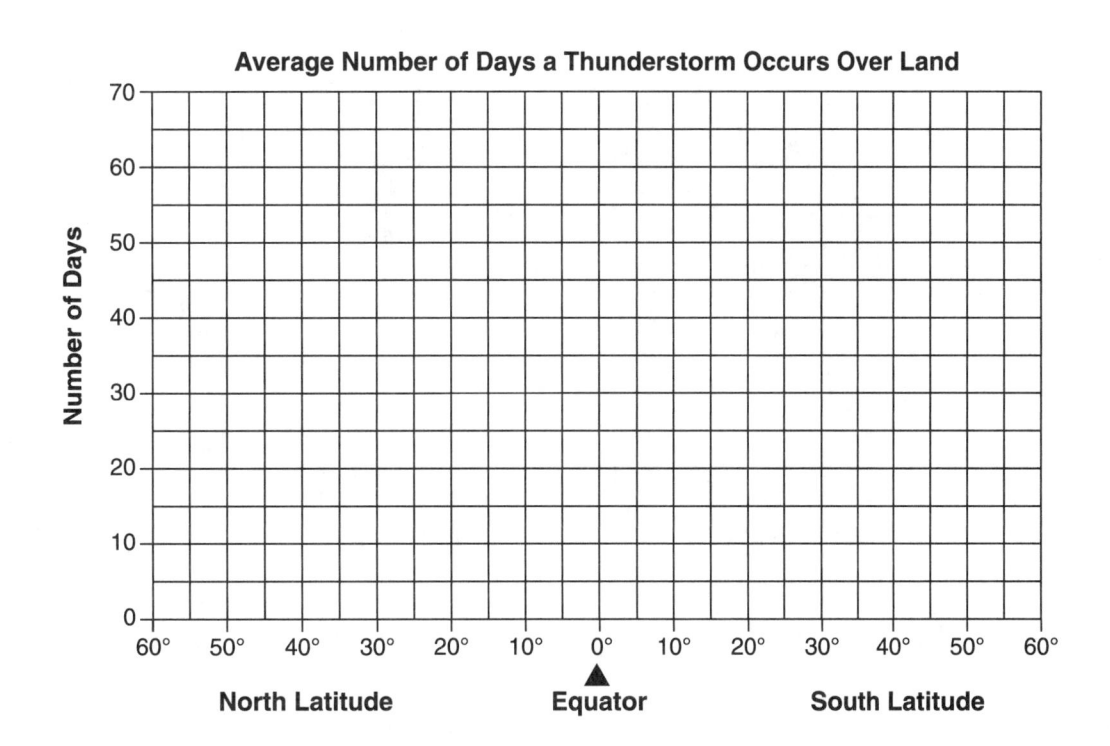

Average Number of Days a Thunderstorm Occurs Over Land

(Grid with y-axis "Number of Days" from 0 to 70, x-axis from 60° North Latitude to 60° South Latitude, with Equator at 0°)

a. On the grid plot with an "**X**" the average number of thunderstorm days per year for each latitude. Connect the centers of the **X**s with a line.

b. State the relationship between latitude and the average number of thunderstorm days per year that occur over land.

_____

_____

31. The data below shows recorded information for a major Atlantic hurricane.

Date	Time	Latitude	Longitude	Maximum Winds (knots)	Air Pressure (mb)
Sept. 10	11:00 a.m.	19° N	59° W	70	989
Sept. 11	11:00 a.m.	22° N	62° W	95	962
Sept. 12	11:00 a.m.	23° N	67° W	105	955
Sept. 13	11:00 a.m.	24° N	72° W	135	921
Sept. 14	11:00 a.m.	26° N	77° W	125	932
Sept. 15	11:00 a.m.	30° N	79° W	110	943

    a. Name the weather instrument used to measure air pressure. _____

    b. Based on air pressure, what would the expected storm surge on September 15? _____

    c. Based on the latitude-longitude data from the data table, use the hurricane symbol to plot the location of the hurricane during this 6-day period. Connect all the symbols with an arrow.

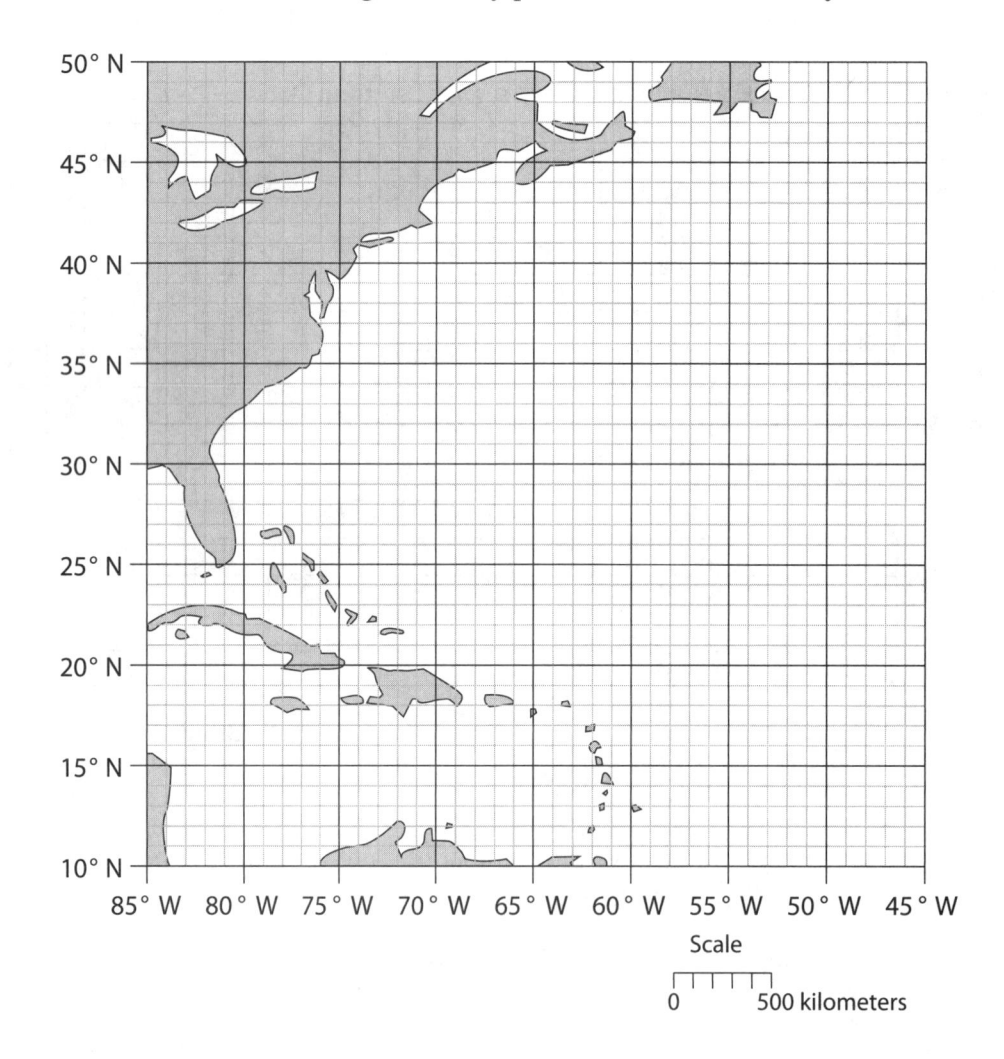

    d. State the general compass direction that the hurricane was moving towards. _____

    e. From September 12 to September 14 the hurricane moved _____ km.

CHAPTER 7 REVIEW

1-4. Base your answers to questions **1** through **4** on the weather station models below which show data that was recorded at four different locations at the same time.

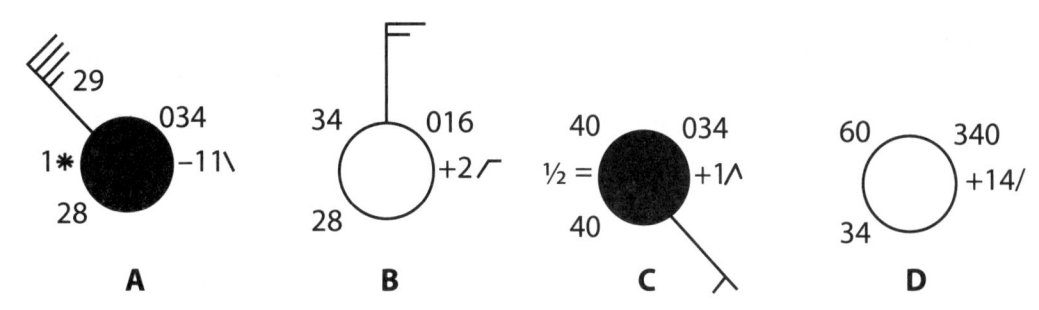

1. Which station has a dewpoint of 34° F?
   (1) A (2) B (3) C (4) D

2. The wind direction at station **C** is from the
   (1) northeast (2) northwest (3) southeast (4) southwest

3. What is the air pressure at station **D**?
   (1) 340.0 mb (2) 934.0 mb (3) 1003.4 mb (4) 1034.0 mb

4. Which station shows that the present air pressure reading is lower than it was 3 hours ago?
   (1) A (2) B (3) C (4) D

5-6. Base your answers to questions **5** and **6** on the weather station model shown below.

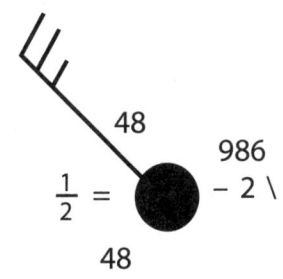

5. Relative to this weather station, a high pressure system is located to the
   (1) northeast (2) northwest (3) southwest (4) southeast

6. Which statement correctly describes the relative humidity at this station?
   (1) The relative humidity is 0% because the cloud cover is 100%.
   (2) The relative humidity is 100% because the air temperature and dewpoint are both 48° F.
   (3) The relative humidity is 98.6% because 986 is the symbol for 98.6%.
   (4) The relative humidity is 50% because ½ is the symbol for 50%.

7. A weather station model is shown below. Which information shown on the station model is most closely associated with measurements from a psychrometer?

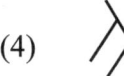

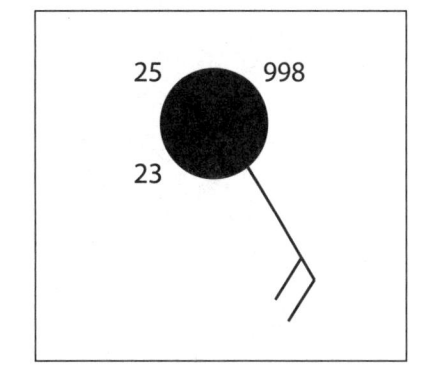

   (1) 23
   (2) 998
   (3) [filled circle symbol]
   (4) [wind barb symbol]

119

8. Rain sometimes turns into ice when it comes in contact with Earth's surface. Which present weather symbol on a station model represents this type of precipitation?

   (1)  ⊙⌣   (2)  ⌐∾   (3) △   (4) ✱

9. A maritime polar air mass approaching New York State would most likely bring
   (1) cool, moist air from the north
   (2) warm, moist air from the south
   (3) cool, dry air from the southeast
   (4) warm, dry air from the southwest

10. What is the source region for a cT air mass that moves into New York?
    (1) southwest United States
    (2) central Canada
    (3) the north Pacific Ocean
    (4) the Gulf of Mexico

11. The diagram shows a cross section of a frontal system. The cloud formation and precipitation shown are caused by: 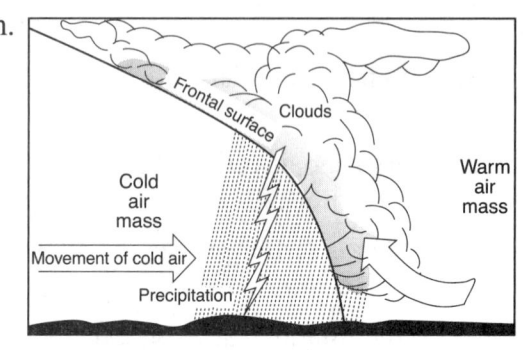
    (1) cold air rising and warming by expansion
    (2) cold air sinking and warming by compression
    (3) warm air rising and cooling by expansion
    (4) warm air sinking and cooling by compression

12. The diagram below represents a cross section of air masses and frontal surfaces along line **AB**. The dashed lines represent precipitation.

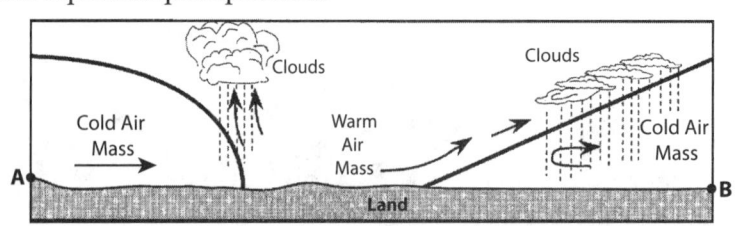

    Which weather map symbols best represents this frontal system?

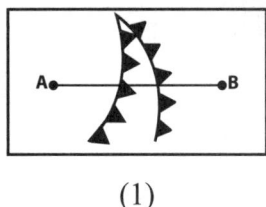

    (1) 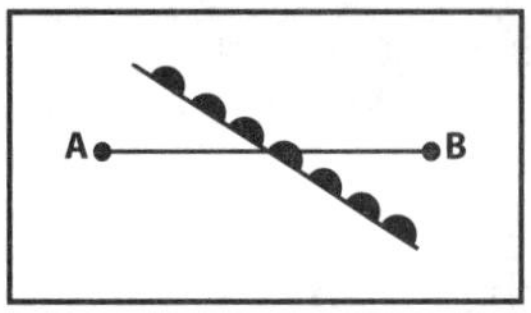   (2) 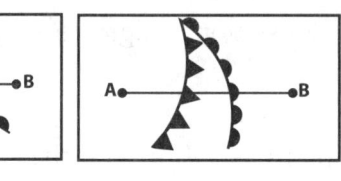   (3) 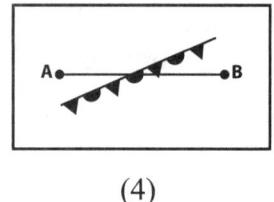   (4)

13. Which map best shows the most probable areas of precipitation associated with the weather systems drawn?

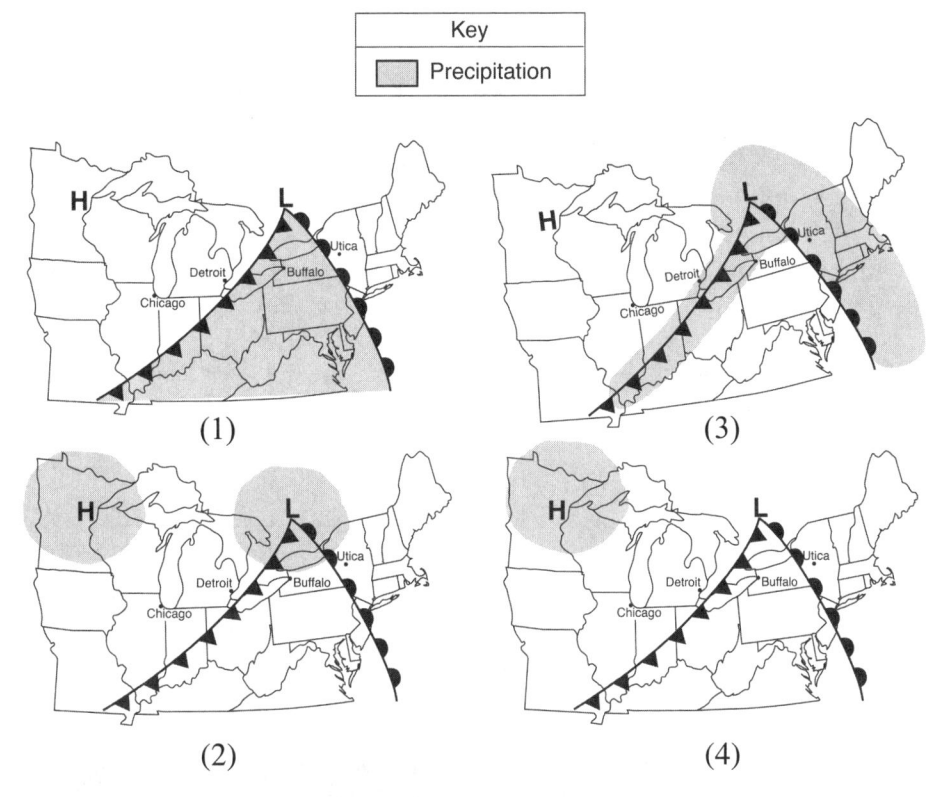

14. In a certain area the air temperature and the dewpoint temperature are approaching the same value, the air pressure is decreasing, and the cloud cover is increasing. What atmospheric change is most likely occurring in this area?

    (1) Warm, moist air is moving into the area.
    (2) Warm, dry air is moving into the area.
    (3) Cold, dry air is moving into the area.
    (4) A cold front has just passed through the area.

15. As a cold front passes a weather station, which changes will be observed?
    (1) Air pressure rises and temperature falls.
    (2) Air pressure falls and temperature rises.
    (3) Both pressure and temperature rise.
    (4) Both pressure and temperature fall.

16. A low-pressure system near Utica, New York, causes heavy precipitation. If this system followed the usual track, which city most likely had the same weather conditions a few hours earlier?
    (1) Plattsburgh   (2) Kingston   (3) Albany   (4) Ithaca

17. A weather station observes that dewpoint and air temperatures are getting further apart and that air pressure is rising. What type of weather is most likely arriving at this station?
    (1) a snowstorm   (2) a warm front   (3) cool, dry air   (4) maritime tropical air

18. Which type of air mass usually contains the most moisture?
    (1) mT   (2) mP   (3) cP   (4) cT

19. Which weather change is most likely indicated by rapidly falling air pressure?
    (1) Humidity is decreasing.
    (2) Temperature is decreasing.
    (3) Skies are clearing.
    (4) A storm is approaching.

20-23. The map below shows a low pressure systems and some atmospheric conditions at weather stations **A**, **B**, and **C**.

20. Which weather station probably has the most unstable weather conditions?
    (1) A (2) B (3) C

21. The current weather at **B** can be described as
    (1) cool with low humidity and high air pressure
    (2) cool with high humidity and low air pressure
    (3) warm with high humidity and low air pressure
    (4) warm with low humidity and high air pressure

22. The atmospheric pressure at the center of the low would most likely be
    (1) 988 mb   (3) 994 mb
    (2) 990 mb   (4) 997 mb

23. Which station has been influenced most recently by the passage of a warm front?
    (1) A   (2) B   (3) C   (4) none of these

24-25. The diagram below shows the satellite image of a Northern Hemisphere hurricane.

24. What is the usual surface wind pattern around the eye of Northern Hemisphere hurricanes?
    (1) clockwise and outward
    (2) clockwise and inward
    (3) counterclockwise and outward
    (4) counterclockwise and inward

25. Which air mass is normally associated with the formation of hurricanes?
    (1) continental tropical
    (2) maritime tropical
    (3) continental polar
    (4) maritime polar

26-28. Use the weather station model below:

26. Interpret this weather station model. Be sure to include units.

present weather	
air temperature	
precipitation amount	
wind speed	

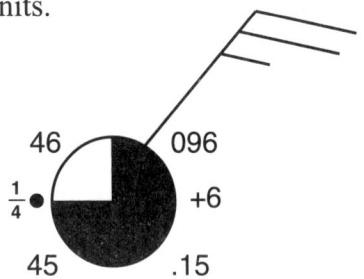

27. Determine the air pressure 3 hours ago at this weather station. _____
Show solution.

28. Why is it raining and not snowing? _____
_____

29-30. Four different weather station models are shown below.

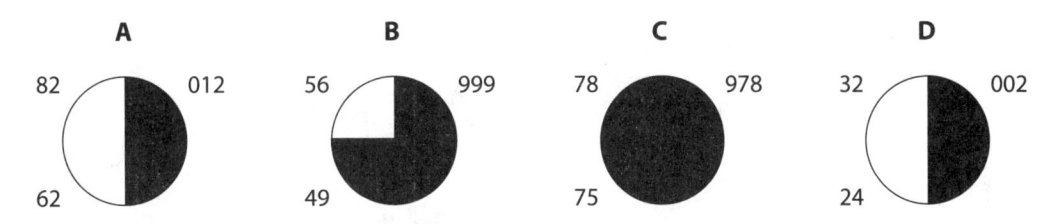

       A   **B**   **C**   **D**

82 ... 012   56 ... 999   78 ... 978   32 ... 002
62   49   75   24

29. List the letters of the four station models in order from highest to lowest air pressure.

_____, _____, _____, _____

30. What evidence indicates that station **C** has the highest relative humidity?
_____

31-35. The diagram shows two weather fronts moving across a portion of Earth's surface. Lines **X** and **Y** represent the frontal boundaries. The large arrows show the general direction the air masses are moving. The smaller arrows show the general direction warm, moist air is moving over the frontal boundaries.

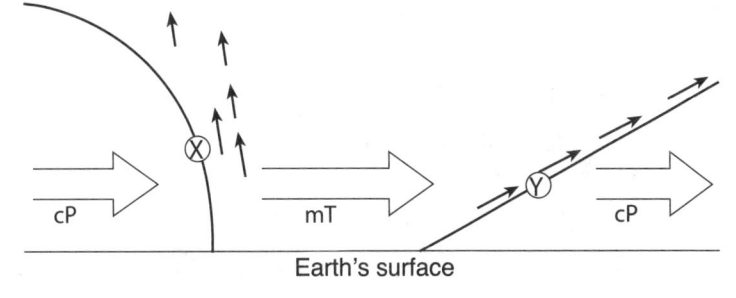

31. Name the type of front represented by letter **Y**. _____

32. Explain why warm, moist air rises. _____

33. What change occurs to the warm, moist air as it rises? _____

34. Draw clouds where they are most likely occurring.

35. What type of front forms when front **X** overtakes front **Y**? _____

36-39. The map below shows three frontal boundaries **A-B**, **B-C**, and **B-D**.

36. Front **A-B** is an occluded front. Draw this symbol on the map.

37. Draw the frontal symbol for **B-C**.

38. Draw the frontal symbol for **B-D**.

39. Write a weather forecast for New York State.

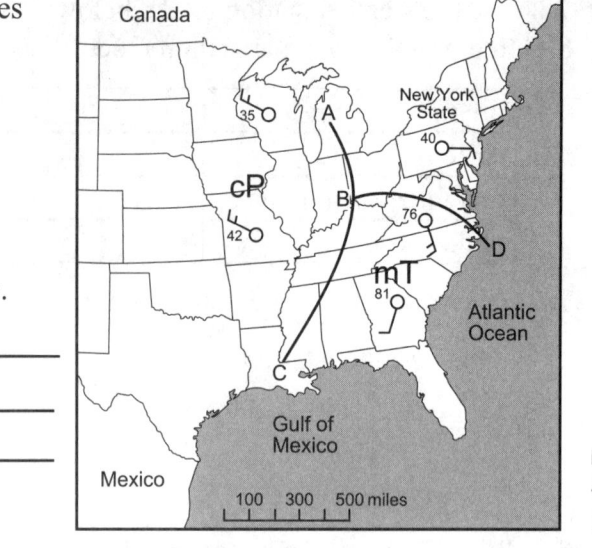

_____

_____

_____

40-45. The weather map below shows a low-pressure system over central United States. Isobars are labeled in millibars.

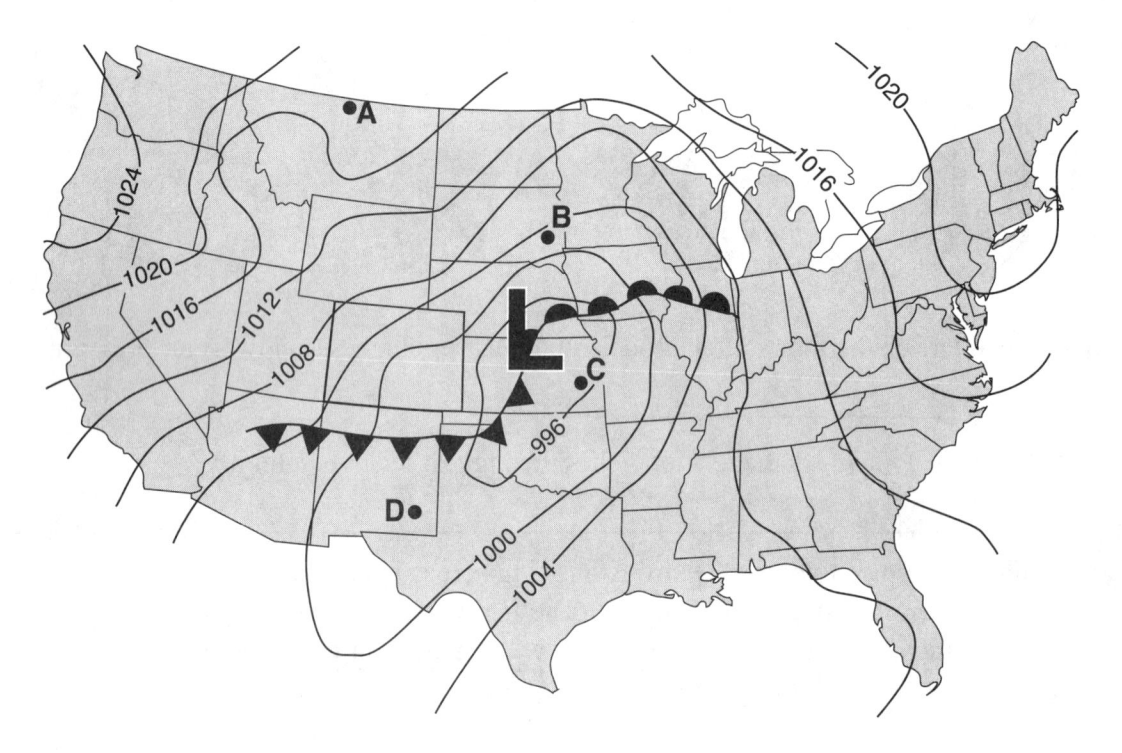

40. Draw four large arrows to show circulation of surface winds associated with the **Low**.

41. Describe the weather associated with the **Low** _____

_____

42. Which location is probably experiencing the strongest winds? _____

How do you know? _____

43. Approximate air pressure for **D** is _____ mb.

44. In the next few hours, **D** will experience (*decreasing*) (*increasing*) temperatures.

45. The center of the **Low** will most likely move toward what direction? _____

124

CHAPTER 8

WATER CYCLE AND CLIMATE
Water Cycle, Soil Water Movement, Factors that Control Climate, Climate Change

THE WATER (HYDROLOGIC) CYCLE

Water on Earth exists in the atmosphere as invisible water vapor and clouds of liquid droplets and ice crystals. Water is found on Earth's surface in oceans, streams, lakes, snowfields, and icecaps. Below Earth's surface, water is stored in the empty spaces between soil particles as subsurface water (groundwater).

Earth has been recycling its water supply for almost four billion years ever since outgassing from volcanoes put water vapor into the atmosphere. **The water cycle** illustrates how water continually moves between the hydrosphere, lithosphere, and the atmosphere. **Evaporation** is the process that causes water from Earth's surface to return to the atmosphere as vapor. Evaporation of surface waters occurs due to heat from insolation and the energy of the wind. **Transpiration** is the process by which green plants release water vapor to the air.

Water vapor in the atmosphere will form clouds when air temperature cools to dewpoint. Precipitation returns water to Earth's surface. Precipitation can **infiltrate** or seep into the ground and become subsurface groundwater. Precipitation can be stored on the surface as ice, snow, or liquid water in puddles and lakes. Water from precipitation can flow on the surface as **runoff**.

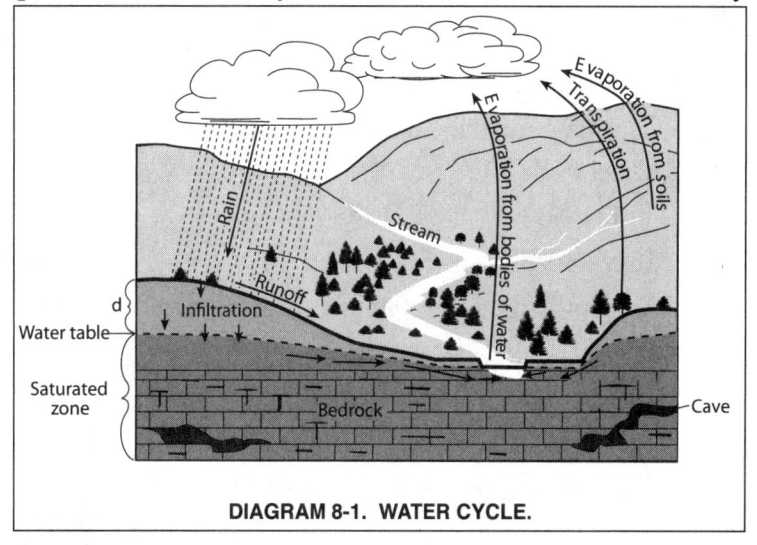

DIAGRAM 8-1. WATER CYCLE.

Runoff occurs when the rate of precipitation is greater than the rate it can be absorbed into the soil. Runoff also occurs if the surface is impermeable, frozen, spaces between soil particles are already saturated (filled) with water, or the slope of the surface is steep and lacking vegetation. Runoff can cause flash floods. Most runoff eventually flows into streams and the ocean. As runoff increases, **stream discharge** (volume of water in the stream) will increase.

Infiltration is the opposite of runoff. Infiltration of precipitation will occur if the soil is porous and permeable, not frozen, unsaturated, covered with vegetation, and the land is flat or gently sloped.

Grasses, trees, and other types of vegetation trap precipitation and slow runoff. Vegetation helps the soil absorb the water. Ground with no vegetation has more runoff than infiltration. **Deforestation** removes vegetation and increases runoff. Roads, parking lots, and buildings are impermeable, causing more runoff.

At Earth's surface, precipitation will....	...if...
...infiltrate the soil	– rate of precipitation is low, – ground is amost level, – soil is not saturated or frozen, – vegetation covers the ground, – soil is porous and permeable
...runoff the surface	– rate of precipation is high, – steep slopes, – soil is saturated or frozen, – vegetation has been removed, – surface is impermeable and/or not porous
...evaporate	– temperature increases, – increase in exposed surface area of water, – it is windy, – low air humidity
...transpire	– surface is vegetated by grasses, bushes, trees

Questions

1. The diagram to the right shows the water cycle.

    Letter **X** is the top of saturated soil (groundwater).

    Explain what is happening at each letter.

    A = _____

    B = _____

    C = _____

    D = _____

    E = _____

    F = _____

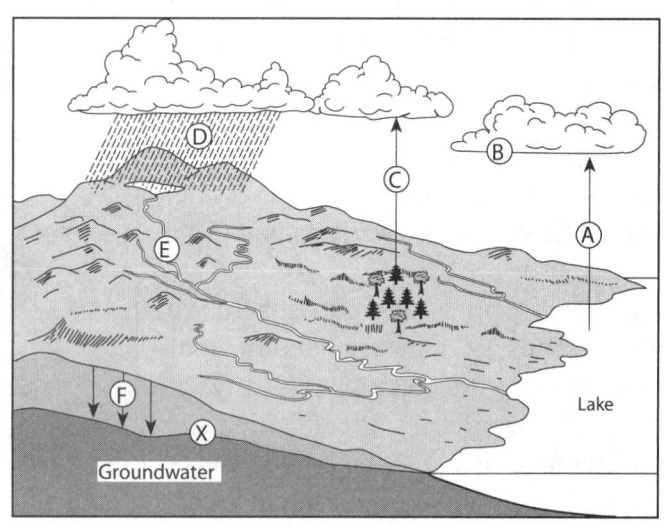

    (Not drawn to scale)

2. Complete each statement by using the term *(decrease)*, *(increase)*, or *(remain the same)*

    a. As insolation increases, evaporation will _____.

    b. As slope of land increases, infiltration will _____.

    c. As slope of land increases, runoff will _____.

    d. As infiltration increases, runoff will _____.

    e. As vegetation is removed from the land, runoff will _____.

    f. As soil becomes saturated, runoff will _____.

    g. When ground is frozen, infiltration will _____.

    h. When it becomes windy, evaporation will _____.

    i. As pavement and parking lots increase, infiltration will _____.

3. On each graph, draw a line to show the relationship between each of the variables.

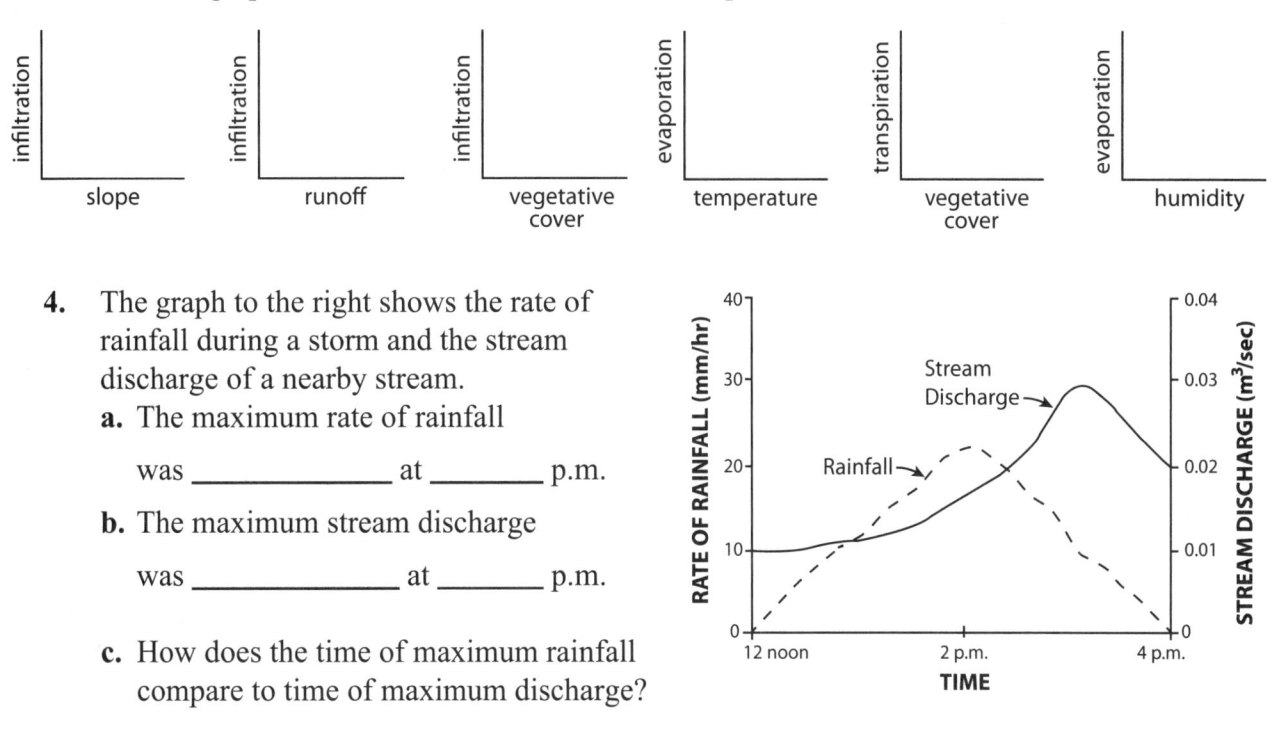

| infiltration vs slope | infiltration vs runoff | infiltration vs vegetative cover | evaporation vs temperature | transpiration vs vegetative cover | evaporation vs humidity |

4. The graph to the right shows the rate of rainfall during a storm and the stream discharge of a nearby stream.
   a. The maximum rate of rainfall was _____ at _____ p.m.
   b. The maximum stream discharge was _____ at _____ p.m.
   c. How does the time of maximum rainfall compare to time of maximum discharge?

   _____

   d. Why? _____

SOIL WATER MOVEMENT

The amount and rate at which water infiltrates and moves through the soil depends on porosity, permeability, and capillarity. **Porosity** is the percentage of open pore space between the soil particles. Subsurface groundwater is stored in these soil pores. As porosity increases, infiltration increases. Porous soil will have soil particles that are round, sorted, and loosely packed. Size of the soil particles does not affect porosity. Large particles have the same porosity as small particles of the same shape and sorting.

Permeability describes the ability of water to enter and flow through the soil. Permeability is greatest for soil which has particles that are large, sorted, round, and loosely packed. As permeability increases, infiltration increases.

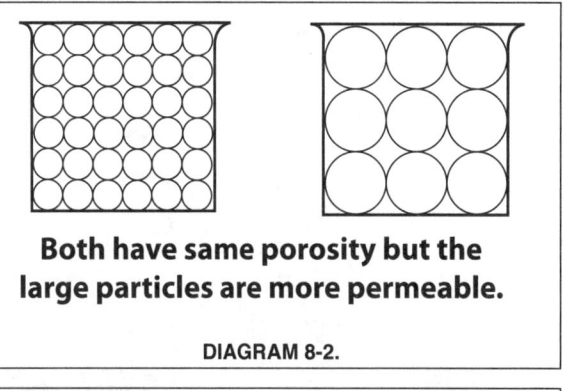

Both have same porosity but the large particles are more permeable.

DIAGRAM 8-2.

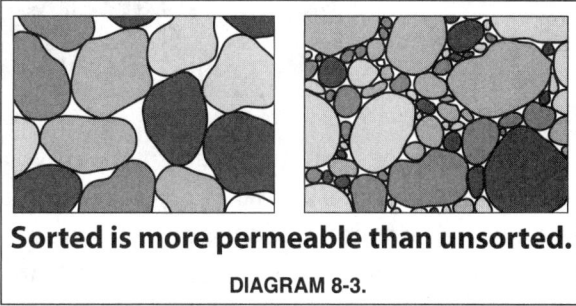

Sorted is more permeable than unsorted.

DIAGRAM 8-3.

127

Capillarity refers to water that is trapped in the upper soil known as the zone of aeration or root zone. Capillary water causes soil to be damp and provides plant roots with water. Capillarity is caused by the attractive force between water molecules and the surfaces of the soil particles. Capillary action causes water to move upwards against the pull of gravity due to this attractive force. **Capillary water** adheres (sticks) to the surface of the soil particles. Capillarity is greatest for small particles since they have a greater surface area. As capillarity increases, infiltration decreases. Small sized particles such as silt and clay have less permeability due to higher capillarity.

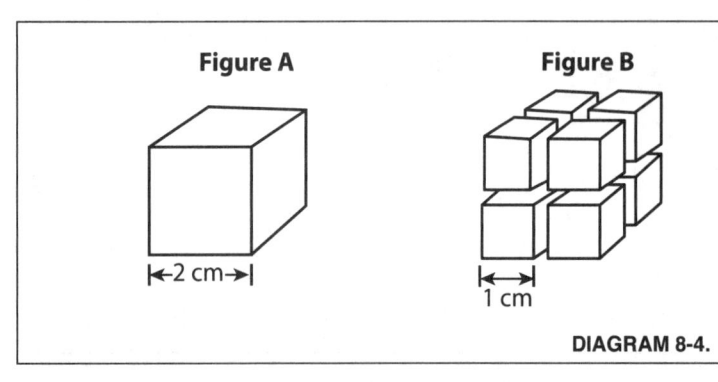

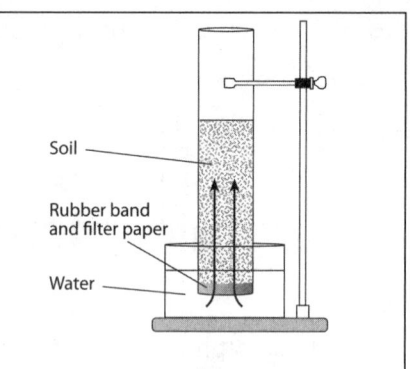

Data Table

Average Soil Particle Diameter (cm)	Height of Water in Column (cm)
0.006	30.0
0.2	8.0
1.0	0.5

Capillarity is the upward movement of water. It is greatest for smaller particles which have more exposed surface area.

Figure A — 2 cm

Figure B — 1 cm

DIAGRAM 8-4.

Infiltration occurs best for soil which is porous, permeable, and has low capillarity. As water infiltrates the soil it will move downward until it reaches impermeable bedrock or saturated soil. In the **zone of saturation** the pores or openings between soil particles are filled with water. Water held in the pores spaces of saturated soil is called **groundwater**. Above the zone of saturation is the **zone of aeration** where pore spaces are mostly empty. Any water held in the zone of aeration is **capillary water**.

The interface or boundary between the zones of aeration and saturation is the **water table**. The depth of the water table varies depending on the amounts of precipitation and infiltration. Water below the water table is **groundwater**.

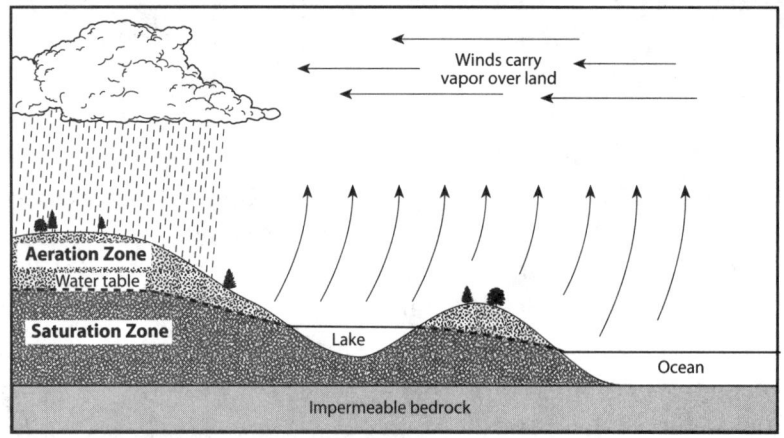

DIAGRAM 8-5.

Groundwater is an important source of water for communities, agriculture, and the surface water found in many rivers and lakes. Adequate groundwater supplies are threatened by droughts, overuse, and a decrease in infiltration as surfaces are made impermeable by buildings and paving.

Clean groundwater is threatened by pollution. These pollutants may come from cesspool waste, household and industrial chemicals, runoff from landfills, and the overuse of pesticides, fertilizers, and herbicides.

Questions

5. For each statement select the soil property described: *(porosity) (permeability) (capillarity)*
   a. Water moves upward within the aeration zone. _____
   b. The empty spaces between soil particles. _____
   c. Infiltrating water adheres to the surfaces of soil particles. _____
   d. Water does not flow through all Earth materials at the same rate. _____
   e. A cubic meter of sand and a cubic meter of pebbles can hold the same amount of water. _____
   f. This property is greatest for small particle sizes. _____
   g. This property is least for small particle sizes. _____

6. Select the graph that shows the relationship between the variables described.

   A   B   C   D

   a. Porosity vs. size of particles: _____
   b. Permeability vs. size of particles: _____
   c. Capillarity vs. size of particles: _____
   d. Infiltration rate vs. permeability of soil: _____
   e. Rate of infiltration vs. particle size: _____
   f. Infiltration vs. height of water table: _____
   g. Precipitation vs. height of water table: _____

7. The diagrams below show four different conditions of the soil below the surface.

   KEY
   ▨ Soil Particles
   ■ Water
   ☐ Pore Space (Air)

   Select the diagram which illustrates the following descriptions.
   a. Saturated soil. _____   d. Zone of saturation. _____
   b. Capillary water. _____   e. Groundwater table. _____
   c. Water adhesion. _____   f. Zone of aeration during drought. _____

8. The diagram shows three human activities (housing, gas station, and road work) taking place at a location.

   a. Label the groundwater table in the diagram.

   b. Which soil zone contains the groundwater? _____

   c. Which soil zone contains capillary water? _____

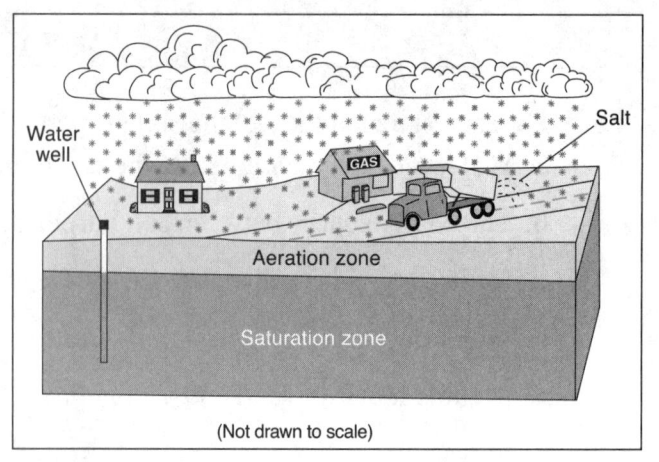

   d. If this area experiences a severe drought, how will the water well be affected?
   _____

   e. Explain *one* house activity that could have a negative impact on the groundwater.
   _____

   f. Explain *one* negative impact that the gas station could have on clean groundwater.
   _____

   g. Explain *one* negative impact that the highway truck could have on clean groundwater.
   _____

9. The diagrams below show four tubes each containing 500 milliliters of sediment.

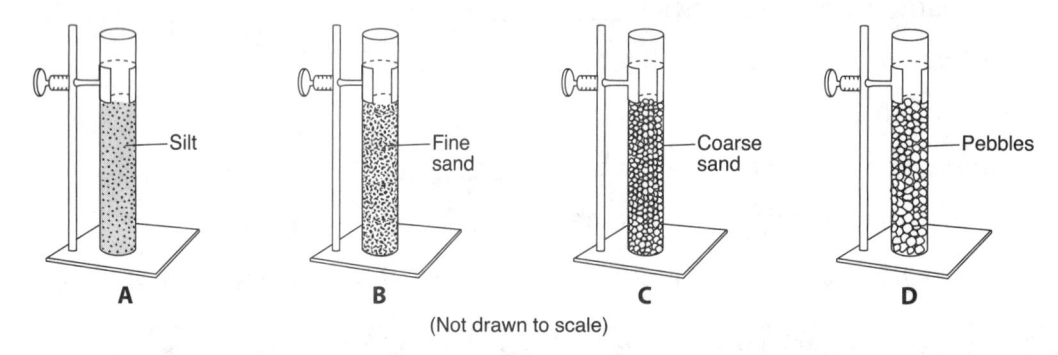

   A — Silt
   B — Fine sand
   C — Coarse sand
   D — Pebbles
   (Not drawn to scale)

   a. Which tube(s) has sediment whose size could be 0.002 cm? _____

   b. Which tube has the greatest permeability? _____
   What change could you make to this tube to make it less permeable? _____
   _____

   c. Which tube would show the greatest capillarity when placed in a pan of water? _____

   d. Which would have the most runoff during a heavy rainfall? _____

10. The diagram below shows some of the processes in the water cycle.

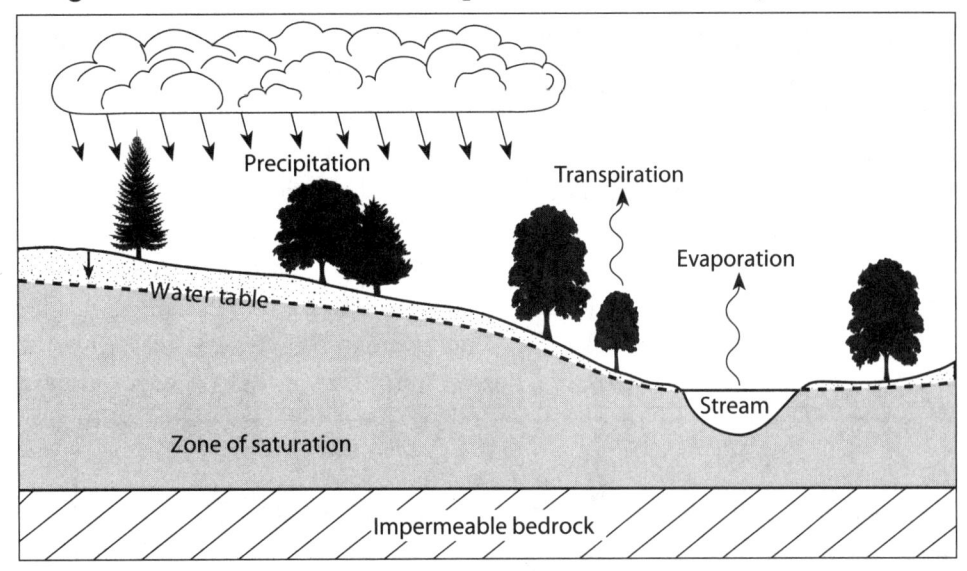

a. Label the zone of aeration.

b. Describe *one* change that would cause more water to evaporate from the stream.
_____

c. Explain the role of the plants in the water cycle. _____
_____

d. Describe *one* surface condition that would cause more runoff to occur. _____
_____

e. Where does condensation occur in this diagram? _____

f. After several days of heavy rainfall, what will happen to the water table?
_____

g. If the water table starts to go down, the level of the stream will _____.

FACTORS THAT CONTROL CLIMATE

Climate is the expected precipitation and temperature for a region over a long period of time. Precipitation and average temperature are the two variables that describe climate. The type of climate in a region depends on the difference between the amount of moisture available from precipitation and the potential need for water to evapotranspire which is controlled by insolation and vegetation. In a **humid climate** the average precipitation is greater than the average potential evaporation. Semi-humid to semi-arid climates have almost equal amounts of precipitation and need for water. In an **arid** (dry) **climate** there is less precipitation than needed.

Many factors affect the climate of a location. These include latitude, planetary winds, location on the continent, ocean currents, elevation, and mountains.

131

LATITUDE AND CLIMATE

Latitude is the most important factor affecting climate. **Tropical climates** are located between the Equator and 30° North and South. Tropical climates have a constantly high temperature because the angle of insolation is always high and duration of insolation has little variation. Tropical climates have no winter and are generally humid.

Temperate mid-latitude climates lie between 30° and 60° North and South. In this climate zone there are seasonal changes in temperature and precipitation. Temperate climates have hot summers and cold winters because the angle and duration of insolation vary during the year.

Polar climates are beyond 60° North and South. Here there are large changes in the duration of insolation and the angle of insolation is always low. Polar climates have very low temperatures. Polar climates have lower precipitation because there is less evaporation at these colder temperatures.

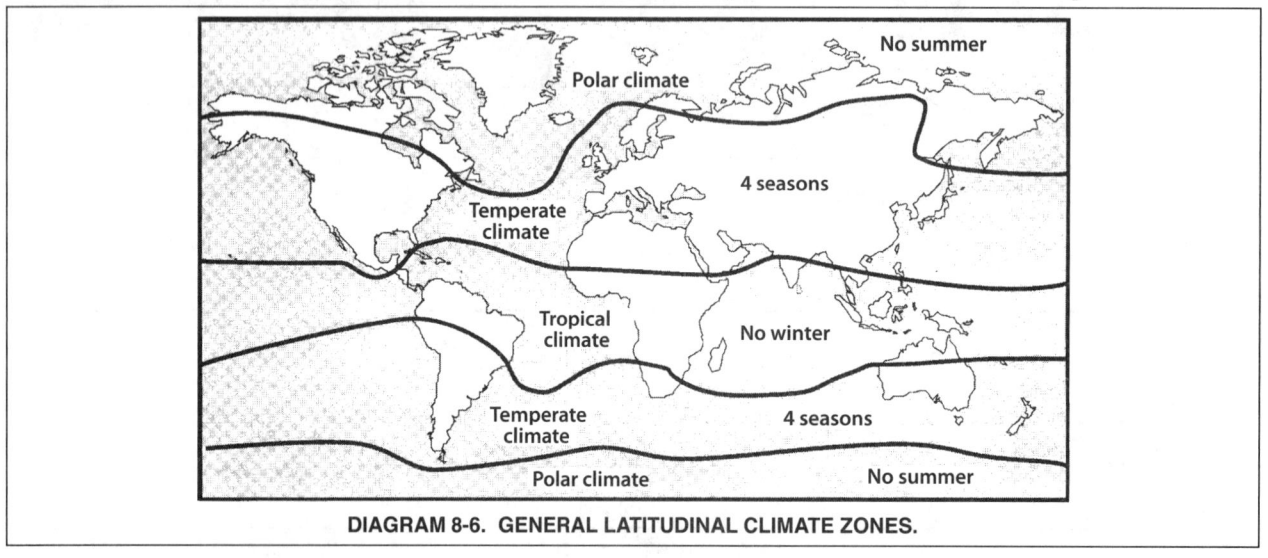

DIAGRAM 8-6. GENERAL LATITUDINAL CLIMATE ZONES.

PLANETARY WIND AND MOISTURE BELTS AND CLIMATE

Planetary winds are another factor that controls climate. If the prevailing wind is from the ocean, it will be humid. If the prevailing wind is from the land, it will be dry.

There is a low pressure belt at the Equator where the air converges. At the Equator the warm, humid air rises, expands, and cools to dewpoint. Cloud formation results in precipitation.

A high pressure belt exists at 30° N and 30° S where air diverges. The air sinks and warms which results in a dry climate. Many of the world's deserts are found at these latitudes. Sinking air is also found at the poles where a cold, dry polar climate occurs.

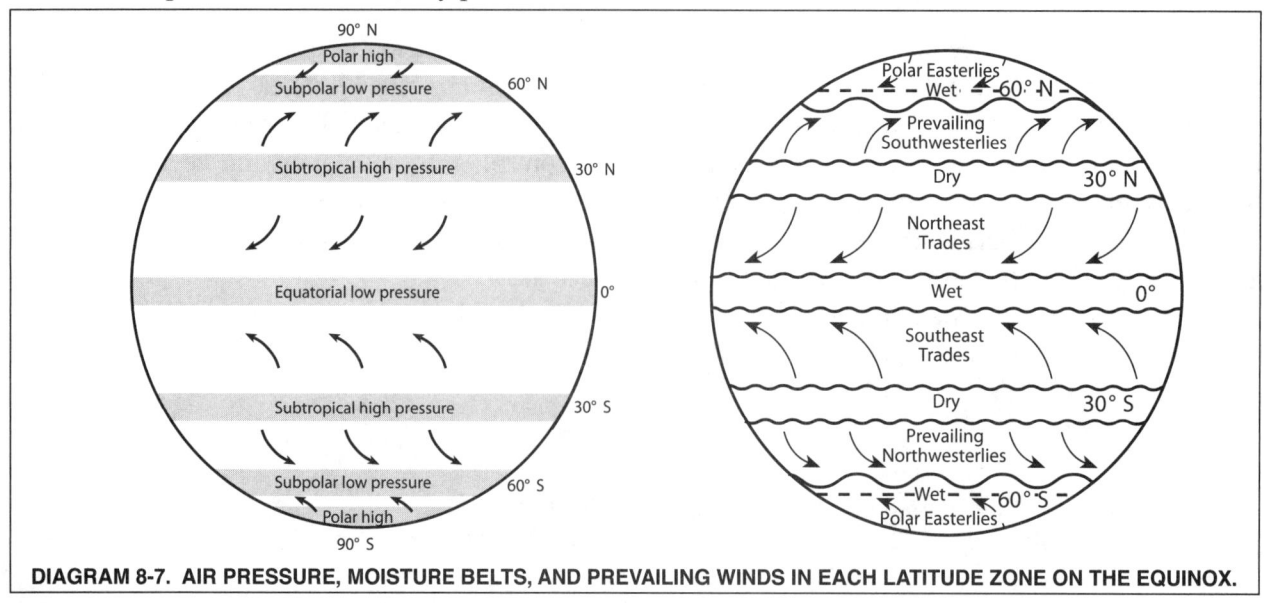

DIAGRAM 8-7. AIR PRESSURE, MOISTURE BELTS, AND PREVAILING WINDS IN EACH LATITUDE ZONE ON THE EQUINOX.

LOCATION ON CONTINENT AND CLIMATE

Another factor influencing climate is location on a continent. Inland areas have a larger variation in temperatures than coastal locations. This is because soil has a lower **specific heat** than water. Soil will heat and cool faster. Coastal areas near the water will heat and cool slowly since water has a high specific heat.

Marine (coastal) climates will have mild seasons, winters are warmer and summers are cooler. Coastal areas are usually more humid than inland areas. Inland locations heat and cool faster. **Continental (inland) climates** have hotter summers and colder winters (severe seasons) compared to coastal locations.

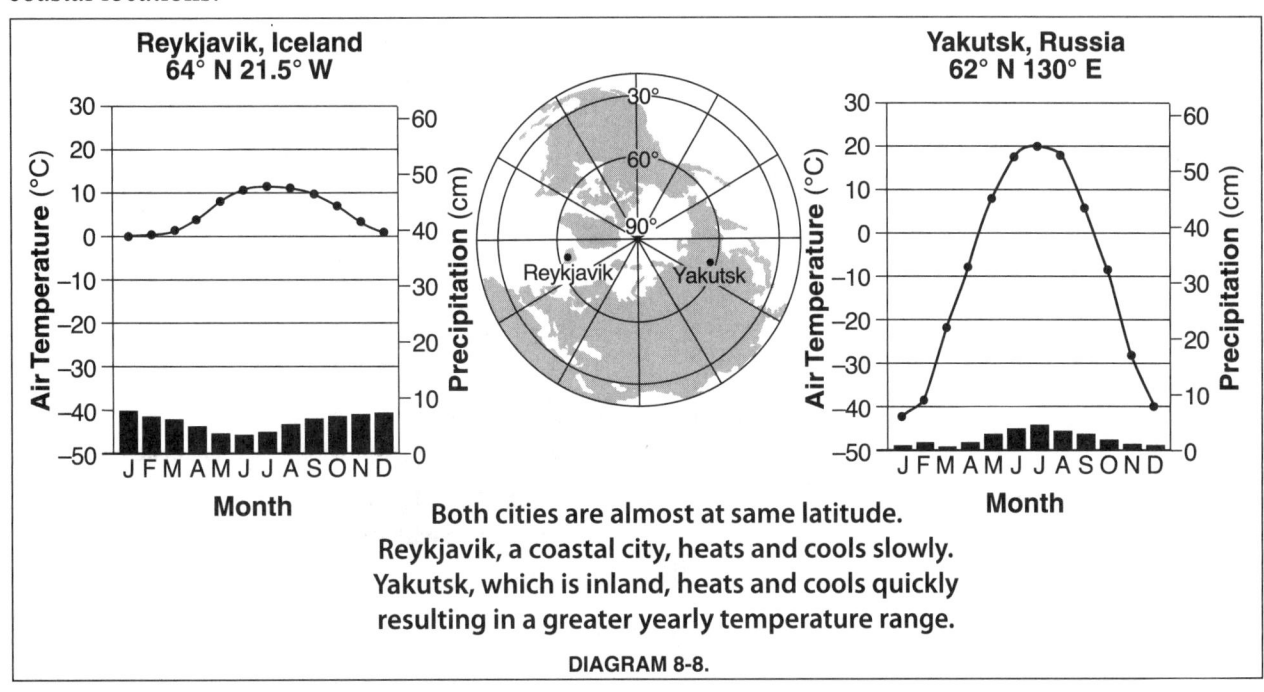

Both cities are almost at same latitude. Reykjavik, a coastal city, heats and cools slowly. Yakutsk, which is inland, heats and cools quickly resulting in a greater yearly temperature range.

DIAGRAM 8-8.

The difference in the heating and cooling of land and water causes monsoons, a seasonal change in local surface winds. For example, in winter India which is a large land area will be colder than the surrounding water causing high pressure to form over land. This results in movement of air from land to water. A dry climate during the winter occurs. In the summer the land heats quickly, resulting in lower pressure over the land causing air to move from the water to the land. When this occurs in the summer, heavy precipitation (monsoons) occurs.

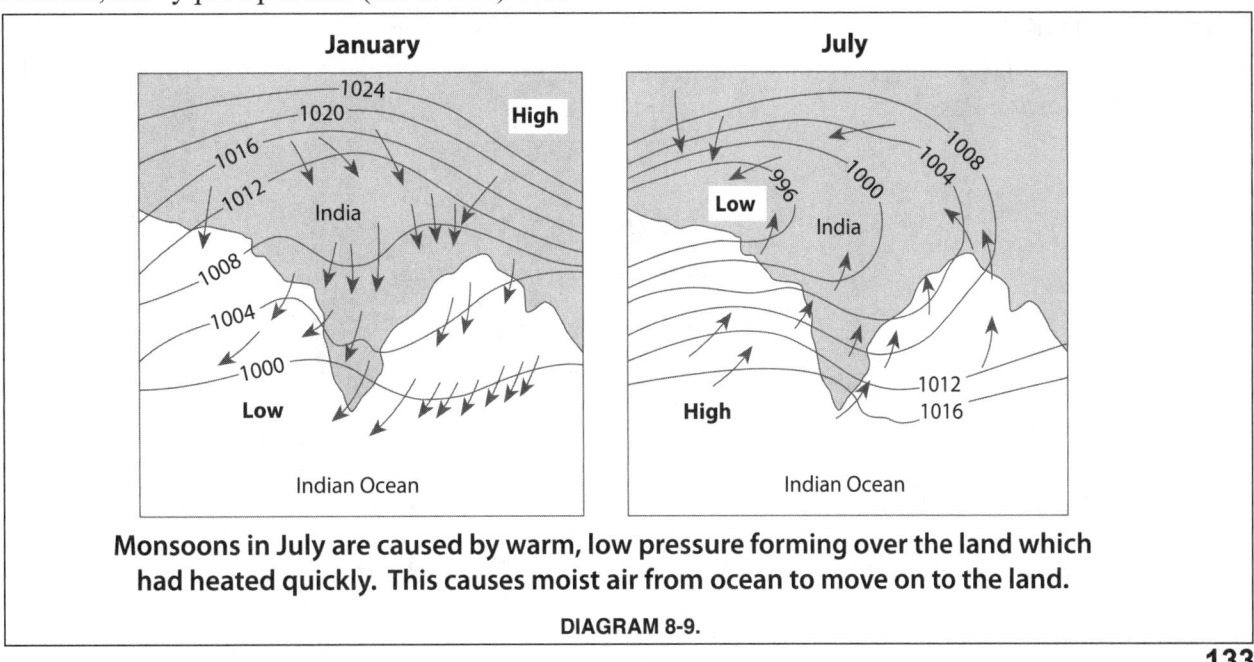

Monsoons in July are caused by warm, low pressure forming over the land which had heated quickly. This causes moist air from ocean to move on to the land.

DIAGRAM 8-9.

133

Ocean Currents and Climate

Surface ocean currents affect climate. Cool ocean currents cause coastal areas to have cooler temperatures. Warm ocean currents cause coastal areas to be warmer.

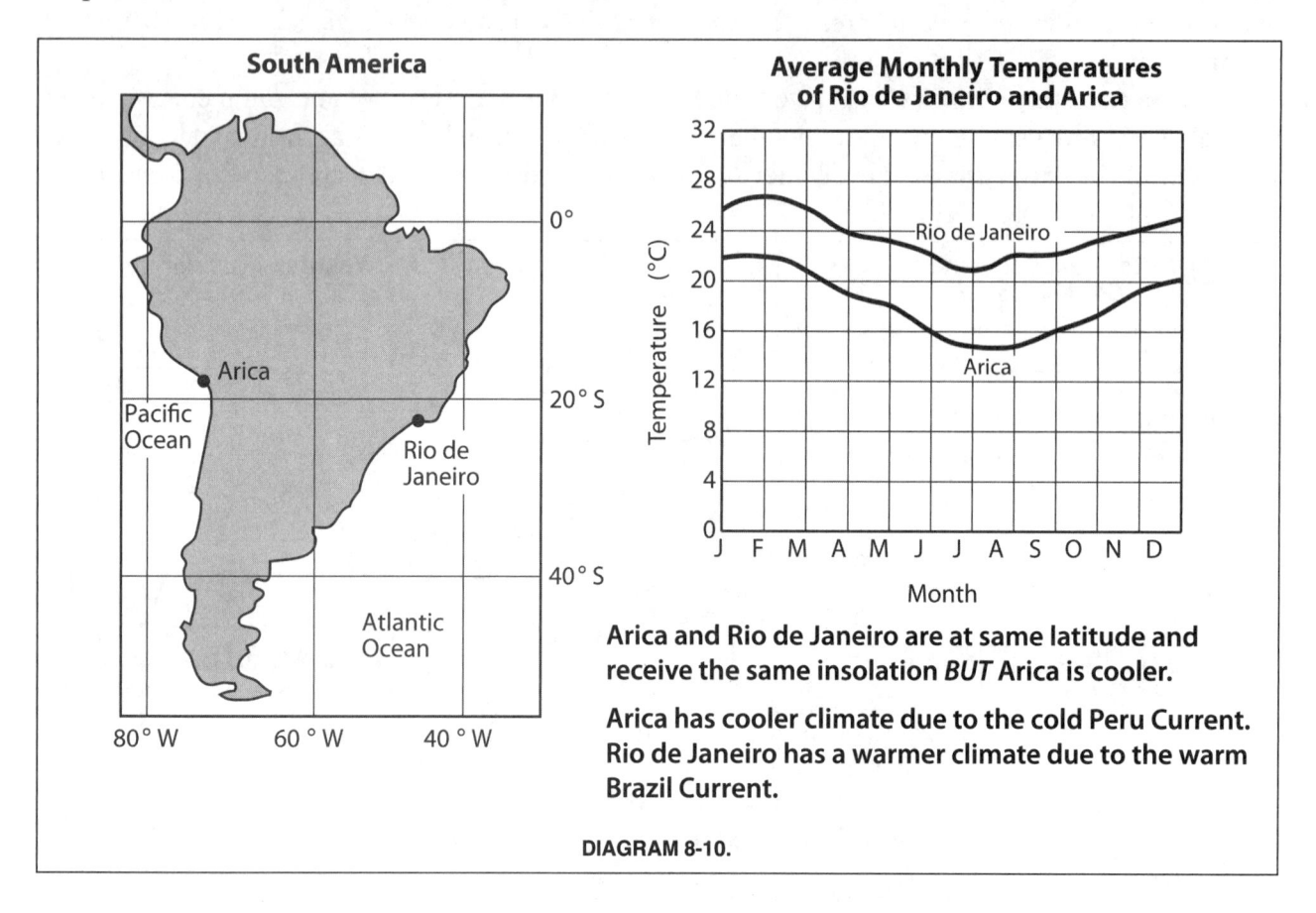

Arica and Rio de Janeiro are at same latitude and receive the same insolation BUT Arica is cooler.

Arica has cooler climate due to the cold Peru Current. Rio de Janeiro has a warmer climate due to the warm Brazil Current.

DIAGRAM 8-10.

Elevation, Mountains and Climate

Elevation above sea level impacts climate. As elevation increases, air expands and cools. This leads to cloud formation and precipitation. Mountain climates are cooler and more humid.

Mountains are barriers to air masses. Mountains such as the Alps and Himalayans block cold northern air from moving southward. Each side of a mountain will have a different climate. The **windward** side of the mountain where air rises will be cooler and more humid than the **leeward** side of the mountain where sinking, dry air warms due to compression. Climates on the leeward side of the mountain are warmer and drier.

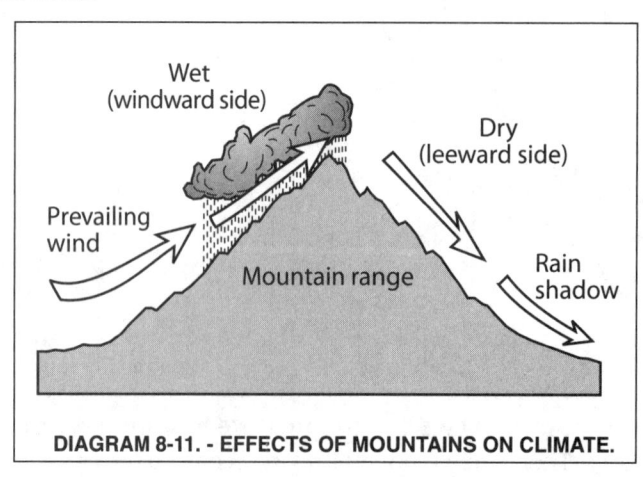

DIAGRAM 8-11. - EFFECTS OF MOUNTAINS ON CLIMATE.

134

Questions

11. On the grids below sketch the graph relationship between the variables.

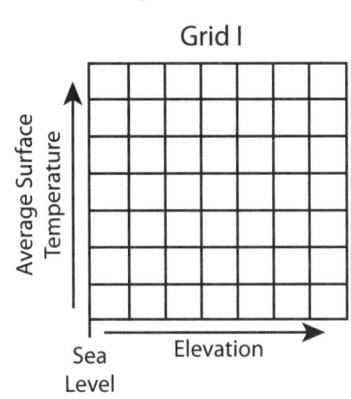

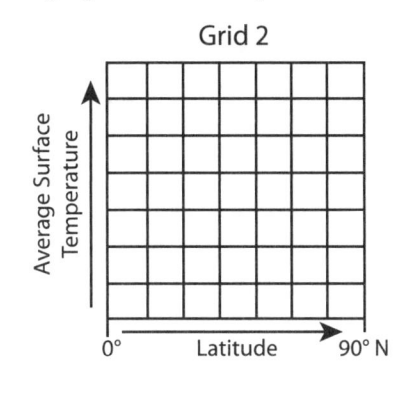

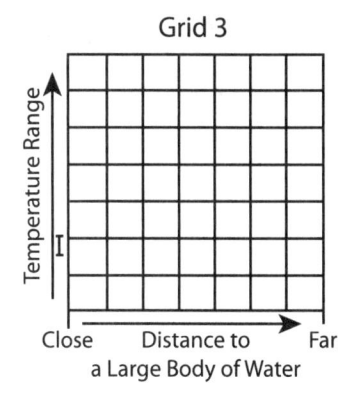

Grid 1: Average Surface Temperature (y) vs. Elevation from Sea Level (x)

Grid 2: Average Surface Temperature (y) vs. Latitude from 0° to 90° N (x)

Grid 3: Temperature Range (y) vs. Distance to a Large Body of Water, Close to Far (x)

12. Below are four climate graphs for four cities, **A**, **B**, **C**, and **D**. The monthly precipitation is shown as a bar graph. The yearly temperature pattern is shown as a line graph.

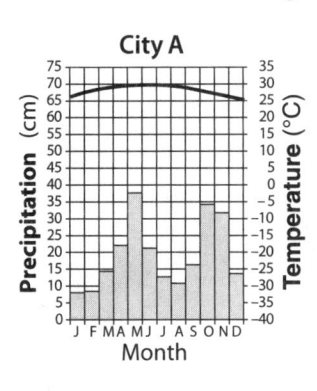

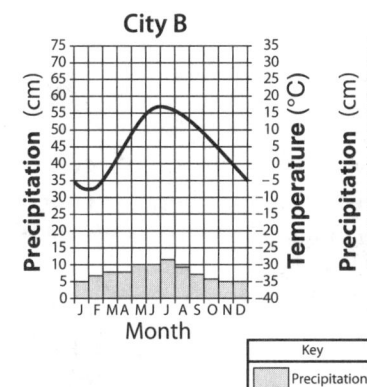

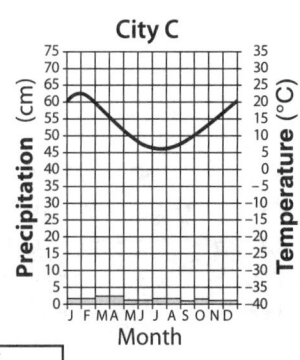

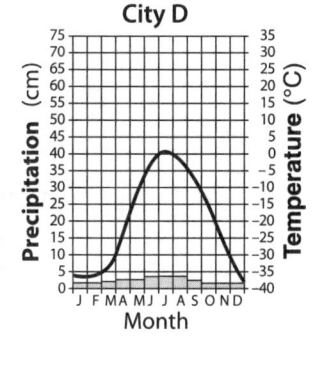

a. Which city is in the Southern Hemisphere? _____ How do you know? _____
_____

b. Which city has the lowest latitude? _____ How do you know? _____
_____

c. Which city has the highest latitude? _____ How do you know? _____
_____

d. City **C** is on the leeward side of a mountain. How do you know this is correct? _____
_____

e. Maximum temperature for city **A** is _____; for city **D** is _____

f. What is the yearly temperature range for city **B**? _____

g. List the cities in order of decreasing precipitation:

_____, _____, _____, _____

135

13. The graph to the right shows the average monthly temperatures for two cities, **A** and **B**, which are both located at 40°N latitude.

   a. How are the temperature ranges for both cities different? _____

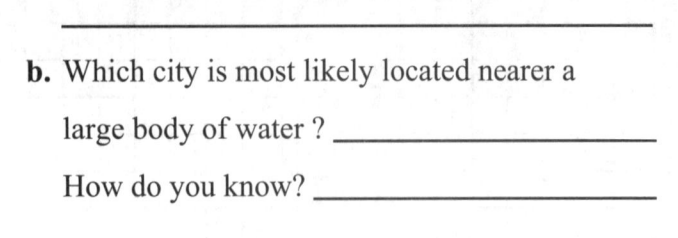

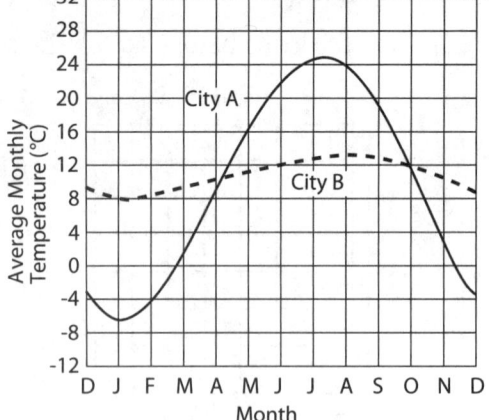

   _____

   b. Which city is most likely located nearer a large body of water? _____

   How do you know? _____

   _____

   _____

   c. Calculate the rate of temperature change for city **A** from February to May in degrees per month. _____ *(Show solution)*

14. Ocean currents have an important impact on the climate of coastal locations. The map below shows the ocean currents in the northern Atlantic Ocean.

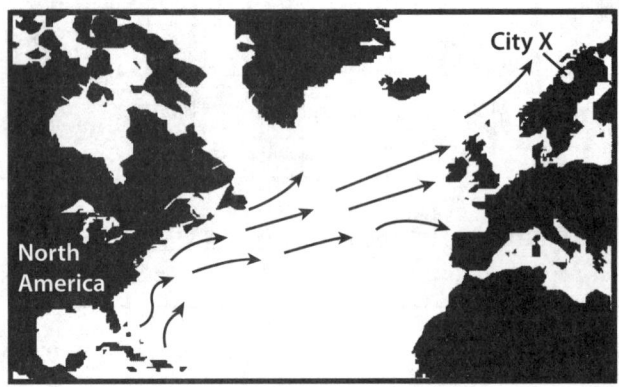

   a. Name the ocean current that directly affects City **X**. _____

   b. How does this ocean current affect the climate of City **X**? _____

   c. On the map label the names of two other ocean current indicated by the arrows.

   d. The ocean currents shown on this map are all *(cold)* *(warm)* currents.

   e. On the map use an arrow to indicate the position and direction of the West Greenland Current. Label the arrow as *(cold)* or *(warm)*.

15-16. This map shows an imaginary continent on Earth. The prevailing winds are shown by the arrows. Locations **A** through **H** are on the continent. All locations are at sea level except location **G** which is located high in the mountain.

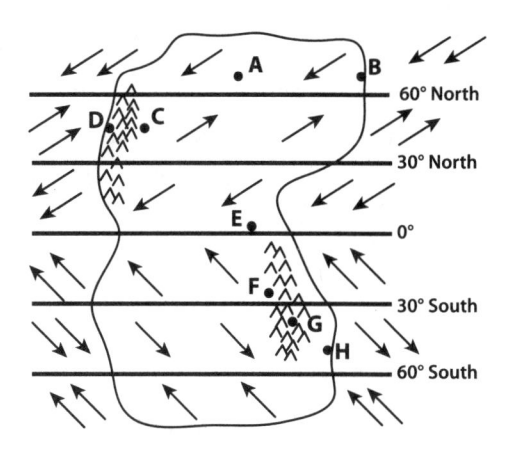

15. Select the location(s) for each description.

a. Highest annual temperature. _____

b. Highest evaporation rate in January. _____

c. Humid climate. _____

d. Arid climate. _____

e. Large seasonal temperature range. _____

f. Constant high temperatures. _____

g. Windward side of mountain. _____

h. Located in a high pressure zone. _____

i. Leeward side of mountain. _____

j. A marine climate. _____

k. A continental climate. _____

l. Affected by southwest winds. _____

m. Affected by northeast winds. _____

n. Atmospheric convergence zone. _____

16. a. State *two* locations which most likely have similar climates. _____ and _____

b. Describe the climate of these two locations. _____

c. What factor causes the similar climate? _____

CLIMATE CHANGE

During Earth's history climate has changed many times. Earth has undergone many periods of warmer and cooler temperatures. The diagram below shows the changes in Earth's temperature during the past 500 million years.

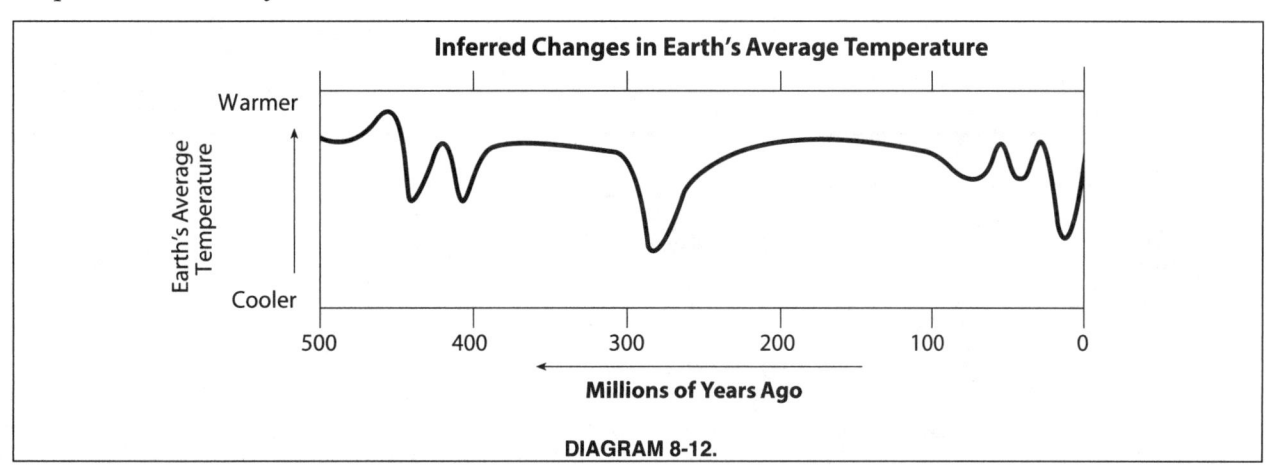

Inferred Changes in Earth's Average Temperature

DIAGRAM 8-12.

Climate change can be caused by changes in Earth's tilt and orbital path, variations in the Sun's output of energy, and dust in the atmosphere from impact events and volcanic eruptions.

Questions

17. If Earth was tilted more than 23.5° how would climate be affected? _____
_____

18. The burning of fossils fuels (coal, oil, natural gas) puts large amounts of the greenhouse gas, carbon dioxide, in the atmosphere. Most scientists believe that this is causing Earth's temperatures to _____ because _____
_____

19. Base your answers on the passage and map below of the volcanic island, Krakatau.

Krakatau

On August 27, 1883, one of the largest volcanic eruptions ever recorded in human history occurred. Krakatau, a volcanic island nearly 800 meters in height, located at 6° S, 105.5° E exploded. Two-thirds of the island was destroyed. Vocanic ash was blasted into the atmosphere to heights between 36 and 48 kilometer. The ash traveled on air currents around the world.

Volcanic Island of Krakatau

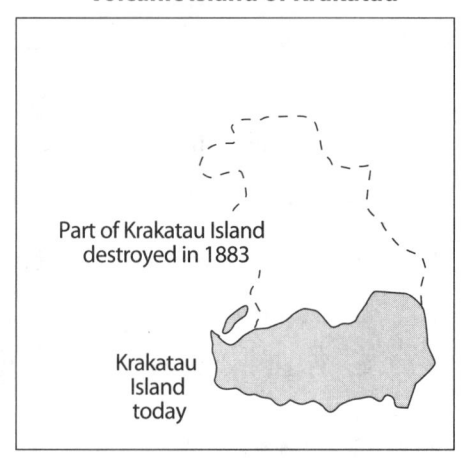

Part of Krakatau Island destroyed in 1883

Krakatau Island today

    a. Name the atmospheric layer that the ash reached. _____
    b. Describe the effect the volcanic ash would have had on worldwide climates.
_____

    c. Why did it have this effect? _____
_____

    d. Krakatau is located on the _____ Plate.

CHAPTER 8 REVIEW

1. The flowchart below shows part of the water cycle.

   Precipitation → Runoff → Ocean → ??? → Water vapor

   Which process should be put in place of the question marks to complete the flowchart?
   (1) condensation (2) deposition (3) evaporation (4) infiltration

2. As surface runoff in a region increases, stream discharge in that region will usually
   (1) decrease (2) increase (3) remain the same

3. Buildings and parking lots completely cover an area that was once an open, grassy field. What factor has most likely increased because of this new land use?
   (1) permeability (2) runoff (3) capillarity (4) local water table

4. Through which sediment does water infiltrate most slowly?
   (1) sand (2) silt (3) clay (4) pebbles

5. When rain falls on a surface, flooding would most likely occur if the
   (1) soil is permeable
   (2) soil is covered with vegetation
   (3) soil pores are filled to capacity
   (4) infiltration rate exceeds precipitation rate

6. Soil with the greatest porosity is composed of particles that are all
   (1) different sizes and shapes
   (2) different sizes and round
   (3) small and angular
   (4) large and round

7. Water moves upward through the soil because of
   (1) capillary action
   (2) permeability
   (3) porosity
   (4) runoff

8. Which property of well-sorted loose material will increase as the particle size decreases?
   (1) capillarity (2) permeability (3) porosity (4) infiltration

9. During a three-week period without rain in July, water continues to flow in a stream on Long Island. The water in the stream is most likely coming from
   (1) the roots of the trees along the stream
   (2) groundwater flowing into the stream
   (3) evaporation
   (4) condensation

10. The diagram represents two identical containers filled with uniform particles. Compared to the container with the larger particles, the container with the smaller particles has
    (1) less permeability and more porosity
    (2) more porosity and more capillarity
    (3) less permeability and more capillarity
    (4) more permeability and more porosity

11. Which graph best shows the relationship between particle size and porosity?

12. Which factor has the greatest effect on the climate of an area?
    (1) distance from Equator
    (2) vegetative cover
    (3) longitude
    (4) clouds

13. Which combination of climate factors generally results in the coldest temperatures?
    (1) low elevation and low latitude
    (2) low elevation and high latitude
    (3) high elevation and high latitude
    (4) high elevation and low latitude

14. What controls the movement of most surface ocean currents?
    (1) density differences at various ocean depths
    (2) prevailing winds
    (3) varying salt content in the ocean
    (4) seismic activity

15. The Gulf Stream and the North Atlantic Current modify the climate in northwestern Europe by making the climate
    (1) warmer and drier
    (2) warmer and more humid
    (3) cooler and more humid
    (4) cooler and drier

16. Which graph best illustrates the temperature changes on adjacent land and water surfaces as they are heated by the Sun from sunrise to noon on the same day?

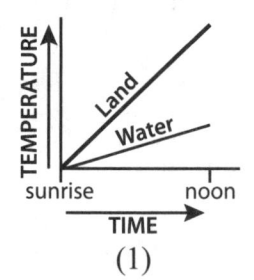

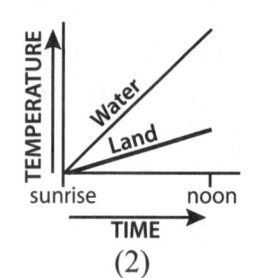

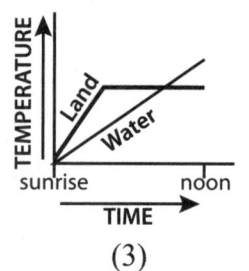

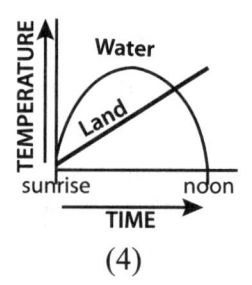

17. The California Current is
    (1) cool and flows north
    (2) cool and flows south
    (3) warm and flows north
    (4) warm and flows south

18. Which coastal location experiences a cooler summer due to ocean currents?
    (1) southeast coast of North America
    (2) southwest coast of South America
    (3) northwest coast of Europe
    (4) northeast coast of Australia

19. The map to the right shows two locations, **A** and **B**, that have the same latitude, elevations, and distance from ocean. Which statement explains why location **A** has a cooler climate than location **B**? 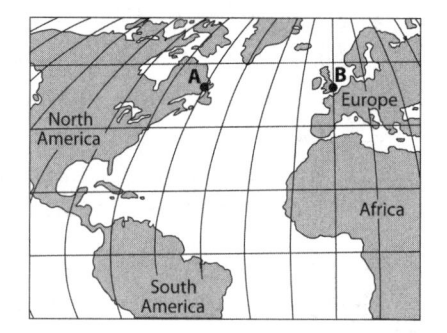
    (1) Location **A** has a longer duration of insolation.
    (2) Location **A** is affected by a cold ocean current.
    (3) Location **A** is further from the Equator.
    (4) Location **A** has less intense insolation each day.

20. The warmest climates on Earth are located near the Equator because the Equatorial regions
    (1) receive mostly high angle insolation
    (2) have the longest daylight hours
    (3) are surrounded by water
    (4) are closest to the Sun

21. What effect does a large body of water usually have on the climate of a nearby landmass?
    (1) The water causes cooler summers and colder winters.
    (2) The water causes cooler summers and warmer winters.
    (3) The water causes hotter summers and warmer winters.
    (4) The water causes hotter summers and colder winters.

22. The map shows two seasonal positions of the polar front jet stream over North America. Which statement best explains why the position of the polar front jet stream varies with the seasons?
(1) Rising air compresses and cools in winter.
(2) Water heats and cools more rapidly than land in the winter.
(3) Prevailing winds reverse direction in the summer.
(4) The vertical ray of the Sun shifts north of the Equator in summer.

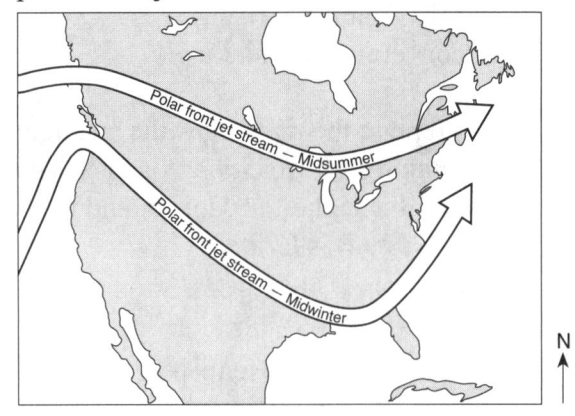

23. The cross section below shows two cities, **A** and **B**, at different elevations.

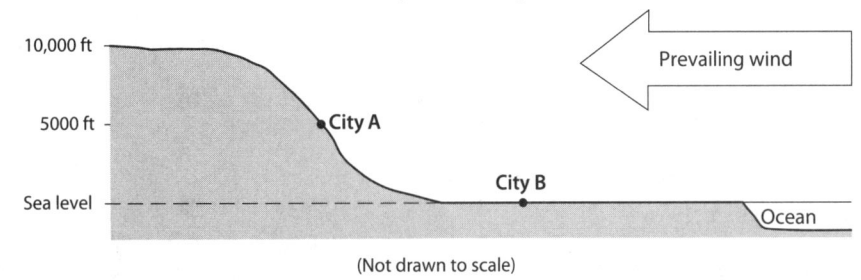

Compared to the yearly temperature and precipitation at city **B**, city **A** most likely has
(1) lower temperature and more precipitation
(2) lower temperature and less precipitation
(3) higher temperature and less precipitation
(4) higher temperature and more precipitation

24. The graph below shows the snow line elevation for different latitudes in the Northern Hemisphere.

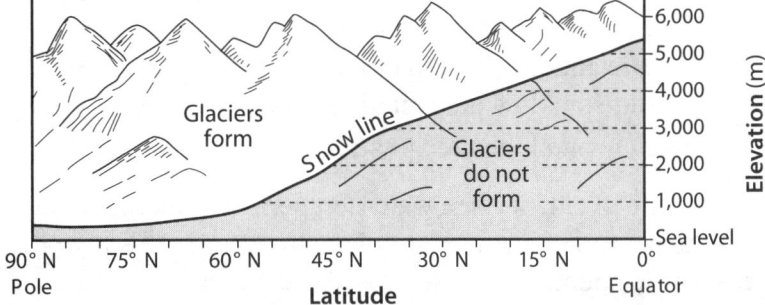

At which location would a glacier most likely form?
(1) 0° latitude at 4,000 m
(2) 15° N latitude at 5,000 m
(3) 30° N latitude at 3,000 m
(4) 45° N latitude at 1,000 m

25. Which graph best shows the average annual amounts of precipitation received at different latitudes on Earth?

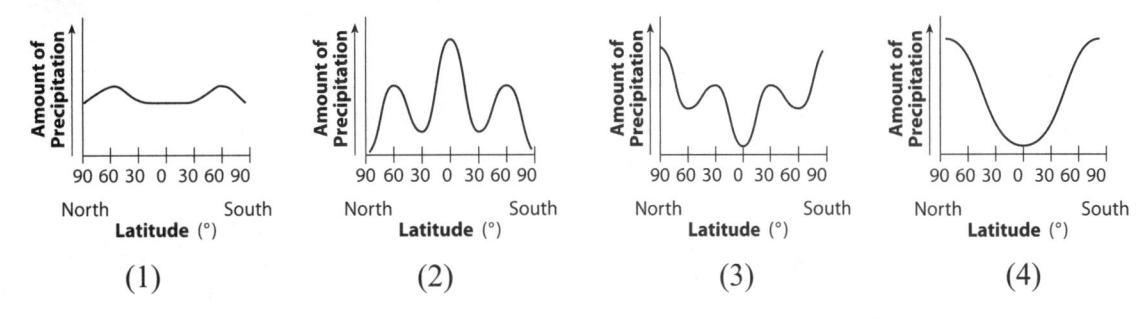

141

26. The planetary winds and moisture belts indicate that dry climates occur where air is
    (1) converging and rising
    (2) converging and sinking
    (3) diverging and rising
    (4) diverging and sinking

27. The graph to the right show the average monthly temperatures of two cities, **A** and **B**. The temperature for city **B** is highest in January and lowest in July because city **B** is located

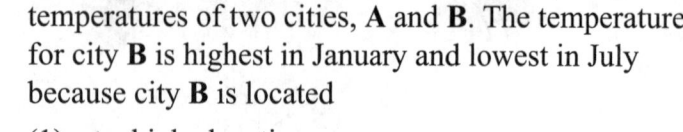

    (1) at a high elevation
    (2) near the ocean
    (3) in the Southern Hemisphere
    (4) near the Arctic Circle

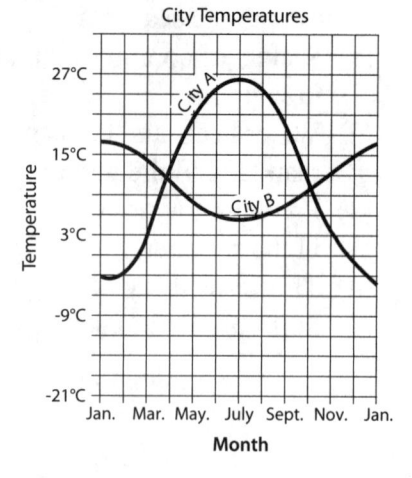

28. The diagram below shows air movement over a mountain. Points **A** and **B** are at the same elevation on opposite sides of the mountain. Compared to the climate at location **A**, the climate at location **B** is

    (1) drier and warmer
    (2) drier and cooler
    (3) more humid and cooler
    (4) more humid and warmer

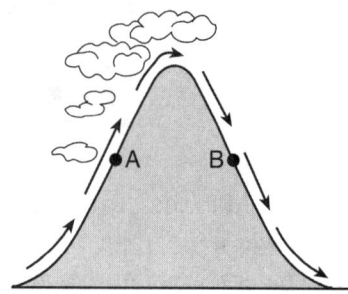

29. Which best describes the climate conditions near the North and South Poles?
    (1) low temperature and low precipitation
    (2) low temperature and high precipitation
    (3) high temperature and low precipitation
    (4) high temperature and high precipitation

30. During an El Nino event, surface water temperatures increase along the west coast of South America. Which weather changes are likely to occur in this region?
    (1) decrease air temperature and decreased precipitation
    (2) decrease air temperature and increased precipitation
    (3) increase air temperature and increased precipitation
    (4) increase air temperature and decreased precipitation

31-33. The diagram below shows Earth's water cycle and some water cycle processes. Letter **A** is on the Earth's surface.

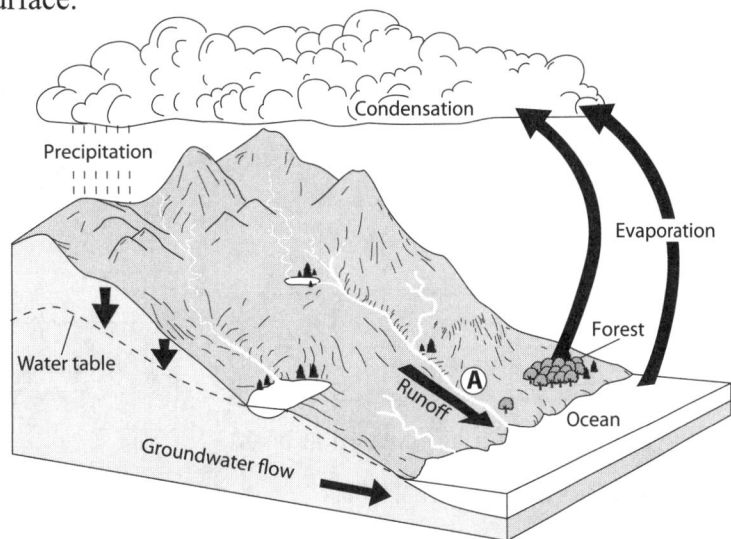

31. Other than evaporation, which other process transfers large amount of water vapor into the air?
    _____

32. Describe one surface condition change at location **A** that would decrease the rate of runoff.
    _____

33. How would the water table be affected if the area experienced a drought? _____
    _____

34-38. The diagrams below show four separate columns, **A** through **D**, filled to the same level with different sediments. The sediments in each tube are of uniform size and shape.

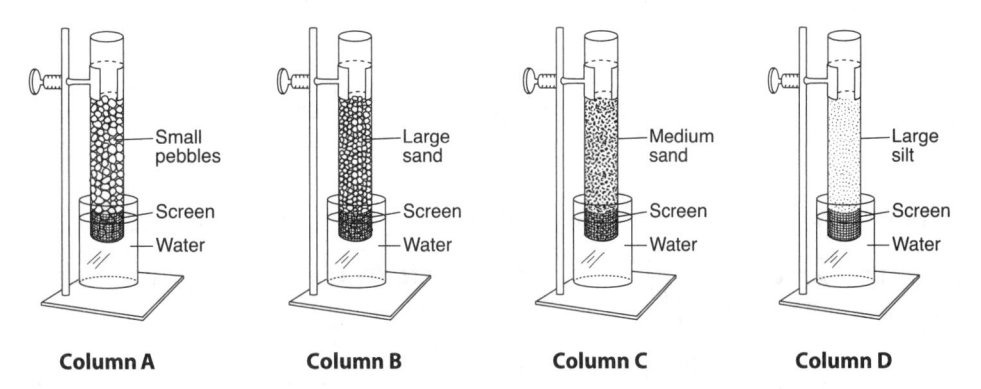

Column A — Small pebbles
Column B — Large sand
Column C — Medium sand
Column D — Large silt

34. Which column contains particles with a diameter of 0.4 cm? _____

35. Which column would allow water to flow through it at the slowest rate? _____

36. After water has flowed through each tube, which column's sediments would retain the most water? _____

37. Describe the relationship between sediment size and porosity that will be observed if water was poured into each column. _____

38. Each beaker is filled with water so the bottom of each tube is submerged in the water. Which tube after six hours will the water rise to the highest level? _____

39-41. The graph to the right shows the average monthly temperatures for a year for city **X** and city **Y**. Both cities are located at the same latitude.

39. What was the annual temperature range for

city **X**? _____ ; city **Y**? _____

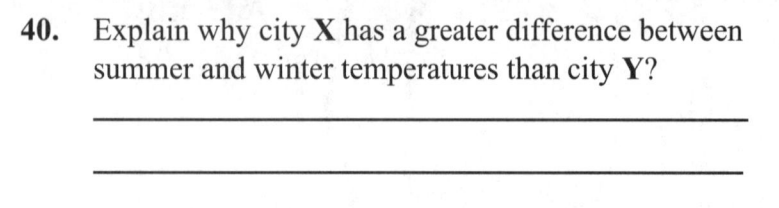

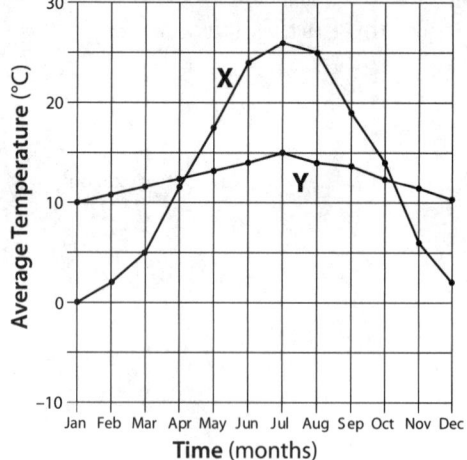

40. Explain why city **X** has a greater difference between summer and winter temperatures than city **Y**?

_____

_____

_____

41. What evidence shown on the graph indicates that both cities **X** and **Y** are in the Northern Hemisphere?

_____

_____

42-44. The map below shows an imaginary continent on Earth. The arrows show the prevailing winds. All locations **A** through **D** are at the same elevation.

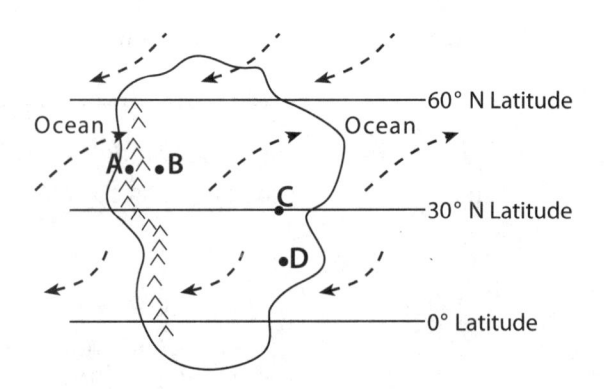

42. Which *two* locations receive the same annual amount of insolation? _____ and _____

43. Identify the *two* locations that have arid climates and explain the factor that causes the arid climate for each one.

   a. location _____ has an arid climate because _____

   b. location _____ has an arid climate because _____

44. Along the 60°N latitude line, the planetary winds are:

   (*diverging and rising*) (*diverging and sinking*) (*converging and sinking*) (*converging and rising*)

45-47. The table below shows the elevation and average annual precipitation at ten weather stations, **A** through **J**, located along a highway that passes over a mountain.

Data Table

Weather Station	Elevation (m)	Average Annual Precipitation (cm)
A	1,350	20
B	1,400	24
C	1,500	50
D	1,740	90
E	2,200	170
F	1,500	140
G	800	122
H	420	60
I	300	40
J	0	65

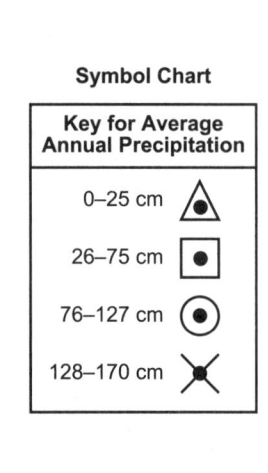

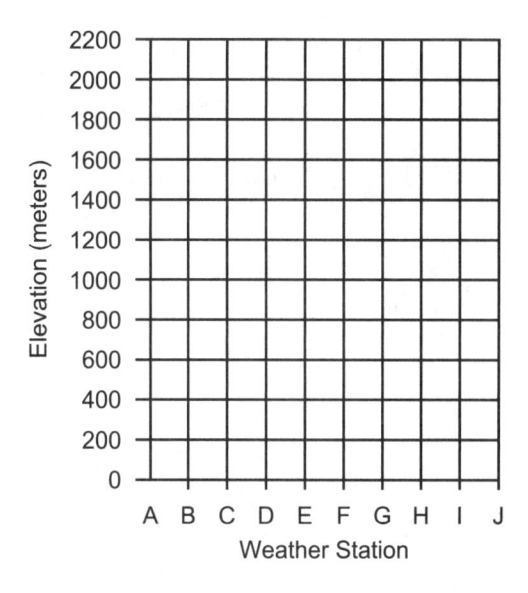

Symbol Chart

Key for Average Annual Precipitation

- 0–25 cm
- 26–75 cm
- 76–127 cm
- 128–170 cm

45. On the grid, plot the data by following these directions:

a. mark the grid with a point showing the elevation of each weather station;

b. surround the data point with the proper symbol from the chart to show the average annual precipitation for the weather station.

46. State the relationship between elevation of weather stations **A** through **E** and the annual average precipitation for these stations.

47. Although station **C** and **F** are at the same elevation, they have different amounts of annual precipitation. Explain why this may occur.

48-50. The map below shows a portion of New York State and Canada. The arrows represent the direction of wind blowing over Lake Ontario for several days one winter.

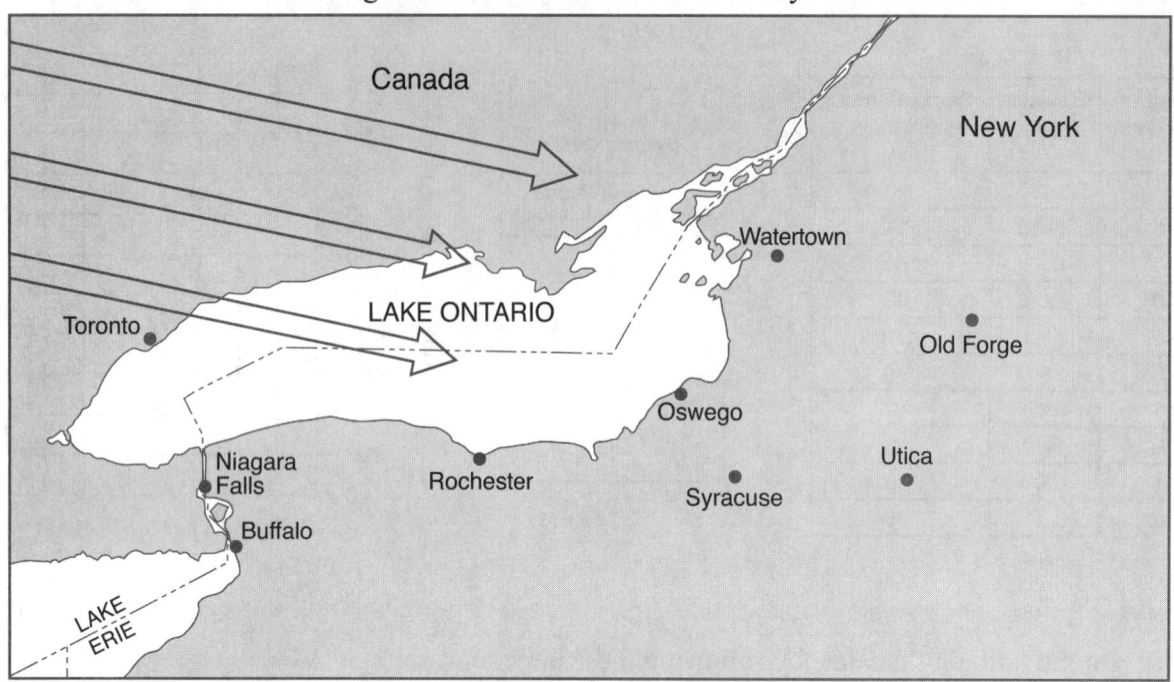

Lake-Effect Snow

During the cold months of the year, the words "lake effect" are very much a part of the weather report in New York State. Snow created by lake effect may represent more than half of the season's snowfall. In order for heavy lake-effect snow to develop, the temperature of the water at the surface of the lake must be higher than the temperature of the air flowing over the lake. The higher the water temperature and the lower the air temperature, the greater the potential for lake-effect snow.

This area of New York Sate that is likely to receive lake-effect snow is often called a "snow belt". It extends along the eastern and southeastern parts of Lake Ontario. Due to the high elevations of the Tug Hill Plateau and the Adirondack Mountains, orographic lifting of the moist air causes heavier snowfall.

48. Why does Oswego, New York usually get more snow than Toronto, Canada? _____

49. Compared to the average winter temperature in Watertown, New York, explain why the average air temperature in Old Forge, New York is colder. _____

50. Explain why the surface of Lake Erie freezes much later in winter than the surrounding land surface. _____

CHAPTER 9

SURFACE PROCESSES
Weathering, Erosion, Deposition

Many forces wear away Earth's surface and change its appearance. **Weathering** breaks up bedrock into smaller pieces. **Erosion** carries away the products of weathering. **Deposition** is the process by which these pieces are placed down. These three processes work together to change the landscape of Earth. Weathering and erosion wear away Earth's surface; deposition builds it up.

WEATHERING

Weathering occurs when rock is exposed to the atmosphere and the hydrosphere. Bedrock is broken down into smaller pieces called **sediment**. Weathering involves the physical and/or chemical break up of rock at or near the Earth's surface.

The products of weathering include sediments, colloids, and ions. Solid **sediments** includes sand, pebbles, cobbles, and boulders. **Colloids** are very small solids that can float in water such as clay and silt. Ions are minerals dissolved in water. These dissolved minerals may make the water salty. The solid particles formed by weathering are classified by their diameter as shown in diagram 9-1.

Physical weathering breaks apart surface bedrock into smaller sediments. **Frost action** (ice wedging) occurs when water gets into a crack in a rock. Water expands when it freezes. The pressure from the expanding ice causes the rock to crack more. This is why potholes form during the winter months. Frost action is common in cold, humid climates where there is the continual cycle of water freezing and melting.

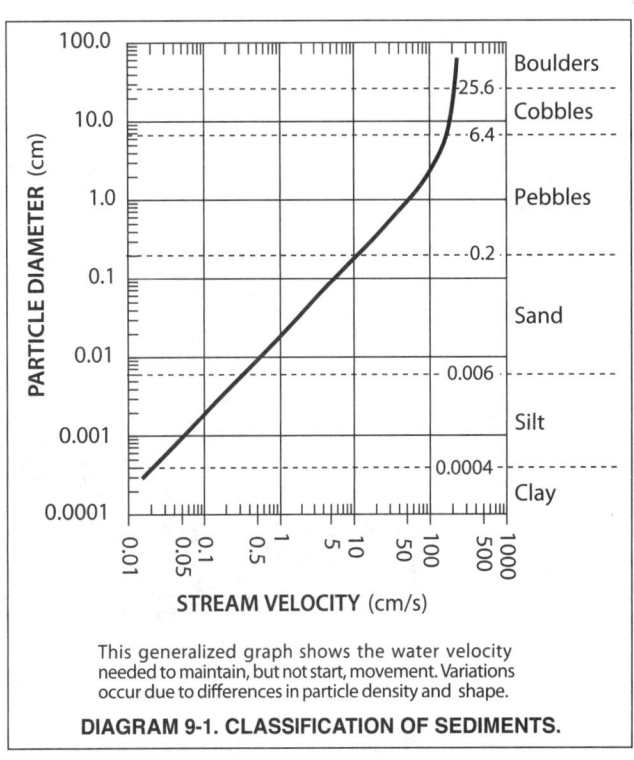

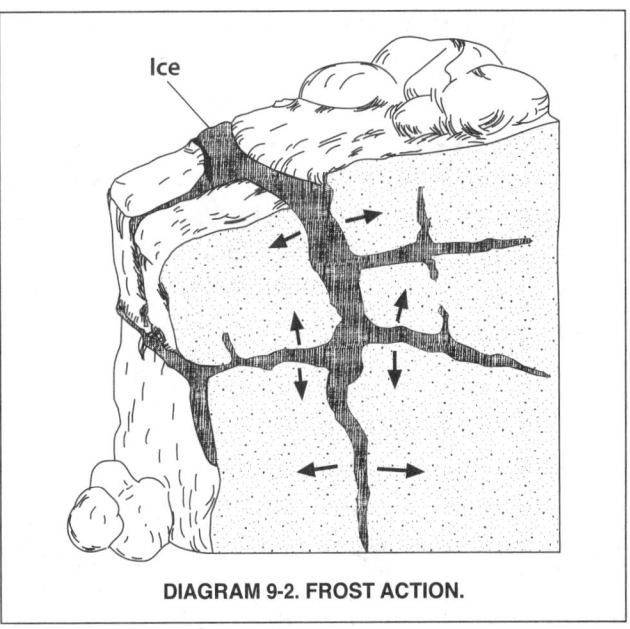

This generalized graph shows the water velocity needed to maintain, but not start, movement. Variations occur due to differences in particle density and shape.

DIAGRAM 9-1. CLASSIFICATION OF SEDIMENTS.

DIAGRAM 9-2. FROST ACTION.

147

Trees and plants cause weathering when their growing roots crack apart bedrock. You may have seen sidewalks that have buckled due to root action.

Another cause of physical weathering is temperature change. Heat during the day causes rocks to expand. When it cools at night, the rocks cool and contract. This constant expansion and contraction will eventually cause rock surfaces to peel and crack.

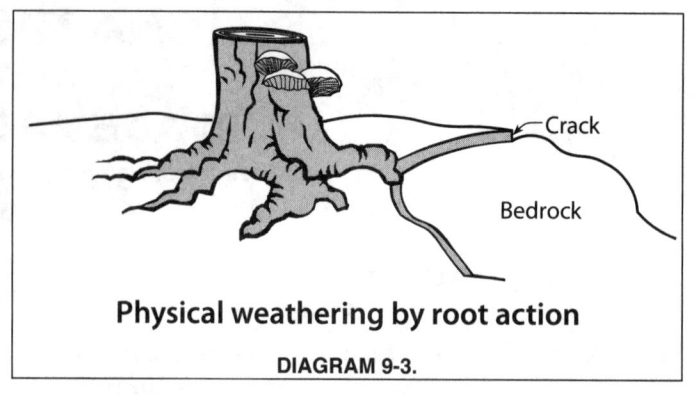

Physical weathering by root action
DIAGRAM 9-3.

Chemical weathering causes rock to chemically change and break apart. A chemically weathered rock may dissolve or change color due to a change in composition. Chemical weathering can be caused by a reaction with oxygen which is called oxidation. Rust and tarnish are examples of oxidation. Acidic water will dissolve limestone bedrock to form caves. Chemical weathering is most common in hot, humid climates.

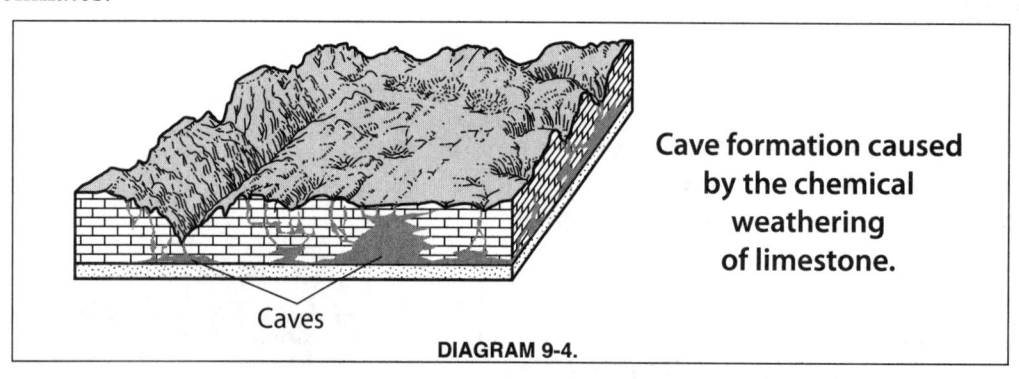

Cave formation caused by the chemical weathering of limestone.
DIAGRAM 9-4.

The rate at which weathering occurs depends on a number of factors. Climate affects the rate and type of weathering. Physical weathering occurs more in cold, humid climates. Chemical weathering occurs best in hot, humid climates.

Some rock types weather more quickly than other types. Sedimentary rocks tend to wear away faster than igneous or metamorphic rocks.

If a rock is exposed to the air and water it will weather faster. Smaller particles have a greater exposed surface area and will weather quickly.

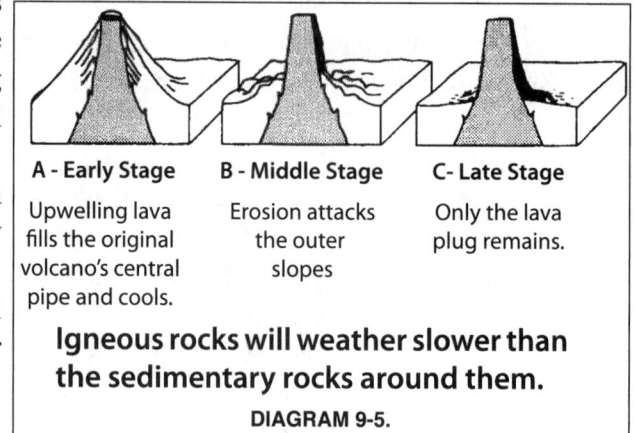

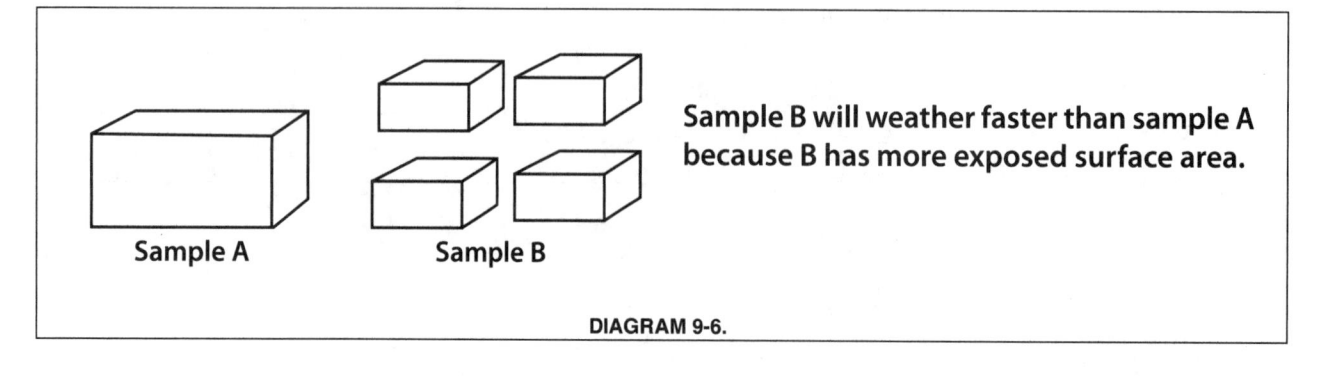

A - Early Stage Upwelling lava fills the original volcano's central pipe and cools.

B - Middle Stage Erosion attacks the outer slopes

C - Late Stage Only the lava plug remains.

Igneous rocks will weather slower than the sedimentary rocks around them.
DIAGRAM 9-5.

Sample A

Sample B

Sample B will weather faster than sample A because B has more exposed surface area.

DIAGRAM 9-6.

Soil is loose material formed by the weathering of rock and the biologic activities of plants and animals. Soil forms when the parent rock is physically and chemically weathered. Plants will root in the loose sediment as it begins to form. Organic matter **(humus)** from plants and animals enriches the soil to form **topsoil**. Subsoil will form below the topsoil by the chemical and physical weathering of topsoil.

Residual soil formed from the bedrock below it and has not been moved. **Transported soil** was moved after it formed. The soil on Long Island was transported there by glaciers.

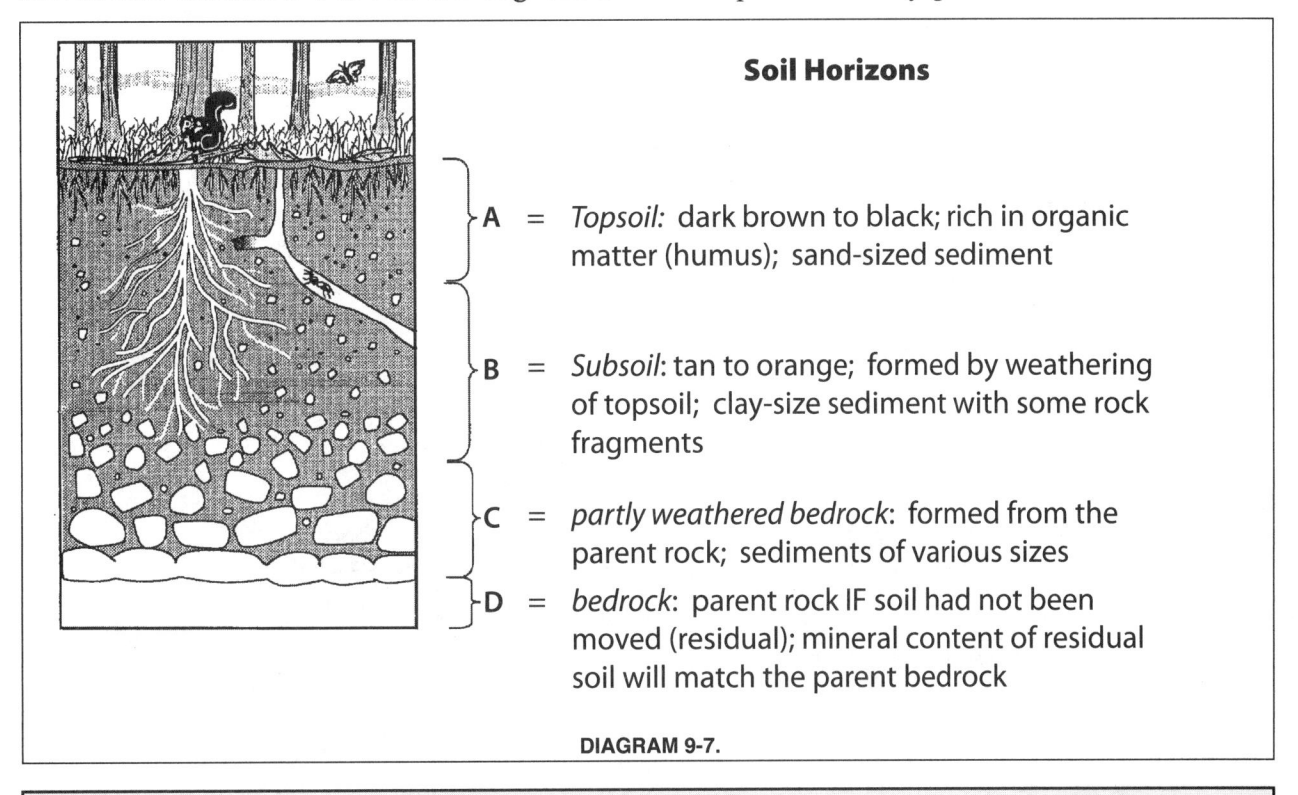

Soil Horizons

A = *Topsoil:* dark brown to black; rich in organic matter (humus); sand-sized sediment

B = *Subsoil:* tan to orange; formed by weathering of topsoil; clay-size sediment with some rock fragments

C = *partly weathered bedrock:* formed from the parent rock; sediments of various sizes

D = *bedrock:* parent rock IF soil had not been moved (residual); mineral content of residual soil will match the parent bedrock

DIAGRAM 9-7.

Questions

1. Use the *ESRT* to name the sediment particle described.

    a. smaller than silt. _____   d. largest sediment. _____

    b. 0.04 cm diameter _____   e. 18.5 cm diameter. _____

    c. 0.0005 cm diameter _____   f. 3.7 cm diameter. _____

2. *Complete* the chart to compare physical and chemical weathering.

	Physical Weathering	**Chemical weathering**
describe . . .		
caused by . . .		
most common in what type of climate?		

149

3. For each statement indicate type of weathering described: *physical* or *chemical*

    a. potholes form _____   **e.** rock surfaces peel apart _____

    b. rock gets discolored. _____   **f.** rock material dissolves in water _____

    c. hot, humid climate _____   **g.** cobbles and boulders form _____

    d. cool, humid climates _____   **h.** bedrock breaks apart _____

4. Describe *three* factors which affect the rate of weathering of rocks.

    a. _____

    b. _____

    c. _____

5. The flow chart below shows a general overview of the processes and substances involved in the weathering of rock.

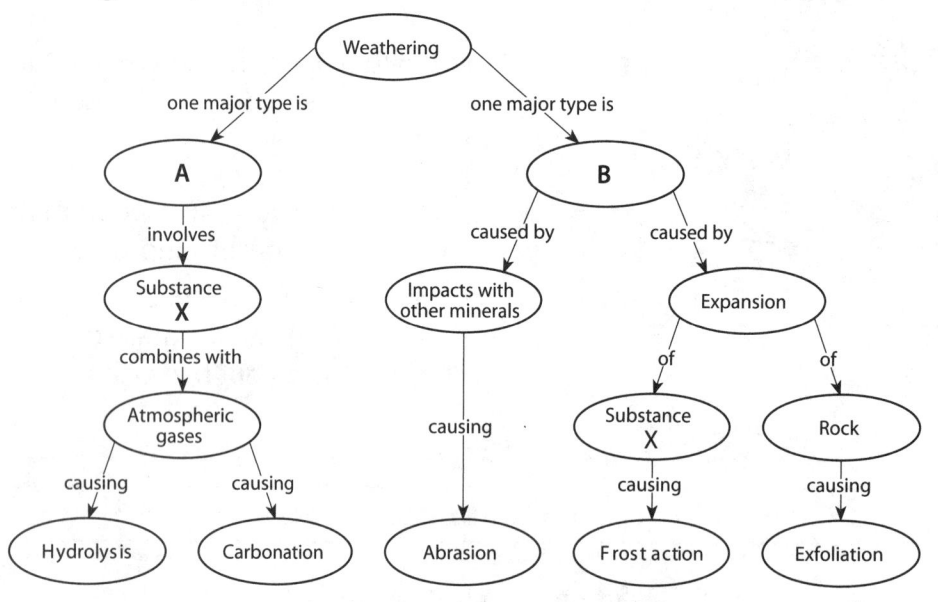

Definitions
Frost action – the breakup of rocks caused by the expansion of substance X
Abrasion – the wearing down of rocks or particles as they rub or bounce against other rocks
Exfoliation – the peeling away of large sheets of loosened material at the surface of a rock
Hydrolysis – the change in a material caused by contact with substance X
Carbonation – the change in a material caused by contact with carbonic acid

    a. The type of weathering at letter **A** is _____.

    b. The type of weathering at letter **B** is _____.

    c. The substance **X** is _____.

    d. Frost action is common in what type of climate? _____

6. The diagram to the right shows the profile of a developing soil.

   a. What processes form soil? _____

   _____

   b. What is the role of organisms in soil formation?

   _____

   c. In the diagram, soil layer "**X**" is the _____.

   d. Which layer is impermeable? _____

   e. Label the "partly weathered rock."

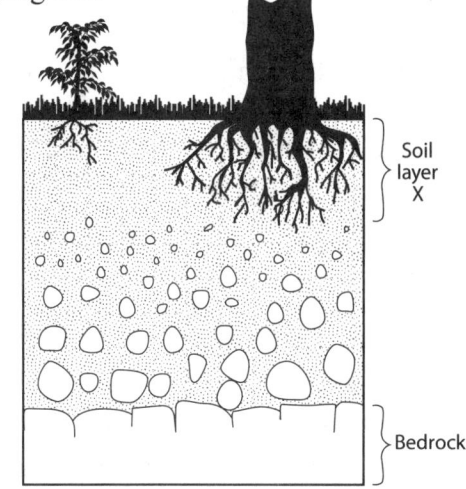

Erosion

Erosion is the process by which sediment is transported to a new location. The transporting agent is called the **agent of erosion**. Running water is the most common and dominant agent of erosion on Earth. Rivers, ocean waves, and runoff from rain and melting snow move a tremendous amount of sediment. Other agents of erosion include glaciers, wind, gravity, and human activities. Human activities, such as construction, building of roads, and mining activities, move large amounts of sediment and change the shape of the land. Each agent of erosion causes unique landscape features.

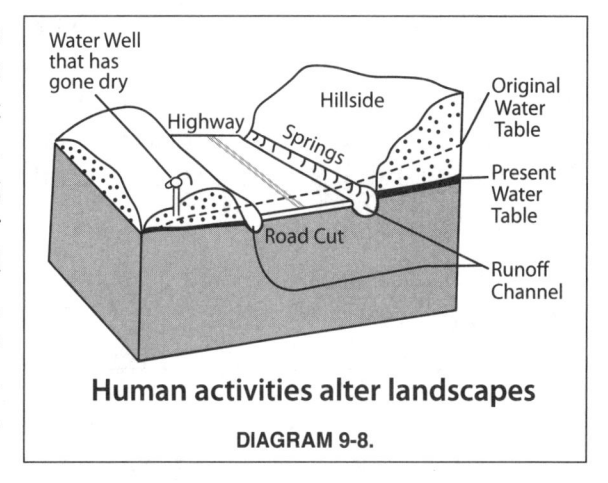

Human activities alter landscapes
DIAGRAM 9-8.

Agent of Erosion: Gravity

Gravity is the driving force behind all the agents of erosion. Gravity can act alone as an agent of erosion by causing sediment to move downhill. This is known as **mass movement**. The sediment deposited at the base of a hill is called **talus**. Talus sediment is unsorted, angular-shaped, and of various sizes. Gravity causes landslides, rock avalanches, and the slow creeping of soil down hill.

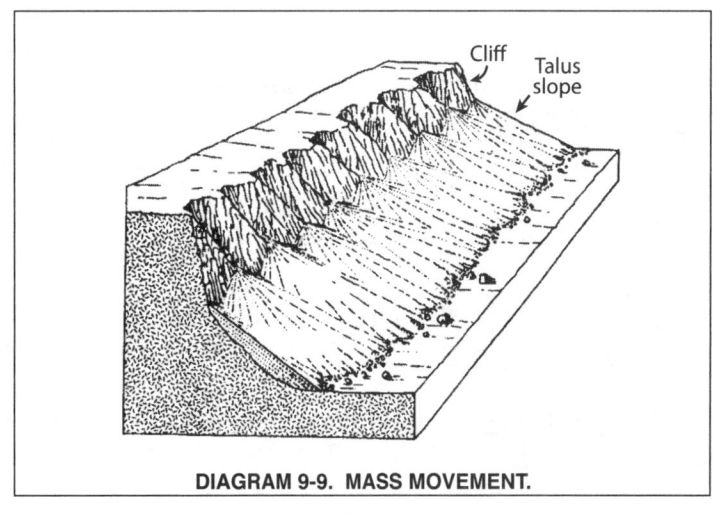

DIAGRAM 9-9. MASS MOVEMENT.

AGENT OF EROSION: WIND

Erosion by wind occurs where soil is loose and unprotected by vegetation. It is most common in arid climates and along beaches. The wind carries small sediment such as sand, silt, and clay. Wind blown sediment abrades, sculptures, and polishes rock surfaces. Rock surfaces may become frosted with small pit marks.

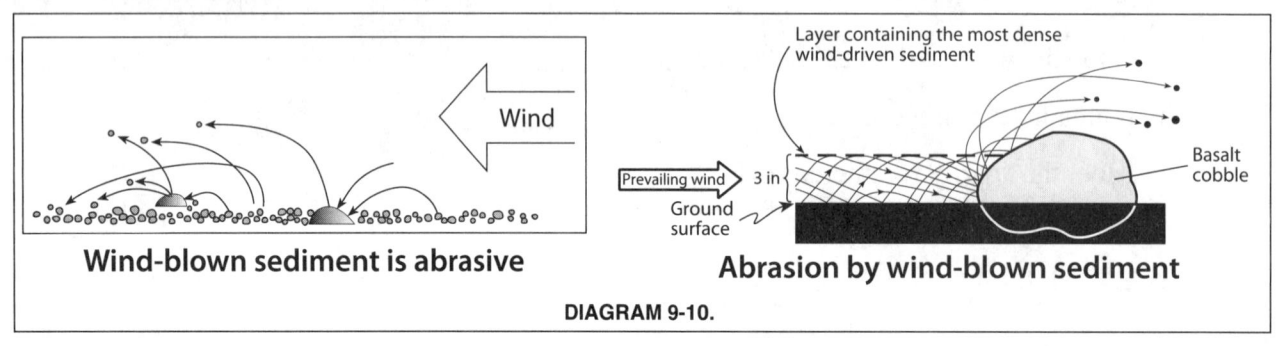

Wind-blown sediment is abrasive

Abrasion by wind-blown sediment

DIAGRAM 9-10.

Wind eroded landscapes are sculptured and angular. **Sand dunes** are hills of sorted, sand-size sediments deposited by the wind. The leeward, protected side will have a steep slope; the windward side of a sand dune will have a gentle slope.

DIAGRAM 9-11.

Questions

7. Talus is a deposit from erosion by gravity.

   a. Where is talus found?_____

   b. Describe the sediment found in talus. _____
   _____

8. The diagram to the right shows a large boulder that was abraded by wind-blown sand.
   Draw an arrow to show the general wind direction for this location.

9. The diagram to the right shows the physical features of a hill slope.

   a. What agent of erosion is acting on this hill? _____

   b. How do you know? _____
   _____
   _____

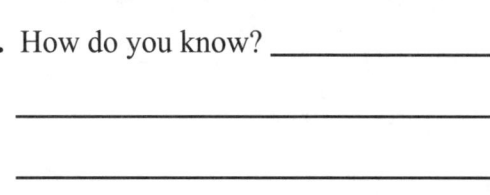

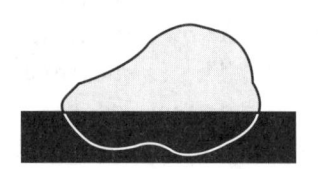

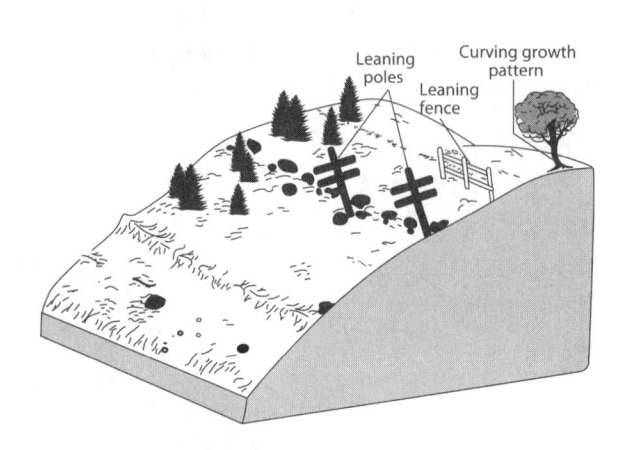

152

Agent of Erosion: Glaciers

Glaciers are moving masses of ice that slowly slide downhill. Glaciers are common in cold, humid climates where snow and ice accumulate year after year.

As glaciers move, they push sediment ahead and carry sediment frozen in the ice. The sediment moved by glaciers is unsorted and angular in shape. The rocks dragged by the glacier cause long **striations** (scratches) in the bedrock that the glacier moves over.

When a glacier begins to melt (retreat) it will leave behind unsorted sediment called **till** and scattered boulders known as **erratics**. Hills of till called **moraines** form along the front and sides of the glacier. Elongated mounds of till called **drumlins** have gentle sloped ends that point in the direction the glacier moved towards.

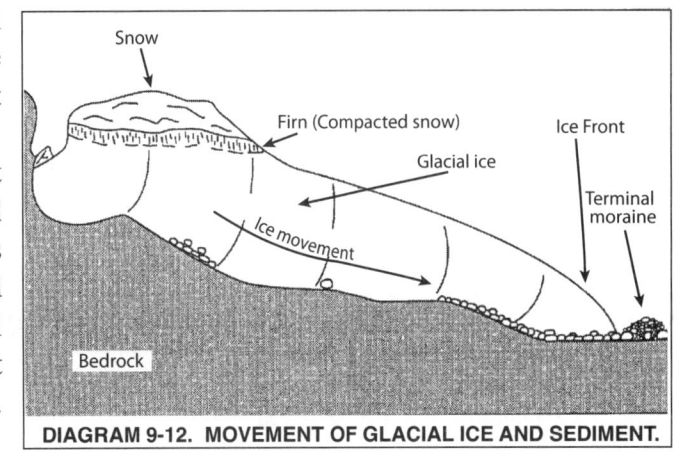

DIAGRAM 9-12. MOVEMENT OF GLACIAL ICE AND SEDIMENT.

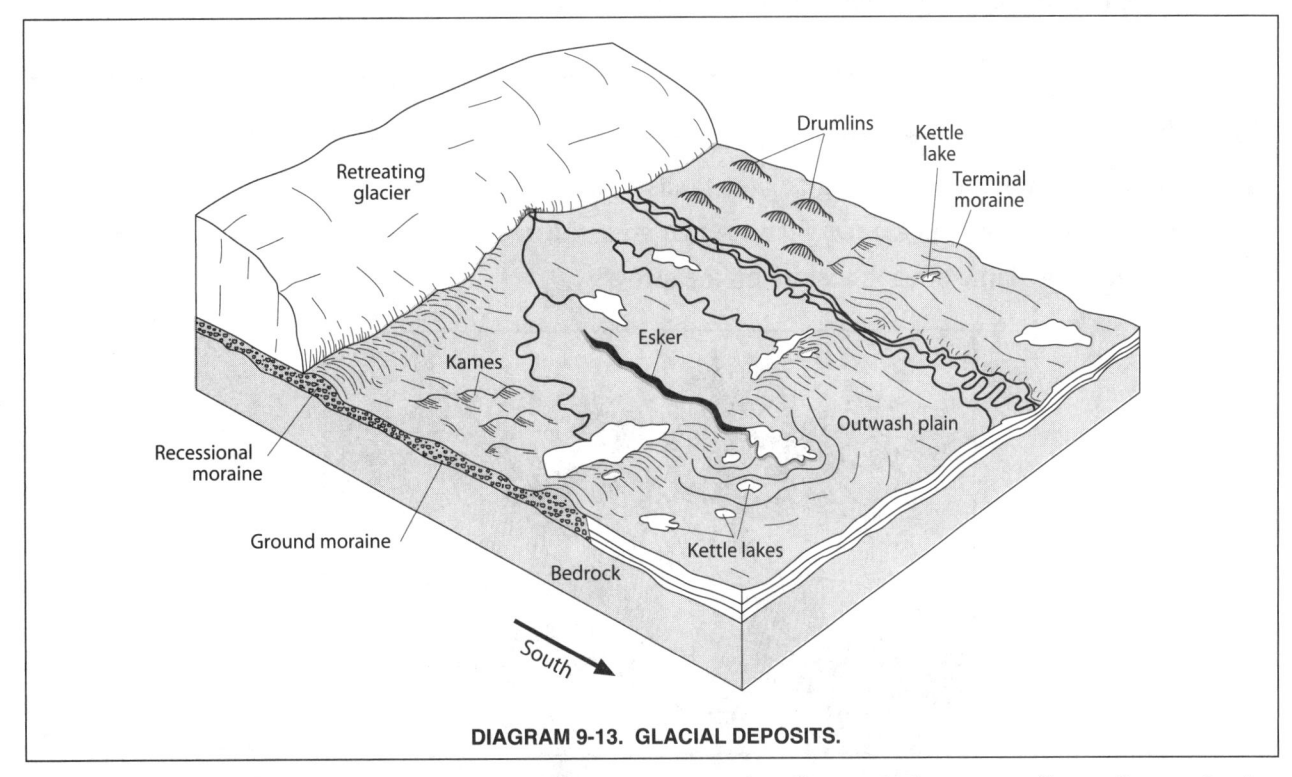

DIAGRAM 9-13. GLACIAL DEPOSITS.

When the glacier begins to melt large amounts of sediment-laden water flows forward. An **outwash plain** of sorted sediment forms. As the glacier retreats large blocks of ice may become buried in the sediment. After the ice has melted a depression called a **kettle hole** forms which can become a lake.

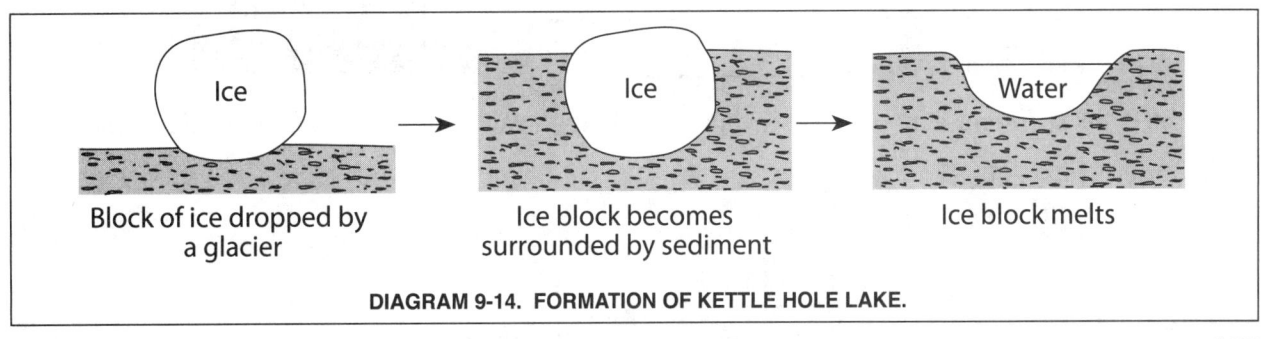

DIAGRAM 9-14. FORMATION OF KETTLE HOLE LAKE.

153

Steep-peaked mountains and U-shaped valleys are common features of a glaciated landscape. The landscape features of glacial erosion and deposition found in many areas of the world indicate that Ice Ages have occurred during Earth's history.

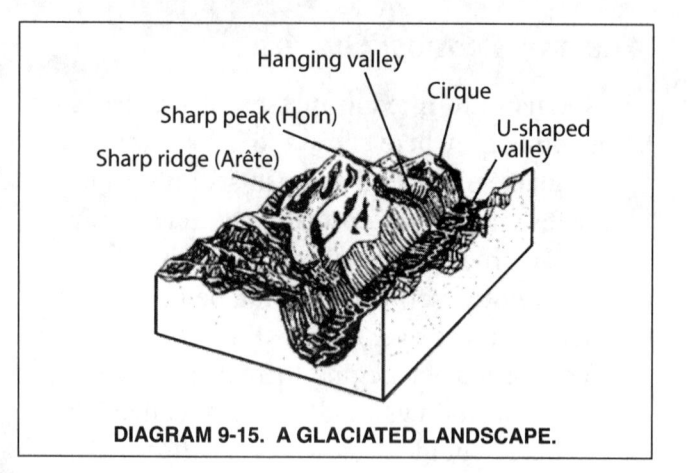

DIAGRAM 9-15. A GLACIATED LANDSCAPE.

Questions

10. Glaciers are most common in climates that are _____ and _____.

11. Explain *two* ways by which glaciers transport sediment.

    a. _____

    b. _____

12. The diagram below shows a glaciated landscape.
    a. Use arrows to indicate *two* features that are evidence of glaciation.
    b. Label the name of each feature on the arrow.

13. The diagram shows the sediment deposited by a retreating glacier.

    a. Describe the sediment deposit. _____
    _____

    b. This sediment is called _____.

    c. This sediment may be deposited in hills along the sides and front of the glacier. These hills are called _____.

154

14. This diagram illustrates a drumlin of sediment deposited by a glacier.
   a. The sediment of the drumlin can be described as (*sorted*) (*unsorted*).
   b. Draw an arrow to show the direction that the glacier was moving toward.

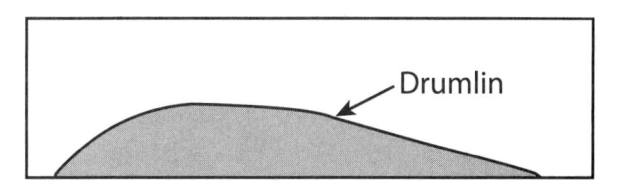

15. The maps below show evidence that New York State was once covered by a glacial ice sheet.

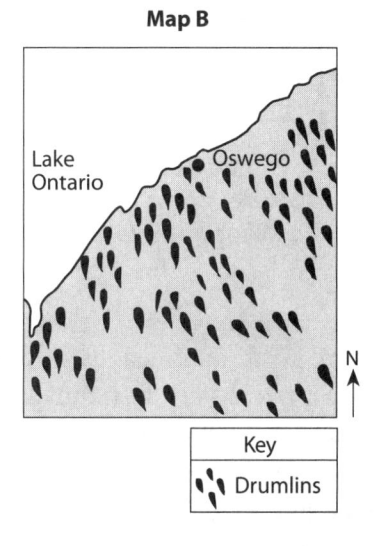

   a. On Map **A** what landscape feature indicates that the glacial ice sheet moved from north to south? _____

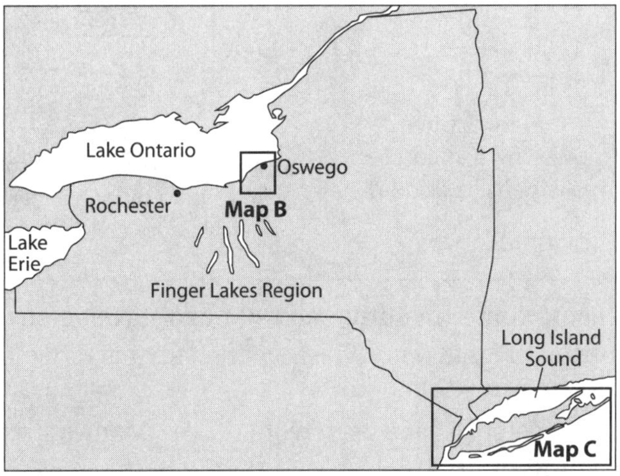

   b. On Map **B**, draw an arrow to show direction that the glacier moved through this region near Oswego.

   c. Drumlins are shown on Map **B**. Describe a drumlin. _____

   d. Part of the landscape of Long Island (Map **C**) is a terminal moraine. Explain how this terminal moraine was formed. _____

AGENT OF EROSION: RUNNING WATER

Running water is the dominant agent of erosion on Earth. It includes the runoff from rain and melting snow as well as streams and rivers.

Streams carry sediment in a variety of ways depending on the size of the sediment. Pebbles, cobbles, and sand are bounced and rolled along the stream bed. This bouncing sediment can abrade the stream bed. Colloids such as silt and clay are carried in suspension. Dissolved minerals are carried in solution.

The faster the stream moves, the larger the amount and sizes of the particles it can carry. As stream velocity increases, erosion increases. Stream velocity will increase if there is an increase in the stream's slope (**gradient**) or an increase in water volume (**stream discharge**). Stream discharge will increase after a storm or when snow melts.

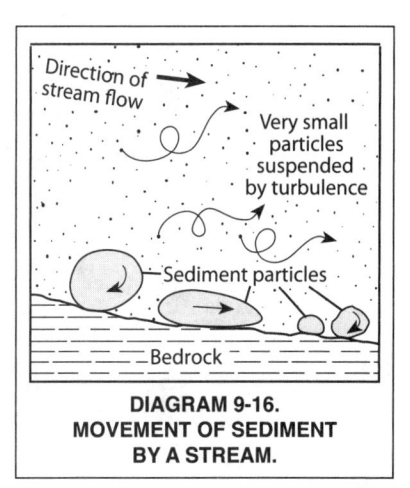

DIAGRAM 9-16.
MOVEMENT OF SEDIMENT BY A STREAM.

155

Development Stages of a Stream

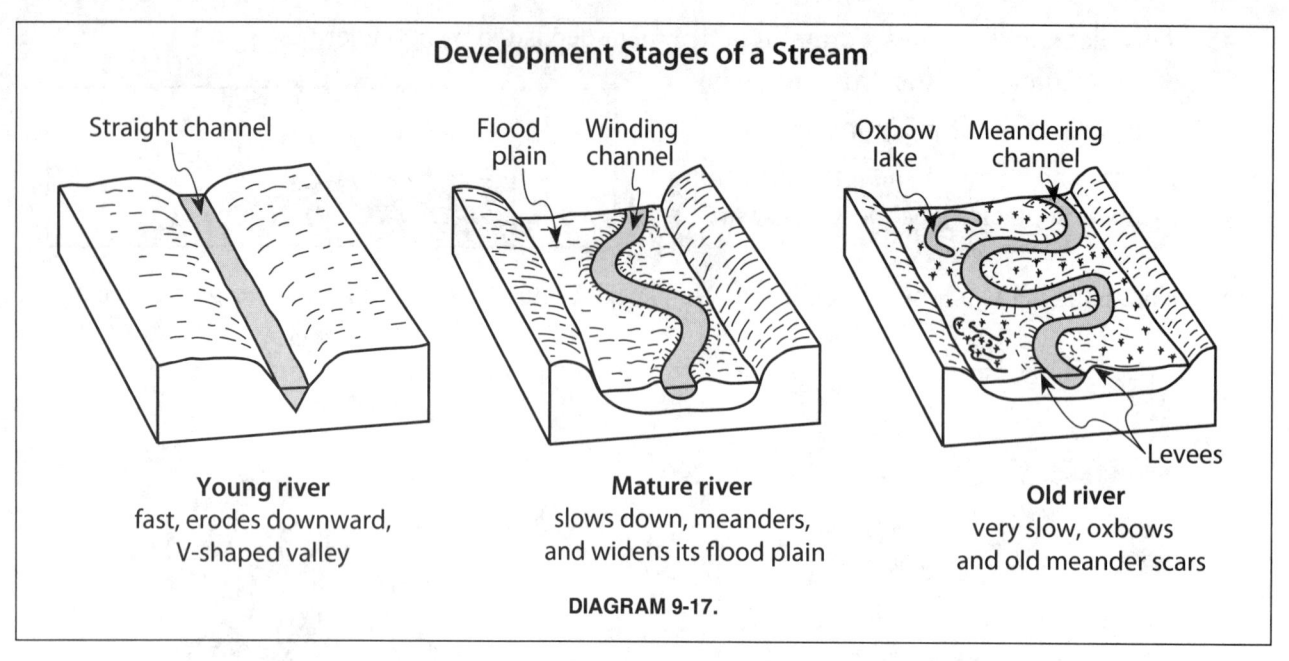

Young river
fast, erodes downward,
V-shaped valley
— Straight channel

Mature river
slows down, meanders,
and widens its flood plain
— Flood plain, Winding channel

Old river
very slow, oxbows
and old meander scars
— Oxbow lake, Meandering channel, Levees

DIAGRAM 9-17.

Streams go through stages of development: youth, maturity, and old age. Young streams have steep slopes, move very fast, and are relatively straight. These fast, young streams erode downward forming steep sided V-shaped valleys.

Mature streams move more slowly since their slopes have been worn down. Mature streams erode laterally or side-to-side. Mature streams **meander** (curve).

Old streams move the slowest and have meander scars called **oxbows**. A stream can become rejuvenated or return to youth if the region is uplifted by tectonic forces. The Colorado River is an example of a rejuvenated stream as indicated by its meanders in steep-walled canyons.

Water on the outside of the meander will move faster than on the inside. As shown in diagram 9-18, erosion will take place on the outside of the meander (**A & D**) and deposition will occur on the inside of the meander (**B & C**) due to the differences in the velocity of the water in the stream. The stream is deeper on the outside of the meander (**A & D**) and shallower on the inside where a sand bar forms (**B & C**).

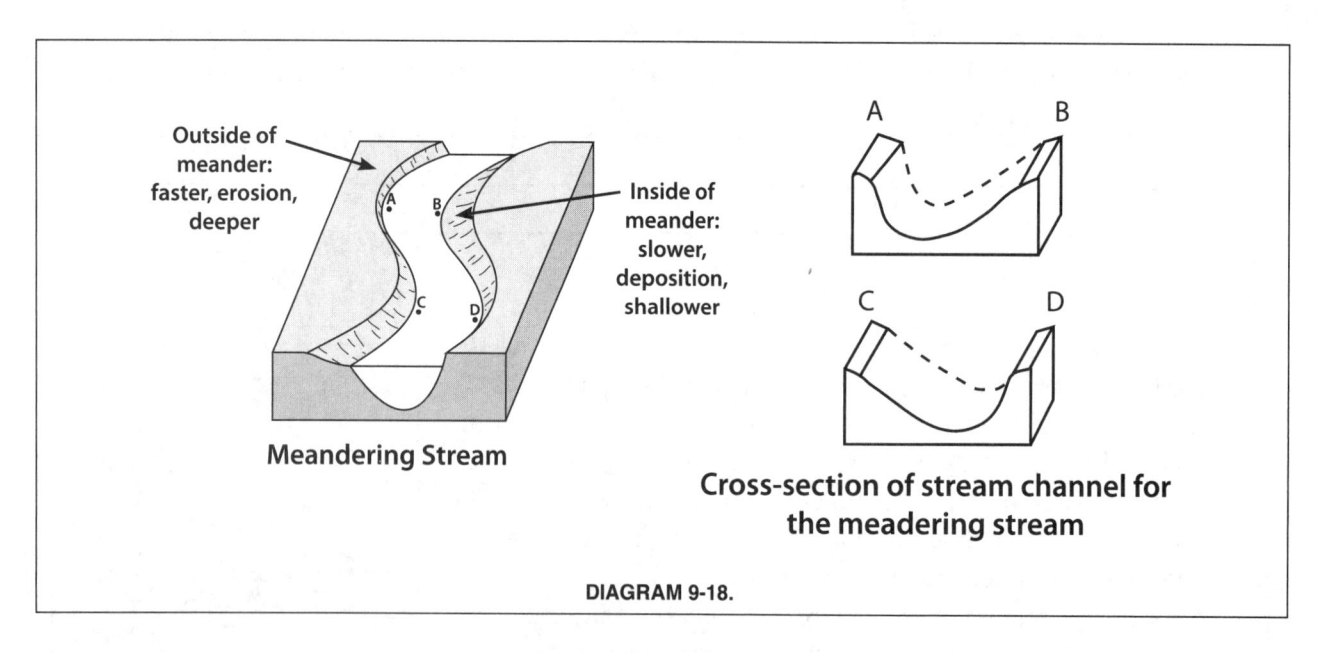

Meandering Stream
Outside of meander: faster, erosion, deeper
Inside of meander: slower, deposition, shallower

Cross-section of stream channel for the meandering stream

DIAGRAM 9-18.

156

Mississippi River watershed

DIAGRAM 9-19.

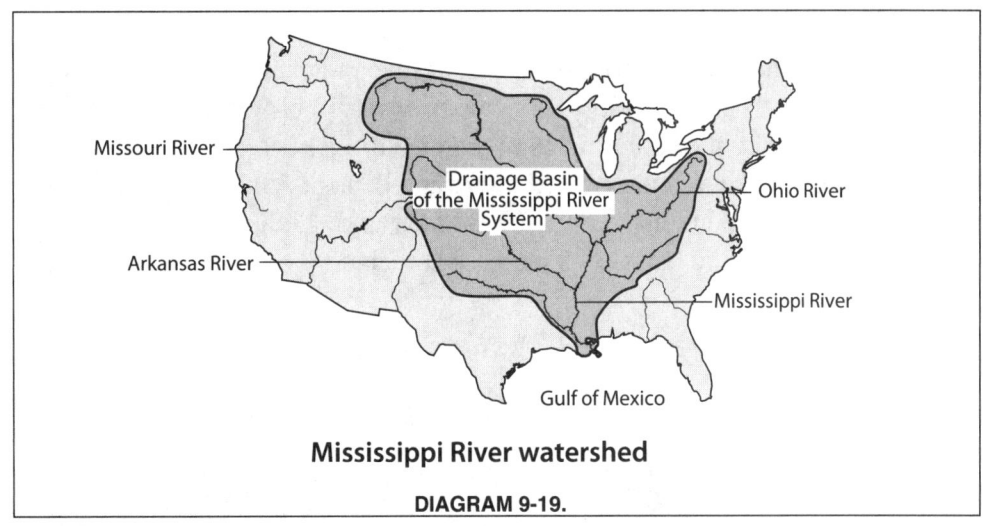

The beginning of the stream is known as the headwaters. The area drained by a stream and all its tributaries is a **watershed**.

When a stream slows down it will deposit its sediment load. These deposits may be found as **sand bars** at the inside of the meander or as a **delta** at the mouth of a river.

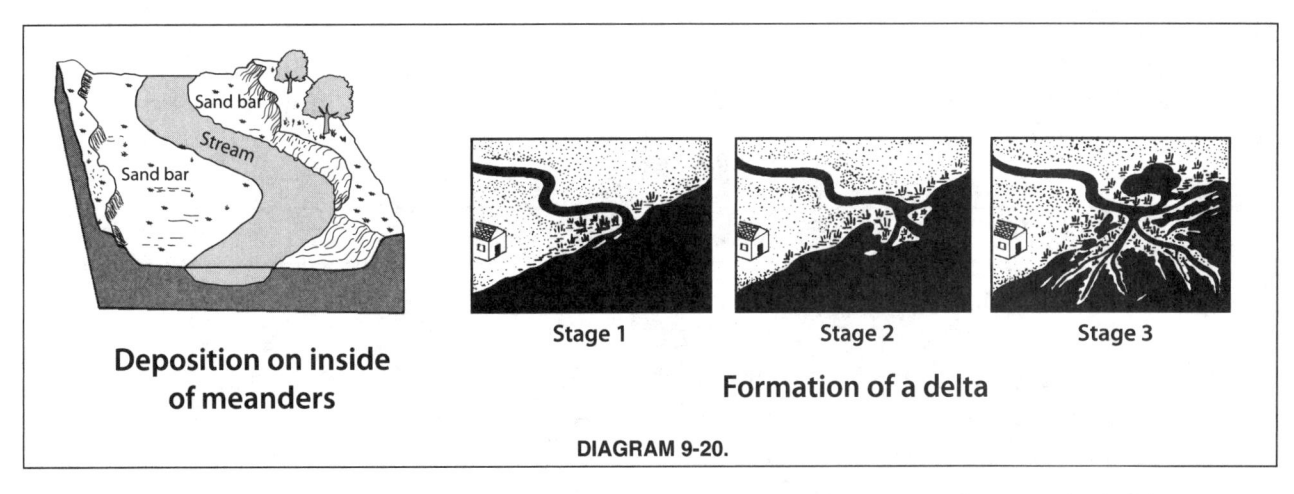

Deposition on inside of meanders

Formation of a delta

Stage 1 Stage 2 Stage 3

DIAGRAM 9-20.

Sediments deposited by running water are rounded, smoothed, and sorted. Deposits left by streams are sorted by size and density. When a stream enters a larger body of water it will slow down and begin to deposit its sediment load. The largest, densest sediment is deposited first.

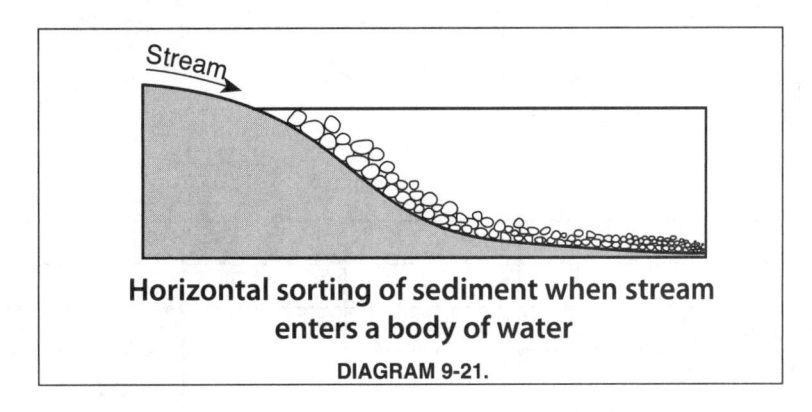

Horizontal sorting of sediment when stream enters a body of water

DIAGRAM 9-21.

AGENT OF EROSION: OCEAN WAVES AND CURRENTS

Ocean waves and longshore currents erode the shoreline and change its shape. Over time a rocky shoreline will become a sandy beach. Waves are caused by wind blowing over the surface of oceans. As the waves continue to strike the shoreline at an angle, a **longshore current** of water parallel to the shoreline develops. This longshore current will move and deposit sediment parallel to the shoreline.

Erosion by waves causes sediment to become rounder and smaller. Deposits from ocean waves and currents include beaches, sand bars, spits, and barrier islands.

Barrier islands protect the mainland from ocean waves and currents

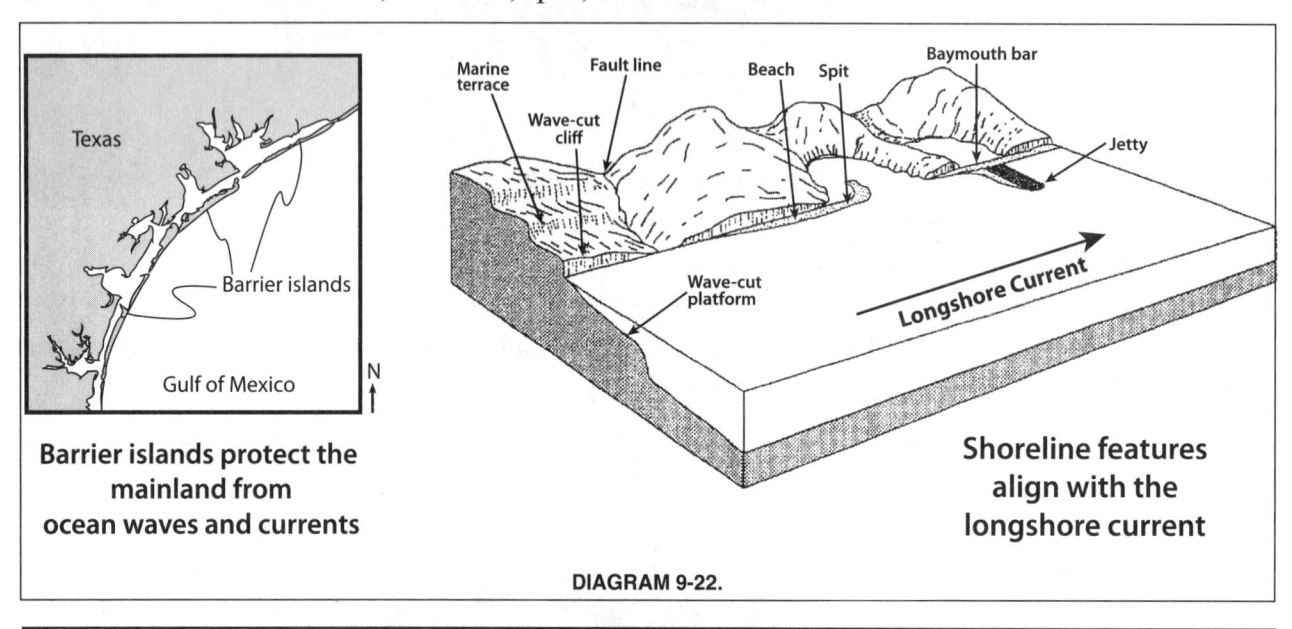

Shoreline features align with the longshore current

DIAGRAM 9-22.

Questions

16. The diagram shows a top view of a mature river. The arrow shows the direction the stream is moving. Points **A** through **D** are locations along the stream channel.
For each statement, place an **X** for the lettered position(s) that the statement applies to. *One example is given.*

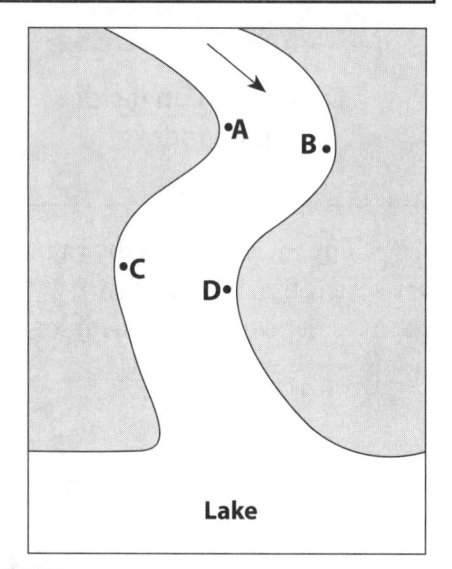

	A	B	C	D
a. fast stream velocity				
b. slow stream velocity				
c. erosion is dominant				
d. deposition is dominant	X			
e. deep				
f. shallow				
g. sandbars form				

158

17. For each statement, select the stage of development for the stream:

(young), (mature), (old)

a. very slow moving. _____   e. V-shape valley. _____

b. moving quickly. _____   f. erodes downward. _____

c. many oxbow scars. _____   g. develops meanders. _____

d. widens its valley. _____   h. relatively straight. _____

18. The diagram below shows some features along a coastline.

a. *Draw an arrow* to show the most likely direction of the longshore current.

b. How do you know? _____

19. For each statement select the agent of erosion that is responsible:

(running water) (wind) (glacier) (oceans waves and currents) (gravity)

a. cold, humid climate _____   i. erratics _____

b. arid climate _____   j. sand bars _____

c. meanders _____   k. oxbow lakes _____

d. V-shaped valley _____   l. U-shaped valley _____

e. mass movement _____   m. drumlins _____

f. landslide _____   n. kettle holes _____

g. striations _____   o. beach _____

h. most dominant agent of erosion on Earth. _____

20. The diagram shows a region of Earth where the landscape has been affected by different agents of erosion. Positions **X** and **Y** are two points along the meander of the stream.

   a. *Label* the delta on the diagram.
   b. *Label* the oxbow lake.
   c. Meanders are evidence that this section is *(young) (mature) (old)*.
   d. The V-shaped valley is a result of erosion by a *(young) (mature) (old)* stream.
   e. Which agent of erosion formed the U-shaped valley? _____
   f. At which position is the stream moving the fastest? *(X) (Y)*
   g. At which position will the stream depth be greater? *(X) (Y)*
   h. Deposition will be occurring at position *(X) (Y)*.

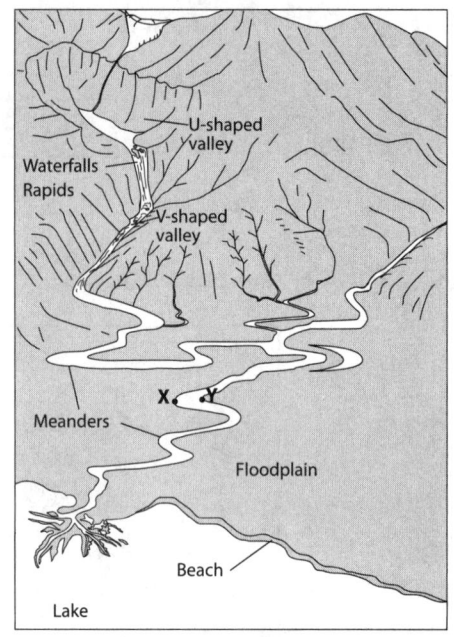

Deposition

Deposition is the process by which the agent of erosion drops its sediment load. This occurs when the agent of erosion slows down. Deposition includes the release of solid sediment (sedimentation) and the release of dissolved minerals (precipitation). Most deposition occurs in water since moving water is the major agent of erosion on Earth. Each agent of erosion causes specific features of deposition to form.

The rate of deposition is affected by three factors: size, shape, and density of the particles. Large sediment will settle faster than smaller sediment. Round sediment will settle faster than flatter sediment. As the density of a particle increases, its rate of deposition will increase. Large, round, dense sediment will be deposited first when the agent of erosion slows down.

Agent of Erosion	Erosional Features	Deposition Features
Gravity	mass movement of loose sediment downhill; *ex:* landslide, soil creep	unsorted, angular sediment, at base of hill; *ex:* talus
Running Water	moving water carves out a stream channel; *ex:* V-shaped valley, meanders, oxbows	sorted, smoothed, and rounded sediment; *ex:* delta, sand bars on inside of meander curves, flood plain
Wind	loose wind-blown sand abrades rock surfaces; *ex:* frosted and pitted rock surfaces, sculptured rocks	sorted, small sediment in a hill; *ex:* sand dunes
Ocean Waves and Currents	waves and longshore currents move sediment along the shoreline	rounded, sorted sediment; *ex:* beaches, sand bars, spits, barrier islands
Glaciers	large mass of compacted ice and snow slowly moves downhill carrying and pushing sediment along; *ex:* U-shaped valleys, striated rock surfaces, kettle and finger lakes	unsorted, angular sediment from clay to boulder size; *ex:* moraines, erratics, drumlins

Questions

21. For each pair, *circle the particle* which will be deposited first
   a. *round particle, flat particle*
   b. *pebble, cobble*
   c. *lead sphere, glass sphere*
   d. *clay, silt*
   e. *granite pebble, basalt pebble*

22. A jar of water with sand, silt, and clay was shaken and then allow to sit quietly. After a few hours the sediments had settled as shown in the diagram to the right.
Explain why this pattern occurred. _____
_____
_____

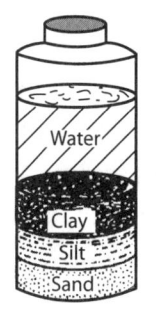

23. The diagram below shows a profile of a stream entering the ocean. The average diameter of the sediment deposited in each area is shown.

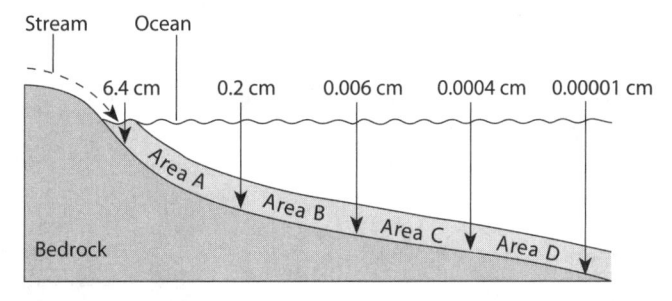

   a. Deposition of the sediments occurred because the steam velocity _____.
   b. State the name of the sediment deposited in each area:
   area **A** :_____; area **B**: _____; area **C**: _____; area **D**: _____
   c. In which area(s) is the water moving at least 50 cm/sec? _____
   d. State the relationship between stream velocity and size of sediment that can be moved.
   _____
   _____

24. *Complete* the following chart for each type of deposit.

Deposit	Describe	sorted or unsorted sediment	Agent of erosion
a. sand dune			
b. delta			
c. moraine			
d. barrier island			
e. talus			
f. till			

161

CHAPTER 9 REVIEW

1. Chemical weathering occurs most rapidly in climates which are
   (1) moist and warm (2) moist and cold (3) dry and cold (4) dry and warm

2. Landscapes will undergo the most physical weathering if the climate is
   (1) dry and hot (2) dry and cold (3) moist and hot (4) moist and cold

3. At high elevations which is the most common form of weathering of rock?
   (1) abrasion of rocks by wind-blown sand   (3) dissolving of minerals into solution
   (2) alternate freezing and melting of water   (4) oxidation of iron-rich minerals

4. A rock will weather faster after it has been crushed because its
   (1) volume has been increased   (3) surface area has increased
   (2) density has been decreased   (4) molecular structure has been altered

5. A sand particle can have a diameter of
   (1) 0.0005 cm (2) 0.005 cm (3) 0.05 cm (4) 0.5 cm

6. Water is a major agent of chemical weathering because water
   (1) cools the surroundings when it evaporates
   (2) dissolves many of the minerals which make up rocks
   (3) has a density of about one gram per cubic centimeter
   (4) has the highest specific heat of all common Earth materials

7. Which event is the best example of erosion?
   (1) breaking apart of shale as a result of water freezing in a crack
   (2) dissolving of limestone by acid rain
   (3) rolling of a pebble along the bottom of a stream
   (4) crumbling of bedrock to form soil

8. Which *two* rock units appear to be the most resistant to weathering and erosion?
   (1) Lockport dolostone and Whirlpool sandstone
   (2) Rochester shale and Albion shale
   (3) Clinton limestone and Queenston shale
   (4) Thorold sandstone and Albion sandstone

9. Solid bedrock is changed to soil primarily by the processes of
   (1) erosion and deposition
   (2) weathering and biologic activity
   (3) infiltration and runoff
   (4) evaporation and transpiration

10. The danger of landslides is greatest when
    (1) a slope is heavily vegetated
    (2) a large amount of precipitation has occurred
    (3) the slope of the land is almost level
    (4) the maximum angle of insolation occurs

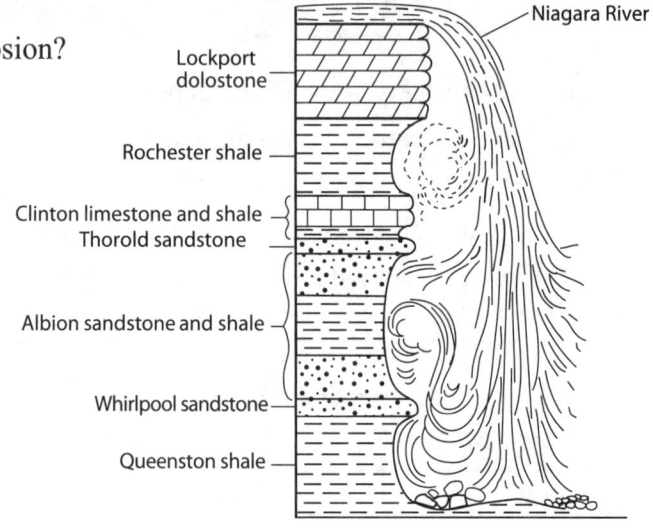

11. Deposits of sediment at the base of a cliff can best be described as
    (1) sorted and round in shape
    (2) sorted and angular in shape
    (3) unsorted and round in shape
    (4) unsorted and angular in shape

12. Sediments in a hill-like deposit are small, well-sorted, and have surface pits that give them a frosted appearance. This deposit was likely transported by
    (1) ocean current   (2) glaciers   (3) gravity   (4) wind

13-15. The diagram shows the edge of a continental glacier that is receding. Letter **R** indicates elongated hills of deposition. The ridge of sediment extending from **X** to **Y** represents a landscape feature.

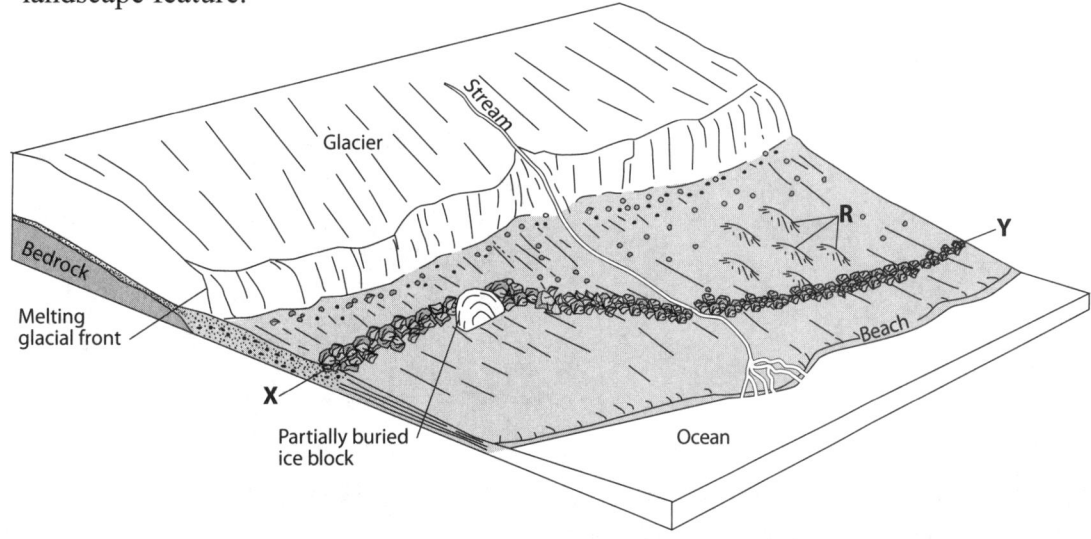

13. The elongated hills labeled **R** are most useful in determining the
    (1) age of the glacier
    (2) direction the glacier moved
    (3) thickness of the glacier
    (4) rate at which the glacier is melting

14. Which feature will most likely form when the partially buried ice block melts?
    (1) drumlin   (2) moraine   (3) kettle hole   (4) sand bar

15. The ridge of sediments from **X** to **Y** can best be described as
    (1) sorted and deposited by ice
    (2) sorted and deposited by meltwater
    (3) unsorted and deposited by ice
    (4) unsorted and deposited by meltwater

16. Glaciers often form parallel scratches and grooves in bedrock because glaciers
    (1) deposit sediment in unsorted piles
    (2) deposit rounded particles in V-shaped valleys
    (3) continually melt and refreeze
    (4) drag loose rocks over the Earth's surface

17. Which landscape feature will most likely form in the mountains by glacial erosion?
    (1) U-shaped valley
    (2) V-shaped valley
    (3) oxbow lake
    (4) sand bar

18. Which graph best illustrates the relationship between the discharge of a river and the particle size that can be transported by the river?

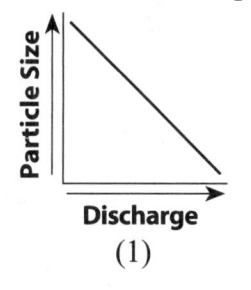

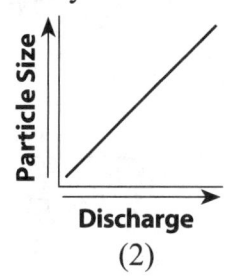

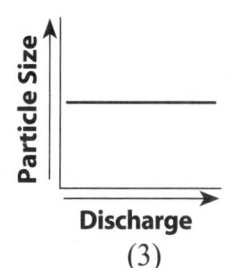

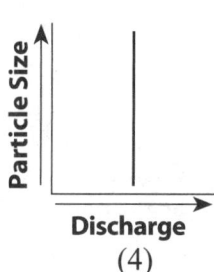

19. A stream flowing at a velocity of 75 centimeters per second can transport
    (1) clay only
    (2) pebbles only
    (3) pebbles, sand, silt and clay, only
    (4) boulders, cobbles, pebbles, sand, silt and clay

20. Rock materials transported in a stream are most likely transported by which methods?
    (1) in solution, only
    (2) in suspension, only
    (3) in solution and in suspension
    (4) in solution, in suspension, and by rolling

21. The map below shows the four watershed regions in New York State labeled **A** through **D**.

    Which letter represents the watershed of the Mohawk and Hudson Rivers?
    (1) A
    (2) B
    (3) C
    (4) D

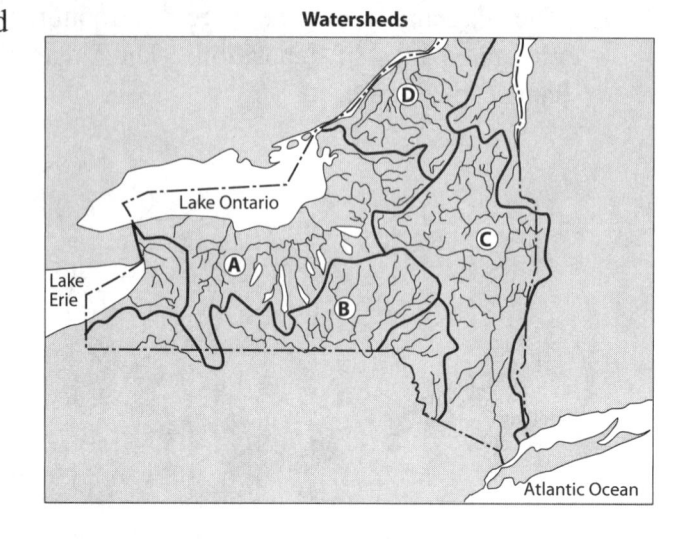

22. Which statement best describes a stream with a steep gradient?
    (1) It flows slowly, producing a V-shaped valley.
    (2) It flows slowly, producing a U-shaped valley.
    (3) It flows rapidly, producing a V-shaped valley.
    (4) It flows rapidly, producing a U-shaped valley.

23. Which statement best describes sediments deposited by glaciers and rivers?
    (1) Both are sorted.
    (2) Both are unsorted.
    (3) Glacial deposits are sorted, river deposits are unsorted.
    (4) Glacial deposits are unsorted, river deposits are sorted.

24. The diagram represents a cross section of a section of a stream. Point **A, B, C, D,** and **E** are located in the stream channel. Which graph best represents the stream velocity at locations **A** through **E**?

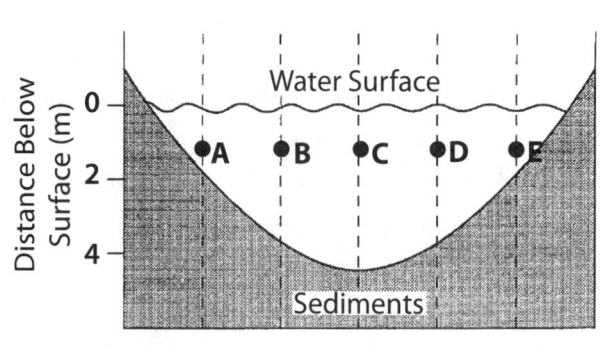

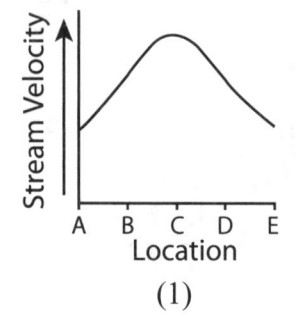

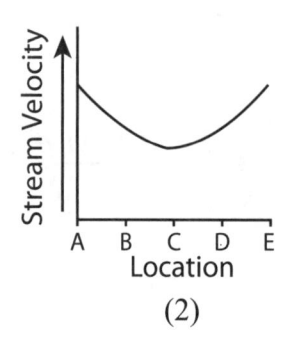

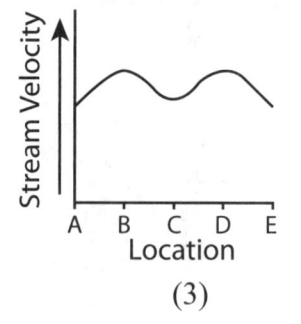

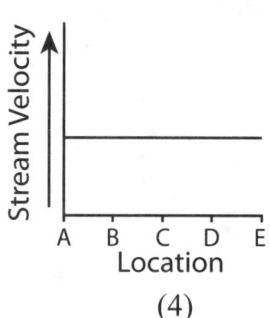

164

25. Why do the particles carried by a river settle to the bottom as the river enters the ocean?
    (1) The density of the ocean water is greater than the river water.
    (2) The kinetic energy of the particles decreases as the particles enter the ocean.
    (3) The velocity of the river increases as it enters the ocean.
    (4) The smaller particles have more surface area.

26. The sand on the beach originated from the weathering of granite bedrock. What mineral is most likely to be found in the sand?
    (1) quartz   (2) calcite   (3) olivine   (4) pyroxene

27. What is the source of most of the dissolved minerals in seawater?
    (1) weathering of seafloor rocks   (3) deep ocean organic sediments
    (2) weathering and erosion of land rocks (4) gases from underwater volcanoes

28. What is the minimum stream velocity needed to move a boulder?
    (1) 10 cm/s   (2) 100 cm/s   (3) 200 cm/s   (4) 500 cm/s

29. Which sized particle will remain suspended the longest as a river enters the ocean?
    (1) pebble   (2) silt   (3) sand   (4) clay

30. The diagram shows a stream flowing towards a lake.

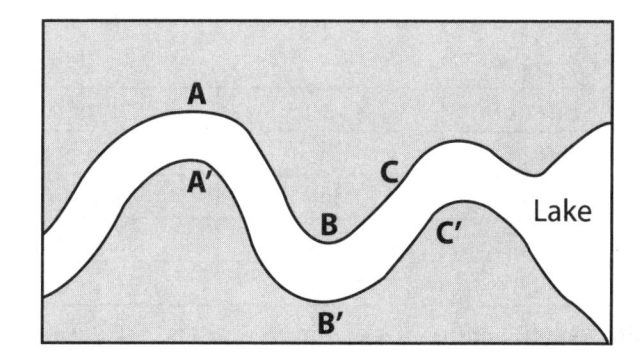

    Which set of diagrams correctly shows the cross sections across the lettered sections of the stream?

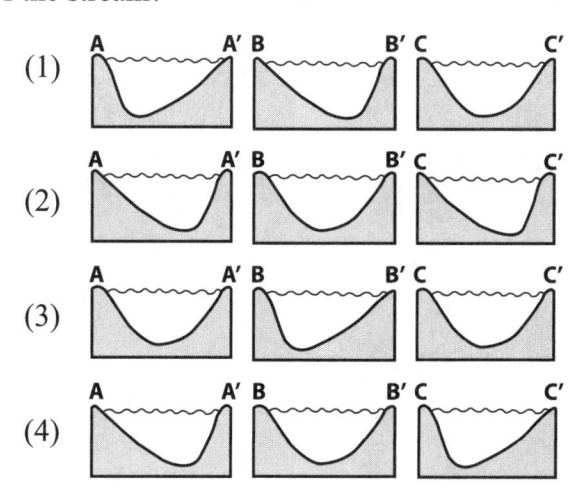

165

31. The diagram shows a meandering stream as it enters a lake. At which points in the stream are erosion and deposition dominant?
    (1) Erosion is dominant at **A** and **D**; deposition is dominant at **B** and **C**
    (2) Erosion is dominant at **B** and **C**; deposition is dominant at **A** and **D**
    (3) Erosion is dominant at **A** and **C**; deposition is dominant at **B** and **D**
    (4) Erosion is dominant at **B** and **D**; deposition is dominant at **A** and **C**

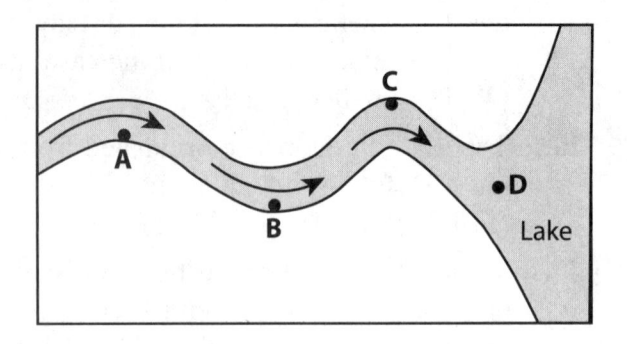

32. Which statement best describes the conditions existing at a stream location where the erosional-depositional system is in dynamic equilibrium?
    (1) More erosion than deposition takes place.
    (2) More deposition than erosion takes place.
    (3) Equal amounts of erosion and deposition take place.
    (4) No erosion or deposition will take place.

33. The four particles shown in the table below are of equal volume and are dropped into a column filled with water.

Particle	Shape	Density
A	flat	2.5 g/cm^3
B	flat	3.0 g/cm^3
C	round	2.5 g/cm^3
D	round	3.0 g/cm^3

    Which particle would usually settle most rapidly?
    (1) A   (2) B   (3) C   (4) D

34. The graph below is incomplete because it does not identify the sediment characteristic (**X**) that would produce the line plotted on the graph.
    Which label should be placed on the horizontal axis to accurately complete the graph?

    (1) Low ⟶ High Particle Density
    (2) Small ⟶ Large Particle Size
    (3) Light ⟶ Heavy Particle Mass
    (4) Round ⟶ Flat Particle Shape

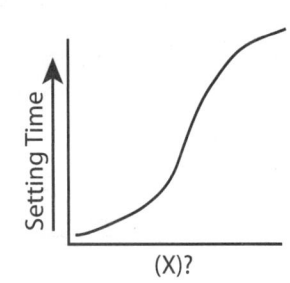

35. A sample of rounded quartz sediment of different particle sizes is dropped into a container of water. Which graph best shows the setting time for these particles?

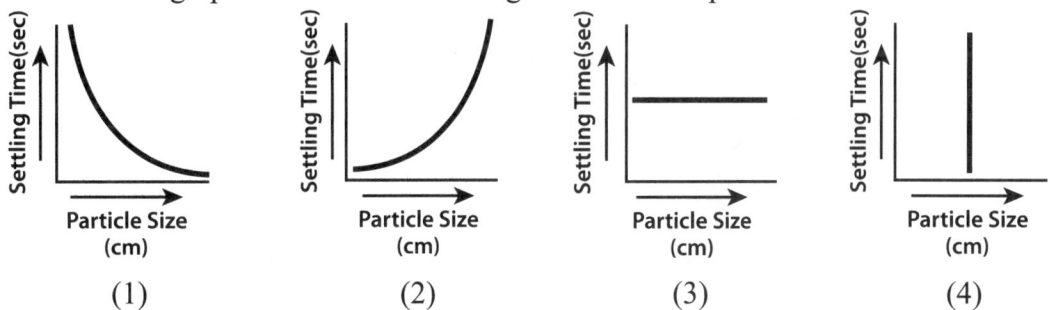

36-37. The diagram below shows a cross-section of sediments deposited by a stream during a three week period.

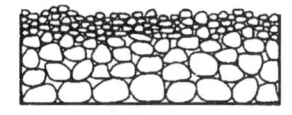

36. During this time period the velocity of the stream decreased. How does the diagram support this statement? _____

37. What would have caused the stream to slow down during this three week period?
_____

38-39. The diagram below illustrates the Rockaway Peninsula located along the south shore of western Long Island.

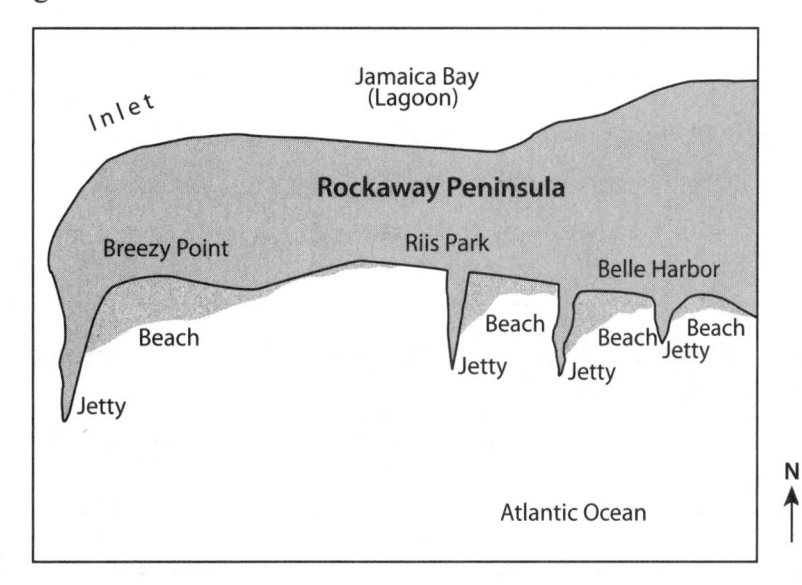

38. *Draw an arrow* to show the direction of the longshore current along the south shore of Long Island.

39. What would be the effect on the shape of the beaches if the jetties were removed?
_____

167

40-45. The graph below shows the effects that average yearly precipitation and temperature have on the type of weathering that will occur in a particular region.

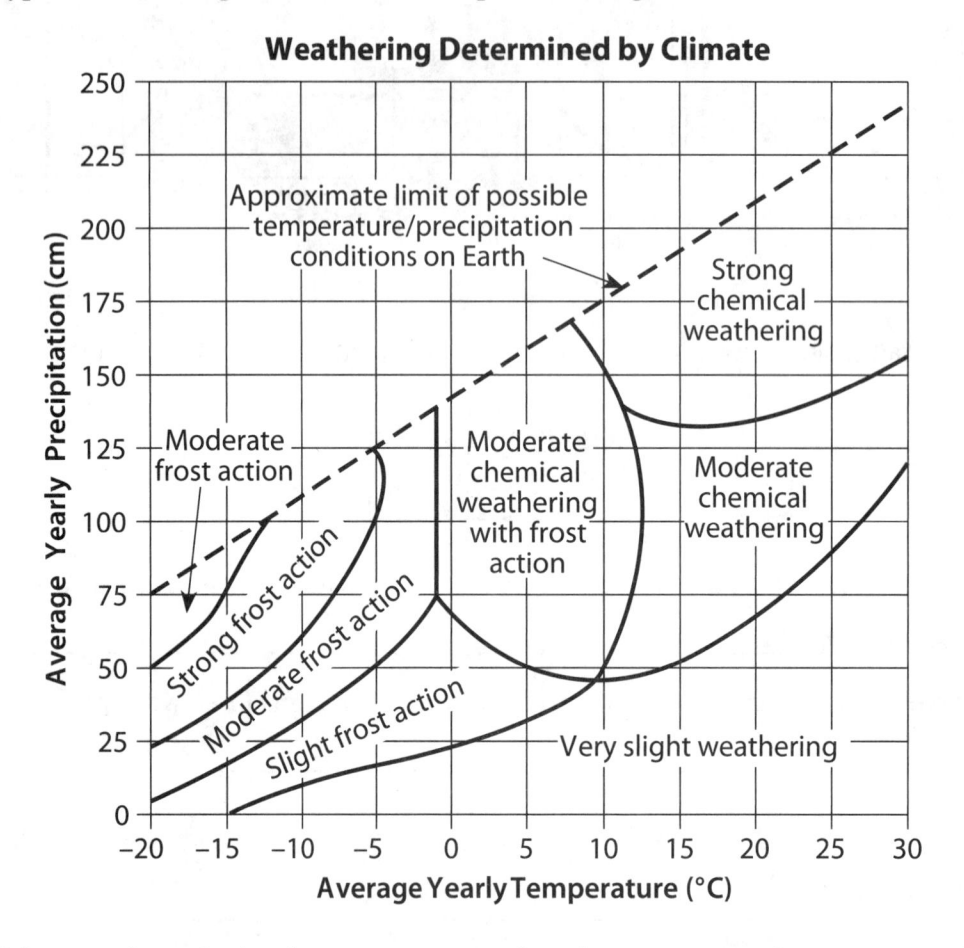

Weathering Determined by Climate

40. Which type of weathering is most common when the average yearly temperature is -5° C and the average yearly precipitation is 75 cm? _____

41. Very slight weathering occurs in climates which are (*dry*) (*humid*).

42. Explain the process of weathering by frost action. _____
_____

43. Frost action will increase when the temperature (*decreases*) (*increases*) and the yearly precipitation (*decreases*) (*increases*).

44. List *two* factors which would cause the amount of chemical weathering to increase.

   a. _____

   b. _____

45. Why does weathering not occur above the dashed line on the graph? _____
_____
_____

168

46-50. The diagram shows a stream flowing into a lake. The arrow shows the direction of streamflow. Points **A** and **B** are locations at the edge of the stream. Line **AB** is a reference line across the stream surface. Line **CD** is a reference line along the lake bottom from the mouth of the stream into the lake. The data table gives the depth of the water and distance from point **A** along line **AB**.

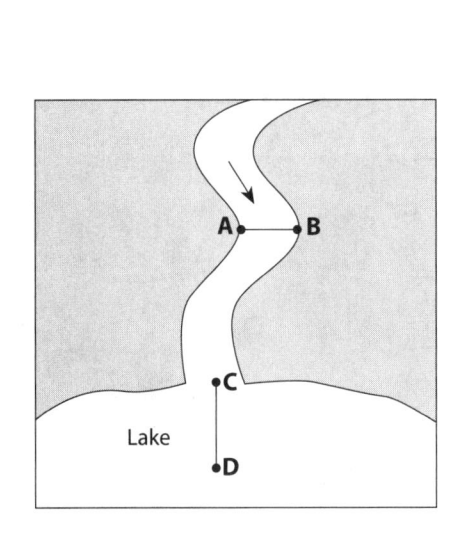

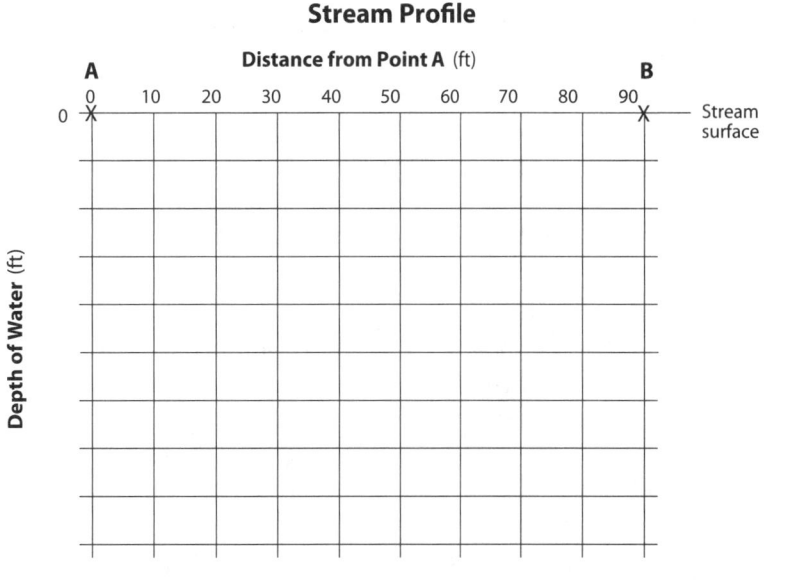

Stream Data Table

	Point A								Point B
Distance from Point A (ft)	0	10	20	30	40	50	60	70	80	90
Depth of Water (ft)	0	2	4	7	11	13	16	17	10	0

46. On the grid above, construct a profile of the depth of water below line **AB** following the directions below.
- Mark an appropriate numerical scale showing equal intervals on the axis labeled "Depth of Water." The zero (0) on the depth of water axis represents the stream surface.
- Using the data table, plot with an **X** the depth of the water at each distance from point **A**. Connect the **X**'s with a smooth, curved line. Points **A** and **B** have been plotted.

47. Based on the map and the data table, explain why the depth of water 20 feet from point **A** is different from the depth of water 20 feet from point **B**. _____

48. The sediments being carried by the stream include clay, pebbles, sand, and silt. List these sediments in the mostly likely order of deposition from point **C** to point **D**.

_____ → _____ → _____ → _____

Point C >>>> -->>>> Point D

49. Sediments of size 0.008 cm are classified as _____.

50. Name a sediment particle that could be held in suspension at **D**. _____

51-55. The reading passage and map below describe the Dust Bowl event of the 1930s in the southern Great Plains.

The Dust Bowl

In the 1930's, several years of drought affected over 100 million acres in the Great Plains from North Dakota to Texas. For several decades before this drought, farmers had plowed the prairie grasses and loosened the soil. When the loose soil became extremely dry from lack of rain, strong winds easily removed huge amounts of soil from the farms. This caused huge dust storms. This region became known as the "Dust Bowl".

In the spring of 1934, a windstorm lasting a day and a half created a dust cloud nearly 2000 kilometers long and caused "muddy rains" in New York State and "black snow" in Vermont. Months later, a Colorado storm carried dust approximately 3 kilometers up into the atmosphere and transported it 3000 kilometers, creating twilight conditions at midday in New York State.

A Portion of the Dust Bowl in the Southern Great Plains

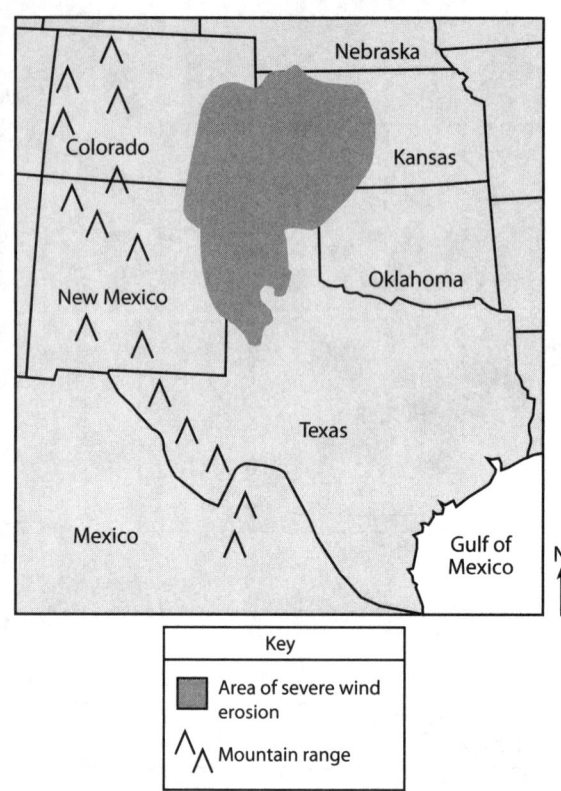

Key:
- Area of severe wind erosion
- Mountain range

51. Identify one human activity that was a major cause of the huge dust storms that formed in the Great Plains during the 1930's. _____

52. Name the layer of the atmosphere in which the dust particles were transported to New York State by the Colorado storm. _____

53. Why did the dust clouds move northeast towards New York and Vermont? _____

54. Explain why the dust clouds were composed mostly of silt and clay particles instead of sand.
_____

55. How did the dust clouds affect insolation? _____

CHAPTER 10

LANDSCAPES AND MAPPING
Landscapes, Classification and Development of Landscapes, Contour Mapping

Landscapes

Long Island was formed by glacial deposition during the last Ice Age. Ocean waves and currents formed Fire Island. The Adirondack Mountains were formed from ancient tectonic events. These are examples of landscapes that have physical features as a result of their formation.

A **landscape** is described by the slope (gradient) of the land, elevation, shape of the land, geologic features, bedrock structure, type of soil, and stream drainage patterns. Landscapes form as a result of weathering, erosion and deposition, major and minor tectonic plate movements, climate, and human activities.

Streams flow downhill and form drainage patterns due to the shape of the land. Streams will flow around resistant bedrock and form valleys through the least resistance rocks. The diagrams below show general stream patterns that result from the shape of the landscape.

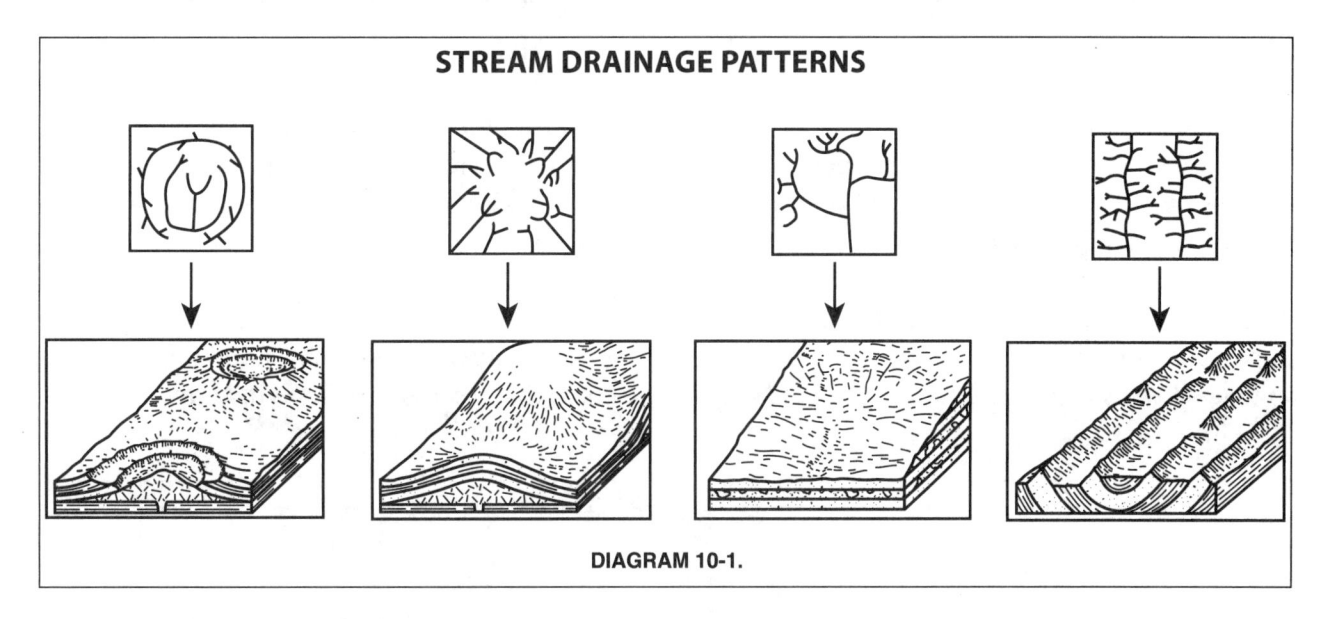

STREAM DRAINAGE PATTERNS

DIAGRAM 10-1.

171

Classification of Landscapes

There are three major landscape regions: mountains, plateaus, and plains. They differ in elevation and bedrock structure.

Landscape	Mountain (highland)	Plateau	Plain (lowland)
elevation (height above sea level)	high relief with steep slopes	high relief; if eroded may have steep slopes	low relief
rock structure	distorted rock layers that are folded, faulted, and/or tilted	horizontal rock layers	horizontal rock or sediment layers
formed by	volcanic activity; tectonic uplift; collisions of plates	uplift of sedimentary rock layers; lava flows	erosion and deposition of sediment
examples	Adirondacks, Alps, Rocky Mountains	Catskills; Colorado Plateau	Atlantic Coastal Plain; Great Plains; Erie-Ontario Lowlands
illustration			

Development of Landscapes

There are several factors that affect landscapes. The shape of the land may change due to **uplifting forces** by volcanoes and plate tectonics. These forces increase the relief of the land. **Leveling forces** such as erosion are destructive and wear away the land. A landscape is in equilibrium when rate of leveling equals the rate of uplift.

Climate is a major factor that alters the landscape. In humid climates, water is the major agent of erosion. Hill slopes will therefore be smooth and rounded. If the climate changes, so will the appearance of the landscape. In arid climates, wind is a major agent of erosion. Landscapes in arid climates have steep, rugged slopes and the bedrock has often been sculptured by wind blown sand.

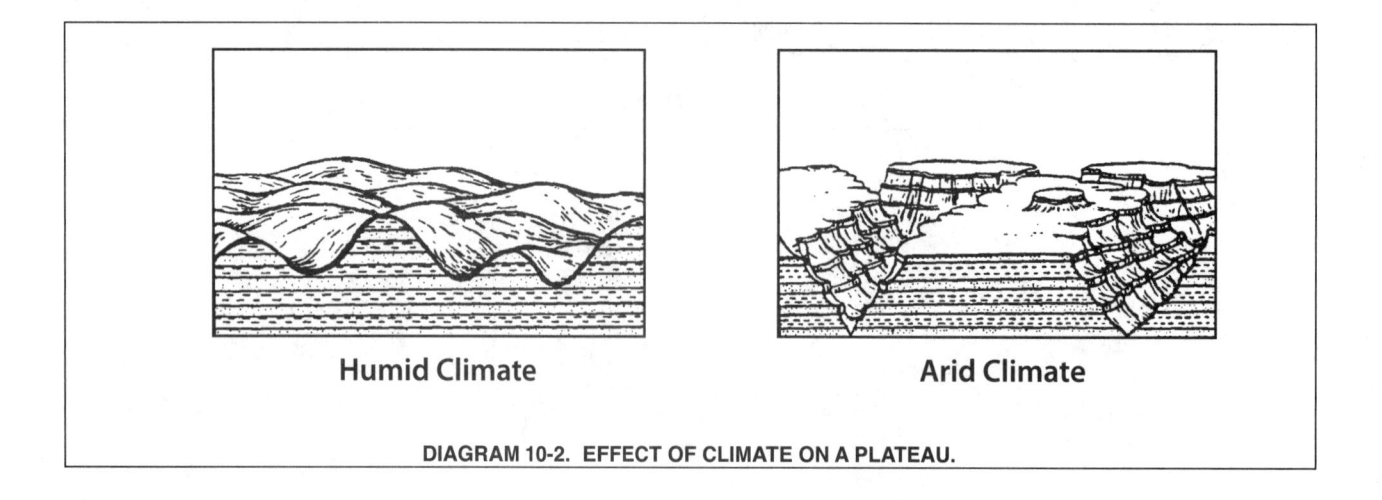

Humid Climate

Arid Climate

DIAGRAM 10-2. EFFECT OF CLIMATE ON A PLATEAU.

The type of bedrock found in an area will determine the features of the landscape. If the rock is resistant to weathering and erosion, steep slopes, ridges, and **escarpments** will form. If the rock is easily weathered and eroded then valleys form.

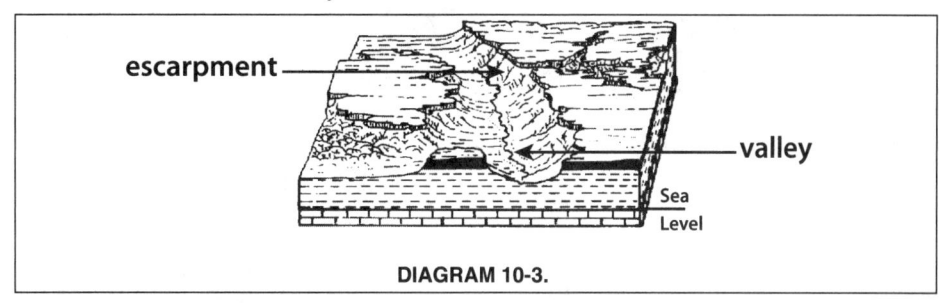

DIAGRAM 10-3.

Human activities can quickly change the landscape. Humans construct roads and buildings. We move soil, farm the land, and remove vegetation. Human activities move more sediment than rivers do!

Landscapes go through stages of development. In young landscapes, uplifting forces are dominant so there are steep slopes and high elevations. Rivers move fast in V-shaped valleys. In mature landscapes, leveling forces become more common. River valleys widen. Old landscapes are dominated by leveling forces so elevations are low and stream velocities are slow. A landscape can be rejuvenated if uplifting forces become dominant again.

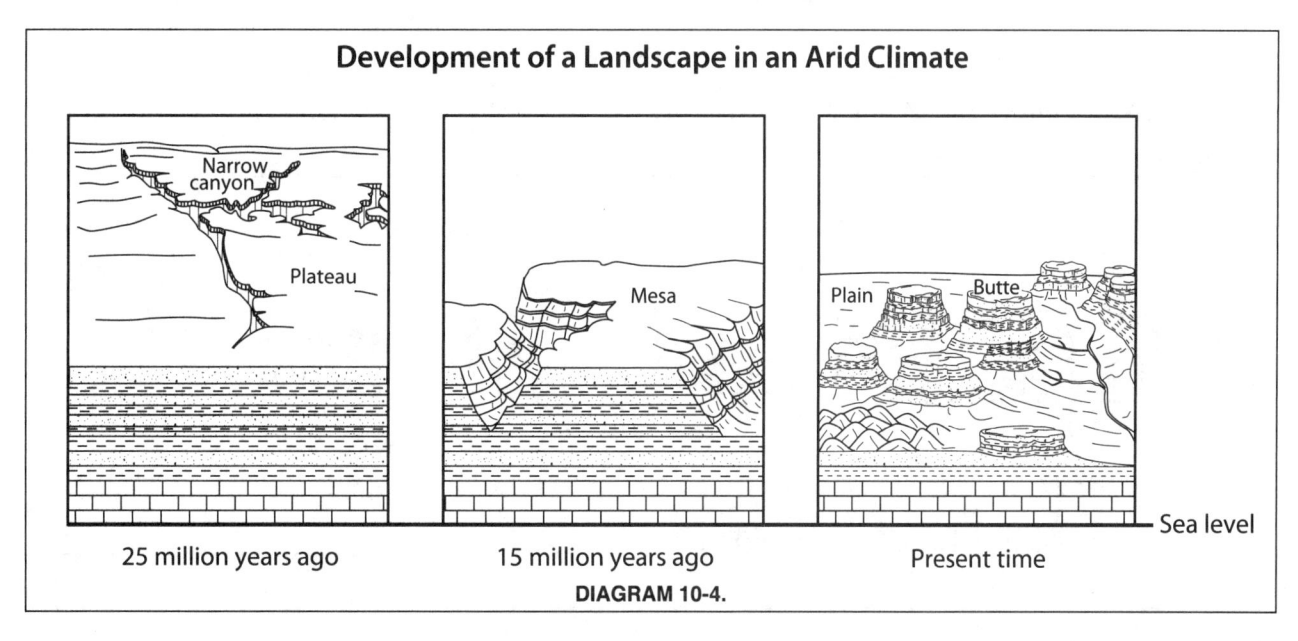

Development of a Landscape in an Arid Climate

25 million years ago | 15 million years ago | Present time

DIAGRAM 10-4.

Questions

1. For each statement select the landscape described: *(mountain) (plateau) (plain)*

    a. horizontal strata of high elevation. _____

    b. folded rock layers. _____

    c. undistorted bedrock of high elevation. _____

    d. low elevation. _____

    e. sediment layers. _____

    f. faulted, tilted strata. _____

2. For each NYS location determine the landscape type: *(mountain) (plateau) (plain)*

   a. The Catskills. _____   f. Riverhead. _____

   b. Hudson River Valley. _____   g. Massena. _____

   c. Adirondacks. _____   h. Finger Lakes. _____

   d. Mt. Marcy. _____   i. Albany. _____

   e. 43°30′ N, 75°45′ W. _____   j. Niagara Falls. _____

3. Describe the change in appearance of this landscape if the climate became more arid.

   _____

   _____

   Why would this occur? _____

4. The *Catskills* are often referred to as the *"Catskill Mountains."* As a student of Earth Science you know this is incorrect.

   a. Explain. _____

   _____

   b. Why would a person think of the Catskills as a "mountain." _____

   _____

5. On the following block diagrams of landscapes, draw arrows to show direction of water flow from the runoff of precipitation.

   a.   b.   c.

6. State the general direction that the Niagara River is flowing. _____

   Explain what you based your answer on. _____

   _____

   _____

7. The cross section below shows the general bedrock structure of an area which has three different landscape regions: **A**, **B**, and **C**.

   (Not drawn to scale)

   Correctly name the landscape type and state evidence supporting your answer.

   A = _____, evidence is _____

   B = _____, evidence is _____

   C = _____, evidence is _____

CONTOUR MAPPING

A **contour map** uses isolines to show elevations, shape of the landscape, steepness of slopes, and stream flow. **Topographic maps** are contour maps that illustrate natural and man-made features.

Contour lines are isolines which connect all the points of the same **elevation** (height above sea level). A location where the exact elevation has been measured is shown on the map by a **benchmark**, for example: ▲• 134. The **contour interval** is the change in elevation from one line to the next. The **map scale** is used to measure the distance from one point to another. **Gradient** is the slope of the land. If the contour lines are far apart, the slope will be gentle. If the lines are close together, the slope will be steep. Gradient is calculated as: $\frac{\text{change in elevation}}{\text{distance}}$ A **profile** is a side view of the landscape.

DIAGRAM 10-5.

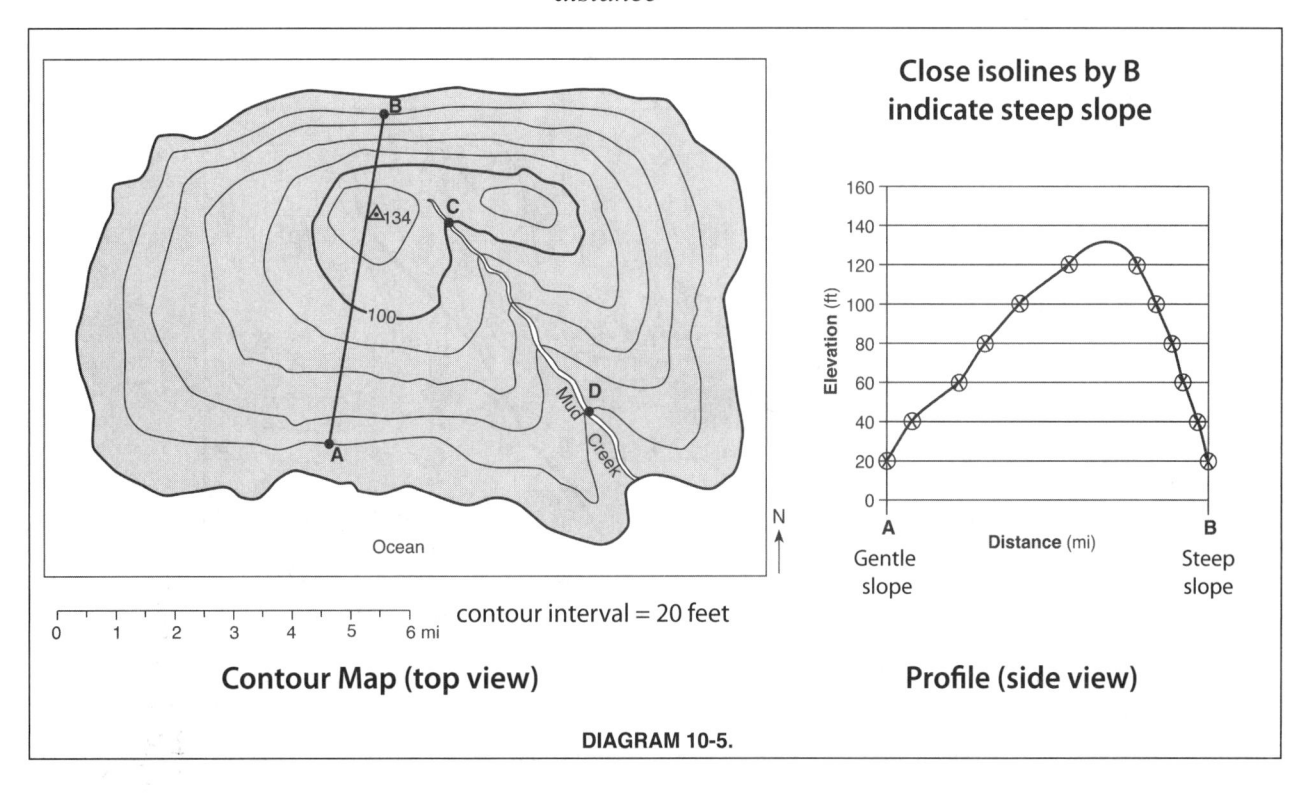

A landscape may have sinkholes or depressions where the elevation goes down. **Hachure marks** on the contour lines indicate a decrease in elevation. The elevation of the first hachure line is the same as the previous regular contour line as shown in the diagrams below.

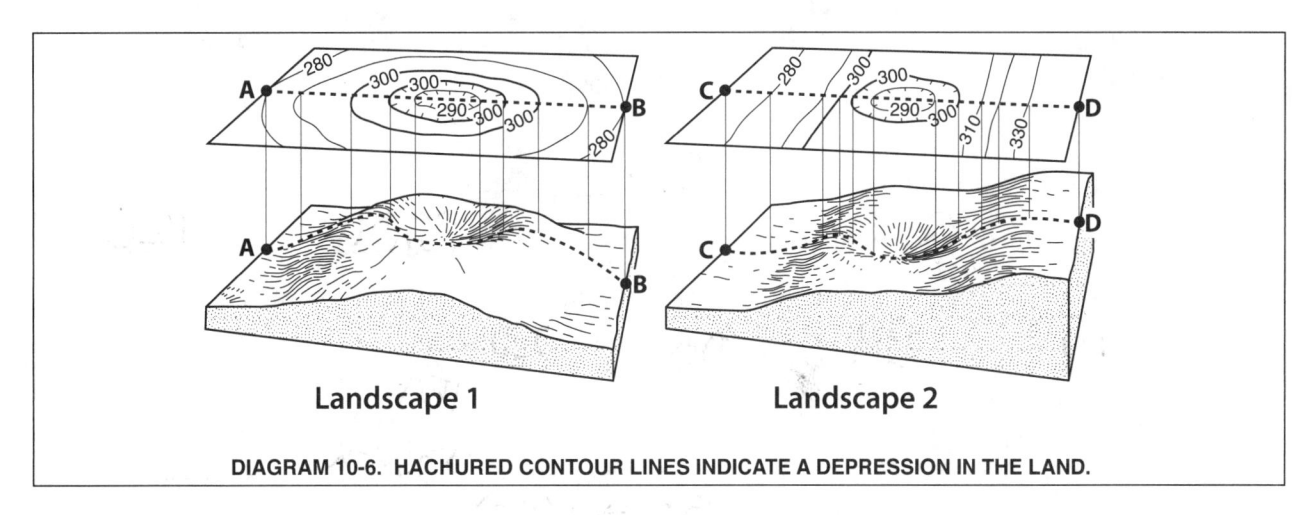

DIAGRAM 10-6. HACHURED CONTOUR LINES INDICATE A DEPRESSION IN THE LAND.

Streams flow downhill in a stream valley. When a contour line crosses a stream, the contour line will bend and point upstream. The stream will flow to lower elevations, in the opposite direction that the contour lines bend.

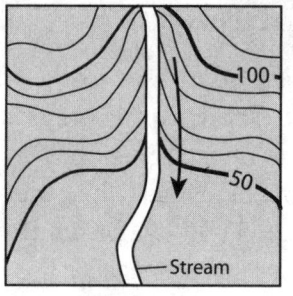

DIAGRAM 10-7.

Questions

8. Use the contour map below. Elevations are in feet.

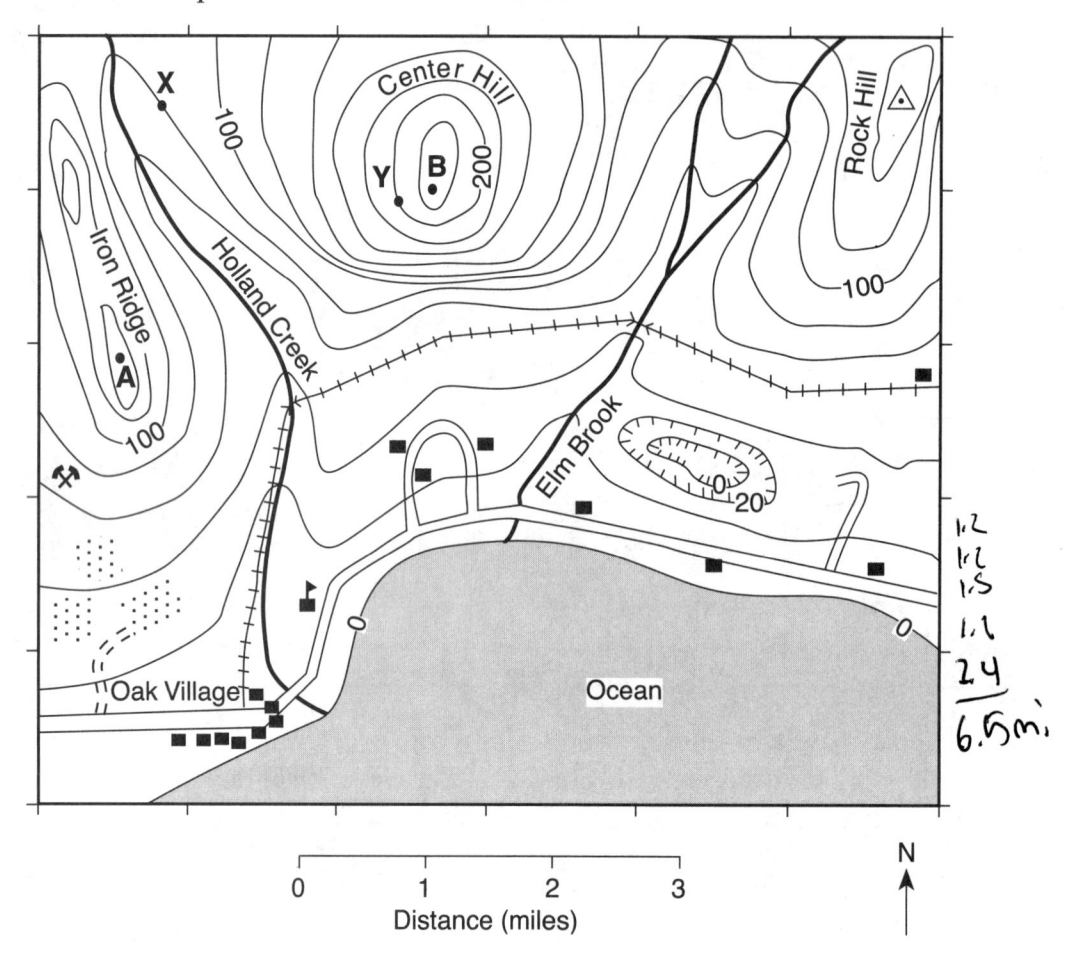

   a. The contour interval of this map is __20__ feet.
   b. Measure the distance to the nearest tenth from: A to B: __1000__; X to Y: __120.00__
   c. Estimate the elevation of the following points to the nearest whole number.
      A: __159__ B: __139__ X: __90__ Y: __200__
   d. What could be the maximum elevation of Rock Hill? __159__
   e. The steepest slopes are on what side of Center Hill? __South__
   f. Compass direction from B to A is __South west__.
   g. Measure the total distance of the railroad track shown on the map. __6.5 mi__

9. Use the contour map below. The elevations are in feet.

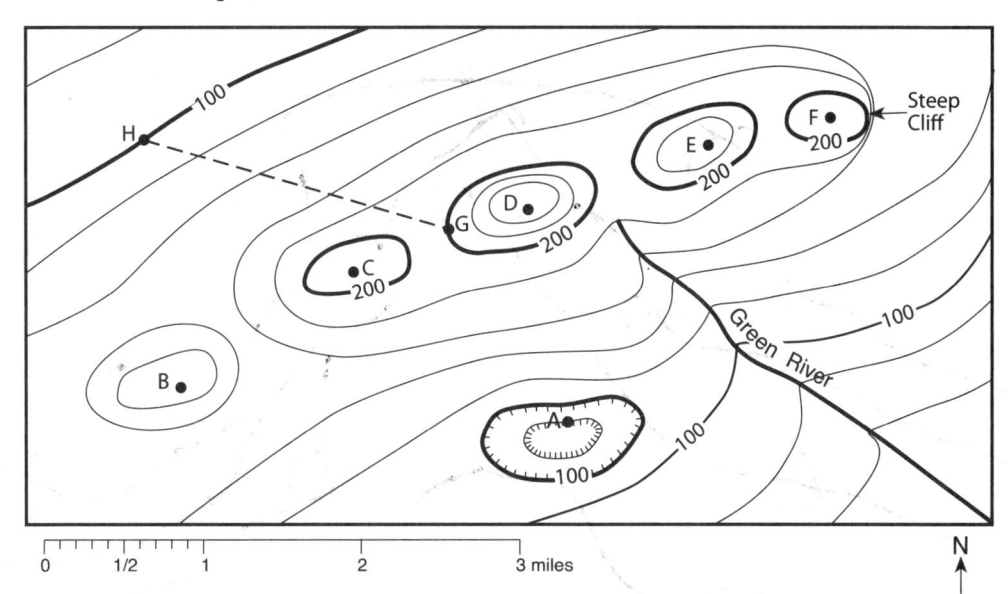

a. The contour interval of this map is __20__ feet.
b. What is the highest possible elevation for B ~~159~~ 149 ; D __259__ ; F __219__
c. The minimum elevation at the center of the depression by A could be __61__.
d. Calculate the gradient between points G and H. __25 ft/m__
   (show solution)

   $200 - 100 = \frac{100}{4} = 25$

e. Calculate the gradient of Green River from its source to the 100 foot contour line. __20 ft__
   (show solution)

   $180 - 100 = \frac{80}{4} = 20$

f. The Green River is flowing southeast. How do you know this is correct? __Contour lines bend upstream.__

177

10. The map below shows the elevation measured in feet for many points in a region. Contour lines for 100 and 120 foot elevations have been drawn.

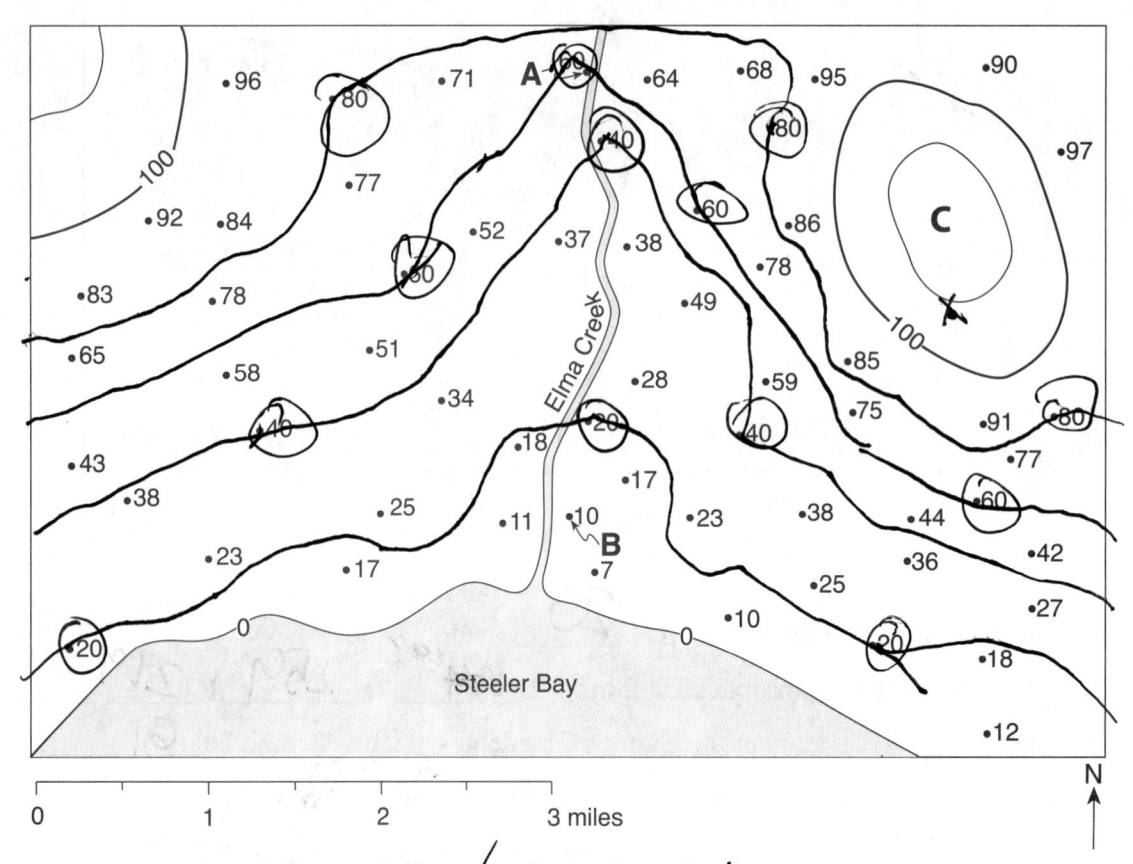

a. On the map draw the 80 foot, 60 foot, 40 foot, and 20 foot contour lines. Label each line.
b. The distance from **A** to **B** is approximately ___50___.
c. Place an "X" near a position that could have an elevation of 118 feet.
d. The maximum elevation for position **C** is ___139___.
e. Elma Creek is flowing in what direction? ___South___
f. Calculate the gradient between **A** and **B**. ___25___
   (Show solution)

To contruct a profile between two points (such as **A** and **B** on the map below) do the following:

1. label clearly the elevation of *EVERY* contour line;
2. use a piece of paper with a straight edge and line the straight edge of the paper along the bottom of the line **A-B**;
3. mark points **A** and **B** on the paper;
4. every time a contour line meets the straight edge place a "tick" mark; label the elevation of that "tick" mark;
5. place the paper with the "tick" marks along the bottom of the profile grid;
6. for each "tick" mark, plot the elevation of that position on the grid;
7. connect the points with a smooth line; if there are two adjacent points with the same elevation do not connect them with a straight line, you must look at the map to determine if the elevation was lower or higher between those two points.

11. Use the contour map below to construct a profile from **A** to **B** and from **X** to **Y**.

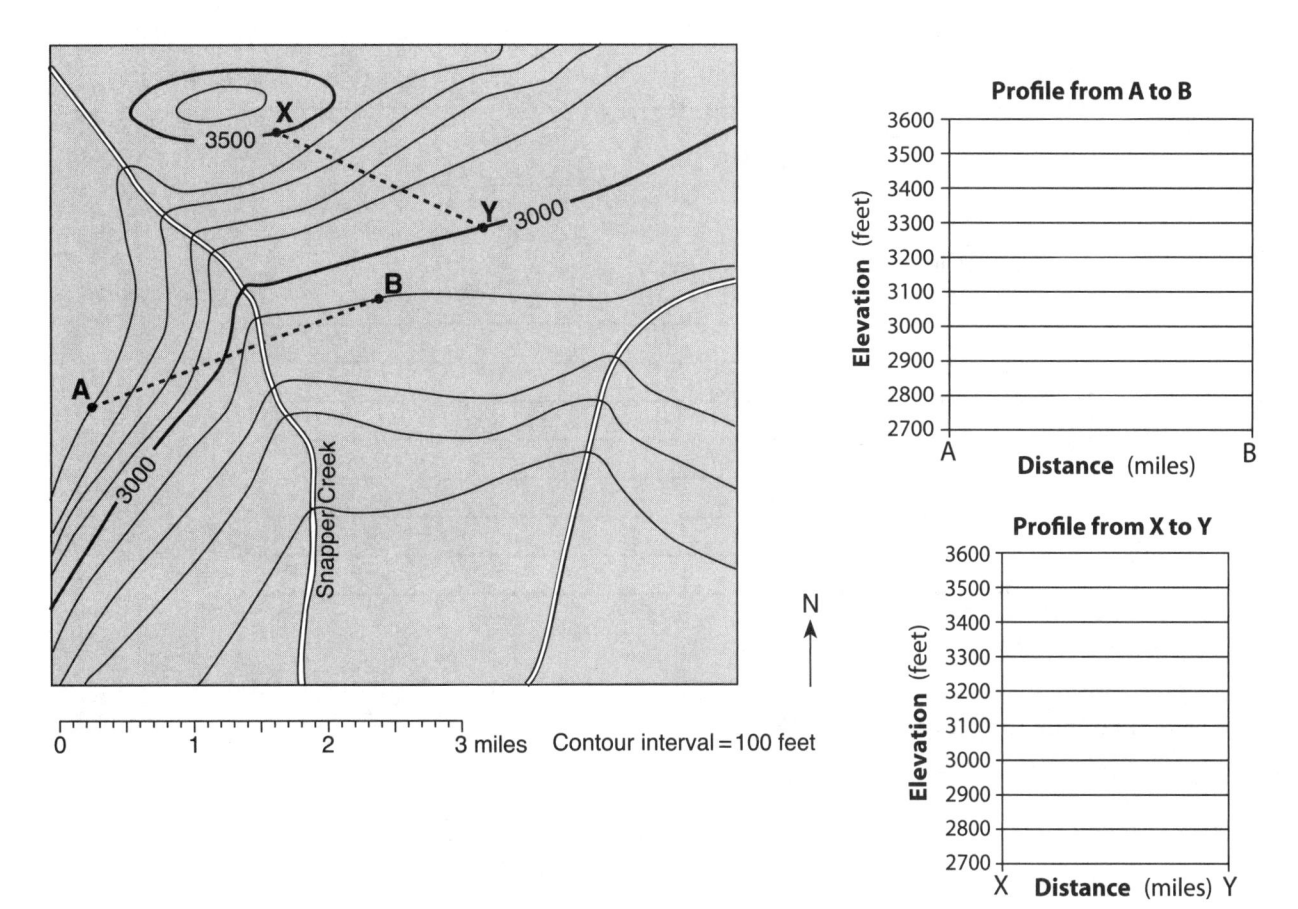

12. Use the contour map below.

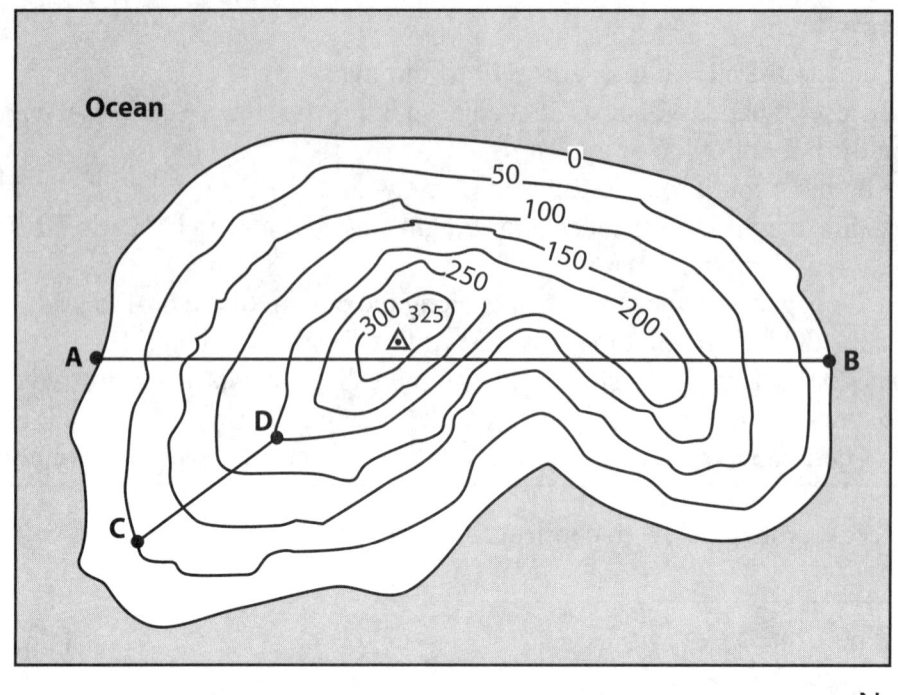

   a. Construct a profile from **A** to **B**.

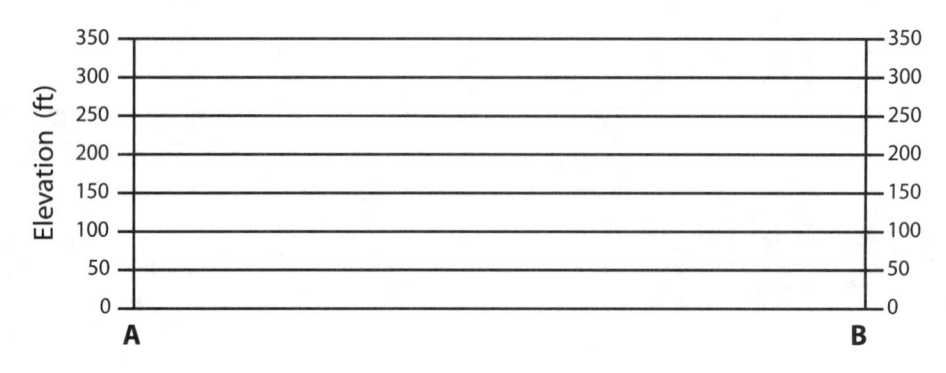

   b. Sketch the profile from **C** to **D**.

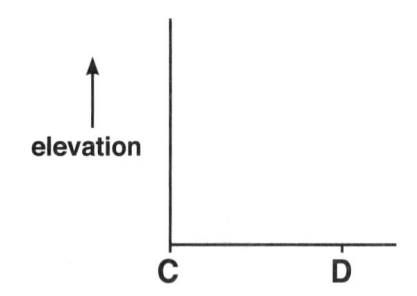

CHAPTER 10 REVIEW

1. A plane traveling in a straight line from Watertown to Utica would fly over what landscape region?
   - (1) Tug Hill Plateau
   - (2) Adirondack Mountains
   - (3) St. Lawrence Lowlands
   - (4) Champlain Lowlands

2. Which sequence shows the order in which landscapes regions are crossed as an airplane flies in a straight line from Albany to Massena, New York?
   - (1) plateau, mountain, plain
   - (2) plateau, plain, mountain
   - (3) mountain, plain, plateau
   - (4) plain, mountain, plain

3. Plattsburgh, New York is located in the
   - (1) Tug Hill Plateau
   - (2) Appalachian Plateau
   - (3) Champlain Lowlands
   - (4) Taconic Mountains

4. The Allegheny Plateau is part of the
   - (1) New England Highlands
   - (2) Interior Lowlands
   - (3) Grenville Highlands
   - (4) Appalachian Uplands

5. The Tug Hill Plateau is part of the
   - (1) Adirondacks
   - (2) Allegheny Uplands
   - (3) Hudson-Mohawk Lowlands
   - (4) Erie-Ontario Lowlands

6. The Hudson River and the Mohawk River flow primarily over what type of landscape region?
   - (1) mountains
   - (2) lowlands
   - (3) plateau
   - (4) highlands

7. Which location in New York is on a plateau?
   - (1) New York City
   - (2) Old Forge
   - (3) Jamestown
   - (4) Rochester

8. The distance in kilometers between Buffalo, NY and Albany, NY is approximately:
   - (1) 90 km
   - (2) 225 km
   - (3) 410 km
   - (4) 810 km

9. The latitude and longitude of Ithaca, NY is approximately
   - (1) 42° N, 76° W
   - (2) 42° 30' N, 76° 30' W
   - (3) 43° 30' N, 77° 30' W
   - (4) 43° N, 77° W

10. The Finger Lakes are located on a
    - (1) plain of sedimentary rocks
    - (2) plateau of sedimentary rocks
    - (3) mountain of sedimentary rocks
    - (4) plateau of metamorphic rocks

11. Which landscape region separates the Adirondacks from the Catskills?
    - (1) Hudson-Mohawk Lowlands
    - (2) Champlain Lowlands
    - (3) Taconic Mountains
    - (4) Tug Hill Plateau

12. Which city in New York is located on a landscape that has distorted rock layers?
    - (1) Niagara Falls
    - (2) Massena
    - (3) Syracuse
    - (4) Old Forge

13. Which factor is most influential in determining the rate of landscape change in a location?
    - (1) prevailing winds
    - (2) sediment size
    - (3) age of bedrock
    - (4) climate

14. Which factor has the *least* influence on the development of a landscape?
    - (1) uplifting forces
    - (2) average rainfall
    - (3) type of bedrock
    - (4) geologic age of the bedrock

15. If the rate of erosion increases while the rate of uplift remains constant, the elevations in the landscape will
    - (1) decrease
    - (2) increase
    - (3) remain the same

16. Continents are divided into different landscape regions based on
    (1) soil composition and sediments
    (2) elevation and bedrock structure
    (3) latitude and climate
    (4) rainfall and temperature

17. Landscapes can be changed the most in the shortest amount of time by
    (1) rivers
    (2) glaciers
    (3) human activities
    (4) wind erosion

18. The chart below describes the relief and bedrock of three different landscape regions.

Landscape Region	Relief	Bedrock
X	Great relief, high peaks, deep valleys	Many types, including igneous and metamorphic rocks, nonhorizontal structure
Y	Moderate to high relief	Flat layers of sedimentary rock or lava flows
Z	Very little relief, low elevations	Many types and structures

Which terms, when substituted for **X**, **Y**, and **Z**, best complete the chart?
(1) X = mountain, Y = plain, Z = plateau
(2) X = plateau, Y = mountain, Z = plain
(3) X = plain, Y = plateau, Z = mountain
(4) X = mountain, Y = plateau, Z = plain

19-20. Use the diagrams below of the topography and bedrock structure of two regions.

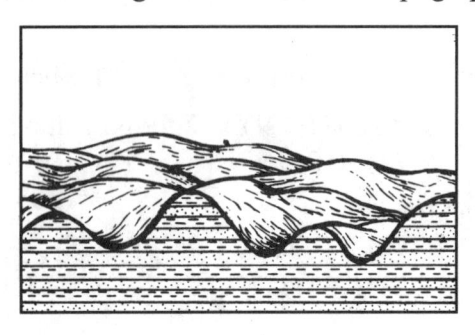

19. These landscapes are both classified as a
    (1) plain   (2) plateau   (3) mountain   (4) lowland

20. Which factor caused the difference in hillslopes in these two regions?
    (1) type of rock
    (2) seismic activity
    (3) time
    (4) climate

21. The diagrams **A**, **B**, and **C** show three different stream drainage patterns.

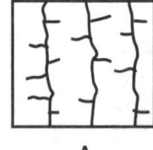

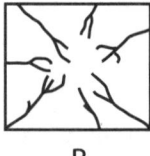

Which factor is primarily responsible for causing these three different drainage patterns?
(1) amount of precipitation
(2) bedrock structure
(3) stream discharge
(4) prevailing winds

22. The diagram illustrates a volcano.

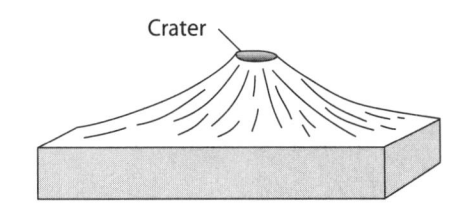

Which diagram shows the stream drainage pattern that most likely formed on the surface of the volcano?

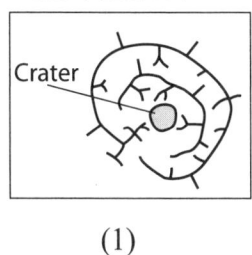

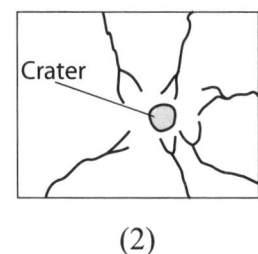

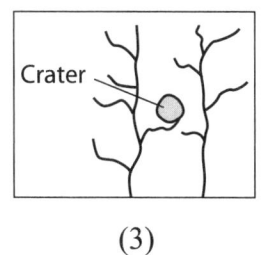

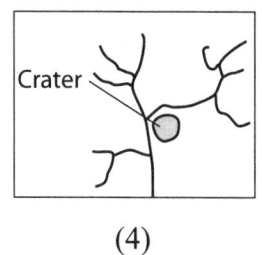

(1)   (2)   (3)   (4)

23. The block diagram shows a portion of Earth's crust.

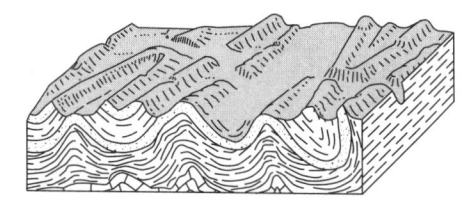

Which stream drainage pattern will most likely occur on this crustal surface?

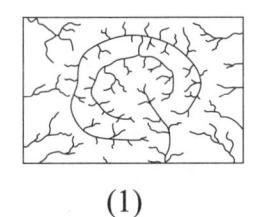

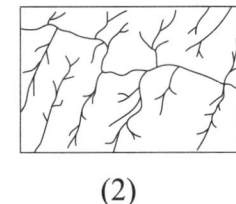

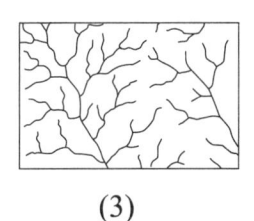

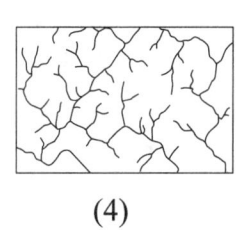

(1)   (2)   (3)   (4)

24-25. The map below shows Jennifer Brook crossing several contour lines and passing points **A** and **B**. Elevation is measured in meters.

24. The approximate gradient between points **A** and **B** is:
    (1) 10 m/km   (3) 40 m/km
    (2) 20 m/km   (4) 80 m/km

25. What is the general direction of streamflow?
    (1) northwest   (3) northeast
    (2) southeast   (4) southwest

26. What is the highest possible elevation for the hill?
    (1) 99 meters
    (2) 189 meters
    (3) 199 meters
    (4) 200 meters

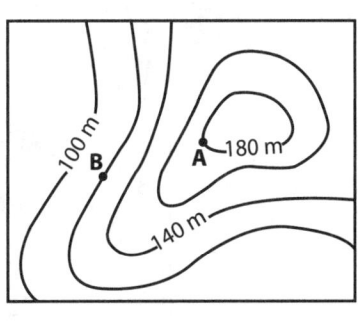

183

27. The contour map below shows elevations recorded in meters.

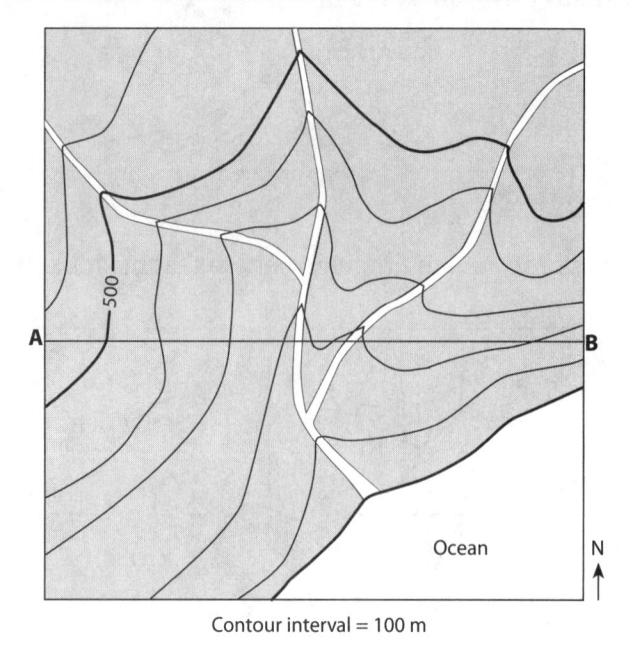

Contour interval = 100 m

Which graph best represents the profile from point **A** to point **B**?

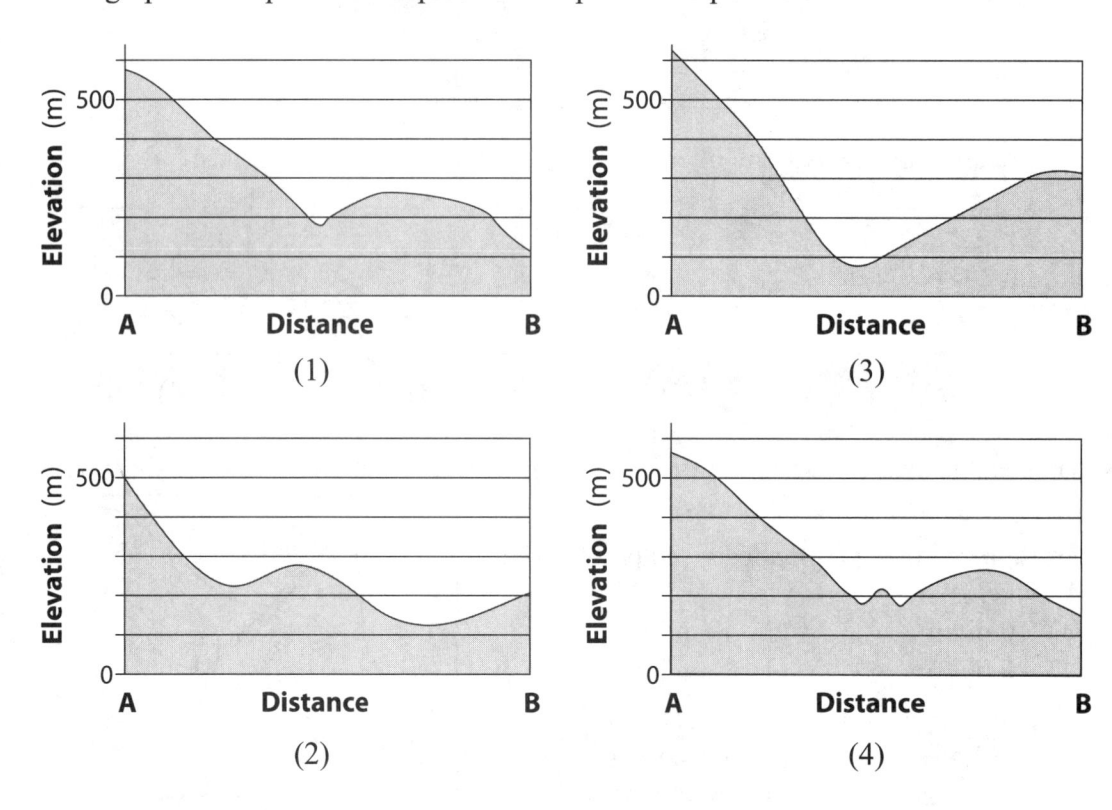

(1)

(2)

(3)

(4)

28. The elevation of the highest contour line on this map is:

(1) 10,500 feet
(2) 10,700 feet
(3) 10.788 feet
(4) 10,800 feet

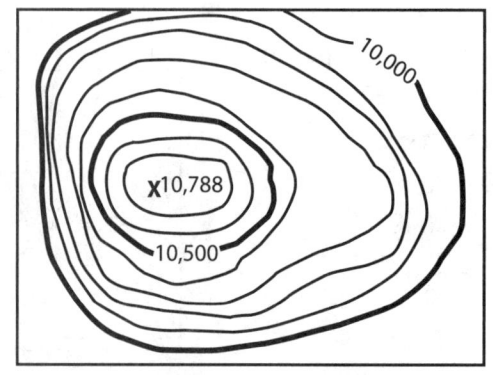

29. Which side of Amethyst Hill is the steepest?
    (1) north
    (2) south
    (3) east
    (4) west

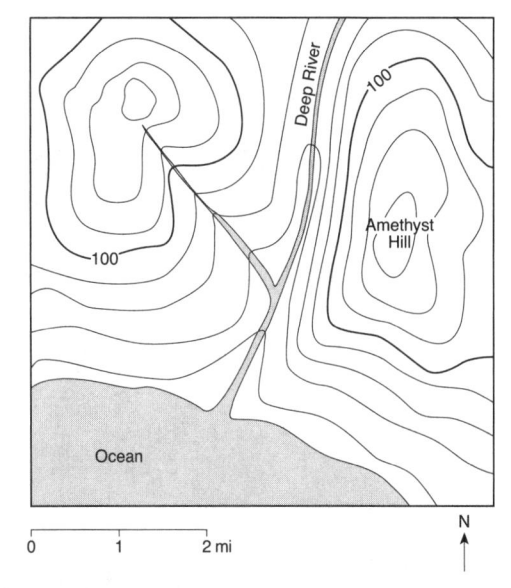

30. What is the contour interval on this map?
    (1) 5 feet   (3) 20 feet
    (2) 10 feet   (4) 25 feet

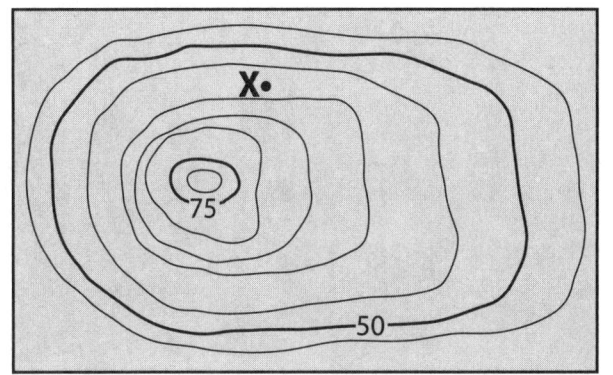

31-33. The block diagram below represents a portion of the Grand Canyon.

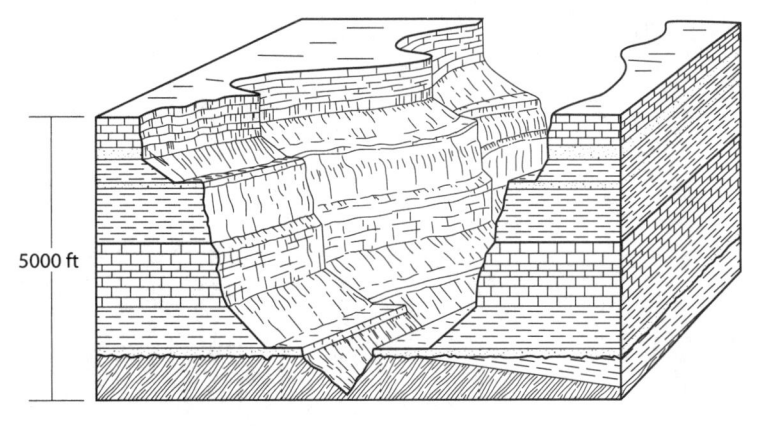

31. This landscape is best classified as a _____.

32. Give *two* characteristics of this landscape region shown in the diagram that supports your answer in #31.

    evidence #1: _____

    evidence #2: _____

33. If the climate in this area became more humid, how would the landscape change?

    _____

185

34. The diagram below shows an unfinished contour map.
Carefully complete the 250 meter and the 500 meter contour lines.

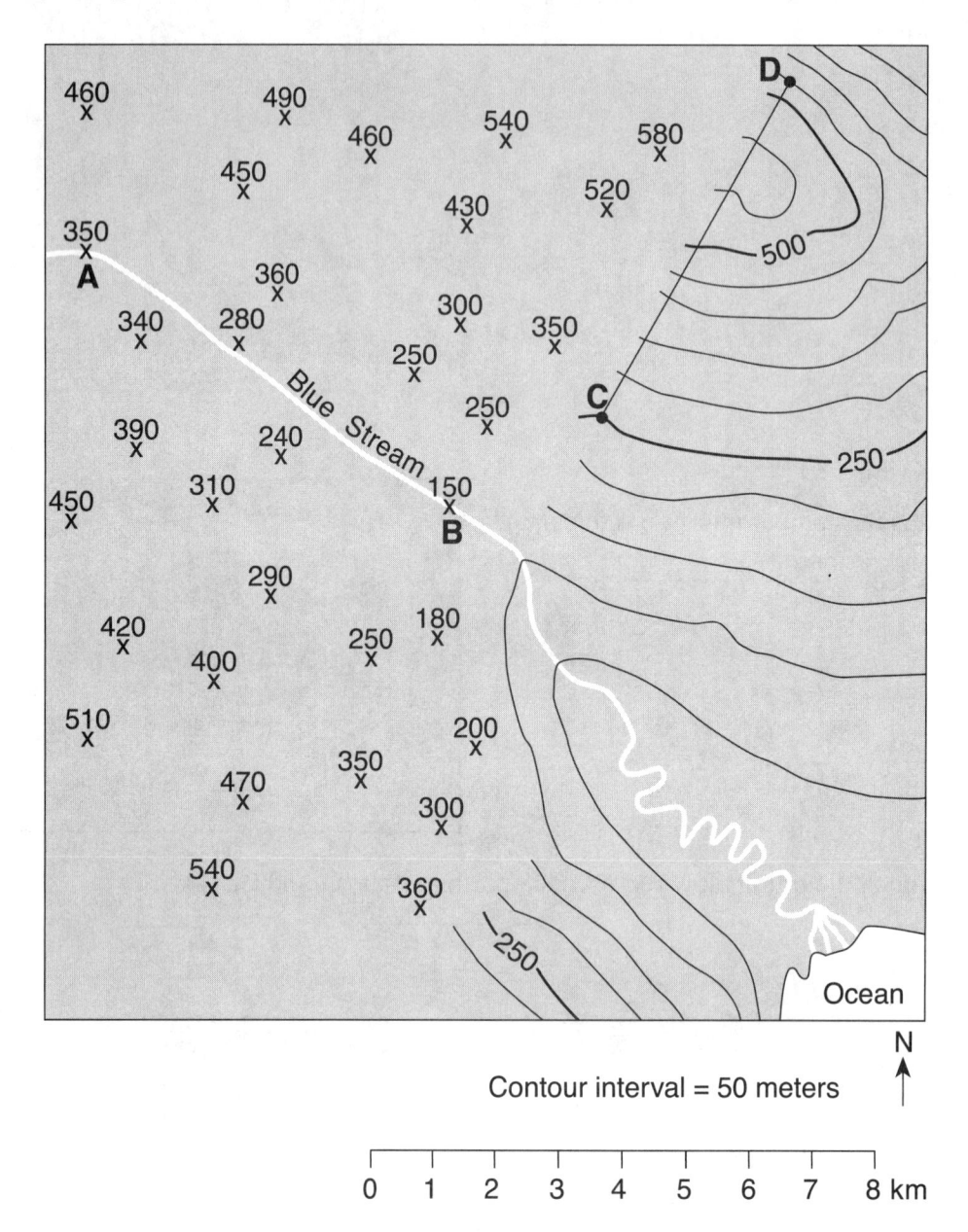

Contour interval = 50 meters

35-38. Use the topographic map shown below. Elevations are in meters.

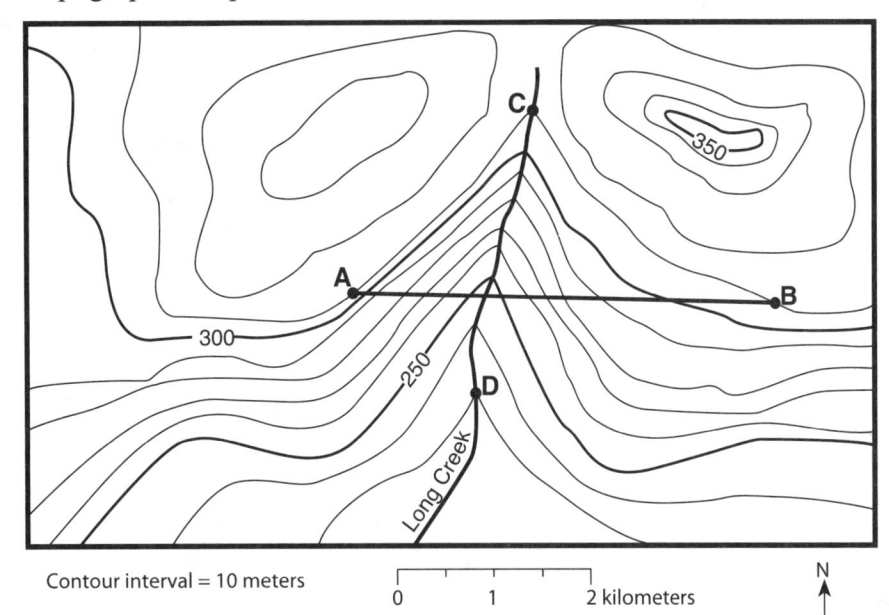

35. On the grid below construct a profile along line **AB**.

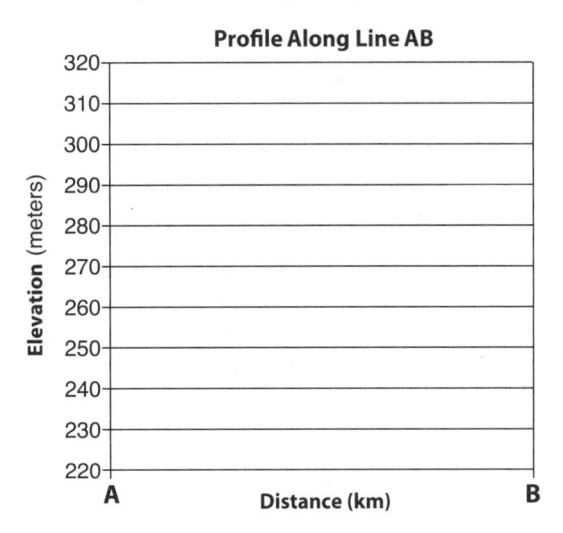

36. Calculate the gradient of Long Creek between points **C** and **D**. _____
(show solution)

37. Explain how the contour lines on the map indicate that Long Creek flows over steep slopes.
_____

38. State the maximum elevation to the nearest whole number of the hill east of **C**. _____

187

39-45. Use the contour map below.

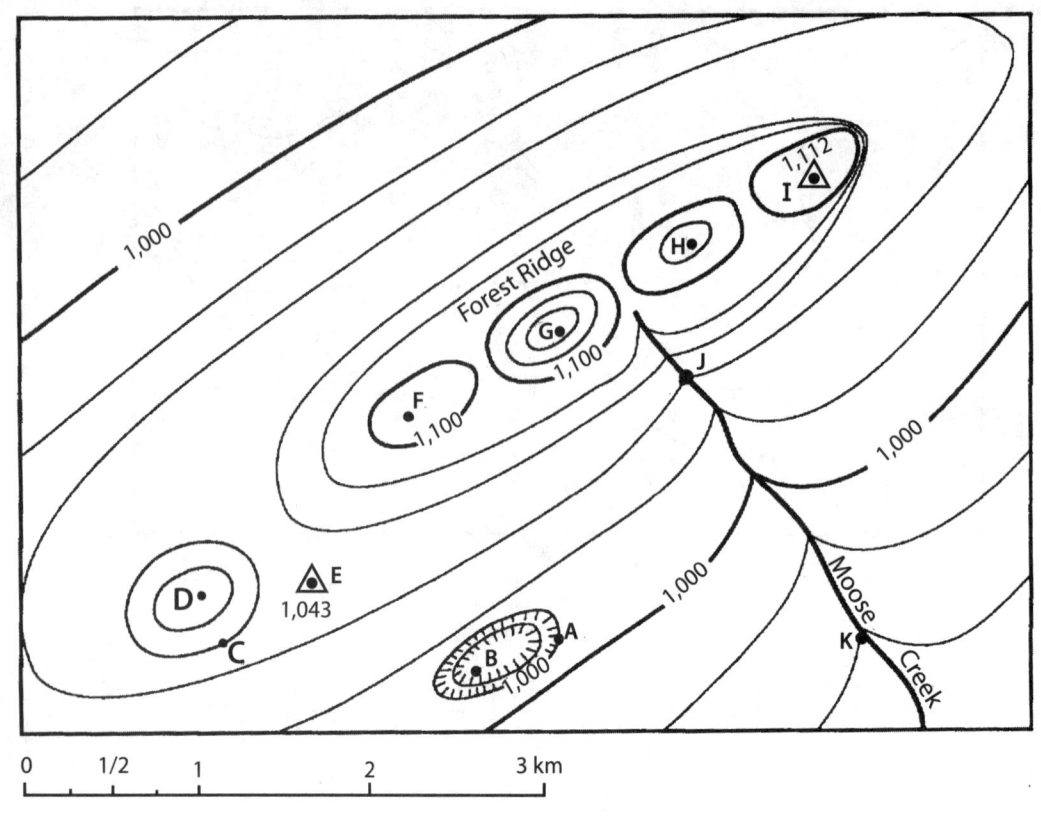

Contour interval = 20 meters
⊥⊥⊥⊥⊥ Hachure lines show depression

39. Which lettered hilltop could have an elevation of 1,145 meters? _____

40. Towards which direction does Moose Creek flow? _____

Explain how you know this. _____

41. What is the *lowest* possible elevation of point **B**? _____

42. What is represented at the symbol by location **E**? _____
_____

43. What is the maximum elevation of hill **F**? _____

44. Calculate the gradient from **J** to **K**. _____
(show solution)

45. State the elevation of point **C**. _____

CHAPTER 11

NATURAL RESOURCES, MINERALS, AND ROCKS
Earth's Natural Resources, Minerals, Rock Cycle, Types of Rocks

EARTH'S NATURAL RESOURCES

Natural resources are the parts of Earth that we use. These resources include rocks that may be used for buildings and roadways, and minerals used in electronics, glass, and jewelry. Soil, sand, water, air, and forests are also natural resources. Fossil fuels such as oil, coal, and natural gas are a valuable natural resource. They formed millions of years ago from the decay of plant matter and microscopic animals.

Earth's resources are classified as renewable or non-renewable. **Renewable** resources can be replaced by nature in our lifetime. Trees, water, and soil are examples of renewable resources. **Non-renewable** resources took millions of years to form and therefore cannot be replaced in our lifetimes, there is a limited supply. Fossil fuels, rocks, minerals, and metals are examples of non-renewable resources.

We depend on Earth's natural resources for our quality of life and global economy. We must conserve resources by using them wisely. We can recycle or reuse them in a different way. We can use alternative materials which are recyclable or renewable; for example, use paper instead of plastic. Alternative energy sources such as solar, wind, geothermal, and hydroelectric are renewable and can replace some of the fossil fuels we currently use.

Questions

1. For each statement, classify the resource as (*renewable*) or (*non-renewable*).

   a. soil. _____   **e.** water. _____

   b. minerals. _____   **f.** coal. _____

   c. sunlight. _____   **g.** trees. _____

   d. aluminum. _____   **h.** oil. _____

2. New York State has many natural resources. For each location state a natural resource that may be mined from that location.

   a. Riverhead. _____

   b. Old Forge. _____

   c. Syracuse. _____
   _____

3. The chart below lists some of the mineral resources we use and the number of years that the resource is estimated to last at our current rate of usage.

Mineral Resources' Future

Mineral Resource	Estimated Supply Time
Salt, magnesium metal	almost infinite
Lime, silicon	thousands of years
Potash, cobalt	200+ years
Manganese ore	200+ years
Iron ore	100 to 200 years
Chromite, feldspar	100 to 200 years
Bauxite (aluminum ore)	50 to 100 years
Phosphate rock, nickel	50 to 100 years
Copper, mercury	less than 50 years
Zinc, lead	less than 50 years

a. Explain why these resources are classified as non-renewable. _____

b. State *one* way that humans can increase the estimated supply time for these resources.

_____

c. Name a mineral which is an ore of iron. _____

4. The data table below shows the uses of salt in the United States.

Uses of Salt in the United States

Salt Usage	Percent	How Used
Water softening	9	Sodium ions from salt replace calcium and magnesium ions in water.
Highways	69	Salt keeps highways free of ice in the winter.
Agriculture	6	Salt is provided for livestock and poultry to balance their diet.
Foods	5	Humans use salt in their diet.
Industry	11	Many industrial processes, such as papermaking, use salt.

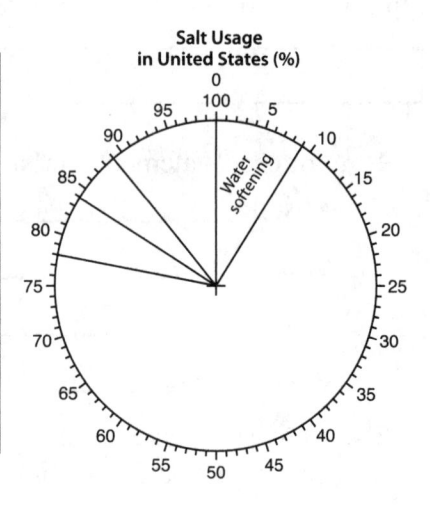

a. Name a mineral which is a source of salt. _____

b. Complete the pie graph for each use of salt. Label each area. *One example is done for you.*

c. For driving safety, icy highways are salted in the winter. Yet this salt is having a negative impact on the environment. Describe one negative effect on the environment from the use of salt on highways. _____

_____

_____

Minerals

Minerals are solid, inorganic compounds made of one or more elements. Minerals form from the cooling and solidification of hot, liquid magma, from the precipitation of dissolved compounds out of water, or the recrystalization of pre-existing minerals.

In Earth's crust, silicon and oxygen are the most common elements by mass. Therefore most minerals are silicates. The silicate compound is made of four oxygen atoms and one silicon atom. This forms the silicon-oxygen tetrahedron, SiO_4.

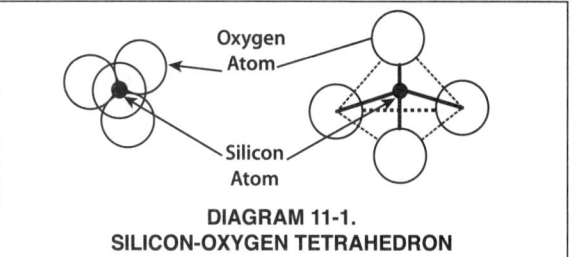

DIAGRAM 11-1. SILICON-OXYGEN TETRAHEDRON

Every mineral has unique physical and chemical characteristics that can be used to identify the mineral. Many of these physical properties are caused by the arrangement of the atoms in the mineral.

mineral property	description
color	physical appearance that is not the most reliable; many minerals have the same color; weathering and impurities alter color
streak	color of the powdered mineral; tested by using a porcelain streak plate (see diagram below)
luster	the reflection of light off the mineral; a metalic luster appears shiny like a metal, a non-metallic luster is dull
hardness	resistance to being scratched by another mineral or common object; a numerical value based on Moh's scale (see chart below)
cleavage	when a mineral breaks along even, smooth surfaces (see diagram below)
fracture	when a mineral breaks unevenly
crystal shape	a definite geometric shape
reaction with acid	bubbling when tested with hydrochloric acid
density	concentration of mass in a unit of volume
other properties that may be used	smell, double refraction, greasy feel

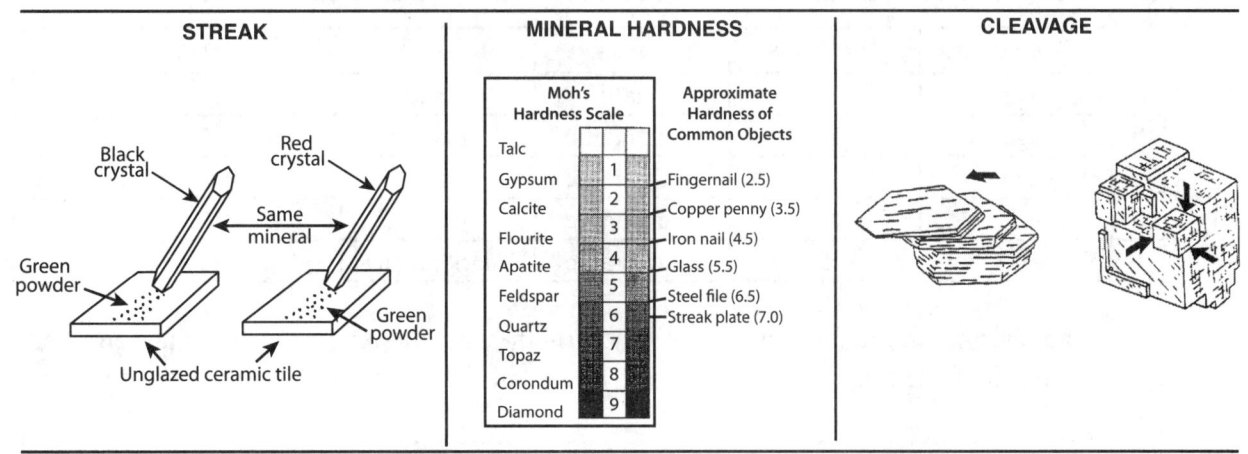

DIAGRAM 11-2. SOME PROPERTIES OF MINERALS.

191

Questions

5. The diagrams below illustrate some of the physical properties of minerals. Name the property shown.

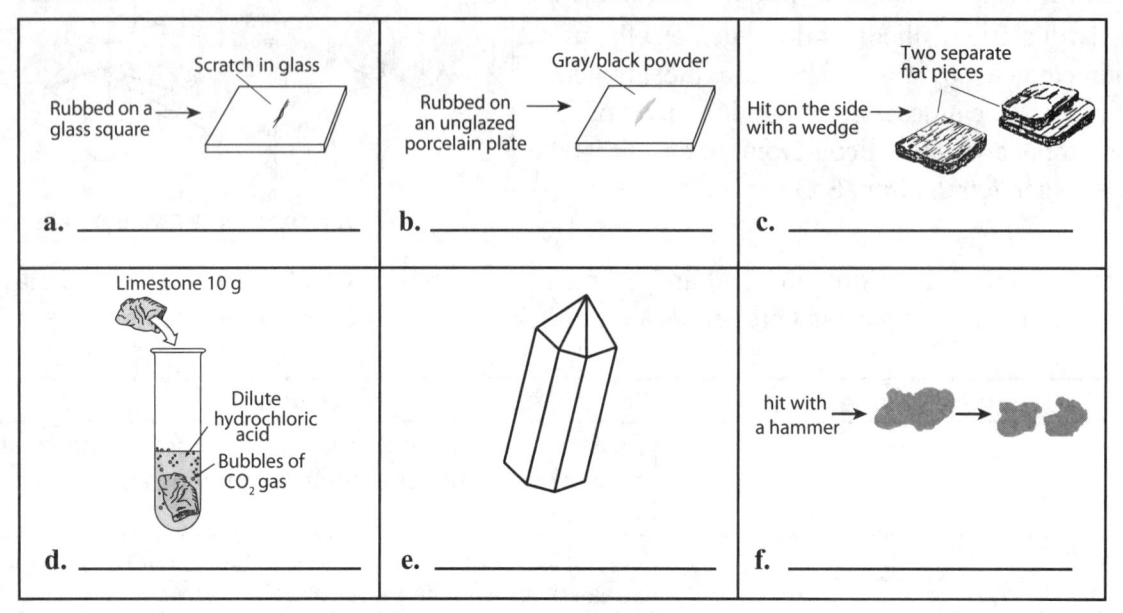

Scratch in glass — Rubbed on a glass square	Gray/black powder — Rubbed on an unglazed porcelain plate	Two separate flat pieces — Hit on the side with a wedge
a. _____	b. _____	c. _____
Limestone 10 g — Dilute hydrochloric acid — Bubbles of CO₂ gas	(crystal shape)	hit with a hammer
d. _____	e. _____	f. _____

6. For each description of physical and/or chemical properties, name the mineral.

   a. fracture, hardness of 7. _____
   b. dark red, can scratch glass. _____
   c. composed of calcium and fluorine. _____
   d. very soft with a greasy feel. _____
   e. a dense cube that can be scratched by a fingernail. _____
   f. cleavage, bubbles with acid. _____
   g. metallic luster, black streak. _____
   h. used as a food additive. _____
   i. red-brown streak. _____

7. The table below provides information about four minerals. Name the mineral.

	Breakage	Hardness	Luster	Color	Mineral
a	cleavage	2.5	metallic	silver	
b	cleavage	2.5	nonmetallic	black	
c	cleavage	3	nonmetallic	colorless	
d	fracture	6.5	nonmetallic	green	
e	cleavage	5	nonmetallic	dark green	

8. The minerals graphite and diamond are made of the same compound. Yet diamond has a hardness of 10, while graphite has a hardness of 2. The difference between these two minerals is caused by _____.

9. A mineral can not be scratched by a copper penny but can be scratched by an iron nail. Its hardness is _____.

10. Construct a bar graph that compares the hardness of the five minerals listed.

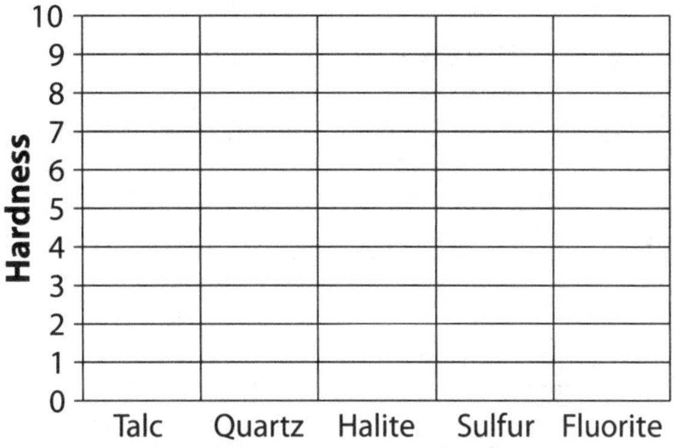
Hardness
10
9
8
7
6
5
4
3
2
1
0
Talc
Quartz
Halite
Sulfur
Fluorite

Rock Cycle

Rocks are solids that are made of minerals. Rocks are classified into three categories based on their origin.

Rock Type	Method of Formation
Igneous Rock	cooling and solidification of hot, molten rock material
Sedimentary Rock	compaction and/or cementation of sediment and/or organic material
Metamorphic Rock	pre-existing rock is changed by intense heat and/or pressure

The **rock cycle** shows how each type of rock is formed and how one rock type can change to another rock type.

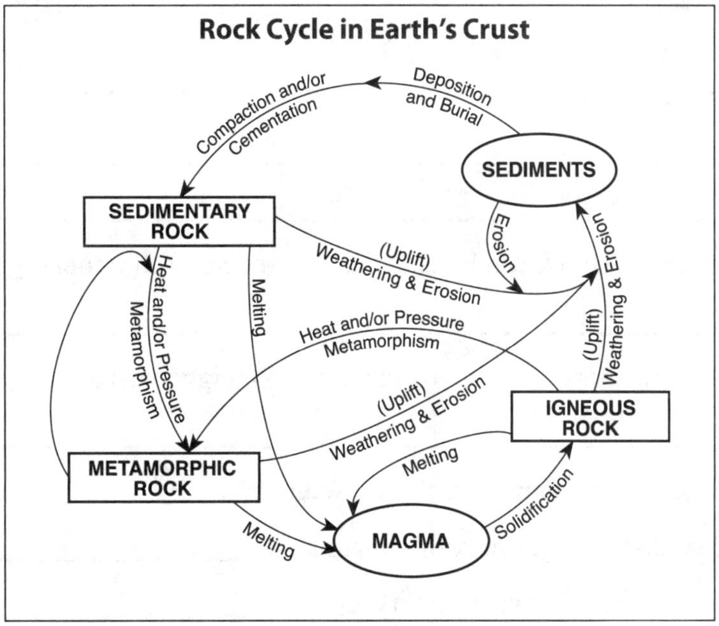
Rock Cycle in Earth's Crust
Compaction and/or Cementation
Deposition and Burial
SEDIMENTS
SEDIMENTARY ROCK
Erosion
(Uplift) Weathering & Erosion
Heat and/or Pressure Metamorphism
Melting
Heat and/or Pressure Metamorphism
(Uplift) Weathering & Erosion
(Uplift) Weathering & Erosion
IGNEOUS ROCK
METAMORPHIC ROCK
Melting
Melting
MAGMA
Solidification

DIAGRAM 11-3.

Questions

11. For each process, name the type of rock that will result.

    a. deposition of sediments. _____

    b. intense heat and pressure. _____

    c. lava cools and hardens. _____

    d. precipitation and deposition of minerals out of seawater. _____

    e. melting and solidification. _____

    f. erosion and deposition. _____

    g. burial and compaction. _____

12. The diagram to the right shows the geologic processes that act continuously on Earth to form different rock types.

    Name the rock type for each numbered circle.

    Rock type **1** = _____

    Rock type **2** = _____

    Rock type **3** = _____

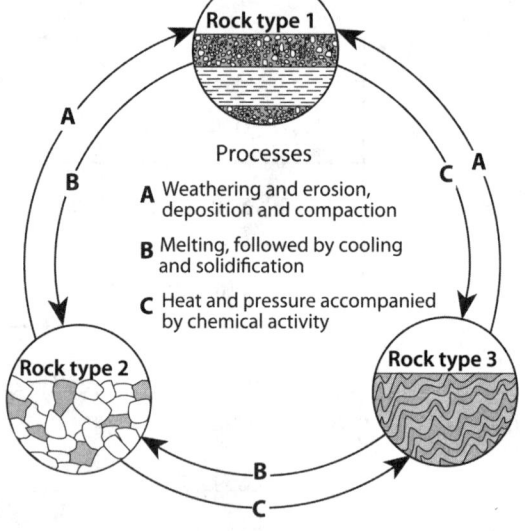

    Processes
    A Weathering and erosion, deposition and compaction
    B Melting, followed by cooling and solidification
    C Heat and pressure accompanied by chemical activity

13. Explain how an igneous rock can be changed to a sedimentary rock. _____
    _____
    _____

14. Explain how an igneous rock can be changed to a metamorphic rock. _____
    _____

15. Explain how a sedimentary rock can be changed to an igneous rock. _____
    _____

16. Name *two* processes that change a rock into sediments. _____ and _____

17. Name the process that changes rock into magma. _____

18. Igneous rocks can form from what rock type(s)? _____

19. Metamorphic rocks form from intense heat and pressure within Earth.

    Does melting form a metamorphic rock? _____ Explain. _____
    _____

ROCK TYPE: IGNEOUS ROCK

Igneous rocks make up about 66% of the rock on Earth. **Igneous rocks** form when hot, molten (liquid) rock material cools and hardens. Molten rock can be found inside Earth as **magma**, which is intrusive (plutonic). Molten rock can also be found on the surface as **lava**, which is extrusive (volcanic). When molten rock cools and solidifies it crystallizes into different minerals, therefore most igneous rocks are **polyminerallic** with intergrown minerals crystals.

The **texture** of an igneous rock refers to the size of the mineral crystals. Texture is determined by the rate of cooling, which depends on where the magma or lava cooled. On Earth's surface, lava will cool very quickly causing a **glassy texture** because crystals did not have time to form. If the lava hardened while steam and gases were bubbling out of it, the igneous rock will have a **vesicular texture** with empty gas pockets. Extrusive igneous rocks have a **fine texture** that is a result of fast cooling, so that mineral crystals did not have time to grow in size.

Intrusive igneous rocks have a **coarse texture**. Magma cools very slowly deep inside Earth, so mineral crystals will have time to form and increase in size. As the cooling time increases, the size of the mineral crystals increases.

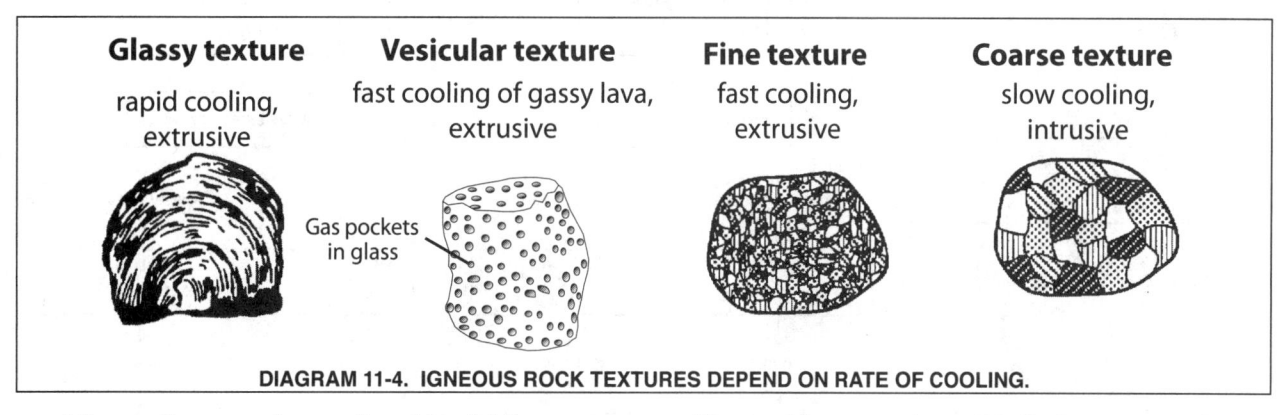

Glassy texture — rapid cooling, extrusive

Vesicular texture — fast cooling of gassy lava, extrusive (Gas pockets in glass)

Fine texture — fast cooling, extrusive

Coarse texture — slow cooling, intrusive

DIAGRAM 11-4. IGNEOUS ROCK TEXTURES DEPEND ON RATE OF COOLING.

Magma/lava can be **mafic** with a high percentage of iron and magnesium. Mafic igneous rocks are dense and dark in color. Magma/lava with a high percentage of aluminum and silicon is **felsic**. Felsic igneous rocks have a low density and are light in color.

Igneous rocks are inorganic. They contain no plant or animal matter and no fossils. Any organic matter or fossils would have been destroyed by the hot liquid magma. The minerals are intergrown (interlocking). Igneous rocks are classified by mineral composition and texture.

Questions

20. Complete the chart to compare three different igneous rocks.

	Basalt	Granite	Rhyolite
texture			
crystal size in mm			
fast or slow cooling?			
environment of formation			
color			
density			
felsic or mafic?			
mineral composition			

21. Dunite is (*polyminerallic*) (*monominerallic*) because _____

22. Texture of pumice is _____ and _____

23. Pegmatite formed by very (*rapid*) (*slow*) cooling of molten rock material.

24. Name the mineral present in the largest quantities in andesite. _____.

25. a. A non-vesicular, coarse, felsic igneous rock is _____.

    b. Igneous rock that is only composed of pyroxene and olivine is _____.

26. Complete parts **A**, **B**, **C**, and **D** of this chart for igneous rocks.

    A = _____

    B = _____

    C = _____

    D = _____

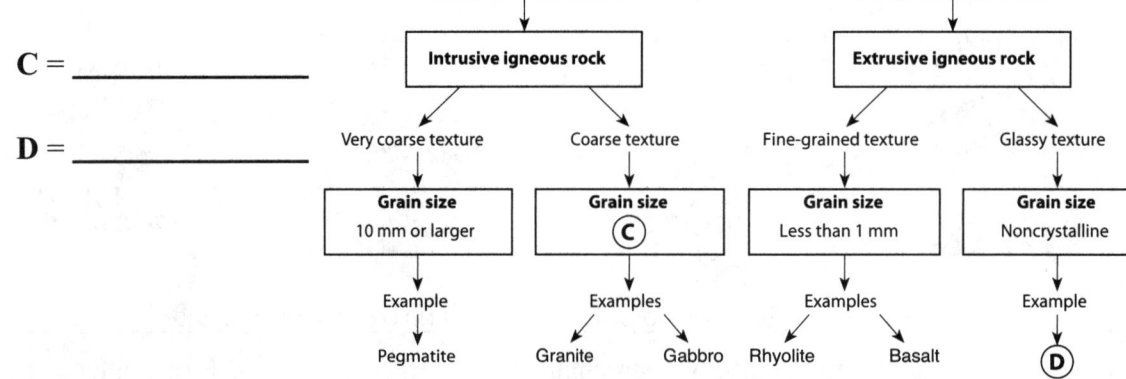

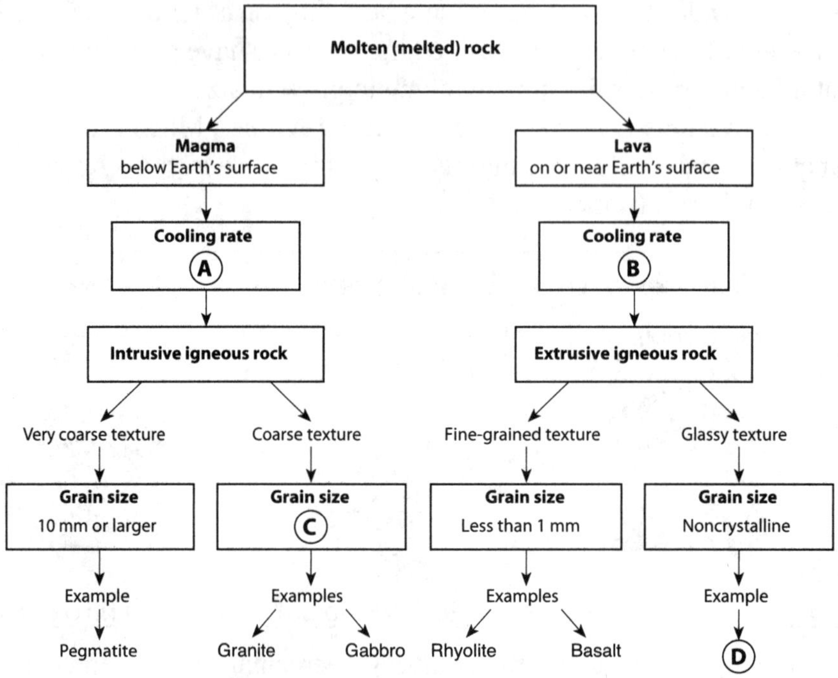

Rock Type: Sedimentary Rock

Sedimentary rocks make up about 8% of Earth's rocks and are commonly found on the Earth's surface. **Sedimentary rocks** are made of solid sediments, which may be organic or inorganic. The rock sediment forms from weathering and erosion. Most sedimentary rock forms on Earth's surface near or in water areas since water is the dominant agent of erosion on Earth. The sediments are compressed, compacted, and/or cemented together.

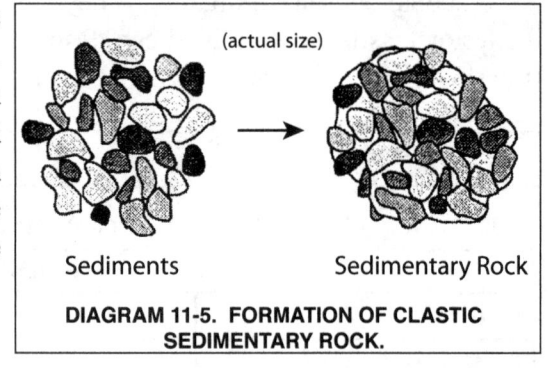

DIAGRAM 11-5. FORMATION OF CLASTIC SEDIMENTARY ROCK.

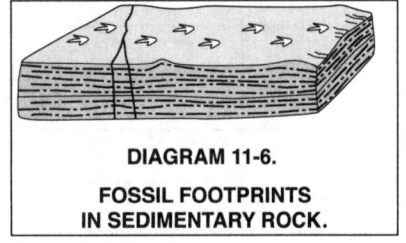

DIAGRAM 11-6. FOSSIL FOOTPRINTS IN SEDIMENTARY ROCK.

Sedimentary rocks may contain fossils and other organic matter. There may be evidence of erosion and deposition in water such as ripple marks and mud cracks.

Sedimentary rocks are classified by the size and type of sediment. **Clastic** sedimentary rocks are made of rock fragments. The sediment may be sorted or unsorted clay, silt, sand, pebbles, cobbles, and/or boulders. The sediments may be in horizontal layers.

Non-clastic sedimentary rocks form as chemical precipitates and evaporites. Dissolved minerals in water settle out when the water evaporates or is still. These rocks are crystalline and usually monominerallic.

Bioclastic sedimentary rocks form by the compaction and cementing of organic sediment. This is the way that coal and limestone form.

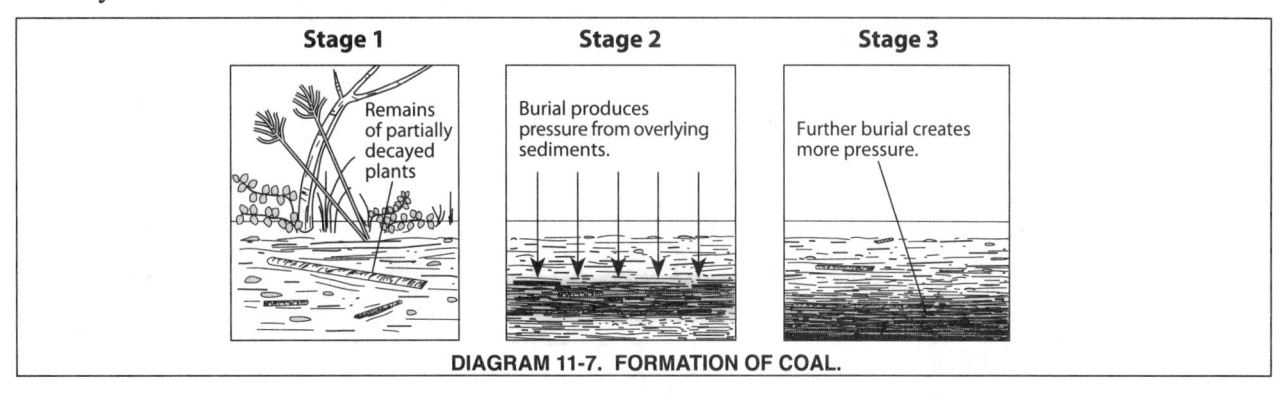

DIAGRAM 11-7. FORMATION OF COAL.

Questions

27. The diagrams below represent some common sedimentary rocks.

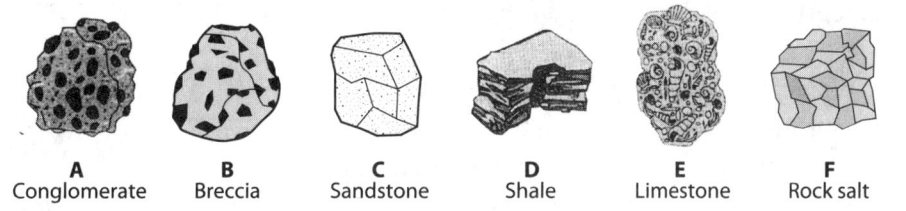

   A. Conglomerate   B. Breccia   C. Sandstone   D. Shale   E. Limestone   F. Rock salt

   Select the letter(s) of the sedimentary rock(s) described.

   a. bioclastic. _____
   b. crystalline. _____
   c. clastic. _____
   d. clay-sized sediment. _____
   e. would bubble in acid. _____
   f. halite. _____
   g. sediment could have been deposit of talus or by a glacier. _____
   h. fragments of rocks. _____
   i. monominerallic. _____
   j. unsorted sediments. _____
   k. sediment size is 0.01 cm. _____
   l. chemical precipitate. _____
   m. made of quartz & feldspar. _____

28. Complete the chart to compare and contrast two different sedimentary rocks.

	siltstone *vs.* shale	conglomerate *vs.* breccia	limestone *vs.* dolostone
compare: (how are they alike?)			
contrast: (how are they diferent?)			

197

29. Two drill-core samples were taken of sedimentary rock layers at *two* different locations 1000 kilometers apart.

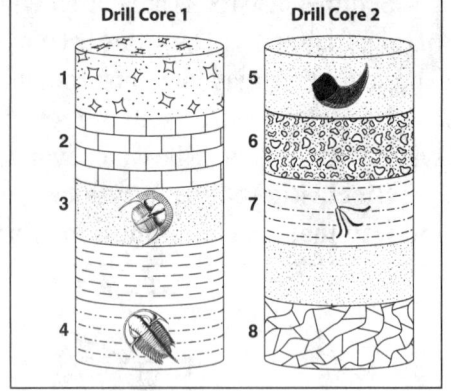

   a. What feature found in some of these rock layers indicate that these are most likely sedimentary rocks? _____

   b. Name the fossil found in layer #3 _____;

   #5 _____

   c. Based on the rock symbols, name the sedimentary rock represented by layers

   #1: _____; #2: _____; #3: _____; #4: _____

   d. Rock layer #8 is the metamorphic rock marble. It formed from the metamorphism of which sedimentary rock? _____

Rock Type: Metamorphic Rock

Metamorphic rocks make up about 26% of the rock on Earth. **Metamorphic rock** forms when pre-existing rock is exposed to extreme heat and/or pressure. The pre-existing rock does not melt. Instead the minerals regrow, recrystallize, and/or rearrange themselves.

There are two types of metamorphism. **Regional metamorphism** is caused by intense heat and pressure affecting a large area. This can occur during mountain building or deep within Earth. **Contact (thermal) metamorphism** occurs in a smaller area that is in direct contact with hot, intruding magma. The zone of contact is the **transition zone**.

Metamorphic rocks are classified by texture. **Foliated metamorphic** rocks have mineral crystal aligned in layers or bands. **Non-foliated** metamorphic rocks are crystalline, the mineral crystals are intergrown and not in layers.

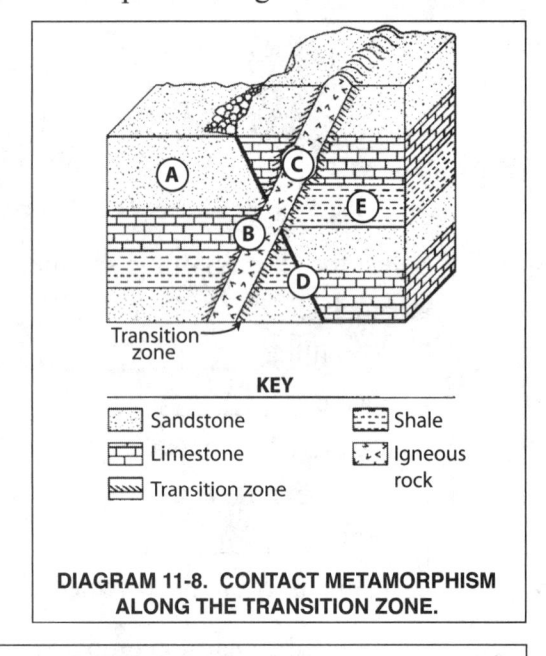

DIAGRAM 11-8. CONTACT METAMORPHISM ALONG THE TRANSITION ZONE.

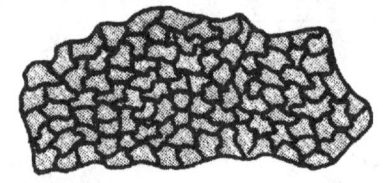

Foliated metamorphic rock

Non-Foliated metamorphic rock

DIAGRAM 11-9. TEXTURES OF METAMORPIC ROCKS.

Questions

30. Name the metamorphic rock described.

  a. banding of minerals. _____

  b. foliated, made of mica only. _____

  c. metamorphism of clay and/or feldspar. _____

  d. alignment of pyroxene, quartz, mica. _____

  e. non-foliated, fine-grained, various minerals. _____

  f. carbon composition. _____

  g. metamorphism of shale. _____

  h. metamorphism of sandstone. _____

31. The geologic cross-section shows various sedimentary rock layers **A**, **B**, **D**, and **E** that were intruded by magma **(C)**.

  a. Contact metamorphism occurs by addition of intense amounts of _____.

  b. Name the metamorphic rock that would form by the contact metamorphism of the sedimentary rocks in layers….

  B: _____ ; D: _____ ; E: _____

32. Select the type of rock each term describes; (*igneous*) (*sedimentary*) (*metamorphic*)

  a. foliated. _____
  f. felsic. _____

  b. plutonic. _____
  g. banding. _____

  c. bioclastic. _____
  h. extrusive. _____

  d. mafic. _____
  i. fragments. _____

  e. mineral alignment. _____
  j. clastic. _____

33. Complete the chart by naming the rock type and specific rock name described.

		Rock Type (igneous, sedimentary, or metamorphic)	Name of Rock
a.	glassy, non-crystalline, black		
b.	gray rock with particles 0.06 cm size compacted together		
c.	fine-grained, intergrown felsic minerals		
d.	many imprints of shells on the rock surface		
e.	minerals separated into bands		
f.	dark color with many small holes		
g.	alignment of mica, quartz, and feldspar with shiny surface		

34. The diagram compares the formation processes and properties for each rock type.

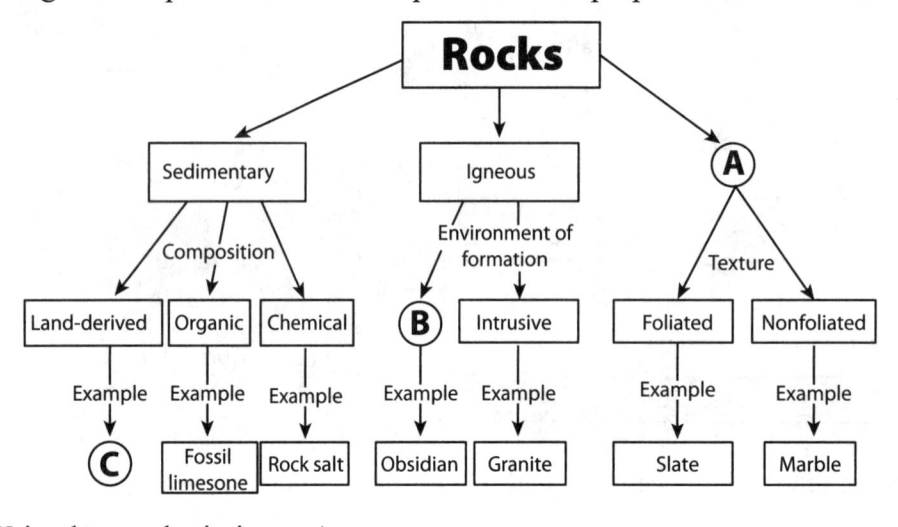

a. Write the word missing at **A**. _____

b. Write the word missing at **B**. _____

c. Write the word missing at **C**. _____

d. An intrusive igneous rock will have (*large*) (*small*) mineral crystals that formed from (*quick*) (*slow*) cooling of molten material.

e. Foliated rocks formed by (*contact*) (*regional*) metamorphism.

f. Another example of an organically formed sedimentary rock is _____.

g. Another example of an intrusive igneous rock is _____.

h. Another example of a foliated rock is _____.

i. Name one mineral present in marble that is not in slate. _____

j. Name *one* observation that a student would make to distinguish obsidian from granite.
_____

35. Rocks are identified as igneous, sedimentary, or metamorphic based on observable physical features. In the chart list characteristics unique to each rock type.

Igneous	Sedimentary	Metamorphic

36. Complete the chart by identifying the rock type and explaining your reason.

	Rock Type (Igneous, Sedimentary, Metamorphic)	Supporting evidence
a.		
b.		
c.		
d.		
e.		
f.		

201

CHAPTER 11 REVIEW

1. Which natural resource is the source of fuel and plastic?
   (1) water (2) petroleum/oil (3) coal (4) uranium

2-5. Base your answers on the chart to the right.

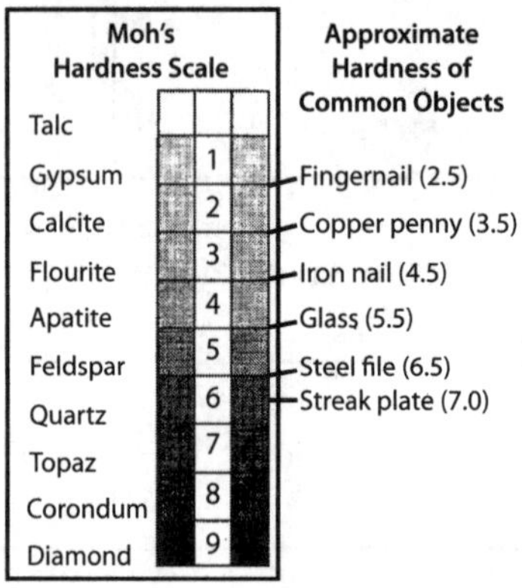

2. Moh's scale of hardness arranges minerals according to their relative
   (1) resistance to breaking
   (2) specific heat
   (3) resistance to scratching
   (4) density

3. Which statement is best supported by the chart?
   (1) An iron nail is made of fluorite.
   (2) Topaz is harder than a steel file.
   (3) A streak plate is made of quartz.
   (4) Apatite is softer than a copper penny.

4. The hardness of the these minerals is most closely related to
   (1) density
   (2) amount of iron in the mineral
   (3) internal arrangement of atoms
   (4) specific heat

5. Which mineral will scratch glass (hardness of 5.5) but not pyrite?
   (1) gypsum (2) fluorite (3) orthoclase (4) quartz

6. Which diagram best represents the silicon-oxygen tetrahedron?

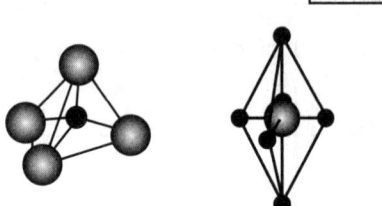

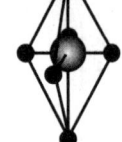

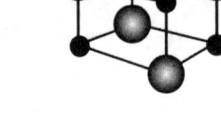

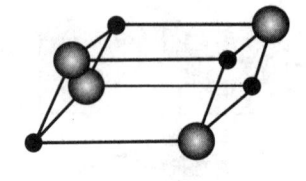

   (1) (2) (3) (4)

7. What causes the characteristic crystal shape and cleavage of the mineral halite as shown in the diagram? 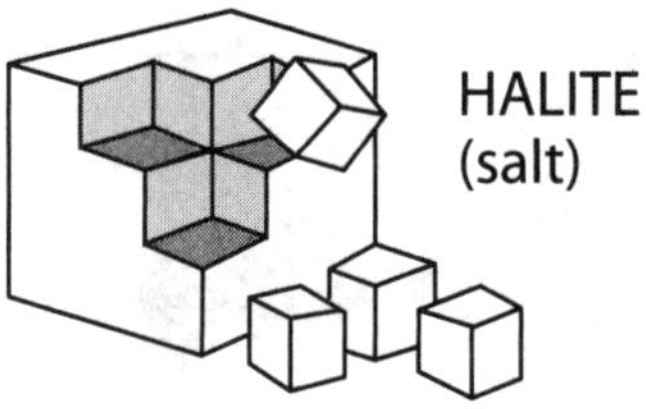
   (1) density of the mineral
   (2) internal arrangement of sodium and chlorine atoms
   (3) amount of metamorphism that occurred
   (4) temperature of the seawater

8. One of the most common minerals in beach sand is quartz. What property of quartz could account for this?
   (1) color (2) fracture (3) luster (4) hardness

202

9. A student created the table below by classifying six minerals into two groups, **A** and **B**, based on a single property Which property was used to classify these minerals?
   (1) color
   (2) luster
   (3) chemical composition
   (4) hardness

Group A	Group B
olivine	pyrite
garnet	galena
calcite	graphite

10. The table below shows some properties of four minerals.

Mineral Variety	Color	Hardness	Luster	Composition
flint	black	7	nonmetallic	SiO_2
chert	gray, brown, or yellow	7	nonmetallic	SiO_2
jasper	red	7	nonmetallic	SiO_2
chalcedony	white or light color	7	nonmetallic	SiO_2

    The minerals listed in the table are varieties of which mineral?
    (1) garnet (2) magnetite (3) quartz (4) olivine

11. Which mineral contains iron, has metallic luster, is hard, and has the same color and streak?
    (1) quartz (2) olivine (3) galena (4) magnetite

12. Which mineral is commonly found in granite?
    (1) quartz (2) magnetite (3) olivine (4) galena

13. Which mineral would most likely weather the fastest?
    (1) pyrite (2) magnetite (3) quartz (4) talc

14. In which group are minerals arranged from least to most resistant to being scratched?
    (1) feldspar, quartz, olivine
    (2) magnetite, galena, quartz
    (3) calcite, pyrite, garnet
    (4) fluorite, halite, talc

15. Rocks are classified as igneous, sedimentary, or metamorphic based primarily on
    (1) texture (2) grain size (3) minerals (4) formation process

16. Which processes change sedimentary rock to igneous rock?
    (1) erosion and deposition
    (2) melting and solidification
    (3) evaporation and condensation
    (4) foliation and recrystalization

17. Rocks that form from fragmental rock particles are classified as
    (1) extrusive igneous
    (2) intrusive igneous
    (3) crystalline sedimentary
    (4) clastic sedimentary

18. Which igneous rock is dark colored, formed by rapid cooling, and is composed of plagioclase feldspar, olivine, and pyroxene?
    (1) obsidian (2) rhyolite (3) gabbro (4) scoria

19. A fine grained rock has a mineral composition that is 50% potassium feldspar, 26% quartz, 13% plagioclase feldspar, 8% biotite, and 3% amphibole. This rock is
    (1) granite (2) rhyolite (3) gabbro (4) basalt

20. Compared to mafic igneous rocks, felsic igneous rocks contain greater amounts of
    (1) quartz (2) iron (3) pyroxene (4) gas pockets

21. Which property is most useful for identifying an igneous rock?
    (1) kind of cement
    (2) mineral composition
    (3) number of minerals present
    (4) types of fossils present

22. Rhyolite and granite are alike in that they are both
    (1) fine-grained   (2) dark-colored   (3) very dense   (4) felsic
23. Which igneous rock has a vesicular texture and mineral composition of quartz and potassium feldspar?
    (1) andesite   (2) pegmatite   (3) pumice   (4) scoria
24. Dolostone is classified as a
    (1) land-derived sedimentary rock
    (2) chemically formed sedimentary rock
    (3) foliated metamorphic rock
    (4) extrusive igneous rock
25. Which rock most likely formed from pebble-sized sediment deposited in shallow water along a shoreline?
    (1) shale   (2) basalt   (3) breccia   (4) conglomerate
26. Particles of which size could form shale?
    (1) 0.2 cm   (2) 0.02 cm   (3) 0.002 cm   (4) 0.0002 cm
27. Large deposits of rock gypsum and rock salt usually form in areas of
    (1) active volcanoes
    (2) continental ice sheet
    (3) mid-ocean ridges
    (4) shallow evaporating seas
28. Which process would form a sedimentary rock?
    (1) melting of rock deep in Earth's crust
    (2) cooling of lava at Earth's surface
    (3) recrystallization of unmelted rock material in Earth's crust
    (4) precipitation of minerals as seawater evaporates
29. Sedimentary rocks of organic origin would most likely form from
    (1) sediments eroded by water
    (2) fragments deposited by glaciers
    (3) shells of marine organisms
    (4) lava flowing out of a volcano
30. What is the main difference between metamorphic rocks and most other rocks?
    (1) Many metamorphic rocks contain only one mineral.
    (2) Many metamorphic rocks have an organic composition.
    (3) Many metamorphic rocks exhibit banding and distortion of structure.
    (4) Many metamorphic rocks have a high amount of silicate minerals.
31. Which rock formed by the recrystallization of unmelted rock material exposed to high pressure and temperature?
    (1) gneiss   (2) granite   (3) dolostone   (4) bituminous coal
32. A plutonic igneous rock composed of pyroxene and olivine would be
    (1) dunite   (2) peridotite   (3) basalt   (4) granite
33. Which observation about an igneous rock would support the inference that the rock cooled slowly underground?
    (1) well-defined mineral layers
    (2) large mineral crystals
    (3) it is 50% potassium feldspar
    (4) light in color and low density
34. Which graph best represents the relative densities of three different igneous rocks?

35. Olivine and pyroxene are commonly found in igneous rocks that are
    (1) felsic with low density
    (2) felsic with high density
    (3) mafic with high density
    (4) mafic with low density

36. Obsidian's glassy texture indicates that it formed
    (1) slowly, deep below Earth's surface
    (2) slowly, on Earth's surface
    (3) quickly, deep below Earth's surface
    (4) quickly, on Earth's surface

37. Which graph best shows the relationship between the time it takes magma to cool and crystal size?

    (1)   (2)   (3)   (4)

38. How do the metamorphic rocks quartzite and schist differ?
    (1) Quartzite contains the mineral quartz and schist does not.
    (2) Quartzite forms by regional metamorphism and schist does not.
    (3) Schist is organically formed and quartzite is not.
    (4) Schist contains the mineral pyroxene and quartzite does not.

39. In New York State, the surface bedrock of the Catskills consists mainly of
    (1) weakly consolidated gravels and sands
    (2) quartzites, dolostones, marbles, and schists
    (3) conglomerates, red sandstones, basalts, and diabase
    (4) limestones, shales, sandstones, and conglomerate

40. Bedrock in the area of Binghamton, New York, consists of
    (1) igneous rock
    (2) sedimentary rock
    (3) tilted volcanic rock
    (4) folded metamorphic rock

41. What change would most likely occur if crustal rock is subducted deep within Earth's crust and subjected to intense heat and pressure but did not melt?
    (1) density will increase
    (2) volume will increase
    (3) mass will increase
    (4) rock becomes more mafic

42-45. The diagrams below represent five different rock samples.

    A: Bands of alternating light and dark intergrown materials
    B: Easily split layers of 0.0001-cm-diameter particles cemented together
    C: Glassy black rock that breaks with a shell-shape fracture
    D: Intergrown 0.5-cm-diameter crystals of various colors
    E: Sand and pebbles cemented together

42. Which sample is composed mostly of clay-sized particles?
    (1) A   (2) B   (3) C   (4) D

43. Which sample formed from lava that cooled rapidly?
    (1) A   (2) E   (3) C   (4) D

44. If sample E were metamorphosed, it would most likely become
    (1) slate   (2) marble   (3) anthracite coal   (4) metaconglomerate

45. Sample D is composed mostly of the minerals potassium feldspar, quartz, and plagioclase feldspar. Sample D is best described as
    (1) felsic   (2) very dense   (3) dark colored   (4) foliated

205

46-47. The diagram is a mineral classification scheme that shows the properties of certain minerals. Letters **A** through **G** represent mineral property zones. Zone **E** represents the presence of all three properties. Assume that glass has a hardness of 5.5.

For *example*, a mineral that is harder than glass, has a metallic luster, but does not have cleavage, would be placed in zone **B**.

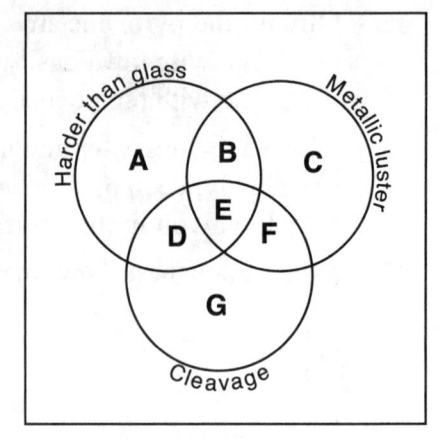

46. In which lettered zone would the mineral potassium feldspar be placed? _____

47. State the name of *one* mineral listed on the "*Properties of Common Minerals*" table that could NOT be placed in any of the zones. _____

48-50. The chart below lists three physical properties of minerals and the definitions. Letters **A**, **B**, and **C** have been left blank.

48. Which physical property of a mineral is represented by letter **A**? _____

49. Write the definition represented at letter **B**.

_____
_____
_____

50. Identify one mineral that could be represented by letter **C**. _____

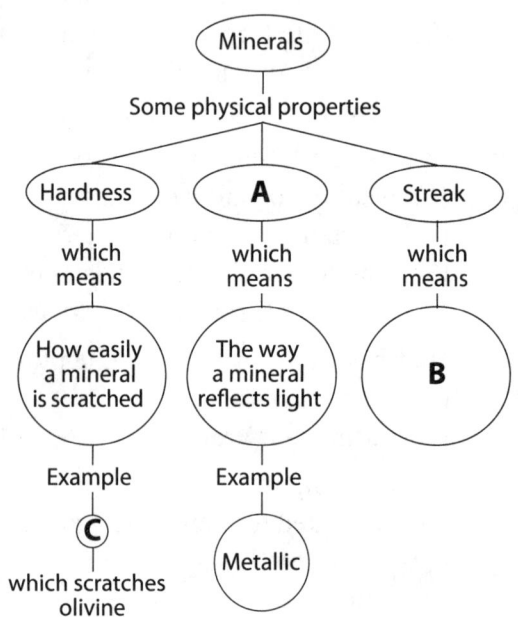

51. A student on a field trip in New York State collected this sample of metamorphic bedrock containing bands of coarse-grained crystals of plagioclase feldspar, pyroxene, quartz, and mica.

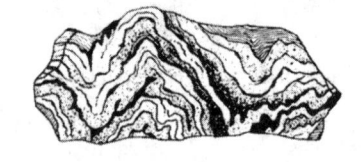

Describe *one* physical feature of this rock sample that identifies it as a metamorphic rock.

_____

206

52-56. The diagram below shows the bedrock structure of a portion of the lithosphere. Letters **A** through **D** are locations in the lithosphere.

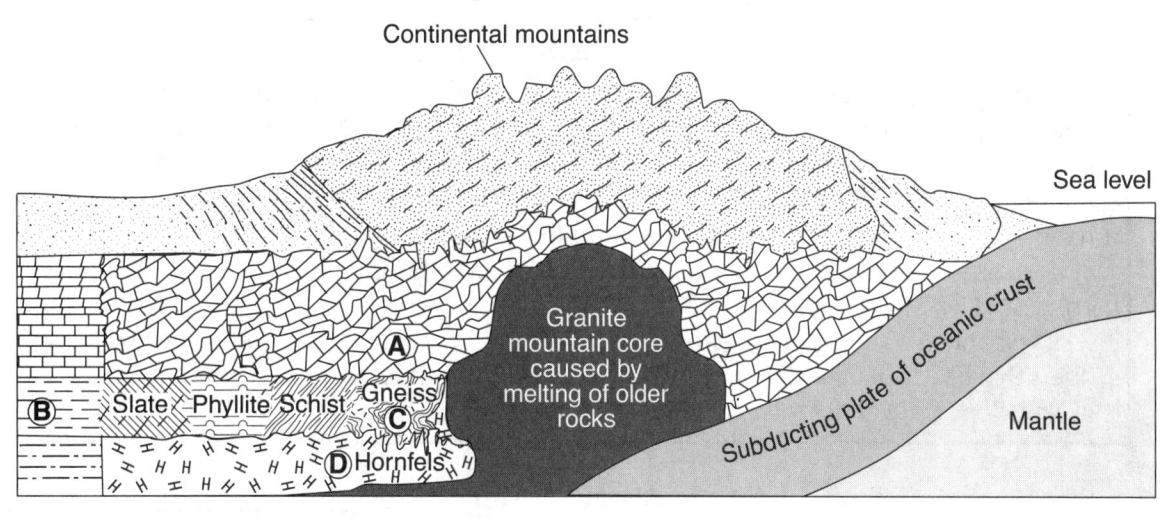

52. Name the metamorphic rock represented by the symbol at **A**. _____

53. Name a mineral found in slate. _____

54. Explain why the types of metamorphic rock change from **B** to **C**. _____
_____

55. Name the igneous rock that is generally found in oceanic crust? _____

56. Why does the oceanic crust subduct under the continental crust when the two plates converge?
_____

207

57-60. Read the passage below about the mineral asbestos.

> **Asbestos**
>
> Asbestos is a general name given to the fibrous varieties of six naturally occurring minerals used in commercial products. Most asbestos minerals are no longer mined due to the discovery during the 1970s that long-term exposure to high concentrations of the long, stiff fibers in asbestos leads to health problems. Workers who produce or handle asbestos products are most at risk, since inhaling high concentrations of the airborne fibers causes the asbestos particles to become trapped in the workers' lungs. This could lead to lung disease and cancer.
>
> Chrysotile is a variety of asbestos that is still mined because it has short, soft, flexible fibers that does not pose the same health threat. Chrysotile is found with other minerals in New York State mines located near 44° 30' N, 74°W.

57. State *one* reason for the decline in asbestos use after 1970. _____

_____

58. Name the New York State landscape region where chrysotile found? _____

59. What causes the physical properties of minerals such as the long, stiff fibers of asbestos?

_____

60. The chemical formula for chrysotile is $Mg_3Si_2O_5(OH)_4$. State the name of the mineral that is most similar in chemical composition. _____

CHAPTER 12

EARTH'S INTERIOR AND THE DYNAMIC CRUST
Earth's Interior, Crustal Movements, Plate Tectonics, Volcanoes, Earthquakes

EARTH'S INTERIOR

As Earth formed the molten materials of the cooling planet separated into four distinct layers due to density differences. Since we have no rocks from Earth's interior, scientists use indirect evidences to infer the conditions inside Earth. The physical structure of Earth's interior is based on the behavior of earthquake (seismic) waves as they travel through Earth. The inferred chemical composition of Earth's cores is based on the composition of nickel-iron meteorites.

As depth into Earth increases, there is an increase in density, pressure, temperature, and gravitational force. The temperature inside Earth increases with depth, due to residual heat left over from Earth's origin, radioactive decay, and increased pressure.

Earth's interior is divided into four layers. The outer, least dense, thinnest layer is the **crust**. The crust is divided into continental crust and oceanic crust. **Continental crust** makes up the land masses. The **oceanic crust** is the rock that makes up the seafloor below the sediment. The oceanic crust is thinner and composed of basalt; the continental crust is thicker and composed of granite.

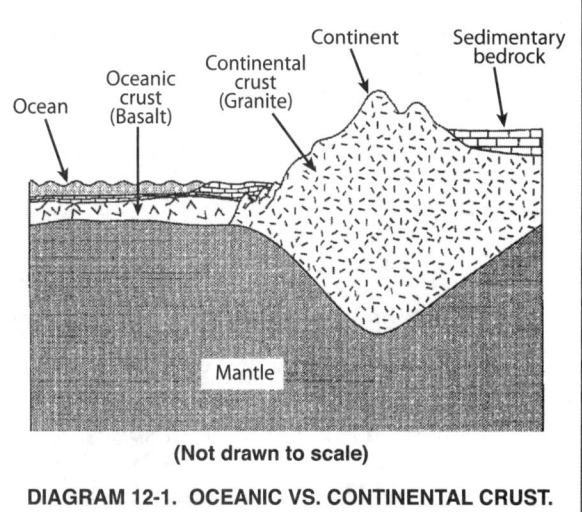

The **Moho** is the interface (boundary) between the crust and the next layer, the mantle. The **mantle** is divided into three regions. The upper part is the thin, **rigid mantle**. Next is the **asthenosphere** which is a gel-like plastic layer that the crust and rigid mantle "float" on. Below this is the solid, **stiffer mantle**. The crust and rigid mantle which lie above the asthenosphere are known as the **lithosphere**.

DIAGRAM 12-1. OCEANIC VS. CONTINENTAL CRUST.

The **outer core** is a liquid because the temperature of the rock is above its melting point. The outer core is made of liquid iron and nickel based on the composition of meteorites. The **inner core** is a very hot solid under extremely high pressure. It is also made of iron and nickel.

The Moon and other planets have interior layers like Earth due to a similar cooling history from the solar nebula.

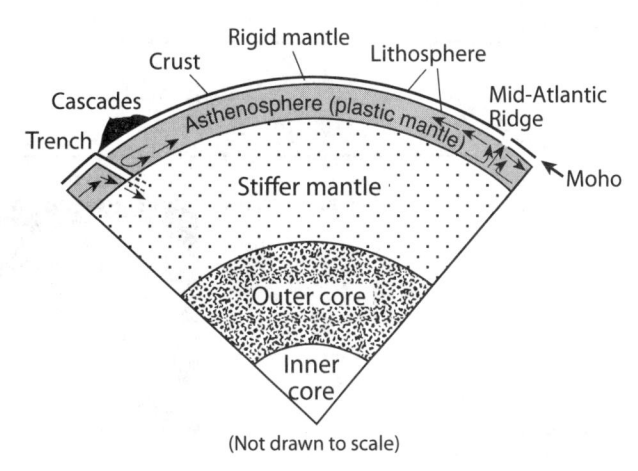

DIAGRAM 12-2. - EARTH'S INTERIOR LAYERS

209

Questions

1. The diagram below represents the zones of Earth's interior identified by letters **A** through **E**. The scale shows depth below Earth's surface in kilometers.

   Zones of Earth's Interior

   | A | B | C | D | E |

   0   1000   2000   3000   4000   5000   6000 km

   Depth Below Earth's Surface

   a. Name the interior layer(s) represented by each letter:

   A: _____ B: _____ C: _____

   D: _____ E: _____

   b. The Moho boundary is located between zones _____ and _____.

   c. The lithosphere includes the _____ and _____.

   d. The approximate thickness of zone **C** is _____ km.

   e. Based on this diagram the approximate radius of Earth is _____ km.

   f. In which lettered zone is the interior temperature 5,500° C? _____

   g. In which lettered zone is the melting point of the rocks 2,000° C? _____

2. For each description name Earth's interior layer which is described.

   a. partially melted rock material: _____

   b. asthenosphere: _____

   c. at depth of 3500 km: _____

   d. pressure of 2.5 million atm.: _____

   e. liquid layer: _____

   f. solid iron and nickel: _____

   g. ocean floor: _____

   h. temperature of 3500° C: _____

   i. rocks melt at 4000° C: _____

   j. density can be 12.9 g/cm³: _____

3. The diagram to the right shows the inferred interior layers of two other planets.

   a. How does Mercury's mantle differ from Earth's? _____

   b. What is the probable reason that planets develop interior layers? _____

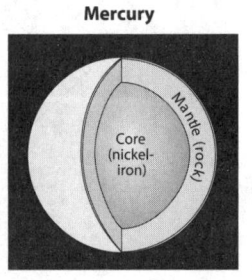

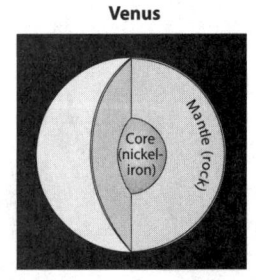

   Mercury — Core (nickel-iron), Mantle (rock)

   Venus — Core (nickel-iron), Mantle (rock)

   (Not drawn to scale)

210

CRUSTAL MOVEMENTS

The physical features of rock layers and locations of fossils indicate that crustal movements have occurred in the past. This is based upon two geologic principles. The first is the **principle of superposition** which states that older rocks are below younger rocks if the crust has not been overturned. The second is the **principle of original horizontality**. This states that sedimentary rocks and extrusive igneous rocks form as horizontal layers.

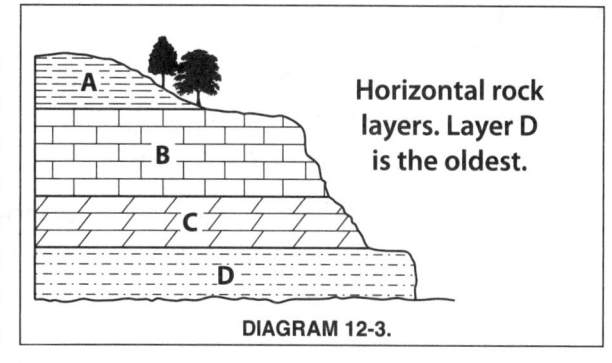

Horizontal rock layers. Layer D is the oldest.

DIAGRAM 12-3.

Many observations support past crustal movements. Geologists find rock layers that have been disturbed and many times the older rocks are above younger rocks. Tilted, faulted, and folded rock layers are evidence of crustal movement. Shallow water marine fossils, such as shark teeth, clams, and corals, have been found high in mountains or deep in the oceans. These displaced fossils indicate that the crust has been uplifted and in other areas the crust has subsided.

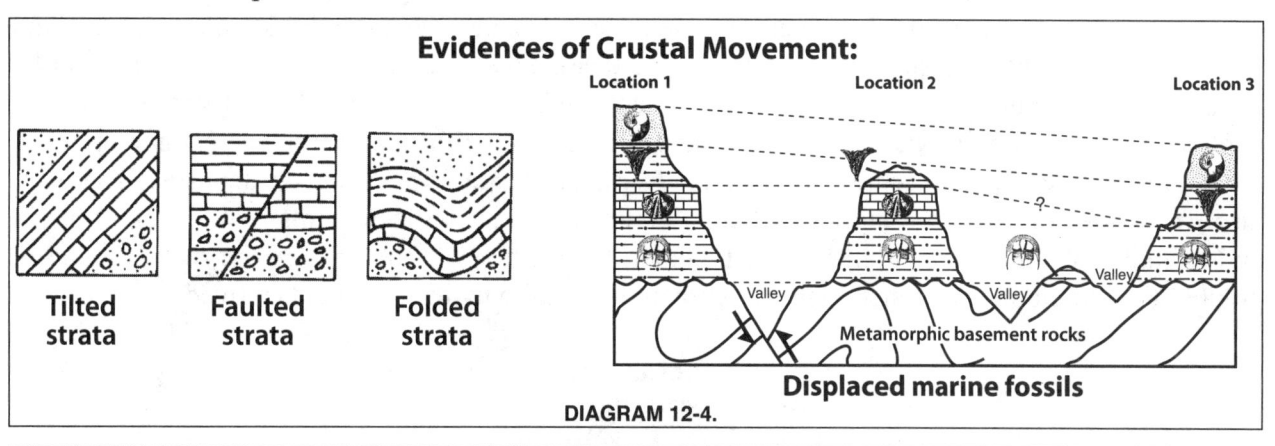

Evidences of Crustal Movement:

Tilted strata | Faulted strata | Folded strata

Displaced marine fossils

DIAGRAM 12-4.

Questions

4. What conclusion can be drawn from finding shallow-water fossils, such as coral, at great ocean depths? _____

5. Coral fossils have also been found in mountains. Why? _____

6. What do tilted and folded rock layers suggest about Earth's crust? _____

7. The diagrams below show cross-sections **A**, **B**, and **C** of exposed bedrock. Select the diagram(s) for each description:

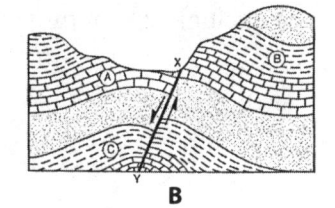

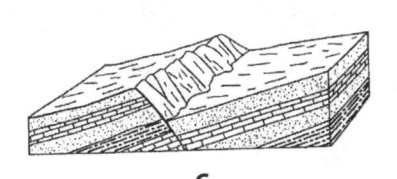

a. horizontal strata: _____
b. faulting: _____
c. tilting: _____
d. folding: _____
e. least crustal movement: _____
f. rock layers in a mountain: _____

211

PLATE TECTONICS

The crust and rigid mantle above the asthenosphere are known as the **lithosphere**. The lithosphere is dynamic; it is in motion. This motion can be slow, as during plate tectonics, or sudden as during earthquakes. The lithosphere moves horizontally, vertically, and laterally.

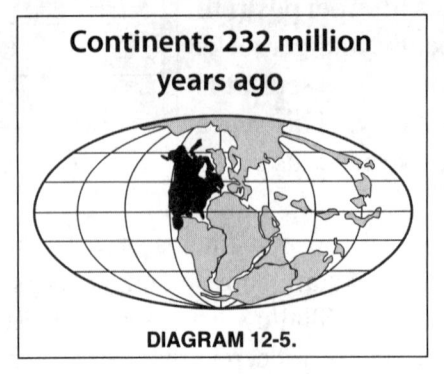

Continents 232 million years ago

DIAGRAM 12-5.

The idea of **continental drift** was proposed in 1912 by Alfred Wegener. He stated that the present continents were once part of a large landmass he called Pangaea. Pangaea broke apart and the land masses drifted apart. The present positions of the continents were different in the past and will be different in the future. One evidence for continental drift is the apparent fit of the shapes of the continents. Another evidence is similar geologic structures on different continents that line up when the continents are placed together. These include glacial striations and mineral deposits. There are also similarities in fossils among the different continents; today the continents have different plants and animals. When the continents are fit together, the magnetized minerals in the rocks line up to point to the same pole. Wegener could not explain how and why the continents moved.

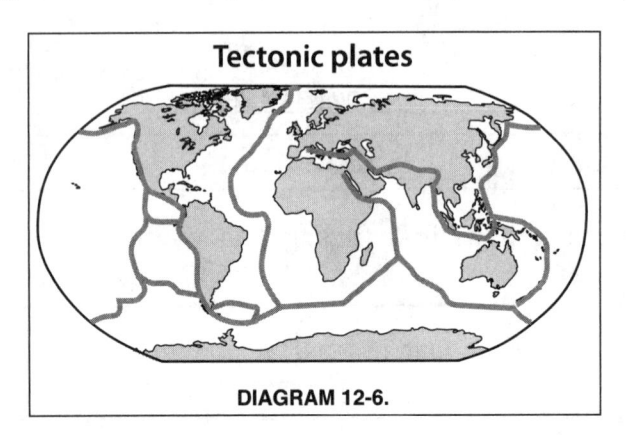

Tectonic plates

DIAGRAM 12-6.

In the 1960's the **theory of plate tectonics** was proposed. This states that Earth's surface is divided into about a dozen, solid, moving pieces called "plates." A **plate** is a piece of the lithosphere: continental and/or oceanic crust and the upper rigid mantle. Plates are in constant motion due to convection currents in the asthenosphere. As the plates move the continents move with them. The edges of plates are geologically active zones of crustal movement where earthquakes, volcanoes, and mountain building occur. The rate and direction of each plate motion varies.

The mechanism for plate tectonics and moving plates is provided by **sea floor spreading**. Crustal plates move atop the asthenosphere due to **mantle convection currents**. New crust is created by volcanic activity at **ridges** and **rift valleys**. This molten, less dense material upwells, spreads out, and pushes older rock away from both sides of the ridge. The youngest oceanic crust is located at the ridges. Evidence for sea floor spreading includes the symmetrical pattern of increasing age of the rock on both sides of ridge and the symmetrical pattern of **polar reversals** on either side of the ridge.

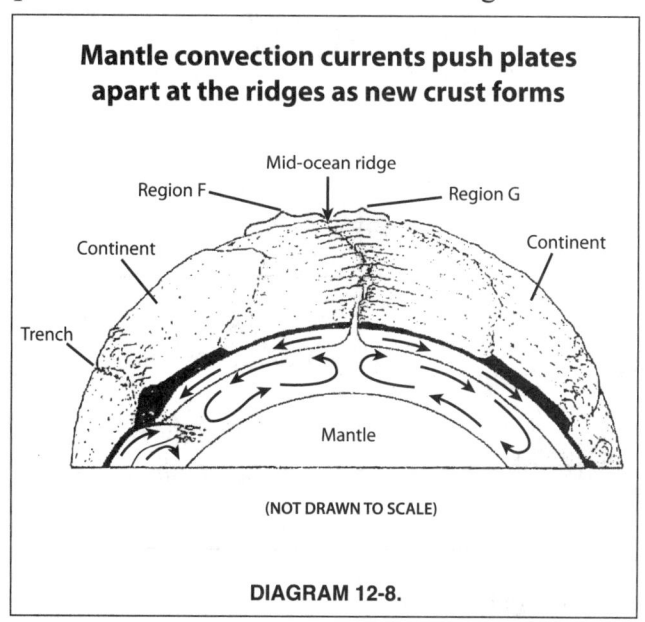

Symmetry of age and magnetism on either side of a ridge

DIAGRAM 12-7.

Mantle convection currents push plates apart at the ridges as new crust forms

DIAGRAM 12-8.

Old crust is destroyed at ocean **trenches** where the dense oceanic crust subducts (sinks) into the mantle, while new crust is created at ridges where magma upwells. This process of subduction and sea floor spreading changes the size and shape of the oceans. The Pacific Ocean is shrinking because deep ocean trenches are subducting and destroying more oceanic crust than the mid-ocean East Pacific Ridge is producing. The Atlantic Ocean is expanding because it has few trenches where crust is destroyed. New ocean floor is created at the Mid-Atlantic ridge, pushing the continents further apart causing the Atlantic Ocean to slowly get larger.

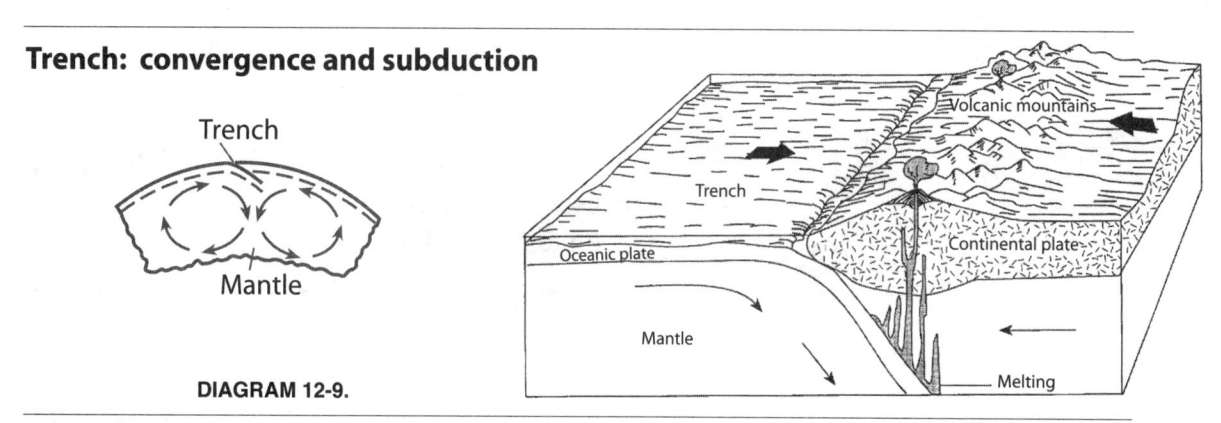

Trench: convergence and subduction

DIAGRAM 12-9.

Mid-ocean ridge: divergence and sea floor spreading

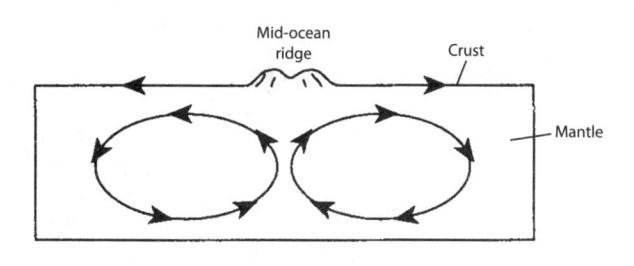

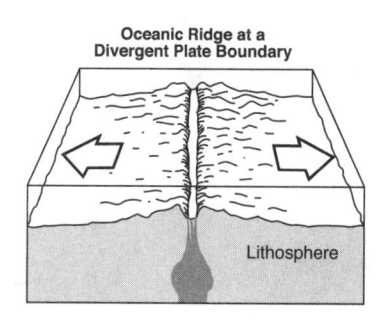

DIAGRAM 12-10.

plate boundary	plate motion	example
![divergence diagram with Plate, Plate, Asthenosphere]	**divergence** • plates move away from each other when heat and magma upwells at ridges and rift valleys. This pushes the plates apart.	• ocean ridges and rift valleys form • *ex:* Mid-Atlantic Ridge, East Pacific Ridge, East African Rift
![transform fault diagram with Plate, Plate, Asthenosphere]	**transform fault** • two plates slide past each other in opposite directions • lateral motion	• transform fault forms • *ex:* San Andreas Fault, Scotia plate
![convergence diagram with Plate, Plate, Asthenosphere]	**convergence** • plates collide without subduction	• mountains form • *ex:* Himalayan Mountains
![subduction diagram with Trench, Oceanic crust, Continental crust, Asthenosphere, Melting (Not drawn to scale)]	**subduction** • denser plate slides under the less dense, overriding plate • basaltic oceanic crust is denser than granitic continental crust	• trench forms • *ex.* Mariana Trench, Peru-Chili Trench

Questions

8. Cite *four* evidences that support the theory that continents were once together.

   a. _____

   b. _____

   c. _____

   d. _____

   _____

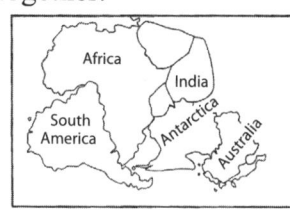

9. What interior layers make up the lithosphere? _____

10. The lithosphere moves above the _____ which is described as _____
    _____.

11. The diagrams below show four different plate boundaries. Select the diagram(s) for each description.

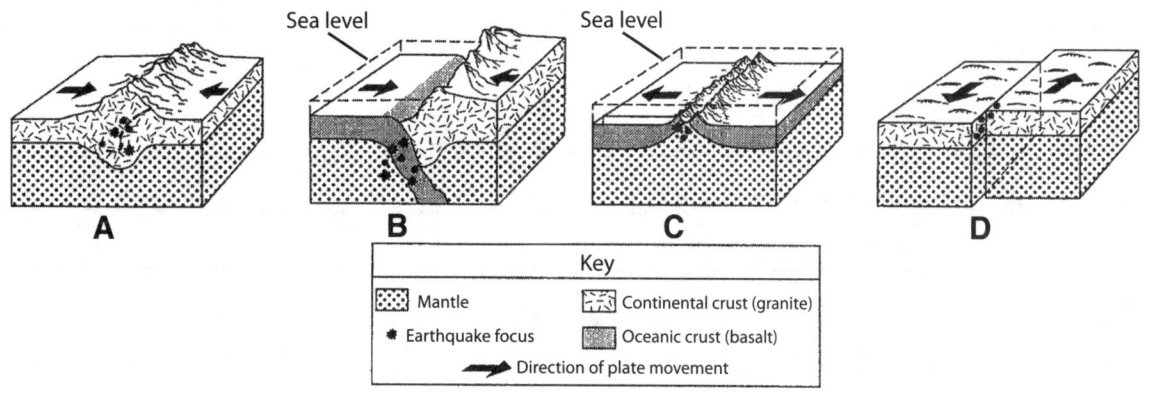

    Key
    - Mantle
    - Earthquake focus
    - Continental crust (granite)
    - Oceanic crust (basalt)
    - Direction of plate movement

    a. convergence: _____
    b. divergence: _____
    c. subduction: _____
    d. crustal collision: _____
    e. new crust forms: _____
    f. old crust is destroyed: _____
    g. transform fault: _____
    h. sea floor spreading: _____

    i. mid-ocean ridge: _____
    j. rift valley: _____
    k. Tonga Trench: _____
    l. Iceland Hot Spot: _____
    m. Indian-Australian & Eurasian Plates: _____
    n. South American & Nazca Plates: _____
    o. Antarctic & Pacific Plates: _____
    p. Scotia & Antarctic Plates: _____

12-13. The diagram shows a cross section of part of Earth's surface and interior. Four areas are labeled: **A, B, C,** and **D**.

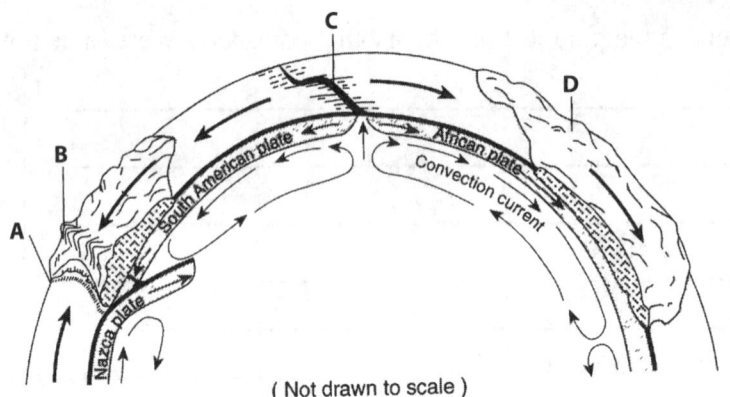

(Not drawn to scale)

12. Select the lettered area(s) that is described.
   a. youngest rock: _____
   b. subduction: _____
   c. trench: _____
   d. ridge: _____
   e. continental landmass: _____
   f. upwelling zone: _____
   g. volcanic mountain chain: _____
   h. granite crust: _____

13. The convection current is in the interior layer known as the _____
   What causes the convection current? _____
   _____

14. Refer to the "*Tectonic Plates*" map on page 5 in the ESRT.
   a. At the Mariana Trench the subducting plate is the _____ Plate and the overriding plate is the _____ Plate.
   b. At the Southeast Indian Ridge the two diverging plates are the _____ Plate and the _____ Plate.
   c. Circle the locations that are at located at a plate boundary:
      (*Hawaii*) (*Iceland*) (*Yellowstone Park*) (*Australia*) (*New York*) (*Bouvet*) (*India*)
   d. State the direction the Cocos Plate is moving towards. _____
   e. The East African Rift is a (*divergent*) (*convergent*) (*transform fault*) (*complex*) plate boundary.
   f. The Tasman Hot Spot is located on the _____ Plate.
   g. Name the plate located at 50° N, 120° E. _____
   h. The boundary between the Scotia Plate and the Antarctic Plate is (*transform fault*) (*divergent*) (*convergent*) (*complex*).

15. The diagram below shows a portion of the oceanic crust near the mid-Atlantic ridge.

The arrows in the diagram represent the magnetic orientation preserved in the igneous rocks. When the igneous rock formed the mineral crystals lined up in the direction of the North Pole. Throughout Earth's history, the north and south poles have switched (reversed) positions. Letters **A** through **E** are locations on the oceanic crust.

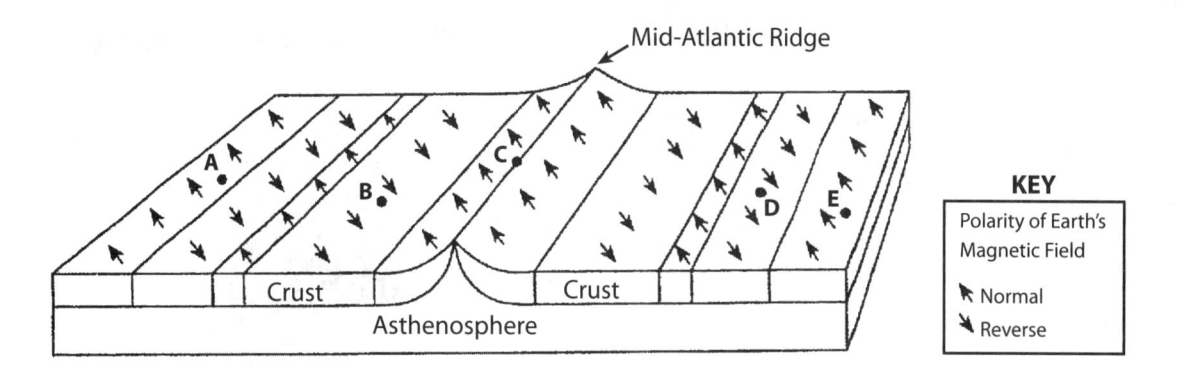

a. Draw arrows to show the mantle convection current for this area.

b. Which location(s) show normal polarity ? _____

c. Which location(s) show reverse polarity? _____

d. Which location has the youngest rock? _____

e. Which *two* locations have rocks of the same age? _____ & _____

f. The highest heat flow would be recorded at letter _____.

g. Name the type of igneous rock found in the crust in this area. _____

h. What causes the symmetry (same pattern) on either side of the ridge? _____

217

VOLCANOES

Volcanoes are weak spots in the crust where molten magma comes to the surface, sometimes pushing up the land to form a domed mountain. Lava that flows out of the vent will harden and form new crust. Most volcanoes are located in specific tectonic zones, such as the Pacific Ocean "ring of fire." These active zones are the boundaries of lithospheric plates where the plates converge or diverge. The plate motion weakens the lithosphere, which fractures and allows magma to reach the surface.

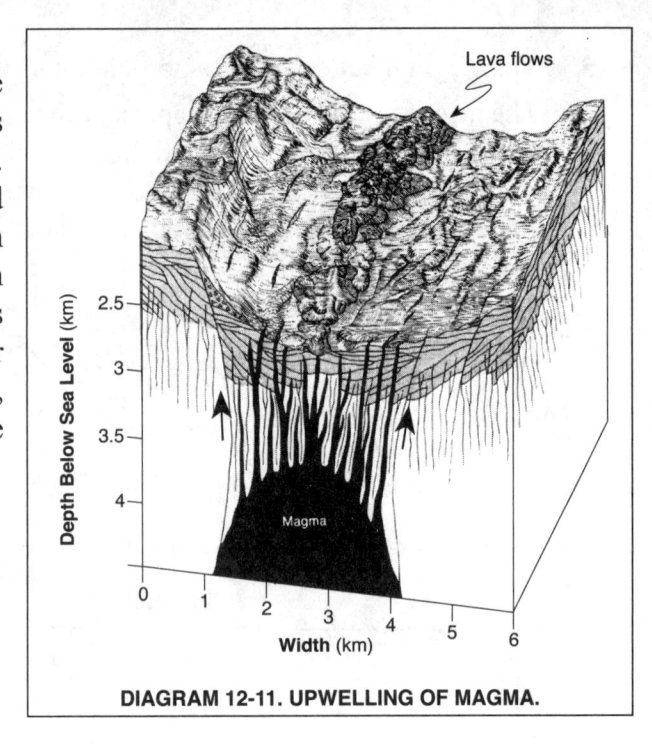

DIAGRAM 12-11. UPWELLING OF MAGMA.

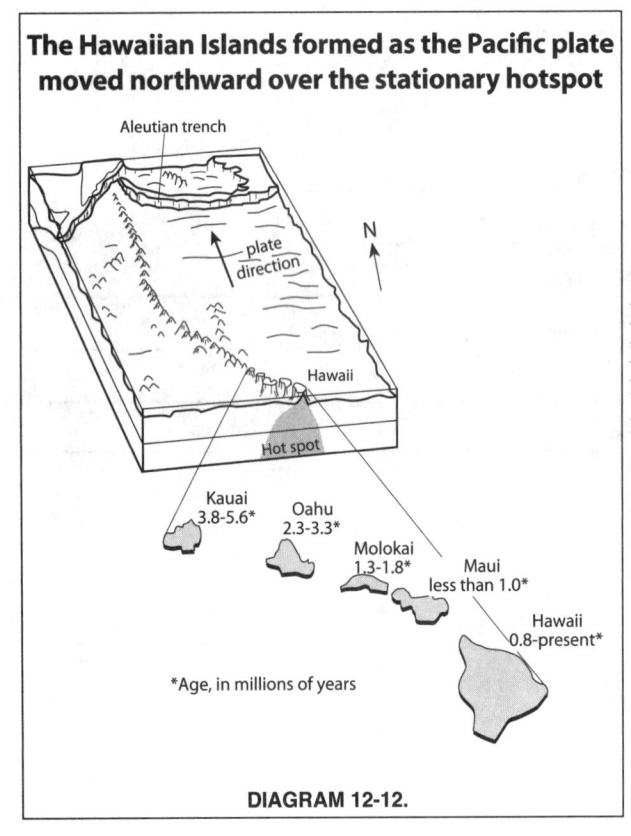

The Hawaiian Islands formed as the Pacific plate moved northward over the stationary hotspot

*Age, in millions of years

DIAGRAM 12-12.

Hot spots are places in the crust above a plume (mass) of magma. This is caused by heat at a specific place in the asthenosphere. Not all hot spots are at plate boundaries. At a hot spot the lava may gently flow out and build up a land mass. As the plate moves, this landmass will move away from the hot spot. The movement of the plate over a hot spot can form a chain of extinct volcanoes like the Hawaii Island Chain. Only the island of Hawaii has active volcanoes because it is over the hot spot.

Volcanoes are very hazardous. Geologists monitor water levels in the craters, changes in elevation, and any change in gases escaping from the volcanoes. Magma contains dissolved gases that are dangerous. Ash, dust, and volcanic bombs can be thrown out of a volcano. Lava flows can bury towns. A cloud of ash and dust can block insolation. In 2010 the ash from a volcanic eruption disrupted air flights across the Atlantic Ocean for weeks. Volcanoes build new land, replenish the soil, and are sources of geothermal energy.

Questions

16. The map below shows the outlines and ages of several volcanic basins (calderas) which formed during the last 16 million years in the northwest United States.

Yellowstone Volcanic Calderas

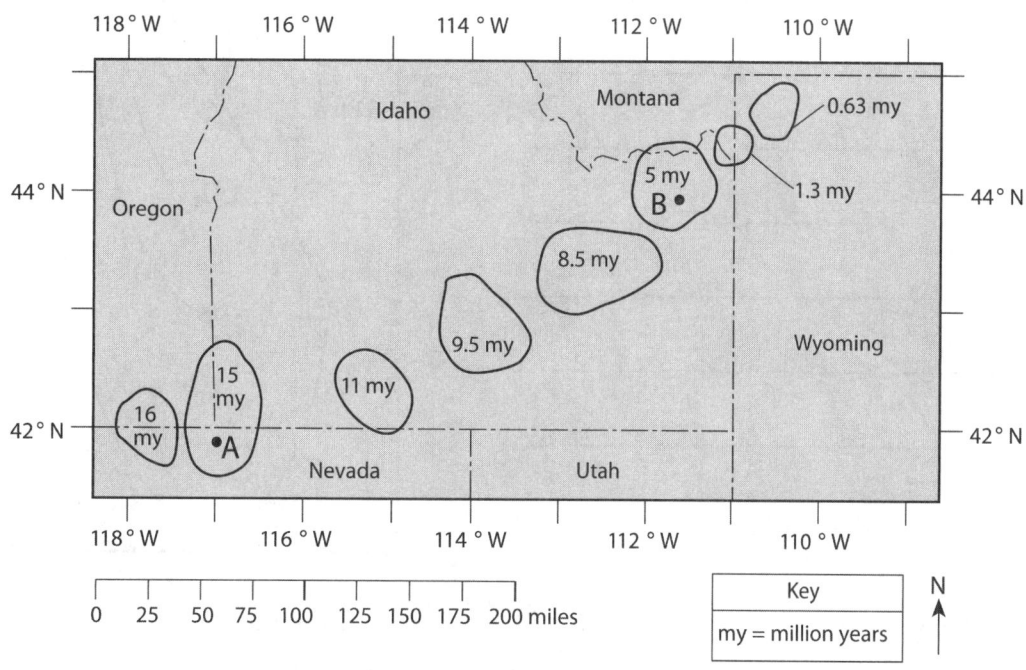

a. Place an "**X**" at the location of the youngest caldera.

b. The ages of the calderas increase toward the _____ direction.

c. What caused this chain of calderas to form? _____

d. The Yellowstone calderas are on the _____ plate.

e. This location (*is*) (*is not*) at the edge of the plate.

f. Draw an arrow to show the direction this plate is moving.

g. The longitude of the center of the 11 m.y. caldera is _____.

h. Calculate the plate motion in miles per million years between points **A** and **B**. _____ (*show solution*)

219

17. On the diagram to the right, the star symbol " ★ " shows the location of a volcano in Iceland. The isolines represent the thickness, in centimeters, of volcanic ash deposited from the eruption. Points **A** and **B** are locations in this area.

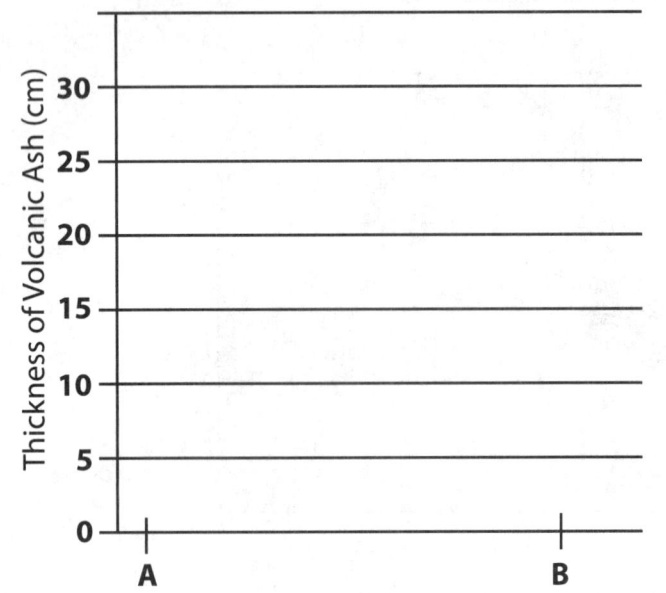

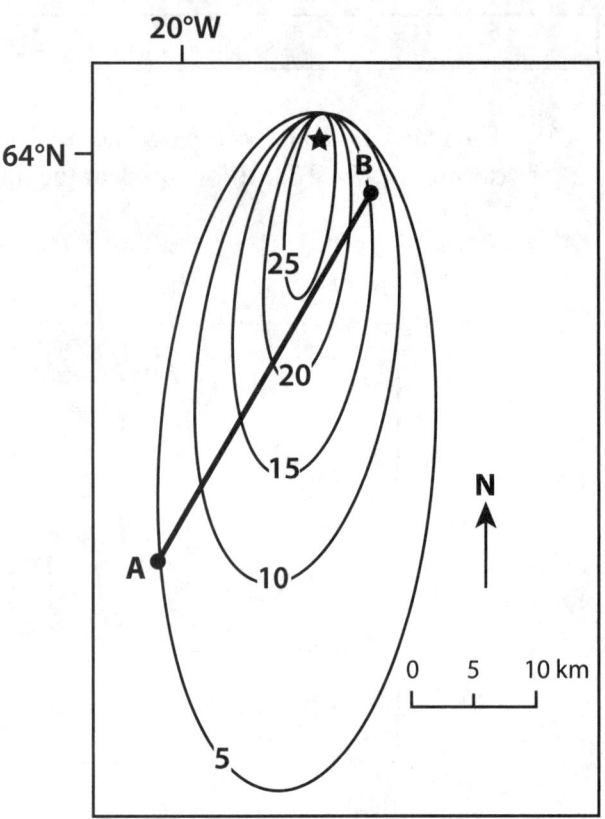

a. On the grid above construct a profile of the ash thickness between **A** and **B**.

b. State *one* factor that could have produced this pattern of ash deposition. _____

_____

c. Explain why volcanoes are likely to occur in Iceland. _____

_____

EARTHQUAKES

There are three major zones of crustal activity. These zones are found along the edges of plate boundaries. These active zones are the Pacific "ring of fire," the Mediterranean-Himalayan zone, and the Mid-Atlantic Ridge. Young mountains, volcanoes, and earthquakes occur most frequently in these zones. The most active zone is the Pacific "ring of fire."

Earthquakes are the natural shaking of the crust and shifting of rock layers. Earthquakes can be caused by volcanic eruptions, the movement of rocks along a fault, or plate movement. The actual place in the crust where the earthquake occurs is the **focus**. The point on Earth's surface above the focus is the **epicenter**. Earthquakes generate vibrations called **seismic waves** that travel throughout Earth. A **seismograph** is the instrument that detects seismic waves. A **seismogram** is the printed recording of the seismic vibrations.

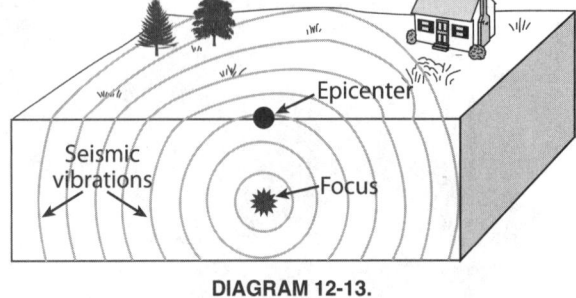

DIAGRAM 12-13.

There are three types of seismic waves. **P-waves** (primary waves) reach the seismograph first. P-waves are compressional waves that travel the fastest and can move through solids, liquids, and gases. **S-waves** (secondary waves) reach the seismograph next. S-waves are shear waves that can travel only through solids.

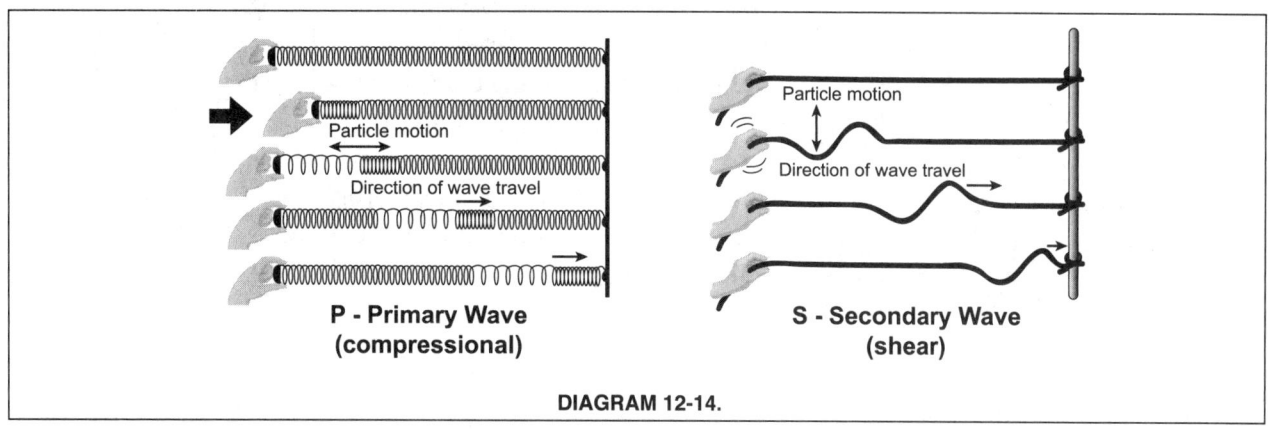

DIAGRAM 12-14.

When an earthquake occurs, both P- and S-waves are received at almost all seismic stations on Earth. Some seismic stations, especially those on the opposite side of Earth from the epicenter, receive only P-waves not S-waves. This occurs because S-waves cannot go through the liquid outer core of Earth. This area is the "P-wave only" zone. Due to the refraction (bending) of seismic waves, there is an area which receives no seismic waves. This is known as the **shadow zone** which is on either side of the "P-wave only" zone.

DIAGRAM 12-15.

The travel time graph for P-waves and S-waves enables geologists to locate the epicenter of the earthquake. It shows the distance P-waves and S-waves travel in a given amount of time. Since P-waves move faster, its travel time is less and it arrives first. The travel time graph lines are not straight because the speed of the waves is not constant. The speed of seismic waves will increase as the rock they travel through becomes denser. Speed increases with depth as the seismic wave goes through the deeper, denser parts of Earth.

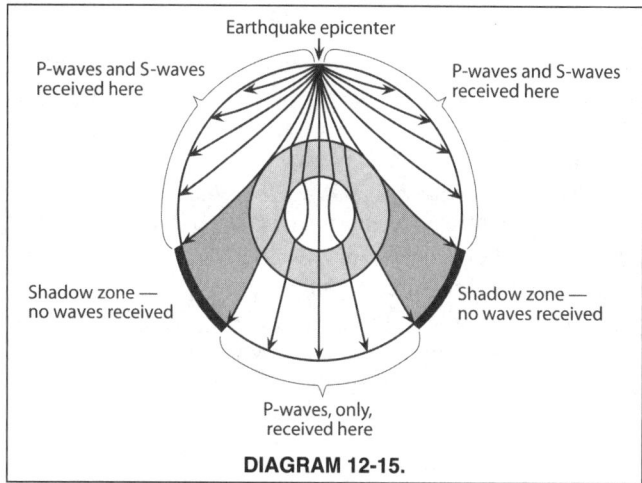

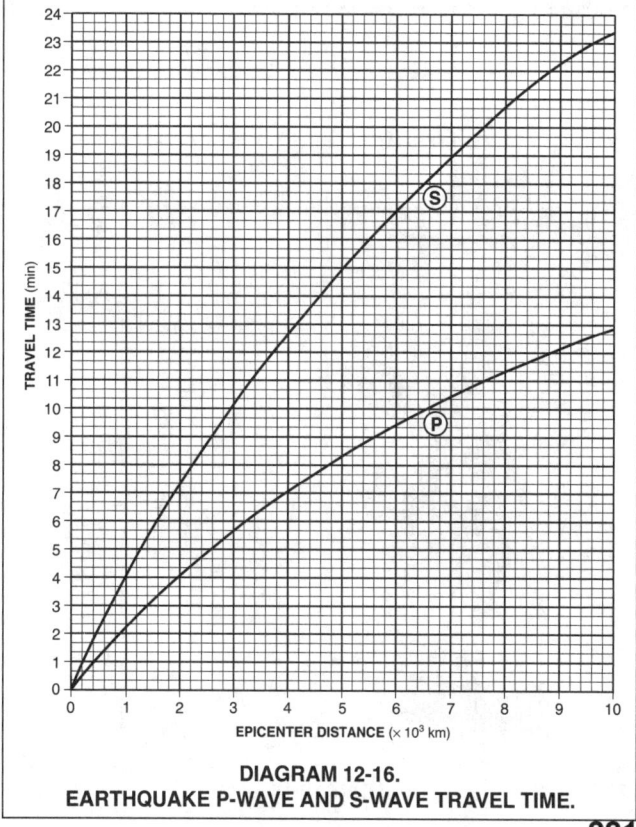

DIAGRAM 12-16.
EARTHQUAKE P-WAVE AND S-WAVE TRAVEL TIME.

The epicenter distance can be determined because as the distance from the epicenter increases, so does the difference between the arrival times of the P- and S-waves. Information from one seismic station can only be used to determine the distance to the epicenter not direction to it.

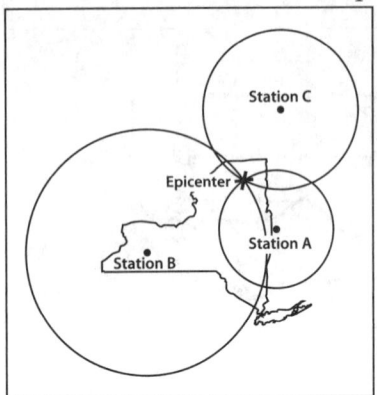

DIAGRAM 12-17.

To locate the epicenter, seismic data from three stations must be used. For each seismogram the distance to the epicenter for the station can be determined. On a map, circles are drawn around each seismic station with the distance to the epicenter as the radius. The intersection of the three circles is the epicenter. On the diagram to the left the epicenter is at point **X**.

DIAGRAM 12-18.

To locate an epicenter do the following:

1. Record the arrival times of the P- and S-waves. Read each seismogram carefully, the time may NOT start at 0 and be sure you know the interval that each hour or minute is divided into. It may be in intervals of 10, 20, or 30;
2. Carefully subtract the arrival times to find the difference in arrival times. Remember there are 60 seconds in each minute and 60 minutes in each hour;
3. Use the "*P-Wave and S-Wave Travel Time*" graph on page 11 of the ESRT to find the distance to the epicenter by using the difference in arrival times. Use a straight edge;
4. Draw a circle with a radius equal to the epicenter distance around each seismic station. When using the map scale, be precise! Measure carefully.
5. The intersection of the three circles is the epicenter. If your circles do not neatly intersect, place your epicenter at the closest location to where they would all appear to meet.

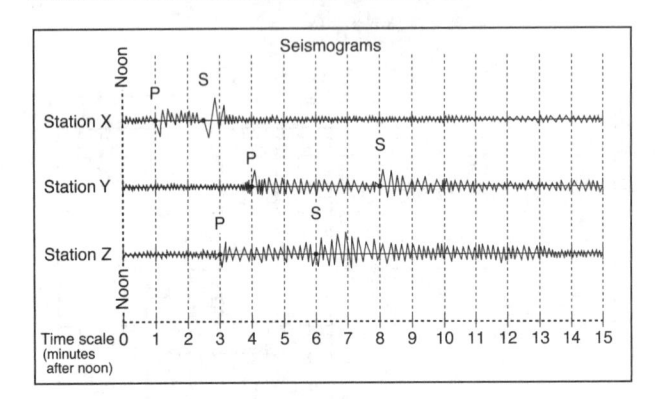

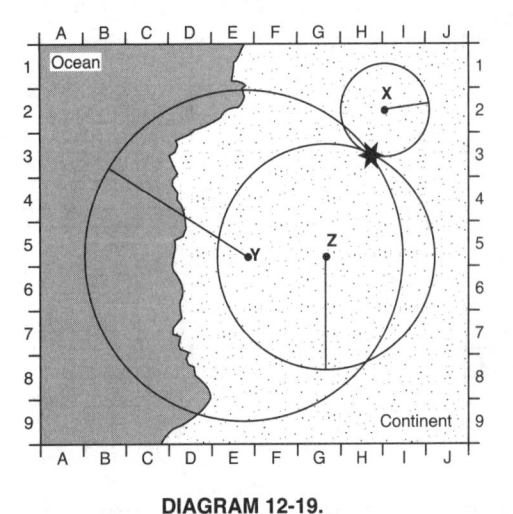

The diagram to the left shows three seismograms recorded from the same earthquake by three seismic stations: **X**, **Y**, and **Z**.

- For station **X**, the P-wave and S-wave arrive close together because this station is the closest to the epicenter. Therefore its circle is smaller.

- Station **Y** has the largest spacing between the P- and S- waves and therefore the largest circle.

- The "✱" is the epicenter where all three circles intersect.

DIAGRAM 12-19.

222

There are two classification systems for earthquakes: the Mercalli scale and the Richter scale. The **Mercalli scale** rates the earthquake by the intensity of the ground motion felt by people and the effect on objects. It describes how the earthquake affected humans, land, and buildings. One earthquake will have different ratings because the effects will differ due to distance from the epicenter. It is a subjective rating. The **Richter scale** rates the magnitude of waves measured by a seismograph. It indicates the energy level of the earthquake.

Earthquakes can have devastating effects on people and the landscape. Severe shaking can damage or destroy buildings and bridges, and break water and natural gas lines. The amount of damage is dependent on the type of ground the structure is on and the length of the vibration time. Aftershocks can occur for days after the earthquake and add to the damage. Landslides can occur especially in areas with many hills and loose soil. **Tsunamis**, seismic sea waves, endanger coastal areas. Earthquakes are difficult to predict. Buildings in earthquake prone areas should be flexible and strong. People should avoid building on steep slopes and near faults. Every person who lives in an earthquake zone should have a plan of action in case an earthquake occurs.

Questions

18. The map below shows the intensity of on earthquake based on the Mercalli scale.

Modified Mercalli Scale

Intensity	Observed Effects
I	Felt by only a few people under very special circumstances
II	Felt by only a few people at rest, especially on the upper floors of buildings
III	Felt noticeably indoors, especially on upper floors of buildings
IV	Felt indoors by many people, outdoors by a few; some awaken
V	Felt by nearly everyone; many awaken; dishes and windows break; plaster cracks
VI	Felt by everyone; many frightened and run outdoors; heavy furniture moves
VII	Everyone runs outdoors; slight to moderate damage in ordinary structures
VIII	Considerable damage in ordinary structures; chimneys and monuments fall
IX	Considerable damage in all structures; ground cracks; underground pipes break
X	Most structures destroyed; rails bend; landslides occur; water splashes over banks
XI	Few structures left standing; bridges destroyed; broad fissures in the ground; underground pipes break
XII	Damage total; waves seen on ground surfaces; objects thrown in air

a. How far was Boston from the epicenter? _____

b. Is it common for an earthquake to occur in this area? _____

Why or why not? _____

c. In which city would heavy furniture have moved? _____

d. In which city was the arrival times of the **P**- and **S**-waves furthest apart? _____

e. What was the travel time of the **P**-wave from the epicenter to Syracuse? _____

223

19. Complete the chart which compares the P-wave and the S-wave.

	P-wave	**S-wave**
also known as		
type of wave..........		
arrives (*1st*) or (*2nd*)?		
materials it can travel through.		

20. The seismogram below shows the arrival of the **P-wave** and **S-wave** from an earthquake at a seismic station.

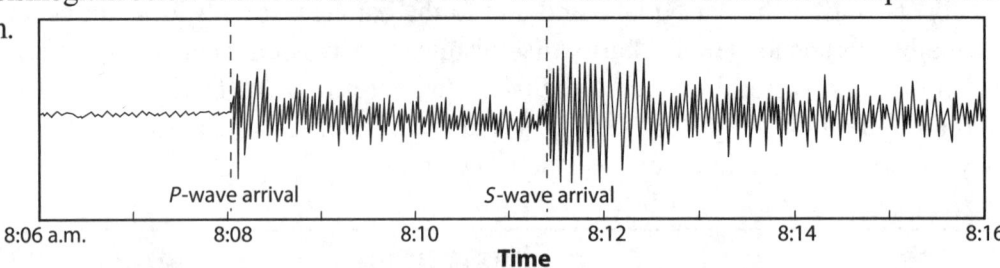

    a. The S-wave arrived at _____ a.m.
    b. Explain why the S-wave arrived after the P-wave. _____
    _____
    c. Determine the distance that this seismic station was from the epicenter. _____
    d. To locate the epicenter what other information do you need? _____
    _____
    e. Explain the relationship between the density of the rock and the speed of the seismic waves. _____
    f. Geologists quickly determined that the epicenter was located on the floor of Indian Ocean. What are they now concerned about? _____

21. Use the "*P-Wave and S-wave Travel Time*" graph in the ESRT to do the following.
    a. How far will an S-wave travel in 10 minutes and 40 seconds? _____
    b. How far will a P-wave travel in 7 minutes and 30 seconds? _____
    c. An earthquake occurred at 10:40:10 p.m. When will the S-wave arrive at a seismic station that is 4400 kilometers away? _____
    d. A seismic station received the P-wave at 6:28:30 a.m. and the S-wave at 6:33:50 a.m. How far is the epicenter? _____
    e. A seismic station 3000 kilometers away from the epicenter received the P-wave at 3:45:40 p.m. At what time did the earthquake occur? _____
    f. An earthquake occurred at 5:29:40 p.m. The P-wave was received at 5:32:50 p.m. How far away is the epicenter? _____
    g. How far apart are the P- and S- waves received at a station 6600 kilometers from the epicenter? _____

22. The data chart gives information collected at four seismic stations, **W, X, Y,** and **Z,** for the same earthquake.

Seismic Station	P-Wave Arrival Time (h:min:s)	S-Wave Arrival Time (h:min:s)	Difference in Arrival Times (h:min:s)	Distance to Epicenter (km)
W	10:50:00	no S-waves arrived	--------	---------
X	10:42:00	10:46:40		
Y	10:39:20		00:02:40	
Z	10:45:40			6200

  a. Fill in the missing data for the chart.

  b. What is a possible reason that station **W** did not receive **S**-waves? _____
  _____

23. The map below shows a portion of the Indian Ocean and surrounding land areas that were affected by a large undersea earthquake that occurred on December 26, 2004. A tsunami was generated by this earthquake. The isolines show the approximate location of the tsunami in half-hour intervals after the earthquake.

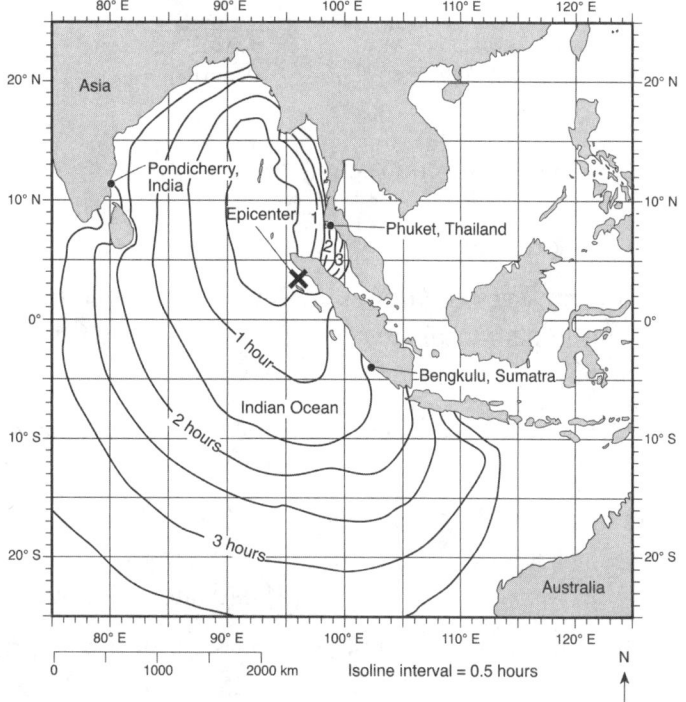

  a. How long did it take the tsunami to reach Pondicherry, India? _____

  b. State the latitude and longitude of the epicenter. _____, _____

  c. This quake occurred along a (*transform*) (*divergent*) (*convergent*) plate boundary.

  d. Name the overriding plate. _____

  e. Calculate the rate of movement of the tsunami that reached Bengkulu, Sumatra in kilometers per hour to the nearest tenth. _____

  (*show solution*)

225

CHAPTER 12 REVIEW

1. Why do the planets in our solar system have a layered interior structure?
   (1) All planets cooled rapidly after they formed.
   (2) The Sun exerts a gravitational force on the planets.
   (3) Each planet is composed of materials of different densities.
   (4) Cosmic dust settled in layers on the planets' surfaces.

2. Theories about the composition of Earth's core is based on the composition of
   (1) meteorites (2) comets (3) ocean crust (4) Moon rocks

3. Scientists have classified Earth's interior into zones based on evidence gained from:
   (1) the Moon's interior   (3) rock from deep drilling
   (2) seismic waves   (4) volcanic eruptions

4. Which graph best illustrates the range of density in each of Earth's interior layers?

   (1)   (2)   (3)   (4)

5. Compared to the oceanic crust, the continental crust is
   (1) thicker, denser, more mafic   (3) thinner, denser, more mafic
   (2) thicker, less dense, more felsic (4) thinner, less dense, more felsic

6. Which temperature and pressure conditions conditions are found in the asthenosphere?
   (1) 1,000° C and 10 million atmospheres   (3) 3,500° C and 0.5 million atmospheres
   (2) 2,000° C and 0.1 million atmospheres (4) 6,000° C and 3.0 million atmospheres

7. Approximately how far below Earth's surface is the outer core-inner core interface?
   (1) 800 km (2) 2,900 km (3) 5,100 km (4) 6,200 km

8. Rocks located 2,900 to 5,100 kilometers below Earth's surface are inferred to be
   (1) iron-rich solid   (3) silicate-rich solid
   (2) iron-rich liquid   (4) silicate-rich liquid

9. The rate of temperature increase in Earth's interior is greatest between depths of
   (1) 250 and 500 km   (3) 2500 and 3500 km
   (2) 1500 and 2500 km   (4) 3500 and 4000 km

10. Which interior layer is characterized by partially melted rock and large-scale convection currents?
    (1) crust (2) lithosphere (3) asthenosphere (4) outer core

11. Which interior layer has a density similar to the densities of the other terrestrial planets?
    (1) crust (2) mantle (3) outer core (4) inner core

12. The temperature at the center of Earth is inferred to be about
    (1) 5,200° C (2) 6,300° C (3) 6,700° C (4) 7,000° C

13. How many millions of years ago were the continents last together?
    (1) 59 m.y.a. (2) 119 m.y.a (3) 232 m.y.a (4) 458 m.y.a

14. Evidence of crustal subsidence is provided by
    (1) zones of igneous activity at mid-ocean ridges
    (2) high heat flow at mid-ocean ridges
    (3) marine fossils found on mountaintops
    (4) shallow water fossils found in rock of the deep ocean floor

15. Which mountain range resulted from the collision of North America and Africa during the late Pennsylvanian Period?
    (1) Allegheny Mountains
    (2) Acadian Mountains
    (3) Taconic Mountains
    (4) Grenville Mountains

16-17. The diagram below shows the magnetic polarity preserved by minerals in the bedrock of oceanic crust near the Mid-Atlantic Ridge. Letters **A**, **B**, **C**, and **D** represent locations in the ocean-floor bedrock.

16. The oldest bedrock is found at location
    (1) A
    (2) B
    (3) C
    (4) D

17. The rocks at which locations formed at a time when the north magnetic pole was located at 90° South?
    (1) A and B
    (2) A and C
    (3) B and D
    (4) C and D

18. Which information indicates that new sea floor forms at an ocean ridge and moves away from the ridge?
    (1) Most volcanoes are located under oceans.
    (2) Seafloor rock is older than continental rock.
    (3) Fossils of marine organisms are found at high elevations.
    (4) The age of the sea floor rocks increases further away from the ocean ridge.

19. Movement of crustal plates is most likely caused by
    (1) Earth's rotation
    (2) tsunamis
    (3) reversals of Earth's magnetic poles
    (4) mantle convection currents

20. Which of the following is a converging plate boundary?
    (1) Southwest Indian Ridge
    (2) Marianas Trench
    (3) East African Rift
    (4) San Andreas Fault

21. What do mid-ocean ridges and hot spots have in common?
    (1) Rising magma and/or heat upwells in these areas.
    (2) They are both located along crustal plate boundaries.
    (3) Local earthquakes occur at great depths here.
    (4) Neither is associated with crustal plate motion.

22. On which plate is the Canary Islands Hot Spot located?
    (1) South American   (2) African   (3) Nazca   (4) Pacific

23. Which location is not at the edge of a plate?
    (1) Easter Island   (2) Galapagos   (3) Hawaii   (4) California coast

24. The interface between the Antarctic Plate and the Pacific Plate is best described as
    (1) converging and located at an ocean ridge
    (2) converging and located at an ocean trench
    (3) diverging and located at an ocean trench
    (4) diverging and located at an ocean ridge

25. What plate boundary is found at 40° S and 80° E?
    (1) East Pacific Ridge
    (2) Southeast Indian Ridge
    (3) Mid-Atlantic Ridge
    (4) Peru-Chili Trench

26. The data table to the right shows the origin depths of all large-magnitude earthquakes over a 20-year period.
    According to this data, most of these earthquakes occurred within Earth's
    (1) lithosphere
    (2) asthenosphere
    (3) stiffer mantle
    (4) outer core

Depth Below Surface	Number of Earthquakes
0-33	27,788
34-100	17,585
101-300	7,329
301-700	3,167

27. Which coastal area is most likely to experience a severe earthquake?
    (1) east coast of North America
    (2) east coast of Australia
    (3) west coast of Africa
    (4) west coast of South America

28. The P-waves from an earthquake can travel through Earth's
    (1) crust, only
    (2) crust and mantle, only
    (3) crust, mantle, and inner core, only
    (4) crust, mantle, outer core, and inner core

29. The seismogram shows the arrival times of P- and S-waves at a seismic station. Approximately how far is this seismic station from the earthquake epicenter?
    (1) 1,650 km
    (2) 1,900 km
    (3) 2,200 km
    (4) 4,100 km

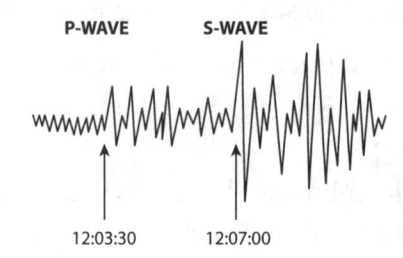

30. The epicenter of an earthquake is 6,000 kilometers from a seismic station. What is the difference in travel time for the P-waves and S-waves?
    (1) 7 min 35 sec   (2) 9 min 20 sec   (3) 13 min 10 sec   (4) 17 min 00 sec

31. An earthquake P-wave arrived at a seismograph at 1:21:40 p.m. The distance to the epicenter is 3,000 kilometers. The earthquake's origin time was
    (1) 1:11:40 p.m.   (2) 1:16:00 p.m.   (3) 1:20:20 p.m.   (4) 1:27:20 p.m.

32. How long will it take for the first S-wave to arrive at a seismic station 4,000 kilometers from the epicenter of an earthquake?
    (1) 5 min 40 sec   (2) 7 min 0 sec   (3) 12 min 40 sec   (4) 13 min 20 sec

33. An earthquake occurred in Massena, New York. For which *two* locations would the P-wave arrival times be approximately the same?
    (1) Rochester and New York City
    (2) Binghamton and Slide Mountain
    (3) Utica and Watertown
    (4) Watertown and Oswego

34. The diagram below shows three circles drawn around three seismic stations. The epicenter is located at point

    (1) G   (3) E
    (2) F   (4) D

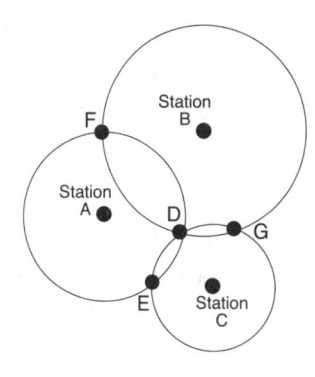

35. What information can be determined by using this seismogram?

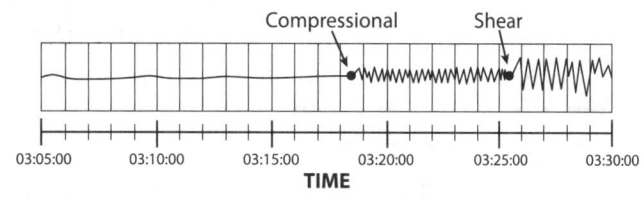

    (1) depth of the earthquake's focus
    (2) direction to the earthquake's focus
    (3) location of earthquake's epicenter
    (4) distance to the earthquake's epicenter

36-39. The diagram below shows the interface between the Indian-Australian Plate and the Fiji Plate.

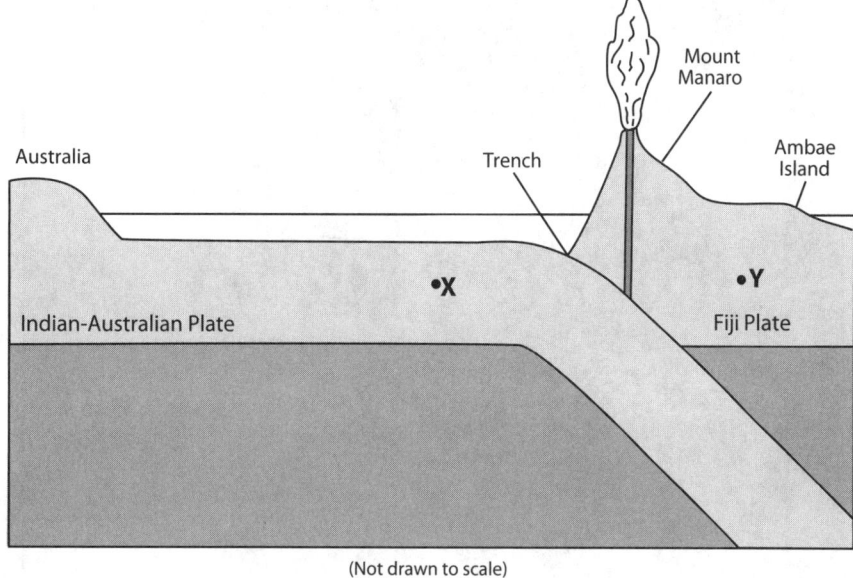

36. This interface is a (*diverging*) (*converging*) plate boundary.

37. Use an arrow to show the movement of the Indian-Australian Plate.

38. Compare the density of the rock at **X** with the density of the rock at **Y**. _____

39. Explain the cause for the Mount Manaro volcano at this boundary. _____

40-42. Base your answers on the following reading and on the map below. The numbers on the map show the predicted relative damage at various locations if a large earthquake occurs along the New Madrid fault system. The higher the number the greater the relative damage.

> **The New Madrid Fault System**
>
> The greatest earthquake risk area east of the Rocky Mountains is along the New Madrid fault system. The New Madrid fault system consists of a series of faults along a weak zone in the continental crust in midwestern United States. Earthquakes occur in the Midwest less often than California, but when they do happen, the damage is spread over a wider area due to the underlying bedrock.
>
> In 1811 and 1812, the New Madrid fault system experienced three major earthquakes. Large areas sank, new lakes formed, the course of the Mississippi River changed, and 150,000 acres of forests were destroyed.

40. On the map draw the **4**, **6**, and **8** isolines indicating relative damage.

41. Using the damage numbers, place an **X** on the map to indicate where the New Madrid fault system is most likely located.

42. An emergency management specialist near the New Madrid region is developing a plan that would save lives and prevent property damage from an earthquake in this area. Describe *two* actions that should be included in this plan.

a. _____

b. _____

43-46. The eruption of Mt. St. Helens in 1980 resulted in the movement of volcanic ash across northwestern United States. The movement of the ash is shown on the map. The times marked along the path indicate the length of time it took the ash cloud to travel to each location.

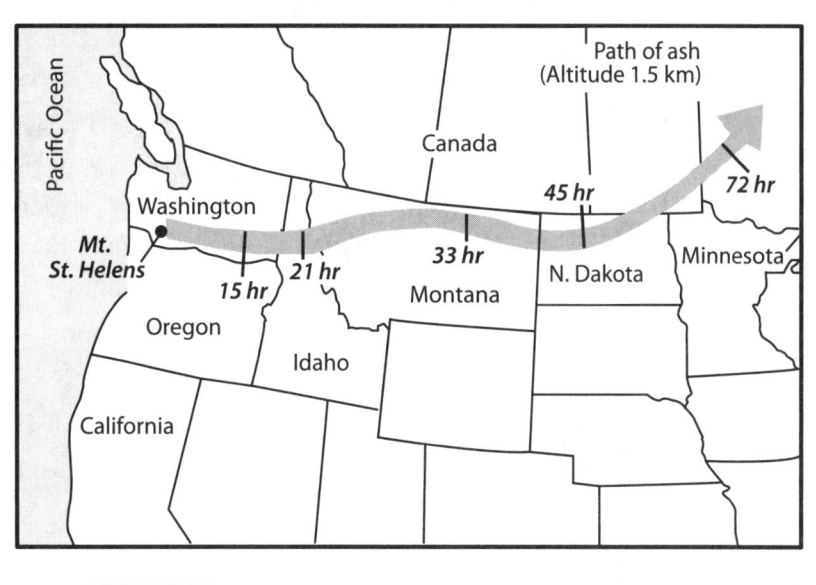

43. Calculate the average rate of movement of the volcanic ash for the first 15 hours. _____
(show solution)

44. State the direction the volcanic ash cloud moved towards. _____

Explain why the ash cloud moved in this direction. _____

_____

45. The ash cloud was located at 1.5 kilometers above Earth's surface. State the name of the atmospheric layer the ash cloud was in. _____

46. As the ash cloud moved, how did it affect the weather of the areas below it? _____

_____

_____

47-50. The diagram below represents three seismograms from the same earthquake for three different seismic stations: **A**, **B**, and **C**.

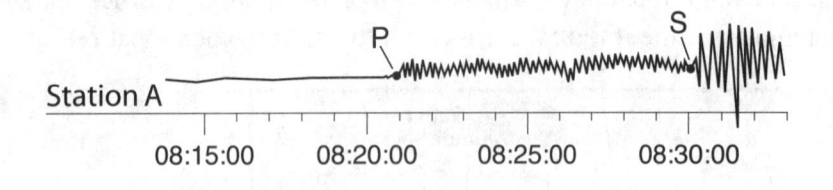

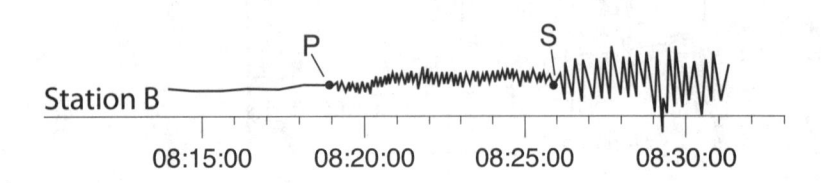

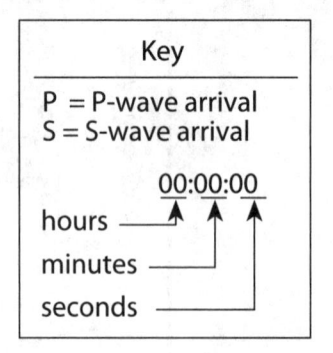

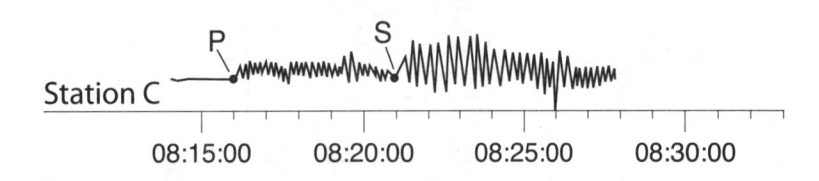

Station A — P arrival at approximately 08:22:00, S arrival at approximately 08:31:00

Station B — P arrival at approximately 08:20:00, S arrival at approximately 08:26:00

Station C — P arrival at approximately 08:17:00, S arrival at approximately 08:20:00

Key
P = P-wave arrival
S = S-wave arrival

00:00:00
hours — minutes — seconds

47. The P-wave arrived at Station **B** at what time? _____

48. Determine the distance that station **C** is from the epicenter. _____

49. Which seismic station is closest to the epicenter? _____

    How do you know? _____

50. How would the seismograms appear if the seismic waves had to pass through the cores of the Earth? _____

_____

232

CHAPTER 13

EARTH'S GEOLOGIC HISTORY
Earth's History, Geologic Time Line and Fossils, Radioactive Dating, Interpreting Geologic History

EARTH'S HISTORY

The oldest rocks found on Earth are estimated to be about 4.2 billion years old. Earth is inferred to be older because the original crustal rocks have been eroded and weathered. Moon rocks and meteorites have been dated to be 4.6 billion years old which is inferred to be the age of Earth when the solar system and Sun formed from the collapse of a giant nebula.

In the beginning, Earth's crust was molten. Most of the gases surrounding this early Earth escaped into space because they were very light elements. After hundreds of millions of years, a solid crust formed and tectonic activity began. Earth's early atmosphere was formed by the outgassing of water vapor, carbon dioxide, nitrogen, and other gasses from Earth's interior through cracks and volcanic eruptions. Earth's oceans formed as a result of precipitation over millions of years from this outgassed water vapor as well as water brought to Earth by collisions with comets. The ocean waters began to absorb the carbon dioxide from the atmosphere. The presence of an early ocean is indicated by sedimentary rocks of marine origin dated to be about 4 billion years old. Weathering of crustal rocks added salts to the ocean.

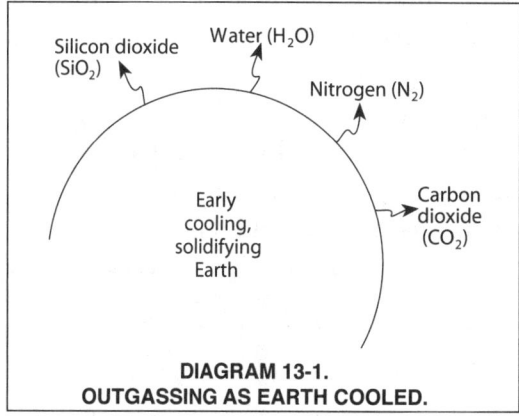

DIAGRAM 13-1. OUTGASSING AS EARTH COOLED.

The development of life caused changes in the composition of Earth's atmosphere. Free oxygen did not form in the atmosphere until oxygen-producing organisms evolved. By 3.3 billions of years ago, one-celled marine organisms called cyanobacteria were present. These photosynthetic organisms often lived in colonies called stromatolites. They absorbed carbon dioxide and released oxygen into the atmosphere.

Questions

1. The age of Earth and solar system is estimated to be _____. Circle this on the diagram to the right.
2. Earth's atmosphere became oxygen-rich in the _____ eon.
3. What process was responsible for the addition of oxygen in the atmosphere? _____
4. Compared to the age of the universe, Earth is (*younger*) (*older*) (*the same age*).
5. How many billions of years after the beginning of the universe did Earth's crust form? _____
6. Life first appeared on land during the _____ era.

233

GEOLOGIC TIME LINE AND FOSSILS

Earth's geologic history is divided into units based on fossil evidence and major geologic events. There are two major **eons**: Precambrian and the present Phanerozoic. The Precambrian eon covered about 85% of Earth's history. It was during this time that Earth's environment evolved to become suitable for modern life forms. During the Phanerozoic eon life forms developed and became complex and diverse.

Eons are divided into **eras** based on the dominant life forms present. Eras end with a major **orogeny** (mountain building event) which changes landscapes and climate and affects life forms. At the end of the Paleozoic and Mesozoic eras major extinctions occurred. Eras are divided into **periods**; periods are divided into **epochs**.

Fossils are evidence of animals and plants that have lived on Earth in the past. A fossil can be the actual remains or a mold, cast, or imprint of the organism. To become a fossil an organism must have had hard parts and/or was quickly buried in sediment. Fossils are found in sedimentary rocks. They provide clues to the environment the organism lived in.

Most of geologic time is devoid of fossil evidence. Fossils are only present in rocks from the last 500 million years. Fossils are difficult to detect in Precambrian rock because the earliest fossils were very small and did not have hard parts. Many of these rocks have been weathered, eroded, or metamorphosed. Most of the fossils represent species that became extinct due to a sudden or drastic environmental change. It is probable that large numbers of past life forms have left no traces in rocks.

The history of life from fossil records shows that the first fossils were simple marine organisms, such as algae, found in rocks over three billion years old. More complex organisms developed (evolved) as time went on.

Questions

7. Place these units of geologic time in order from largest to smallest.
   (*eon*) (*epoch*) (*era*) (*period*): _____, _____, _____, _____

8. State our present geologic time.
   eon:_____ era: _____ period:_____ epoch: _____

9. Complete the chart based on the "*Bedrock Geology of New York State*" map in the ESRT.

	Eon	Period	rock type
a. Allegheny Plateau			
b. shoreline of Lake Ontario			
c. Syracuse	Phanerozoic		
d. Mt. Marcy			
e. Jamestown			sedimentary
f. 44° 55' N, 74° W			
g. 41° 05' N, 74° 05' W			
h. New York City			

10. On a field trip to upstate New York, a student finds the following fossil assemblage, *Stylonurus*, *Manticoceras*, and *Phacops*, in the surface bedrock.

   a. What geologic period are these rocks from? _____

   b. What landscape region did the student visit? _____

11. Name the geologic eon described.
   a. longest period of time: _____
   b. atmospheric oxygen developed: _____
   c. life forms became diverse: _____
   d. eon we are in now: _____
   e. oldest known rocks: _____
   f. late Archean: _____

12. The strip below shows the relative lengths of the geologic periods of the Phanerozoic.

   | Paleozoic | Mesozoic | Cenozoic |

   Positions: 1, 2, 3, 4 (Present day)

   Select the number or numbers that apply for each statement.

   a. invertebrates are dominant. _____   c. humans. _____

   b. advance of last continental ice. _____   d. earliest dinosaurs. _____

13. Complete the chart.

	Era	Period (include late, middle, early)
a. earliest mammals	Mesozoic	
b. Earth's first forests		
c. earliest flowering plants		early Cretaceous
d. large carnivorous mammals		
e. abundant amphibians		
f. 367 million years ago		
g. 10 million years ago		
h. Taconian orogeny		
i. initial opening of the Atlantic Ocean		
j. no New York State rock record		
k. fossil: *Centroceras*		
l. fossil: *Tetragraptus*		

14. Complete the chart for each fossil illustrated.

	fossil name	general fossil group name	geologic period
a.		dinosaur	
b.			
c.			
d.			
e.			
f.			

g. Which two fossils are approximately the same age? _____ and _____

h. The oldest fossil is _____; the youngest fossil is _____

15. Use the ESRT to answer these questions.
    a. Cretaceous is part of the _____ era.
    b. The Period that began 299 million of years ago was the _____.
    c. The shortest Mesozoic Period is _____.
    d. The epoch that ended 23 m.y.a. is the _____.
    e. The Ordovician lasted for _____ million of years.
    f. The Cambrian index fossil is _____.
    g. *Maclurites* fossil is classified as a (*trilobite*) (*coral*) (*gastropod*) (*brachiopod*).
    h. The Taconic orogeny occurred during which period? _____
    i. The Alleghenian orogeny was caused by the _____.
    j. The New York State landscape regions caused by the Grenville orogeny are the _____ and _____.
    k. Name a crinoid fossil from the Silurian. _____

RADIOACTIVE DATING

The age of a rock, fossil, or material can be estimated by using **radioactive elements**. **Radioactive dating** compares the amount of the radioactive element to the amount of the stable element (decay product) found in the sample. Most elements exist in more than one form called isotopes. **Isotopes** of an element have the same number of protons but different number of neutrons and therefore a different atomic mass. Some isotopes are unstable. The unstable, **radioactive isotope** "disintegrates" (changes) into a stable element.

A radioactive element's **half-life** is the time it takes for one-half of the radioactive element to "decay" (change) to a stable element. Half-life is a constant value, nothing will change its value. Half-lives range from billions of years to seconds. Radioactive elements with short half-lives, such as C-14, are used for dating recent organic materials and rocks that are less than 50,000 years old from the Cenozoic era. Those with longer half-lives, such as K-40, are useful for dating older rocks.

number of half-lives	% (fraction) of radioactive element A	% (fraction) of stable decay product B
0	100 (1)	0 (0)
1	50 (½)	50 (½)
2	25 (¼)	75 (¾)
3	12.5 (⅛)	87.5 (⅞)
4	6.25 (1/16)	93.75 (15/16)

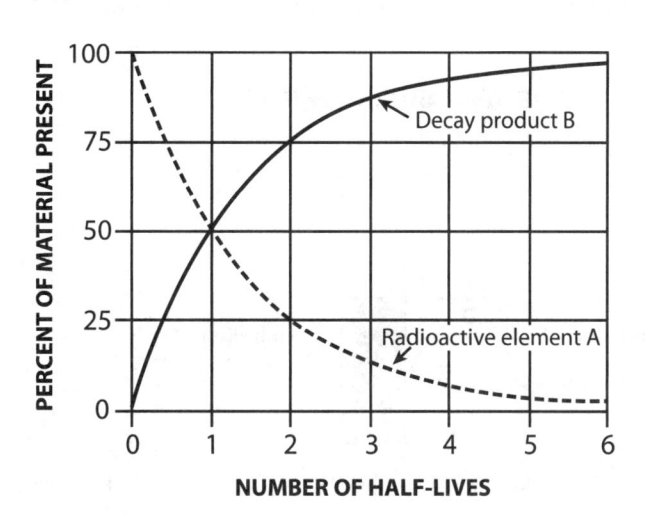

**DIAGRAM 13-2.
RADIOACTIVE ELEMENT vs. STABLE DECAY PRODUCT.**

237

Questions

16. Which radioactive isotope can not be used to date Moon rocks? _____
    Why not? _____
    _____

17. How much of 20 g of K-40 remains after 3.9 billion yrs? _____

18. A fossil bone has 100 g of C-14 and 700g of N-14. How old is the bone? _____

19. A rock has only one-fourth of its original K-40. How old is the rock? _____

20. Carbon-14 can not be used to date fossils from the Paleozoic. Why not? _____
    _____

21. Name the radioactive isotope described.
    a. stable product is Ca-40: _____
    b. shortest half-life: _____
    c. half-life is more than Earth's age: _____
    d. longest half-life: _____

22. The diagram below shows the "decay" of a radioactive element into a stable element for one half-life.

Radioactive Decay Model

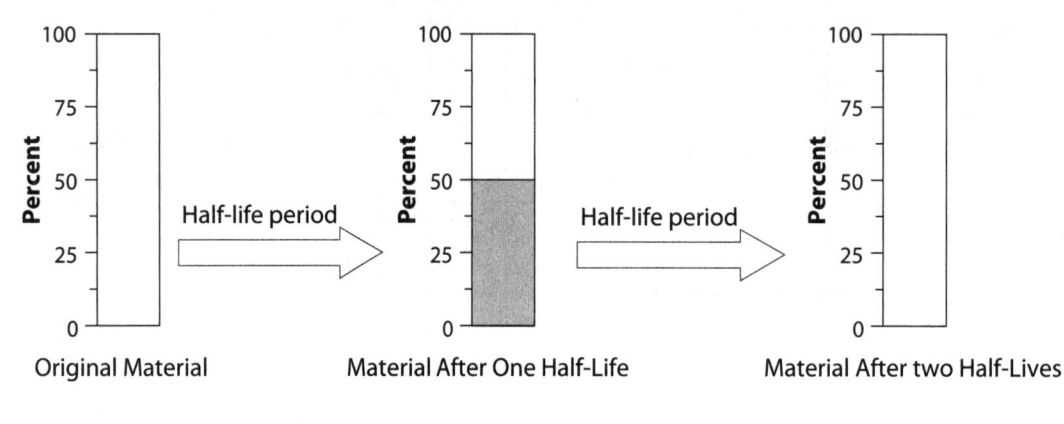

Original Material → Half-life period → Material After One Half-Life → Half-life period → Material After two Half-Lives

Key:
- Radioactive element (white)
- Stable decay element (shaded)

a. Complete the last diagram for the "material after two half-lives" by shading in the amount of stable decay element after two half-lives.

b. If the radioactive element in this model had been Uranium-238, how much time will have passed for one half-life? _____

238

23. The graph below shows the "decay" of 50 grams of a radioactive isotope.

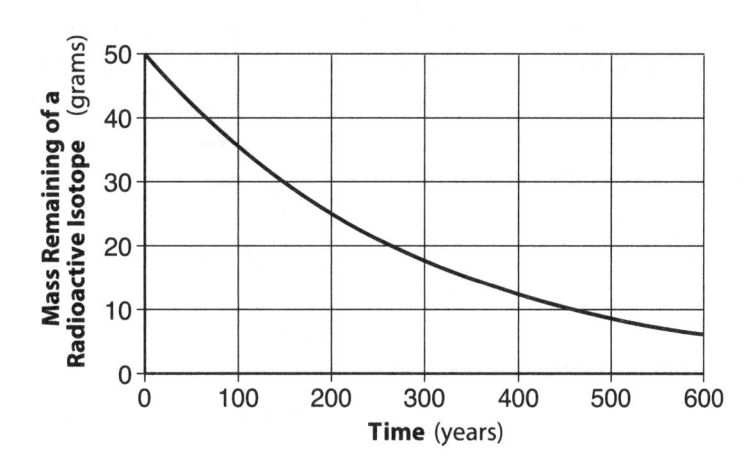

   a. What is this isotope's half-life? _____

   b. If this isotope was in a rock that was metamorphosed, how would the half-life be affected?
   _____

   c. After 350 years, how much of the 50 grams of the isotope will remain? _____

   d. On the graph, draw a dashed line for the "stable decay product." Label the line.

24. The data chart to the below shows the decay of radioactive carbon-14.

Radioactive Decay of Carbon-14

Number of Half-Lives	Percentage of Original Carbon-14 Remaining	Time (years)
0	100	0
1	50	5700
2	25	11,400
3	12.5	17,100
4	6.25	
5	3.125	28,500
6	1.562	34,200

   a. Complete the data chart for the time for 4 half-lives of carbon-14.

   b. Plot the data with an "**X**" for each half-life and percentage of original carbon-14. Connect the "**Xs**" with a smooth curve.

239

Interpreting Geologic History

Geologic history involves the study of geologic events which have occurred in the past. Earthquakes, erosion, deposition, uplift, folding, rock formation, volcanic activity, and metamorphism are examples of **geologic events**. Geologic history is preserved in Earth's rocks and is interpreted by observations of the composition, structure, position, and fossil content of the rock layers.

The process of **relative dating** is the sequencing of geologic events and rocks from oldest to most recent. The **principle of superposition** states that the bottom layer in a series of undisturbed sedimentary rock layers is the oldest. The **principle of original horizontality** states that rock layers should be horizontal. If not, then geologic events such as folding, faulting, and tilting occurred after the rock formed.

Rock layers are older than faults, joints (cracks), mineral veins, intrusions, and folds that go through them. In sedimentary rocks, the sediments are older than the sedimentary rock and the mineral cement that holds the rock together.

Igneous Intrusion

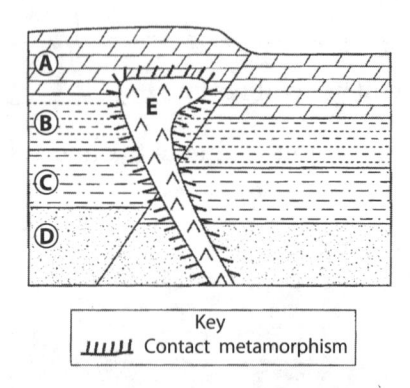

An **intrusion** is magma that cuts across rock layers. It will contact metamorphose the rock layers it touches.

- Intrusions are younger than any feature it cuts across and the rock it metamorphoses.
- This intrusion (**E**) is younger than rock layers (**A,B,C,D**) and the fault.

Igneous Extrusion

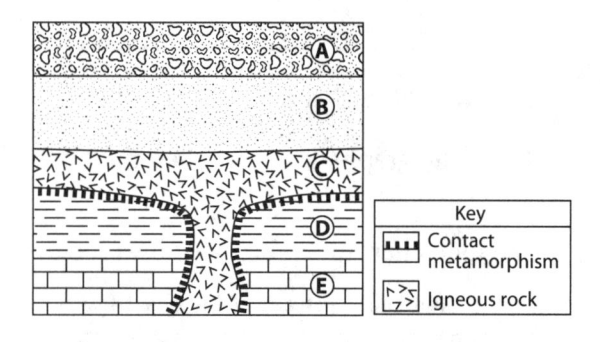

An **extrusion** of lava will cut across rock layers and flow on the surface. An extrusion contact metamorphoses the rock it cuts across and the rock below it.

- This is a buried extrusion (**C**).
- Rock layers **A** and **B** formed after the extrusion (**C**).
- This extrusion (**C**) is older than **A** and **B** but younger than **D** and **E**.

DIAGRAM 13-3.

Examples of Relative Dating of Geologic Events

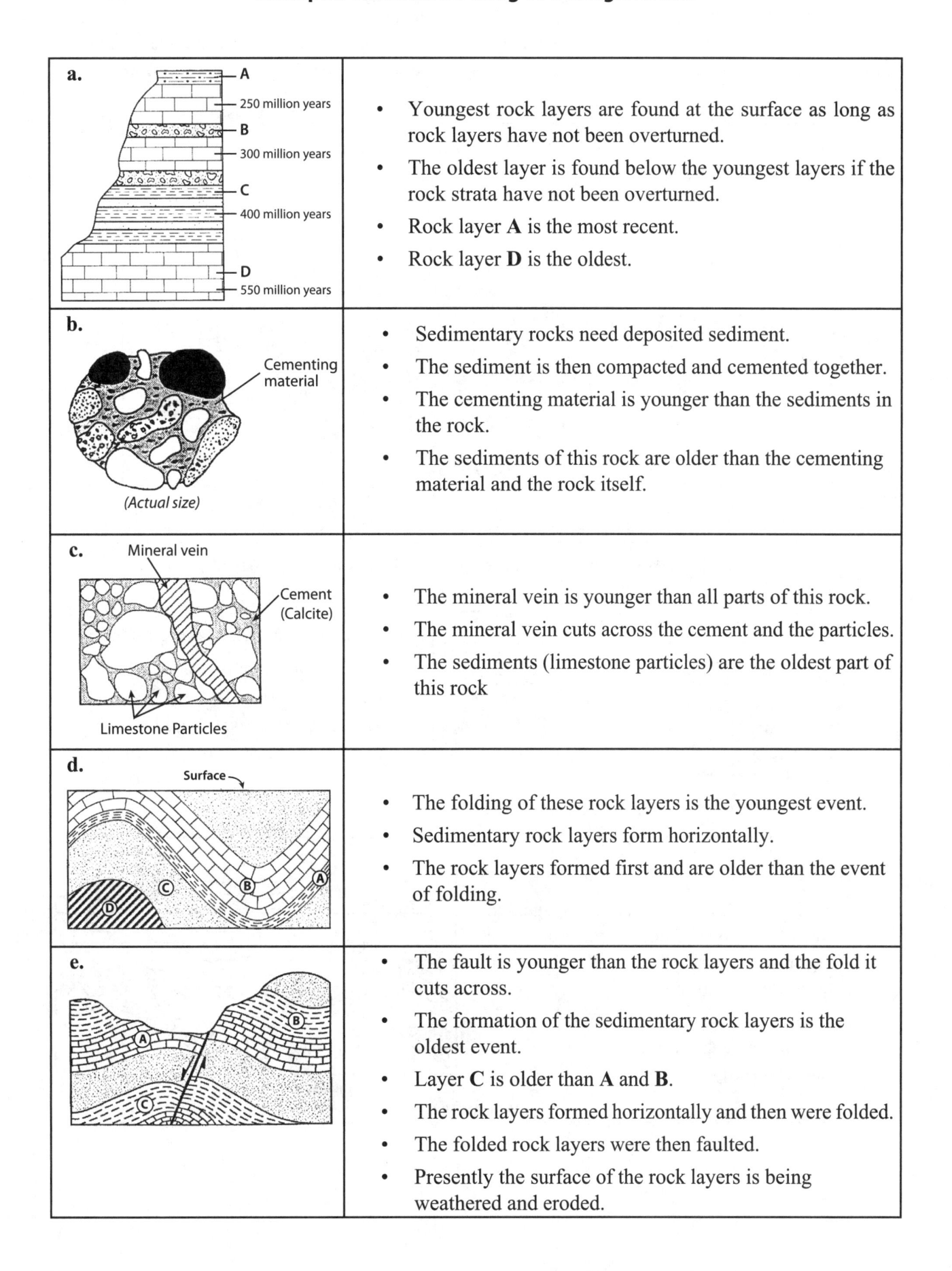

a. (diagram showing rock layers A at 250 million years, B at 300 million years, C at 400 million years, D at 550 million years)	• Youngest rock layers are found at the surface as long as rock layers have not been overturned. • The oldest layer is found below the youngest layers if the rock strata have not been overturned. • Rock layer **A** is the most recent. • Rock layer **D** is the oldest.
b. (diagram of sediment with cementing material, actual size)	• Sedimentary rocks need deposited sediment. • The sediment is then compacted and cemented together. • The cementing material is younger than the sediments in the rock. • The sediments of this rock are older than the cementing material and the rock itself.
c. (diagram showing mineral vein, cement (calcite), and limestone particles)	• The mineral vein is younger than all parts of this rock. • The mineral vein cuts across the cement and the particles. • The sediments (limestone particles) are the oldest part of this rock
d. (diagram of folded rock layers A, B, C, D with surface)	• The folding of these rock layers is the youngest event. • Sedimentary rock layers form horizontally. • The rock layers formed first and are older than the event of folding.
e. (diagram showing folded and faulted rock layers A, B, C)	• The fault is younger than the rock layers and the fold it cuts across. • The formation of the sedimentary rock layers is the oldest event. • Layer **C** is older than **A** and **B**. • The rock layers formed horizontally and then were folded. • The folded rock layers were then faulted. • Presently the surface of the rock layers is being weathered and eroded.

Unconformities are buried erosional surfaces which cause gaps in the rock record. Unconformities occur when rock layers are destroyed by erosion. They formed when crustal uplift exposed the rock layers to the forces of weathering and erosion. Subsequent subsidence (sinking) of the area may result in deposition and formation of new rock layers on top of the erosional surface (unconformity).

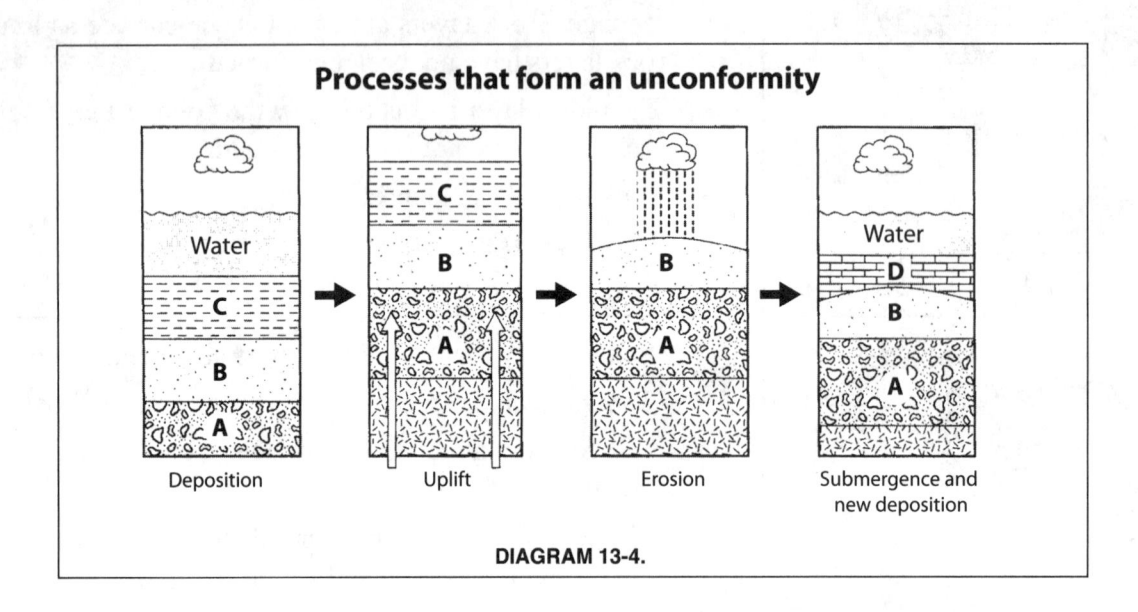

Processes that form an unconformity

Deposition → Uplift → Erosion → Submergence and new deposition

DIAGRAM 13-4.

Examples of geologic histories that include an unconformity.

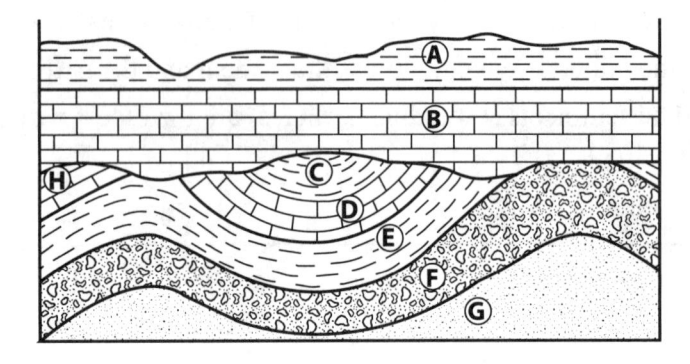

- Layers **G,F,E,D,C,H** formed.
- Layers **G,F,E,D,C,H** were folded.
- Uplift and erosion destroyed the top of the folds.
- Submergence in a water environment resulted in formation of sedimentary rock layers **B** then **A** above the unconformity.

- Using the geologic time line on page 8 of the ESRT, you will notice that Ordovician age rock is missing between the Cambrian and Silurian layer.
- Erosion destroyed the Ordovician rocks, leaving a gap in the rock record (an unconformity).

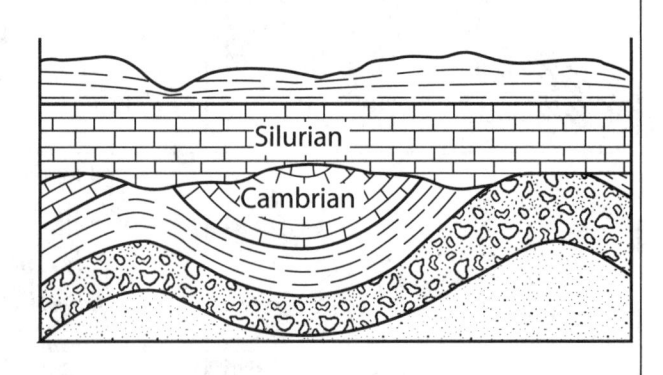

DIAGRAM 13-5.

Events in geologic history can often be placed in order of relative age by using fossil evidence. Some life forms existed for only specific periods of time. These organisms which lived over a large geographic range for a short-period of geologic time are called **index fossils**. The diagram to the right gives some examples of Paleozoic index fossils. Pages 8 and 9 of the ESRT list more examples.

Many times geologists will **correlate** (match) rock layers from one location to another as being of the same age. Methods of correlation include similarity in rock type or structure, similarity in position in the rock column, and similarity in index fossils. Layers of volcanic ash are good time markers because ash is quickly deposited over large surfaces of rock after a volcanic eruption.

BEDROCK AGE	INDEX FOSSIL
Mississippian	Spirifer
Devonian	Mucrospirifer
Silurian	Eospirifer
Ordovician	Michelinoceras

DIAGRAM 13-6.

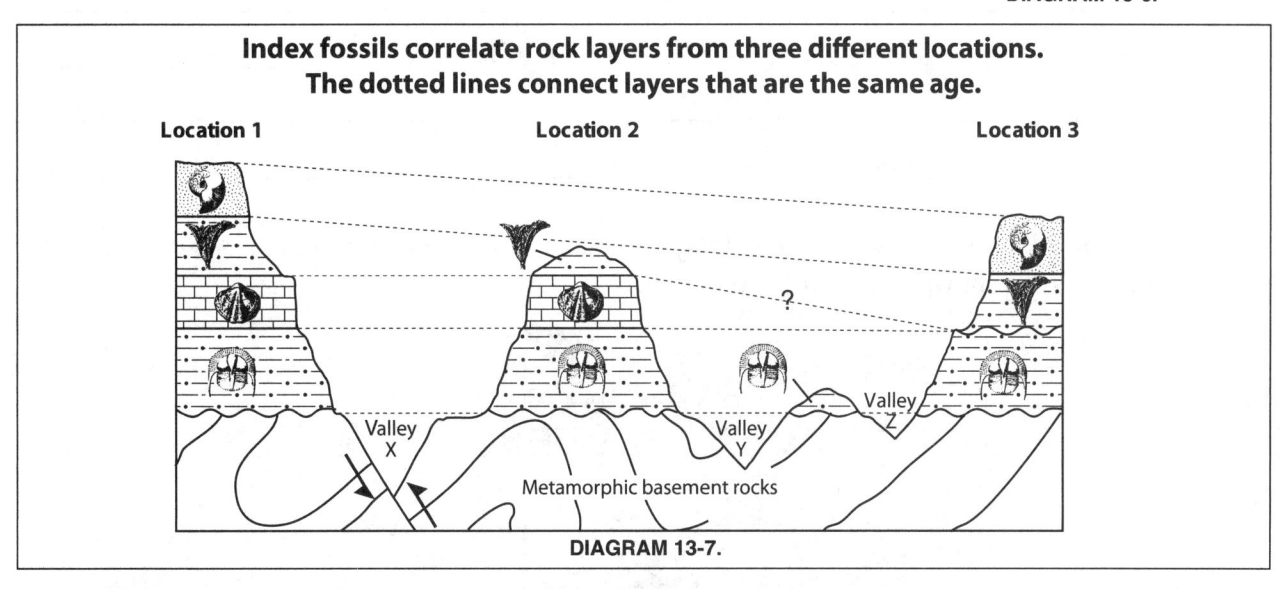

Index fossils correlate rock layers from three different locations. The dotted lines connect layers that are the same age.

DIAGRAM 13-7.

Questions

25. The diagram below shows two outcrops separated by 15 kilometers.

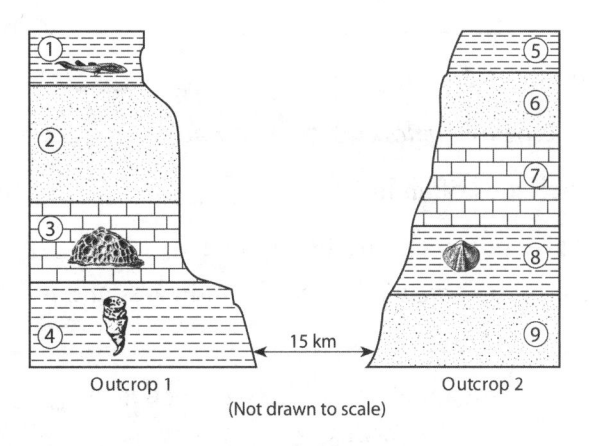

    a. In layer **4**, the fossil name is _____.

    b. Name the rock that makes up layers **4** and **8**. _____

    c. Layers **4** and **8** both formed during the _____ Period. How do you know this? _____

    d. Of these two outcrops, the oldest rock layer is # _____.

243

26. The geologic cross-section below shows bedrock layers **A** through **D**.

   a. Which letter represents the most recent rock layer? _____

   b. Line **XY** represents a _____ which is older than layer(s) _____.

   c. An unconformity exists under layer _____.

   d. How does an unconformity form? _____
   _____

   e. Why is there no layer **B** above layer **A** to the left of letter **X**? _____
   _____

   f. Name the rock that makes up layer **C**? _____

27. The block diagram below shows rock layers that have not been overturned. A New York State index fossil is shown in one of the rock layers.

   a. State the evidence that supports the inference that the fault is older than the rhyolite. _____
   _____

   b. The index fossil shown lived during the _____ period.

   c. This index fossil is a type of (*trilobite*) (*ammonoid*) (*coral*) (*graptolite*).

   d. Describe *two* characteristics of an index fossil. _____ and _____

   e. Describe the crystal size of mineral in the rhyolite. _____

   f. Explain what the size of the crystals in rhyolite indicates about the rate of cooling. _____
   _____

   g. Identify the metamorphic rock that most likely formed at **A**. _____

   h. What type of metamorphism occurred at **A**. _____

   i. What caused the type of valley shown on the surface? _____
   _____

   j. What type of crustal movement occurred below the unconformity? _____

   k. Compared to the fault, the unconformity is (*younger*) (*older*) (*same age*).

244

28. A geologic cross section is shown below. The rock types are indicated in the key.

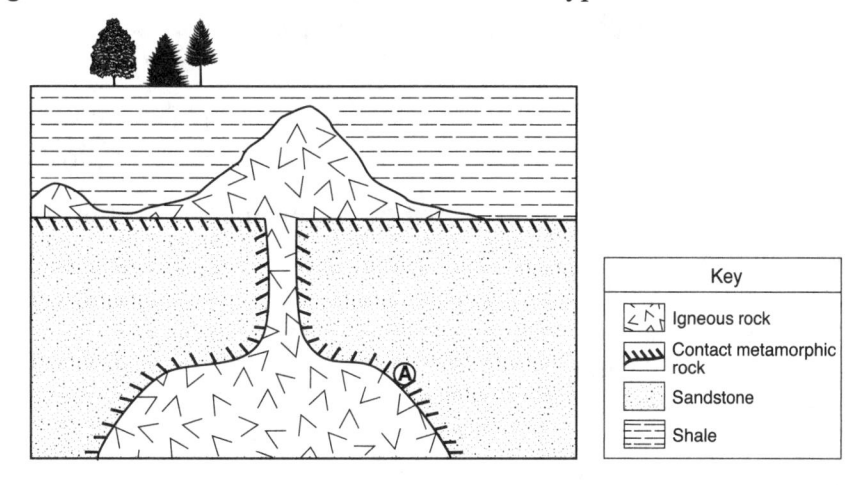

    a. The igneous rock is classified as an (*extrusion*) (*intrusion*) because _____
_____

    b. The youngest rock layer is the _____.

    c. The oldest rock layer is the _____.

    d. Fossils would not be found in the _____.

    e. Name the metamorphic rock that would form at the transition zone **A**. _____

29. The diagram below shows the rock structure of a region of Earth's crust.

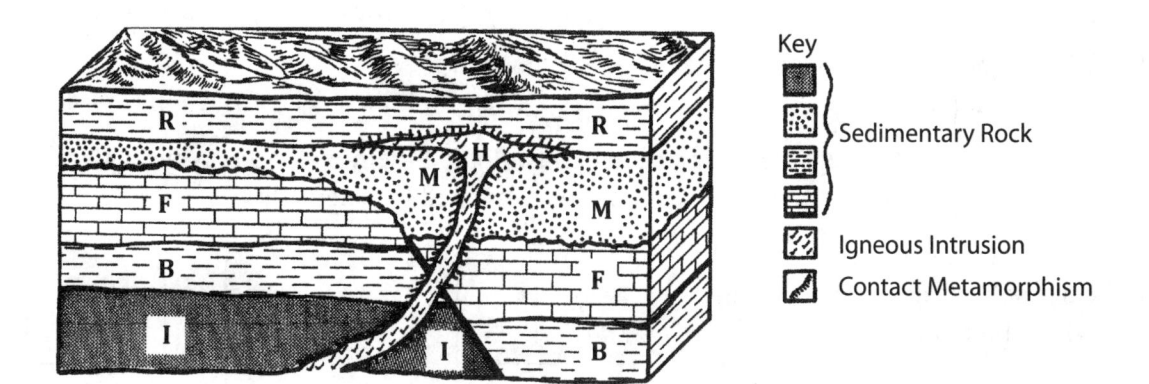

Place the rock layers and geologic features in order from oldest to most recent:
B, F, I, M, R, intrusion (H), fault, unconformity

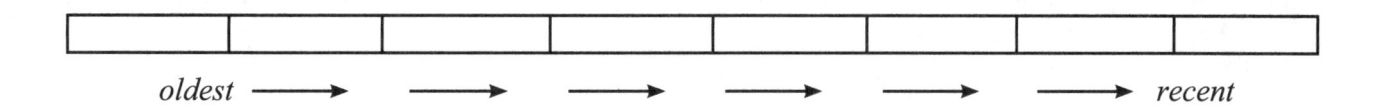

oldest ⟶ ⟶ ⟶ ⟶ ⟶ ⟶ recent

245

CHAPTER 13 REVIEW

1. Which atmospheric gas formed as a direct result of the appearance of certain life forms?
   (1) oxygen   (2) nitrogen   (3) helium   (4) hydrogen

2. Which geologic event occurred in New York State at about the same time as the extinction of dinosaurs and ammonoids?
   (1) formation of the Catskill Delta
   (2) deposition of sands and clays underlying Long Island
   (3) initial opening of the Atlantic Ocean
   (4) advance and retreat of the last continental ice sheet

3. Which New York landscape region is composed primarily of Cretaceous through Pleistocene unconsolidated sediments?
   (1) Champlain Lowlands   (3) Erie-Ontario Lowlands
   (2) Hudson-Mohawk Lowlands   (4) Atlantic Coastal Plain

4. The division of Earth's geologic history into units the time of eons, eras, periods, and epochs is based on
   (1) absolute dating techniques   (3) climate changes
   (2) fossil evidence   (4) seismic data

5. According to fossil evidence, most plants and animals that lived on Earth in the geologic past have
   (1) survived until the present   (3) become extinct
   (2) been discovered and identified   (4) lived on land

6. Which life form appeared first?
   (1) trilobite   (2) fish   (3) graptolites   (4) stromatolites

7. Which group of organisms are still in existence today?
   (1) brachiopods   (2) eurypterids   (3) graptolites   (4) trilobites

8. The diagram below represents a portion of a geologic timeline. Letters **A** through **D** represent time intervals between the labeled events, as estimated by scientists.

   Fossil evidence indicates that the earliest birds developed during which time interval?
   (1) A
   (2) B
   (3) C
   (4) D

9. The time line below represents the geologic history of Earth. At what letter did oceanic oxygen begin to enter the atmosphere?

   (1) A   (2) B   (3) C   (4) D

246

10. Which index fossil has been found in Ordovician-age bedrock?

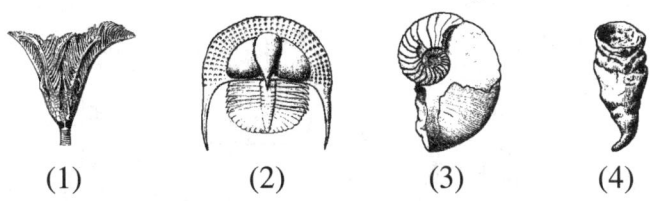

    (1)   (2)   (3)   (4)

11. Which group of organisms had the shortest record of life on Earth?

    (1) eurypterids   (2) graptolites   (3) birds   (4) placoderm fish

12. According to fossil evidence, which sequence show the order in which these life forms first appeared on Earth?

    (1) reptiles ⟶ amphibians ⟶ insects ⟶ fish
    (2) insects ⟶ fish ⟶ reptiles ⟶ amphibians
    (3) amphibians ⟶ reptiles ⟶ fish ⟶ insects
    (4) fish ⟶ insects ⟶ amphibians ⟶ reptiles

13. Which is the correct fossil sequence from oldest to youngest?

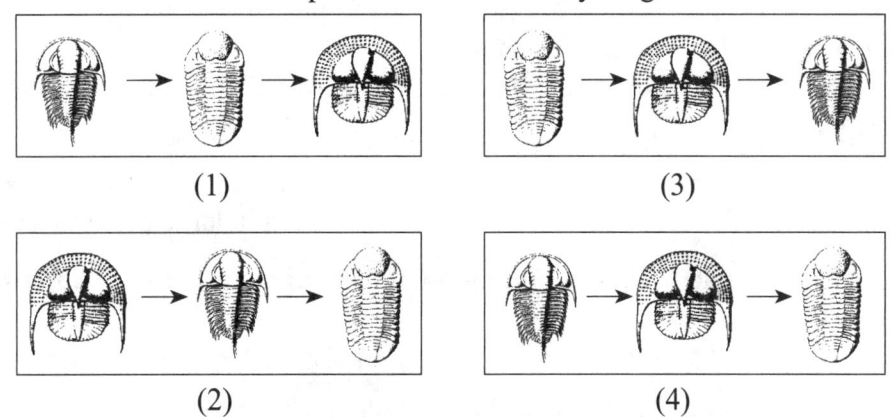

    (1)   (3)

    (2)   (4)

14. Which two types of organisms survived the mass extinction that occurred at the end of the Permian?
    (1) trilobites and nautiloids   (3) placoderm fish and graptolites
    (2) corals and vascular plants   (4) gastropods and eurypterids

15. Living corals are found in warm, shallow seas. Coral fossils have been found in sedimentary rocks of Alaska. These findings suggest that
    (1) coral can grow in cold climates
    (2) ocean current carried coral to Alaska
    (3) Alaska once had a tropical marine environment
    (4) Alaska's cold climate fossilized the coral

16. Brachiopod fossils were found in a layer of limestone rock. In which type of environment did the limestone layer form?
    (1) shallow marine   (2) tropical forest   (3) coastal plain   (4) interior grassland

17. Approximately how long ago were the Taconic Mountains uplifted?

    (1) 540 million years ago   (3) 310 million years ago
    (2) 450 million years ago   (4) 120 million years ago

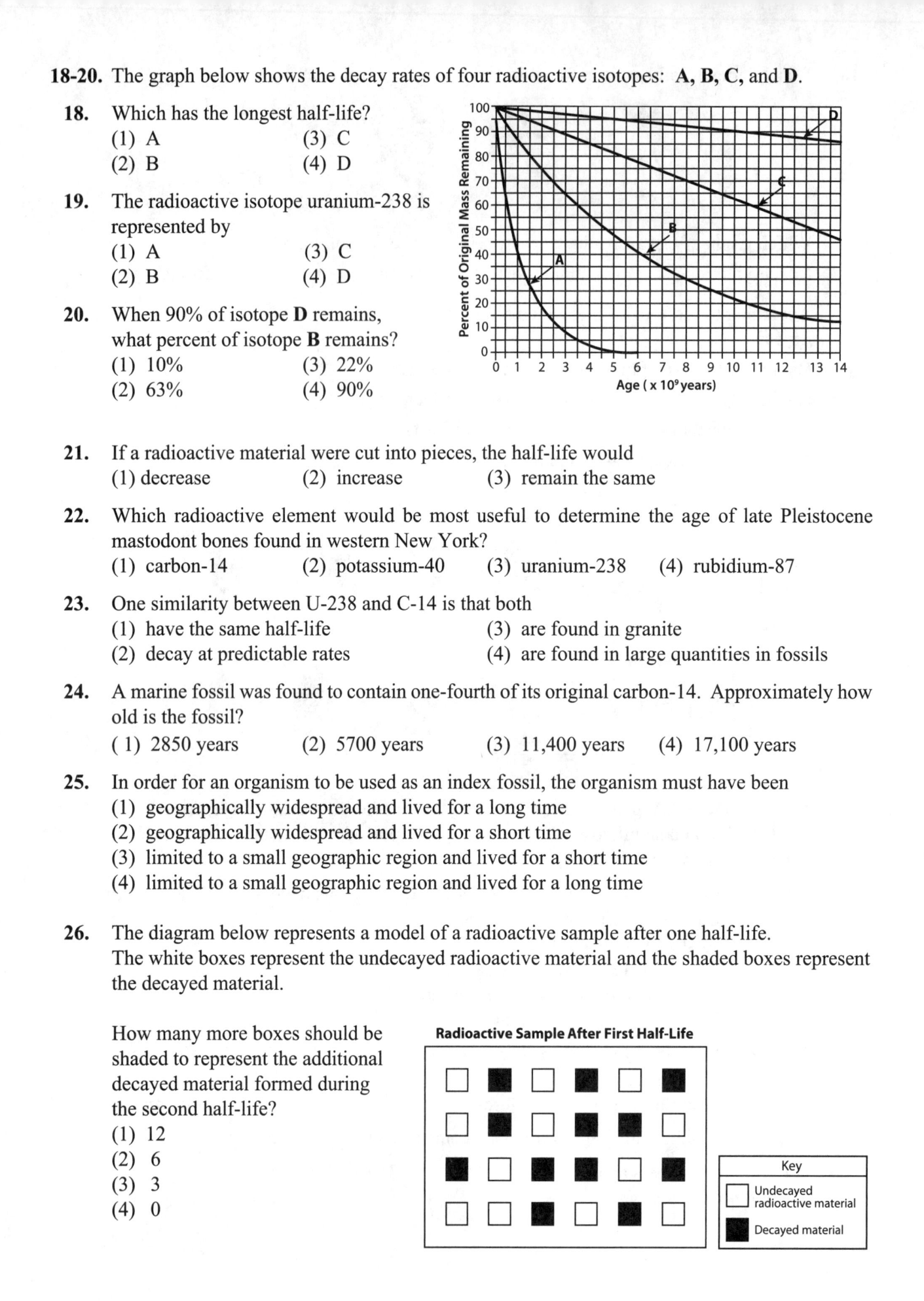

18-20. The graph below shows the decay rates of four radioactive isotopes: **A, B, C,** and **D**.

18. Which has the longest half-life?
    (1) A   (3) C
    (2) B   (4) D

19. The radioactive isotope uranium-238 is represented by
    (1) A   (3) C
    (2) B   (4) D

20. When 90% of isotope **D** remains, what percent of isotope **B** remains?
    (1) 10%   (3) 22%
    (2) 63%   (4) 90%

21. If a radioactive material were cut into pieces, the half-life would
    (1) decrease   (2) increase   (3) remain the same

22. Which radioactive element would be most useful to determine the age of late Pleistocene mastodont bones found in western New York?
    (1) carbon-14   (2) potassium-40   (3) uranium-238   (4) rubidium-87

23. One similarity between U-238 and C-14 is that both
    (1) have the same half-life   (3) are found in granite
    (2) decay at predictable rates   (4) are found in large quantities in fossils

24. A marine fossil was found to contain one-fourth of its original carbon-14. Approximately how old is the fossil?
    (1) 2850 years   (2) 5700 years   (3) 11,400 years   (4) 17,100 years

25. In order for an organism to be used as an index fossil, the organism must have been
    (1) geographically widespread and lived for a long time
    (2) geographically widespread and lived for a short time
    (3) limited to a small geographic region and lived for a short time
    (4) limited to a small geographic region and lived for a long time

26. The diagram below represents a model of a radioactive sample after one half-life. The white boxes represent the undecayed radioactive material and the shaded boxes represent the decayed material.

    How many more boxes should be shaded to represent the additional decayed material formed during the second half-life?
    (1) 12
    (2) 6
    (3) 3
    (4) 0

27. The cross-sections below show the exposed bedrock at two different locations that are 200 km apart. The rock layers have not been overturned.

Rock layer **X** at location **B** is most likely the same relative age as which rock layer at location **A**?

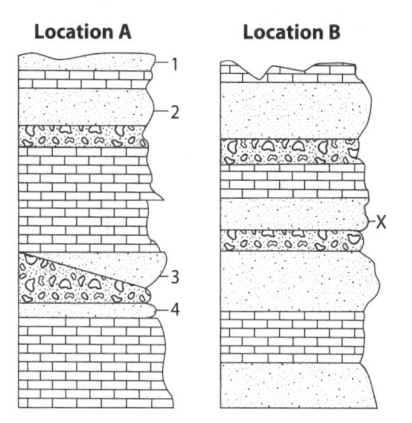

(1) 1 (3) 3
(2) 2 (4) 4

28. The cross-sections below represent three widely separated outcrops of exposed bedrock. Letters **A**, **B**, **C**, and **D** represent fossils found in the rock layers.

Which fossil appears to have the best characteristics of an index fossil?

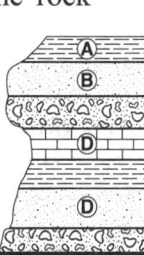

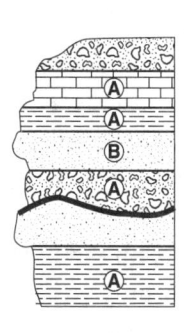

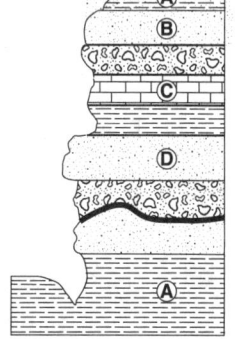

(1) A (3) C
(2) B (4) D

29. How are index fossils and volcanic ash deposits similar?
(1) Both can be dated with carbon-14.
(2) Both normally occur in sedimentary rock.
(3) Both resist chemical weathering.
(4) Both can serve as geologic time markers.

30. The two diagrams below represent two geologic cross-sections. Layers of igneous rock and contact metamorphism zones are shown.

Which statement best describes a difference between igneous rock **A** and igneous rock **B**?

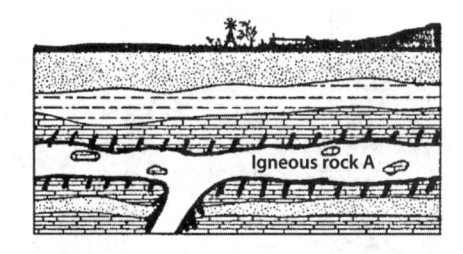

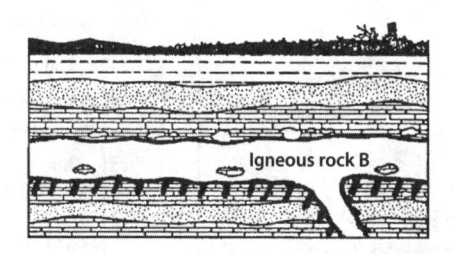

(1) **A** was extrusive; **B** was intrusive
(2) **A** cooled quickly; **B** cooled slowly
(3) **A** formed after the limestone above it; **B** formed before the limestone above it
(4) **A** was exposed to weathering; **B** was not exposed

31. The diagram below shows a cross-section of a portion of Earth's crust where no overturning of rock layers has occurred. Two rock layers **A** and **B** are labeled.
    Which statement is correct?

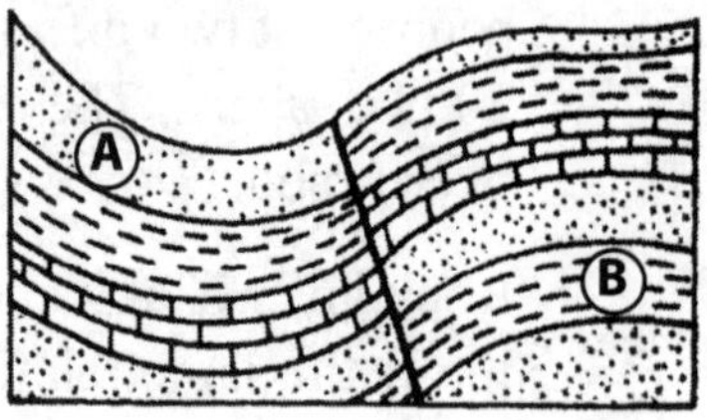

    (1) Folding occurred before the formation of **A** and **B**.
    (2) Faulting occurred before folding of the rock layers.
    (3) Faulting occurred before formation of rock layer **A**.
    (4) Folding occurred after formation of rock layers **A** and **B**.

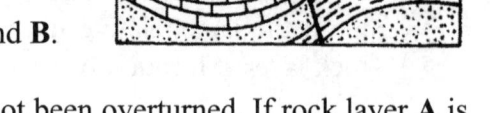

32. Use the cross-section below of an eroded fold that has not been overturned. If rock layer **A** is of Devonian age, rock layer **E** could be of what age?

    (1) Triassic
    (2) Carboniferous
    (3) Cambrian
    (4) Paleogene

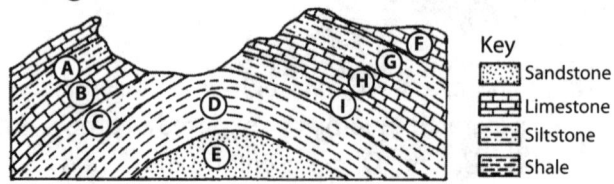

    Key
    Sandstone
    Limestone
    Siltstone
    Shale

33. The diagram represents a cross-section of a series of rock layers of different geologic ages. Which statement is true regarding the order of these rock layers?
    (1) The oldest is on the bottom.
    (2) An unconformity exists between the layers.
    (3) The layers have been overturned.
    (4) The Permian layer has been totally eroded.

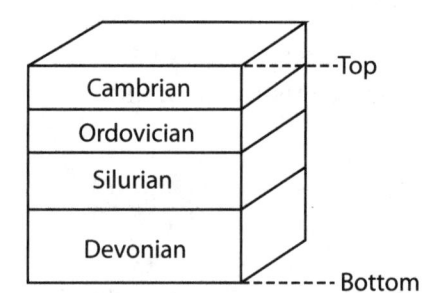

34. An igneous intrusion is 50 million years old. What is the probable age of the rock immediately surrounding the intrusion?

    (1) 10 million years   (2) 25 million years   (3) 40 million years   (4) 60 million years

35-36. Three rock outcrops **A**, **B**, and **C** are shown below. Overturning has not occurred.

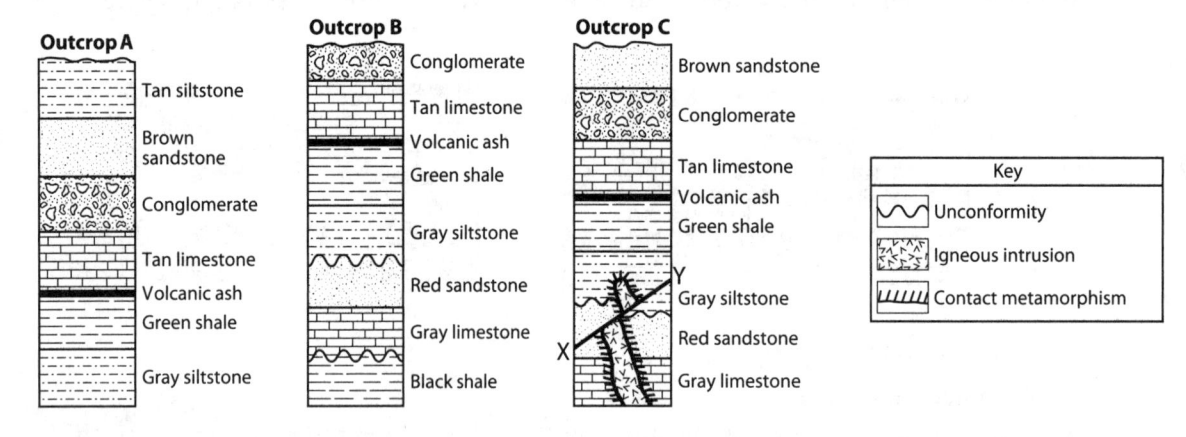

    Key
    Unconformity
    Igneous intrusion
    Contact metamorphism

35. Which sedimentary rock shown in the outcrops is the youngest?
    (1) black shale   (2) conglomerate   (3) tan siltstone   (4) brown sandstone

36. What is the youngest geologic feature in the three bottom layers of outcrop C?
    (1) fault   (2) intrusion   (3) unconformity   (4) contact metamorphism

37-40. The block diagrams below represent three widely separated outcrops. All rock layers are sedimentary. No overturning has occurred. Layers labeled with the same letter are the same age.

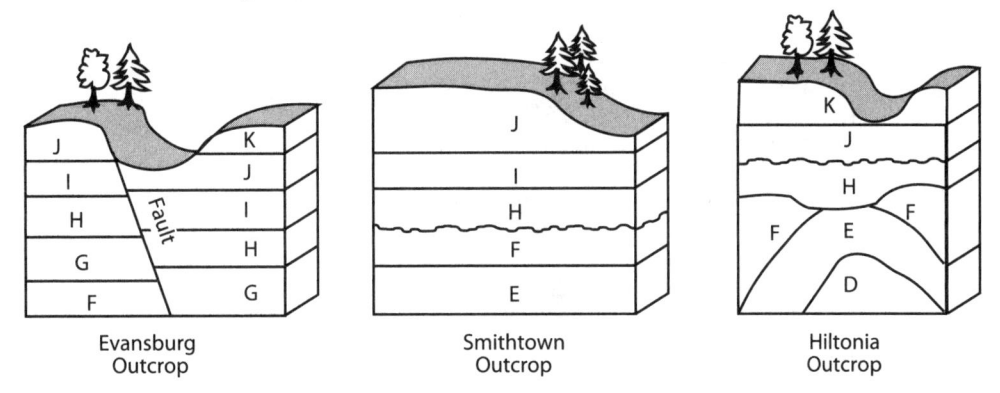

37. Which of the geologic processes that affected layer **F** happened first?

(1) deposition of sediments in layer **F**
(2) erosion of the surface of layer **F**
(3) folding of layer **F**
(4) faulting of layer **F**

38. The fault in Evansburg Outcrop is younger than

(1) G, only
(2) J, only
(3) G and J, only
(4) F, G, H, I, and J

39. Which would provide the most reliable evidence for the idea that layer **J** was deposited at the same time in each location?

(1) the percentage of mineral cement in each **J** layer
(2) the thickness of each **J** layer
(3) the mineral composition of each **J** layer
(4) the fossils in each **J** layer

40. Which order of events occurred at the Hiltonia Outcrop between the formation of layer **F** and the beginning of the formation of layer **H**?

(1) uplift ⟶ erosion ⟶ folding ⟶ deposition
(2) folding ⟶ uplift ⟶ erosion ⟶ subsidence
(3) subsidence ⟶ erosion ⟶ deposition ⟶ folding
(4) folding ⟶ erosion ⟶ faulting

41-43. The cross-section below shows rock layers that have not been overturned. Erosion has exposed the intrusion at the surface.

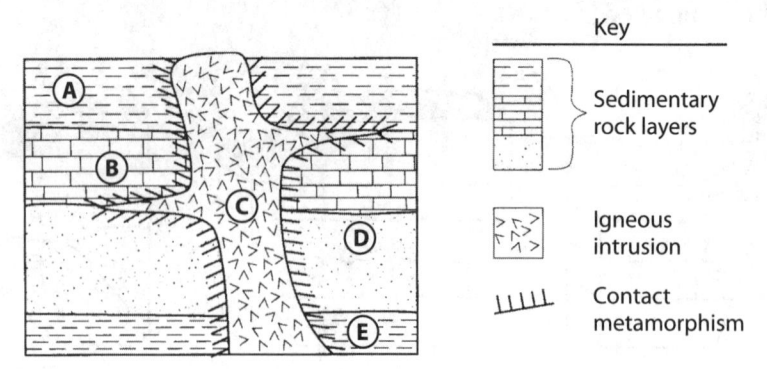

Key
- Sedimentary rock layers
- Igneous intrusion
- Contact metamorphism

41. Using letters **A** through **E**, list the rock units in order from oldest to youngest.

_____, _____, _____, _____, _____

42. State the name of the sediment that was compacted to form rock unit **A**. _____

43. State *one* observation about the mineral crystals in igneous rock **C** that would provide evidence that igneous rock **C** is an intrusive igneous rock. _____

44-49. Use the following map of New York State and the ESRT. Letters **A, B, C, D,** and **E** are surface locations.

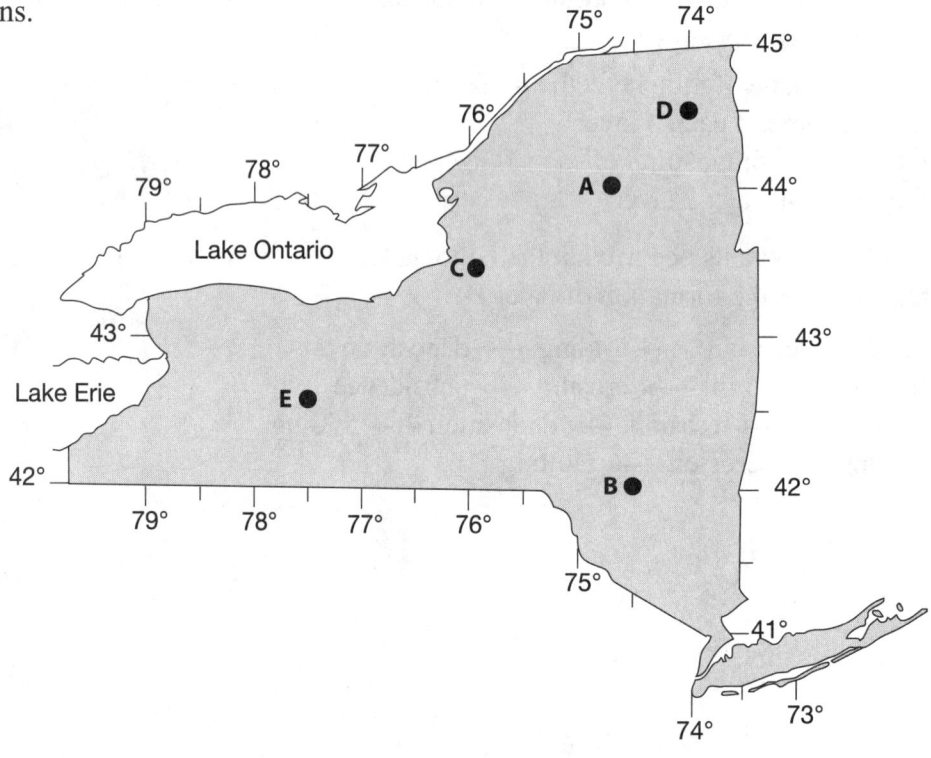

44. From **C** to **E** the age of the rocks *(decreases)* *(increases)* *(stay the same)*.
45. The oldest rock surface is found at letter(s)? _____
46. Which letter is located on the youngest rock surface? _____
47. Name the gastropod fossil that you could find at **B**. _____
48. The Geologic Period of rocks at **C** is _____
49. The type of rock at location **D** is _____

50-56. The diagrams below represent two bedrock outcrops, **I** and **II**, found several kilometers apart in New York State. Rock layers are lettered **A** through **F**.

Drawings represent specific index fossils.

50. In outcrop **II** circle the rock layer that is missing in outcrop **I**.

51. During which geologic time period was rock layer **C** deposited? _____

52. Name the sedimentary rock that is layer **F**. _____

53. How did layer **B** form? _____

54. Compared to layer **D**, layer **A** is (*younger*) (*older*) (*same age*). How do you know?
_____

55. Explain why carbon-14 can not be used to find the age of these index fossils.
_____

56. Name the fossil shown in layer **C**. _____

57-61. The cross sections below represent three widely separated outcrops. Index fossils are found in some of the rock layers.

57. Line **XY** represents a(an) _____; Line **GH** is a(an) _____

58. Circle the index fossil from the early Devonian period in outcrop III.

59. For outcrop **III**, list in order from oldest to youngest the relative age of the layers **A, B, C, D,** line **GH**, and line **XY**.

_____, _____, _____, _____, _____, _____
oldest   youngest

60. Place a letter **B** on another rock layer that is the same age as layer **B** in outcrop **III**.

What is your evidence? _____

61. What is unusual about outcrop **I**? _____

253

62-65. The geologic cross section below shows rock structure of a region of Earth's crust. Letters **A** through **H** are rock units. Lines **J-J'** and **K-K'** are interfaces within the cross section. Rock layers **A**, **B**, and **C** have not been overturned.

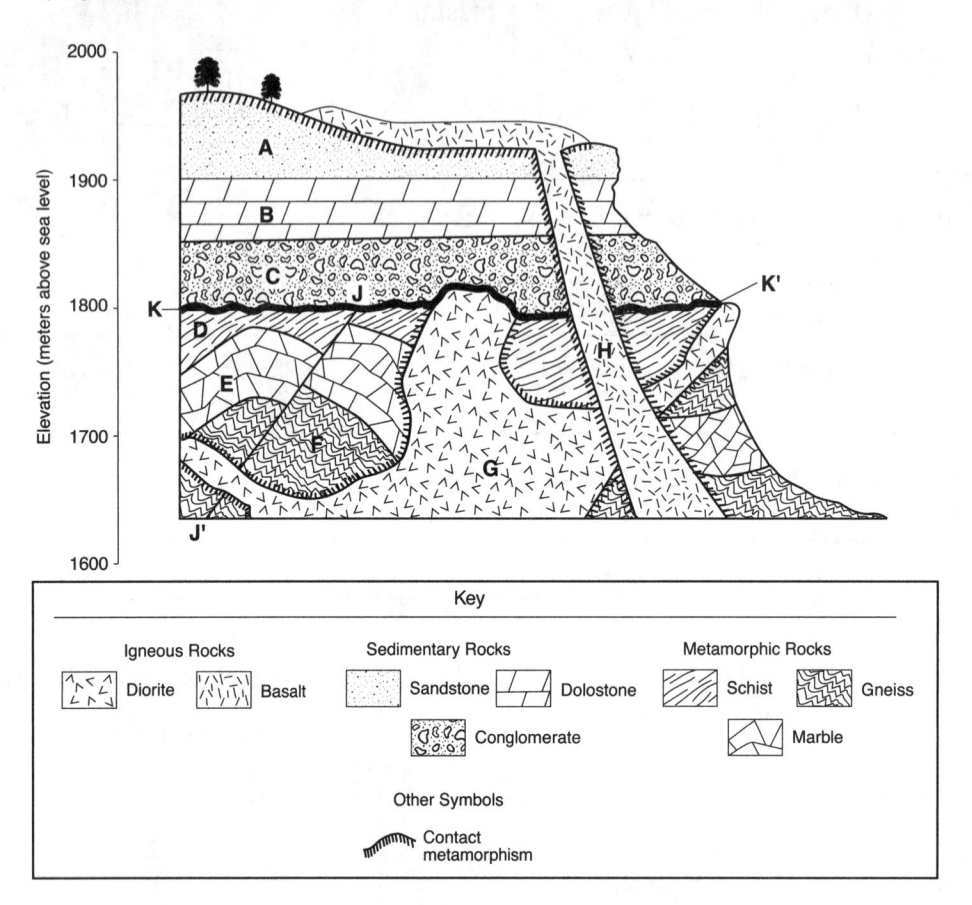

62. List the following geologic events from oldest to youngest:

*extrusion **H**, intrusion **G**, fault **J-J'**, unconformity **K-K'**, folding of **D,E,F***

_____, _____, _____, _____, _____

oldest   *youngest*

63. List the rock layers **A, B, C, D, E, F, G, H** in order from oldest to youngest.

_____, _____, _____, _____, _____, _____, _____, _____

oldest   *youngest*

64. Rock layers **D** and **E** were affected by *(contact)* *(regional)* metamorphism.

65. What was the most likely rock that **E** was metamorphosed from? _____

_____

66. Which igneous rock in this diagram has the largest mineral crystals? _____

Why? _____

CHAPTER 14

EARTH SCIENCE REFERENCE TABLES (ESRT)

PAGE 1

RADIOACTIVE DECAY DATA

- Geologists use radioactive isotopes to determine the age of fossils and rocks.
- The first column lists the **radioactive isotopes** from shortest to longest half-life.
- The second column lists the elements the isotope **disintegrates** (changes or decays) to.
- The third column gives the **half-life**, this is the time it takes for one-half of the isotope to change. Half-life is a constant, unchangeable value.
- C-14 has the shortest half-life, 5700 years; C-14 is used to date organic remains from living organisms from the past 50,000 years.
- Rb-87 has the longest half-life of 49 billion years. Since Earth is only 4.6 billion years old only a small amount of Rb-87 has disintegrated.

Questions

1. Which radioactive isotope has a half-life approximately the same age as Earth? _____

2. Carbon-14 can not be used to date dinosaur bones. Why not? _____

3. Which radioactive isotope disintegrates to Pb-206? _____; to Ca-40? _____

4. Plant pollen is preserved in sediment from the last continental ice. Which radioactive isotope is best to use to determine the age of this pollen? _____

5. A piece of wood that originally contained 100 grams of C-14 now only has 25 grams. How many years ago was this part of a living tree? _____

6. A rock contains only one-fourth of its original K-40. How old is the rock? _____

7. If 80 grams of original K-40 has been disintegrating to Ar-40 for 3.9×10^9 years. How many grams of K-40 is left? _____

8. As the amount of radioactive isotope decreases, the half-life will _____.

SPECIFIC HEATS OF COMMON MATERIALS
- The materials are listed in order from highest to lowest specific heat.
- **Specific heat** is a measure of how many joules of energy are needed to raise the temperature of one gram of the material 1°C.
- Specific heat indicates how quickly a material heats up and cools off.
- A material with a high specific heat will heat and cool slowly.
- A material with a low specific heat will heat and cool quickly.

Questions

9. How many joules of energy are required to heat 10 grams of iron 1°C? _____
10. Compared to liquid water, ice will heat (*slower*) (*quicker*).
11. On a sunny day which will heat more quickly? (*lake*) (*plowed farmland*)
12. Compared to dry air, humid air will heat and cool (*slower*) (*quicker*).

EQUATIONS
- This chart lists four commonly used equations in Earth Science.
- **Eccentricity** is a measure of how elliptical an orbit is. As the value of eccentricity increases, the orbit becomes more elliptical.
- Eccentricity is used in the Astronomy unit to describe the shape of an orbit.
- Eccentricity has NO units and is always rounded to the thousandth place.
- The eccentricities of the objects in the solar system are found on page 15 of the ESRT.
- **Gradient** is the rate of change in a value over distance in a given area (field).
- The gradient equation is used in the Geology unit to describe the steepness of the slope of the land. It can also be used in the Meteorology unit to determine the change in air pressure or temperature within a given distance.
- In the gradient equation, the field value can be elevation, temperature, or air pressure. Be sure to subtract the value of one point from the other point.
- In the gradient equation distance could be a given value in the problem or measured on a map or diagram. When measuring carefully put a "tick" mark on one point and another "tick" mark on the other point. Determine the distance between the two points using the given scale.
- Gradient is rounded to the nearest tenth. It always includes units such as feet/mile, meters/kilometer, °C/mile, millibars/kilometer.
- **Rate of change** is similar to gradient except it is the change in a value in a given amount of time.
- Examples of rate of change are change in temperature per minute or change in air pressure per hour.
- Rate of change is rounded to the nearest tenth and must include the correct units.
- **Density** is the amount of mass in a unit of volume.
- Density is calculated as mass divided by volume. It is usually rounded to the nearest tenth. The unit of density is g/cm³ or g/mL.
- Density equation is used in the Meteorology unit to determine the density of a parcel (volume) of air or in the Geology unit to determine the density of rocks and minerals.

Questions

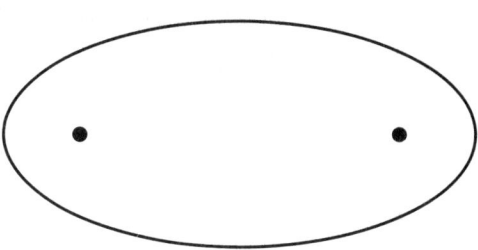

13. Calculate the eccentricity of the ellipse below. _____
    Show solution.

14. A hot air balloon traveled upwards for 500 feet. During this ascent the temperature changed from 24°C to 18°C. Calculate the temperature gradient. _____
    (*Show solution to the nearest hundredth*)

15. The diagram shows the temperature change in an area. Calculate the temperature gradient between points **X** and **Y**. _____
    (*Show solution*)

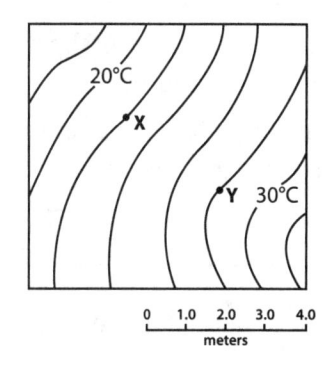

16. At 9 a.m. the air temperature was 45°F. At 1 p.m. the temperature was 57°F. Calculate the rate of temperature change during this time period. _____
    (*Show solution*)

17. A rock has a mass of 51.8 grams and a volume of 20.6 mL. Calculate its density.
    _____ (*show solution*).

PROPERTIES OF WATER

- Heat energy is gained (added) when ice melts or water vaporizes (evaporates). It takes more energy to evaporate water.
- Heat energy is released (removed) when water freezes or vapor condenses to a liquid.
- More energy is released when water vapor condenses into water droplets.
- Water is densest as a liquid at 3.98° C.

257

Questions

18. When changing liquid water to a solid, heat energy will be (*gained*) (*released*).

19. When cloud droplets form, heat energy is (*gained*) (*released*) to the atmosphere.

20. How many joules of heat energy are need to vaporize 50 grams of water? _____

21. Approximately how much more energy is needed to vaporize one gram water than to melt one gram ice? _____

22. Explain how the density of water changes as its temperature increases from 0° C to 10° C.
_____
_____

Average Chemical Composition of Crust, Hydrosphere and Troposphere

- The most common elements of planet Earth are listed by name and chemical symbols in the first column.
- These common elements are listed by their percentage in Earth's **crust** (solid upper rock layer), Earth's **hydrosphere** (water), and Earth's **troposphere** (the lower layer of the atmosphere).
- The common elements in the crust are given as percentage by mass and by volume. The common elements in the hydrosphere and atmosphere are given as percentage by volume.
- There are over 92 elements found on Earth. "Other" refers to elements that are in such small amounts that there is no need to name them.

Questions

23. Which element is found in all three areas: crust, hydrosphere, and troposphere? _____
24. The most abundant element in the waters of Earth is _____
25. The percent by volume of magnesium (Mg) in the crust is _____
26. The percent by mass of aluminum (Al) in the crust is _____

Page 2

Generalized Landscape Regions of New York State

- This map shows the landscape regions of NYS and surrounding areas.
- The "Key" in the bottom center is used to determine the type of boundary between the regions.
- The map scale and compass directions are in the bottom right corner.
- Lowlands are low elevation **plains** of horizontal rock or sediment. They are commonly found along lakes, rivers, and coastlines.
- **Plateaus** have high elevation and horizontal rock layers (strata).
- **Mountains** (highlands) have high elevation and the rock layers have been disturbed by tilting, folding, and/or faulting.
- This landscape map is often used with the map on page 3; both are the same size.

Questions

27. In NYS, the Catskills are a (*mountain*) (*plateau*) (*plain*).
28. Within the United States, the Allegheny Plateau is part of the _____.
29. The Interior Lowlands (*are*) (*are not*) part of NYS.
30. Name the three landscape regions in NYS that are part of the New England Province.
    _____, _____, _____
31. Name the landscape region at 43°N and 77°W: _____

PAGE 3

GENERALIZED BEDROCK GEOLOGY OF NEW YORK STATE

- The major waterways (rivers, lakes, seas) and cities of NYS are named as well as the adjoining states. The capital of NYS is Albany.
- Latitude (°N) and longitude (°W) are shown along the outline of NYS. Remember that each degree of latitude and longitude is divided into sixty minutes.
- Latitude in NYS ranges from 40° 30′N to 45°N.
- Longitude ranges from about 72°W to 79°45′W.
- A map scale in miles and kilometers and the compass directions are given.
- The different map patterns (symbols) refer to the age of the surface bedrock by geologic period or era written in capital letters in the key.
- For each geologic period/era the specific type of rock and its rock classification (igneous, metamorphic, sedimentary) is written in lower case letters.
- The geologic periods and eras are listed in the key from most recent to oldest. Use page 8 and 9 of the ESRT to give you further information on these geologic times.
- This map can be superimposed over the map on page 2 to determine the landscape of a given location in NYS.

Questions

32. Name the state that adjoins NYS at 44°N? _____
33. Name the landscape region that Mt. Marcy is located in. _____
34. Name the landscape region that Rochester is located in. _____
35. Name the geologic age of the surface rock at Watertown. _____
36. Name the rock type for surface bedrock at Jamestown. _____
37. Name the geologic age of the surface rock in the Catskills. _____
38. For Utica, the latitude is _____ and the longitude is _____
39. Measure the distance in kilometers from Watertown to Kingston. _____
40. Measure in miles the length of the Genesee River in NYS. _____
41. What is the compass direction from Binghamton to Plattsburgh. _____
42. Name the location/city with the youngest rock? _____ oldest? _____

43. The elevation of Lake Ontario is _____.
44. Name the three minerals found in Silurian NYS rock. _____, _____, _____
45. Regional metamorphism occurred in NYS about _____ millions years ago.

PAGE 4

SURFACE OCEAN CURRENTS

- Earth's surface ocean waters circulate (move) between the tropics and the poles.
- The "Key" indicates that *warm ocean currents* are solid, black arrows and *cool ocean currents* are open, white arrows.
- Due to the Coriolis effect caused by Earth's rotation most of the ocean currents in the Northern Hemisphere curve to the right and to the left in the Southern Hemisphere.
- Names of the major landmasses and oceans are given.
- Latitude and longitude are along the margins of this map.
- Major latitude lines are shown along the right margin (Arctic Circle, Tropic of Cancer, Equator, Tropic of Capricorn, and Antarctic Circle).
- The Prime Meridian is 0° longitude. The International Date Line is 180° longitude.
- The Western Hemisphere is to the left of the Prime Meridian and stops at the International Dateline. On this map most of the Eastern Hemisphere is to the left of the International Dateline.

Questions

46. The Kamchatka Current is (*warm*) (*cold*) and travels (*northward* (*southward*).
47. The Peru Current is (*warm*) (*cold*) and travels (*northward*) (*southward*).
48. Name the ocean current along the southeast coast of Africa. _____
49. Name the current at 60°S, 100°E. _____
50. Name the warm ocean current in the South Atlantic Ocean. _____

PAGE 5

TECTONIC PLATES

- Earth's **lithosphere** (crust and rigid mantle) is "broken into" more than a dozen **plates** which move and shift slowly.
- This map locates and names the major plates of Earth and shows their direction of movement.
- The "Key" for this map shows the symbols used for each type of plate boundary.
- A **transform plate boundary** occurs where two plates are moving alongside each other.
- A **divergent plate boundary** is where two plates are moving apart. The plates are diverging (separating) along ridges where new crust is being formed by upwelling of magma.
- A **convergent plate boundary** occurs where two plates are coming together with the denser plate subducting under the less dense plate. Many of these occur along trenches.

- Some plate boundaries are still being studied by geologists and are labeled as complex or uncertain.
- **Hot spots** are locations where magma upwells towards the surface.
- Latitude and longitude values are given along the margin of the map similar to page 4.

Questions

51. Name the type of plate boundary between the Nazca Plate and the South American Plate. _____

52. At the Mariana Trench, the Pacific Plate is the (*subducting*) (*overriding*) plate.

53. The Canary Islands Hot Spot is on the _____ Plate.

54. The Easter Island Hot Spot is located on the _____.

55. The Indian-Australian Plate is generally moving (*northward*) (*southward*).

56. The San Andreas Fault is a _____ plate boundary.

57. State the direction the Arabian Plate is moving towards. _____

58. Name the plate located at 50°S, 100°W. _____

59. South America is slowly moving (*eastward*) (*westward*).

60. Name a complex or uncertain plate boundary. _____

PAGE 6

ROCK CYCLE IN EARTH'S CRUST

- This diagram illustrates the processes that form each of the three rock types.
- The diagram also shows how one rock type changes to another rock type. Follow the arrows for the steps of formation.
- **Sediments** are the loose rock material needed to form sedimentary rock.
- **Magma** is the liquid rock material within the Earth that is needed to form igneous rock.

Questions

61. Metamorphism is caused by _____ and/or _____.

62. Name the *two* processes that form sediments. _____ and _____

63. When a rock melts it will become _____.

64. What *two* processes occur during the uplift of land surfaces? _____ & _____

65. Solidification of _____ will form _____ rock.

66. Metamorphic rock must first undergo _____ in order to become sedimentary rock.

67. After sediments are deposited and buried they will then be _____ and/or _____ in order to form a sedimentary rock.

261

Relationship of Transported Particle Size to Water Velocity

- This graph shows how fast water must be moving (*stream velocity in cm/second*) to move a specific sized sediment (*particle diameter in cm*).
- Both axes of the graph are numbered logarithmically, the increments are not equal. For example, on the horizontal X-axis, the first number is 0.01 (cm/s), the next line is 0.02, then 0.03, and so on until 1. Then the numbers go by ones, then by tens, then by hundreds.
- Along the right side of the graph the names of each sediment type are given. The dash lines separate the range of sizes for each. Clay is less than 0.0004 cm, silt is 0.0004-0.006 cm, sand is 0.006-0.2 cm, pebbles are 0.2-6.4 cm, and cobbles are 6.4-25.6 cm. Any sediment larger than 25.6 cm is classified as a boulder.
- As the stream's velocity increases, the size of the particle it can move increases.
- Remember if a stream can move a pebble it can also move sand, silt, and clay.
- Use a straight edge to accurately use this graph.

Questions

68. A particle with a diameter of 0.09 cm is classified as a _____
69. A particle with a diameter of 0.004 cm is classified as a _____
70. A stream which is moving at 30 cm/s can move which particles? _____
71. A stream which is moving at 5 cm/s can not move particles which are larger than _____ cm.
72. To move a particle which has a 10.0 cm diameter the stream must be moving at least _____ cm/s.

Scheme for Igneous Rock Identification

- This is divided into three sections: igneous rocks, characteristics, and mineral composition.
- The top section gives the names of seventeen igneous rocks. These igneous rocks are described by their **environment of formation**: intrusive or extrusive.
- Igneous rocks are described by their texture which is the crystal size of the minerals.
- **Extrusive (volcanic)** igneous rocks have crystal sizes less than 1 mm or are **non-crystalline** (no mineral crystals are visible). Their textures are fine to glassy. Cooling of lava on Earth's surface occurs rapidly so mineral crystals do not have time to grow.
- **Intrusive (plutonic)** igneous rocks have mineral crystal sizes greater than 1 mm giving them a coarse to very coarse texture. Cooling of magma inside Earth occurs slowly so mineral crystals will grow larger.
- Some extrusive igneous rocks may show evidence of gas pockets (**vesicular**).
- These seventeen igneous rocks are grouped by their characteristics of color, density, and composition.
- The lighter colored igneous rocks have a lower density and are **felsic**, rich in silicon (Si) and aluminum (Al)
- The darker colored igneous rocks have a higher density and are **mafic**, rich in iron (Fe) and magnesium (Mg)
- The mineral composition of igneous rocks is recorded as percent by volume. This chart shows that there is a range in the percent of each mineral in the igneous rocks.
- When given the name of the igneous rock, read across to get its texture and read down to get its characteristics and mineral composition.

Questions

73. If mineral crystals are 1 mm to 10 mm in size the texture is _____.
74. Intrusive igneous rocks (*can*) (*can not*) be vesicular.
75. Volcanic igneous rocks can have a _____ or _____ texture.
76. The environment of formation for diorite is _____; for scoria is _____
77. State the texture of pumice: _____; peridotite: _____; pegmatite: _____
78. Name a felsic, coarse textured igneous rock. _____
79. Name a low density, fine textured, vesicular igneous rock. _____
80. In granite the minimum amount of amphibole can be _____%, maximum amount _____%
81. The range in amount of olivine in basalt is _____ to _____ %.
82. The only monominerallic igneous rock is _____, it is made of the mineral _____
83. Name the minerals found in andesite: _____
84. Complete this information for the igneous rock rhyolite:

    a. environment of formation: _____   e. color: _____
    b. crystal size: _____   f. density: _____
    c. texture: _____   g. mafic or felsic? _____
    d. mineral composition: _____

PAGE 7

SCHEME FOR SEDIMENTARY ROCK IDENTIFICATION

- There are two sections to this chart. The top section is for sedimentary rocks that are made of *inorganic land-derived* sediments. The bottom section is for the sedimentary rocks that form *chemically and/or organically*.
- Sedimentary rocks are described by texture, grain size, and composition.
- The texture of sedimentary rocks depends on the type of particle it is made of. **Clastic** refers to inorganic rock fragments or sediments (clay, silt, sand, pebble, cobbles, boulders). **Crystalline** sedimentary rocks are made of a mineral that had settled out of water as a precipitate or evaporite. **Bioclastic** sedimentary rocks are made of organic sediments such as plant remains or shell pieces.
- Grain size for clastic sedimentary rocks depends on the type of sediment particle in the rock. Remember the size of each sediment is also given on page 6 of the ESRT.
- The inorganic land-derived sedimentary rocks are polyminerallic, composed of a variety of minerals.
- The chemically and/or organically formed sedimentary rocks are monominerallic, composed of one type of mineral.
- The "comments" column describes the fragments in the rock or how the rock formed.
- Each sedimentary rock name has a map symbol that is used on diagrams of cross-sections of rock layers.

Questions

85. Describe the *two* ways that limestone can form. **a.** _____
   b. _____

86. Explain how an evaporite such as rock salt forms. _____
_____

87. If limestone is tested with laboratory acid it will "bubble." Why? _____
_____

88. Name the sedimentary rock described.
   a. angular rock fragments. _____
   b. particles are 0.0005 cm. _____
   c. rounded pebbles in clay. _____
   d. carbon. _____
   e. bioclastic. _____
   f. [image] _____

SCHEME FOR METAMORPHIC ROCK IDENTIFICATION

- This chart is divided into two sections based on the texture of the metamorphic rock.
- **Foliated** metamorphic rocks have mineral alignment or banding of minerals.
- **Non-foliated** metamorphic rocks do not have minerals that are aligned or banded.
- Grain size describes the size of the mineral crystals in the rock.
- The minerals common to metamorphic rocks are listed.
- The type of metamorphism that formed each metamorphic rock is described.
- Foliated metamorphic rocks formed from regional metamorphism due to intense heat and pressure. The intensity of the metamorphism (heat and pressure) increases from slate to gneiss.
- Non-foliated metamorphic rocks formed from **regional metamorphism** or from **contact metamorphism** with hot intruding magma or lava.
- The "comments" column describes special appearances of the metamorphic rock and/or the pre-existing rock that it formed from.
- Next to each rock name is the map symbol that is used on diagrams of cross-sections of rock layers.

Questions

89. The metamorphic rock formed only from contact metamorphism is _____

90. The degree of metamorphism on phyllite compared to schist was (*more*) (*less*).

91. Slate can be classified as monominerallic. Why? _____

92. A coarse-grained, non-foliated metamorphic rock is _____

93. Name the metamorphic rock described:

   a. banding of minerals. _____

   b. made of quartz. _____

   c. high-grade regional metamorphism. _____

   d. fine-grained, minerals aligned. _____

   e. metamorphism of shale. _____

   f. metamorphism of dolostone. _____

PAGE 8 & 9

GEOLOGIC HISTORY OF NEW YORK STATE

- Pages 8 and 9 of the ESRT are one chart that is on two pages. Be sure to line up both pages if you are given separate pages.

- This chart shows the divisions of geologic history, the descriptions of life forms during each time, and the geologic events affecting New York State.

- On the far left is time in million years ago. The most recent time is on the top, the oldest time is on the bottom of the chart. (1000 million years = 1 billion years)

- The time scale at the bottom is 4600 million years ago (m.y.a.), which is the estimated time of the origin of Earth and solar system. 4600 million years is equal to 4.6 billion years (b.y.).

- The time scale at the top is 0 million years ago, which is the present.

- There are two main **eons**, the Precambrian and the Phanerozoic. We are currently in the Phanerozoic eon.

- The information about the Precambrian is in the shaded area starting on the left side of the chart. The Precambrian takes up almost 85% of time on Earth, yet we know very little about it. Many of the rocks from the Precambrian have been weathered and eroded. The organisms from this time were tiny and soft bodied so there are few preserved fossils.

- The Precambrian eon is divided into two sub-eons: the Archean and the Proterozoic. These sub-eons are divided into early, middle, and late.

- The boundary between the Precambrian and the Phanerozoic is approximately 542 million years ago. This boundary is marked by a thick solid line.

- The current Phanerozoic eon is divided into three eras: Paleozoic, Mesozoic, and Cenozoic. Each era is further divided into periods, which are divided into epochs.

- Another time line in million years ago is to the right of the epochs. These numbers can be used to state when a period or an era began (bottom number) and when it ended (top number). For example the Triassic period began 251 million years ago and ended 200 million years ago. Therefore the Triassic lasted for 51 million years.

- The column headed "Life on Earth" describes the highlights of life on Earth for that time period. For each description read straight across to the left to find the epoch, period, and era when it occurred. For example, earliest birds occurred during the late Jurassic period, which is part of the Mesozoic era, which is part of the Phanerozoic eon.

- "NY Rock Record" column shows which time periods have sediment or rock found in New York. The solid black line represents when there is a rock record. When that line is broken or missing, there is no rock record in NYS for this time period. Rock records may be missing due to weathering and erosion. A gap in the rock record is called an **unconformity**.
- "Time Distribution of Fossils" uses lines to show the span of time when each animal and plant group existed on Earth. Organisms that have become extinct do not have lines that extend to the top of the chart. For example, trilobites went extinct at end of Paleozoic, 251 million years ago.
- The letters on these lines indicate a specific index fossil that lived during that time. Names and drawings of the index fossils are along the bottom of the chart. They are NOT listed by age.
- An **index fossil** is an organism that lived for a short period of time, over a large geographic area. Index fossils are used to estimate how old a rock is.
- For the column "Important Geologic Events in New York" carefully read across to find the time period when the event occurred.
- An **orogeny** refers to major mountain building events. An orogeny is indicated by the "mountain" symbol:
- "Inferred Positions of Earth's Landmasses" illustrates the positions of the continents during the Phanerozoic Eon. North America is the darker area. From these diagrams we can see that the landmasses have "drifted" throughout geologic time and changed their position relative to each other and relative to latitude and longitude.

Questions

94. The Silurian is a period of the _____ era and the _____ eon.
95. The Paleozoic began _____ m.y.a. and ended _____ m.y.a.
96. The Cretaceous lasted for _____ million years.
97. Oceanic oxygen begins to enter the atmosphere _____ m.y.a.
98. Name the geologic periods in which there is no rock record in NYS. _____
99. Name the abundant life form of the Early Permian. _____
100. Earliest grasses occurred in the _____ epoch.
101. Index fossil *Ctenocrinus* lived during the (*early*) (*middle*) (*late*) _____ period.
102. Name the oldest trilobite index fossil. _____
103. Compared to corals, gastropods appeared (*earlier*) (*later*) (*at the same time*).
104. Dome-like uplift of the Adirondacks occurred during the _____ period.
105. North America began to "drift away" from Africa _____ m.y.a.
106. The Grenville Orogeny occurred during the _____ eon.
107. The oldest known rocks on Earth are about _____ m.y.a.
108. Name the oldest coral index fossil. _____

PAGE 10

INFERRED PROPERTIES OF THE EARTH'S INTERIOR

- This chart is title "inferred" because no one has drilled further than 13 kilometers into Earth.
- The properties of Earth's interior layers are inferred from observations of the behavior of seismic waves as they travel through Earth and the composition of meteorites.
- The top diagram illustrates the relative thicknesses of the interior layers and their densities.
- The **crust** is the top dark, bold line. Examples of crustal features from Pacific Ocean eastward to the middle of the Atlantic Ocean are the trench, Cascades, and Mid-Atlantic Ridge.
- The crust and rigid mantle make up the **lithosphere**.
- The arrows in the asthenosphere show the **mantle convection currents** which are responsible for the movement of the lithospheric "plates."
- The range in density of each interior layer is given to the right. The crust is divided into continental (land) crust and oceanic (sea floor) crust. Oceanic crust is denser.
- The **outer core** and **inner core** are composed of iron and nickel.
- All diagrams on this page share the same horizontal X-axis: depth in kilometers.
- The pressure in Earth's interior is measured in millions of atmospheres. One atmosphere is the normal air pressure we feel at sea level.
- The temperature graph shows the inferred interior temperature at each depth and the melting points of the rocks at each depth.
- In the outer core the temperature of the rocks are greater than the melting point, therefore the rocks in this layer are a liquid. Even though the outer core is a hot, liquid this is not the source of volcanic magma or lava.
- As with all graphs use a straight edge to get precise answers.
- The vertical dashed lines correspond to the margins of each interior layer.

Questions

109. According to this diagram the mantle is subdivide into three sections: the _____ mantle, the _____ mantle, and the _____ mantle.

110. As depth towards Earth's center increases, the density of the rock _____

111. As depth towards Earth's center increases, the pressure _____ and interior temperature _____.

112. The depth of Earth's center is approximately _____ kilometers.

113. Rocks with a density of 4.8 g/cm³ are found in which layer? _____

114. Granitic rock is found in the _____ crust.

115. The interior cores of Earth are rich in the elements _____ and _____.

116. Name the interior layer with the following characteristic:

   a. depth of 2400 km: _____
   b. temperature of 5500°C: _____
   c. density of 11.6 g/cm³: _____
   d. a liquid: _____
   e. pressure of 3.6 million atm: _____
   f. melting point is 4000°C: _____

Page 11

Earthquake P-Wave and S-Wave Travel Time

- **Travel time** is the amount of time it takes for the P-wave or the S-wave to move a given distance.
- The travel time axis is labeled by minutes. Each minute is divided into smaller increments of 20 seconds.
- **Epicenter** is the surface location above where the earthquake occurred in the crust.
- The epicenter distance axis is labeled by every 1000 (10^3) kilometers, which is divided into smaller increments of 200 km.
- As epicenter distance increases, the travel time of the seismic wave increases.
- As epicenter distance increases, the travel time difference between P- and S-wave increases.
- To find the P-wave or the S-wave travel time for a given distance, follow the line up from the distance until you reach the graph line, and go across to the left for the time.
- To find the distance a P- or S-wave travels in a given time, locate the time on the Y-axis, follow it across until you reach the graph line, and go down to find the distance.
- Use this graph when given a seismogram of the P- and S-wave information as recorded by a seismograph.
- If given a seismogram subtract the P-wave arrival time from the S-wave arrival time. Once you have this number, take a piece of paper and mark off the time difference using the Y-axis of the chart. Slide the paper between the S- and the P-wave graph lines until the spacing matches. Now read down to get the distance to the epicenter. You can check your answer by counting the lines between the S-and the P-wave lines, each line is 20 seconds.

Questions

117. How far will a P-wave travel in 8 minutes and 20 seconds? _____

118. If the P-wave and S-wave arrive 6 minutes and 40 seconds apart, the distance to the epicenter is _____ kilometers.

119. The time it takes the S-wave to travel 5.8×10^3 kilometers is _____

120. Based on this seismogram, the distance to epicenter was _____

    P-Wave   P-Wave
    12:03:30 12:07:00

Page 12

Dewpoint(°C) and Relative Humidity(%)

- Dewpoint temperature and relative humidity indicate the amount of moisture present in the air.
- Both are determined by using a psychrometer and these charts.
- The dry-bulb temperature is the air temperature.

- The wet-bulb temperature is affected by the amount of evaporation that is taking place. Evaporation is a cooling process, therefore the wet-bulb temperatures is always less than the dry bulb/air temperature.
- When the relative humidity is low, the air is dry and more evaporation will occur. The wet-bulb temperature will be lower.
- Dewpoint is affected by relative humidity. When air is drier the dewpoint is lower; when it it humid the dewpoint is closer to the air temperature.
- To use both charts, temperatures must be in Celsius (°C).
- You must subtract the wet-bulb temperature from the dry-bulb temperature.
- Go across the row for the given dry-bulb temperature and the down the column of the difference between the wet-bulb and dry-bulb to find dewpoint or relative humidity.
- If you are given a problem where the dry-bulb temperature is an odd number, go in between the two even numbers to estimate the answer.

Questions

121. What is the relative humidity when the air temperature is 22°C and the wet-bulb temperature is 19°C ? _____

122. What is the dewpoint temperature when the dry-bulb temperature is 7°C and the wet-bulb temperature is 4°C ? _____

123. What is the dewpoint when the air temperature is 14°C and the relative humidity is 60%? _____

124. What was the wet-bulb temperature when the air temperature was 18°C and the dewpoint was 13°C? _____

PAGE 13

TEMPERATURE

- There are three different temperature scales: Fahrenheit (°F), Celsius (°C) and Kelvin (K).
- Each temperature scale has different increments. The Fahrenheit scale is two degrees between each line. The Celsius and Kelvin scales are one degree between each line.
- The temperatures for water boils, room temperature, and water freezes are given.
- Use a straight edge to convert from one temperature scale to the other. Remember to check how far apart each increment is and be aware of the negative values.

Questions

125. Room temperature is _____°F or _____°C.
126. 295 K is _____ °F or _____°C.
127. -10°C is _____ °F or _____ K.
128. Water freezes at _____ K.

PRESSURE

- For this chart, **pressure** refers to the force of air pushing down.
- Barometric air pressure can be measured in millibars (mb) or inches (in) of mercury (Hg).
- The spacing on the millibar scale is 1.0 mb between each line.
- The spacing on the inch scale is 0.01 between each line. Air pressure in inches is recorded to the nearest hundredth.
- **One atmosphere** is the average air pressure at sea level.
- When using this chart, check your numbers carefully to be sure you are reading them correctly on the scale. Use a straight edge to do conversions from millibars to inches.

Questions

129. One atmosphere is equal to _____ mb or _____ inches of Hg.

130. 30.24 in = _____ mb; 983.0 mb = _____ in

KEY TO WEATHER MAP SYMBOLS

- The National Weather Service uses these symbols on their weather maps.
- There are more than 600 weather stations in the United States which measure data for each weather variable. This information is recorded on a **station model**. An example of a station model is shown.
- The interpretation for this sample station model is given. The variables are always put in the same place, except for wind direction.
- **Air temperature** and **dewpoint** are recorded in °F. Only the number is written. Do not write the unit °F.
- **Visibility** is the distance in miles that one can clearly see to at that location. It is written as a whole number or as a fraction. No unit is written.
- **Present weather** describes any weather event that is occurring at that location. The symbol is placed to the right of the visibility. A key for the symbols used is given.
- Winds are named for the direction they are coming from. A north wind is blowing from the north. **Wind direction** is shown by a line pointing to that compass direction.
- **Wind speed** is measured in knots (1 knot = 1.15 mile/hour). The wind speed is indicated by "feathers" or short lines extending off the wind direction line.
- Each whole "feather" or line equals 10 knots; a half "feather" or half line equals 5 knots. An indented half-line equals 5 knots. A darkened triangular "flag" represents 50 knots. Add these "feathers" together to get the wind speed.
- If there is no wind direction line then it is calm.
- The station model circle is filled in to show the amount of **cloud cover**. If it is not filled in then skies are clear.
- Barometric air pressure is written as a code in three digits.
- When decoding pressure place a decimal between the last two numbers. Then place a "10" in front if the first digit is a 0 to a 4. Place a "9" in front if the first digit is a 5 or greater. *Examples*: "499" = 1049.9 mb; "704" = 970.4 mb
- When decoding air pressure you can use the "*Pressure*" chart to check your answer.

- **Barometric trend** indicates if the air pressure has risen or fallen during the past three hours.
- To decode barometric trend place a decimal between the two numbers. If the pressure has been rising, a "+" or line sloping upward (/) is drawn. A "–" or a line sloping downward (\ ) indicates that the pressure has fallen. A straight line (–)means that the pressure has been steady, it has not changed. Example: "- 45" means air pressure has fallen 4.5 mb in past three hours.
- **Precipitation** is written in decimal form. This is the amount of precipitation in inches that has fallen at that location in the past six hours.
- An **air mass** controls the weather for a location. The abbreviations for each type is given. Notice that the second letter is capitalized.
- **Continental air masses** have dry air; **maritime air masses** have moist air.
- **Tropical air masses** are warm; **polar air masses** are cold; **arctic air masses** are very cold.
- A **front** is the boundary of an approaching air mass. At a front the weather will change as the new air mass moves to the location. The symbols used on a weather map for each type of front is given. The shaded symbols along the frontal line point in the direction the front is moving towards.
- On a National Weather map the location of hurricanes and tornados are shown by the given symbols. These symbols will be used when tracking the path of these severe storms.

Questions

131. Decode these air pressures: 723 = _____ mb; 394 = _____ mb; 049 = _____ mb

132. Two station model are shown. Decode the weather variables. Include units

station model	28 006 1* ● +02 25 .2	46 096 ¼ ● –6 45 .15
sky conditions		
air temperature		
dewpoint		
wind direction		
wind speed		
visibility		
present weather		
precipitation amount		
air pressure		
air pressure trend		

133. Interpret the weather map below. The numbers represent each of the three fronts.

   a. Name each front type: 1 = _____; 2 = _____; 3 = _____

   b. Front **#3** is moving *(towards) (away from)* Jackson, MS.

   c. The air in the **cP** air mass is *(cold) (warm)* and *(dry) (moist)*.

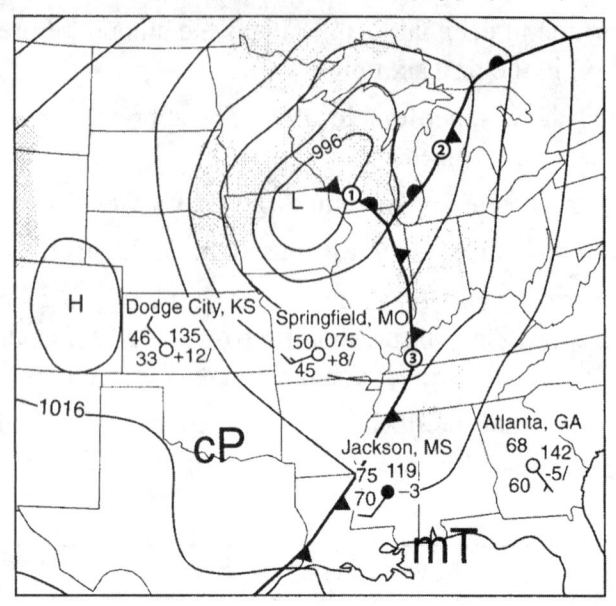

PAGE 14

Selected Properties of Earth's Atmosphere

- Earth's atmosphere is separated into four distinct layers based on temperature patterns: troposphere, stratosphere, mesosphere, and thermosphere.
- The boundaries between each layer is a "pause" and indicated by a dashed horizontal line.
- **Altitude** is the elevation above sea level; it is given in both kilometers and miles. The kilometer scale has increments of 10; the mile scale has increments of 5.
- The layer closest to Earth's surface is the troposphere. It is separated from the stratosphere by the **tropopause**. The stratosphere is separated from the mesosphere by the **stratopause**; the mesosphere is separated from the thermosphere by the **mesopause**.
- The temperature scale is in °C. The solid graph line shows the pattern of temperature change within each layer.
- The temperature in the troposphere decreases with altitude due to adiabatic cooling.
- The temperature in the stratosphere increases due to the absorption of ultraviolet radiation by ozone.
- The air pressure within each layer is given in atmospheres(atm). Normal air pressure at sea level is 1.0 atm. The pressure scale is in increments of 0.25 atm.
- The concentration of water vapor is given in grams per cubic meter (g/m^3). Most of the water vapor is concentrated in the troposphere, closest to Earth's surface.

Questions

134. Name the atmospheric layer found at: 65 miles: _____; 38 km : _____

135. Name the feature found at 80 kilometers. _____

136. As altitude increases in the mesosphere the temperature *(decreases) (increases)*.

137. The temperature of the stratopause is _____

138. The temperature at 40 miles is _____

139. The air pressure at the tropopause is _____

140. The concentration of water vapor at 10 km is _____

PLANETARY WINDS AND MOISTURE BELTS IN THE TROPOSPHERE

- The diagram shows the general movement of air in the troposphere and the global moisture belts. Latitude is indicated for every 30°.
- The prevailing planetary winds curve to the right in the Northern Hemisphere and to the left in the Southern Hemisphere. This is known as the Coriolis Effect and is caused by Earth's rotation.
- This diagram shows the planetary winds and moisture belts for the equinoxes. As the Sun's intense vertical ray shifts with the seasons so will these winds and moisture belts. In the summer they will shift northward; in the winter they will shift southward.
- The diagram uses dashed arrows to show the prevailing winds in each latitude zone. They are named for direction the wind is coming from.
- The solid arrows show the convection currents in the troposphere. Differences in temperature and density causes air to rise and sink.
- **Jet streams** are rapidly moving easterly winds in the upper troposphere. They are shown by a circle with an (X) inside it.
- Subtropical jet streams occur at about 30°N and S.
- Polar jet streams occur at about 60°N and S.
- At the Equator and at 60°N and S the air converges and rises. This results in cloud formation so these are wet zones of moisture.
- At about 30°N and S cool, dense air sinks and diverges. The sinking air will warm, increasing its ability to hold on to moisture. These latitude zones are dry. This is where the world's deserts are located. This also occurs at both Poles where polar deserts occur.
- At about 60° N and S latitude the polar front occurs. Along the polar front cool arctic air meets with warmer air from the mid-latitudes forming storm systems.

Questions

141. Name the planetary wind found at 55°S : _____; 38° N: _____

142. Name the jet stream found at 60° S. _____

143. Converging air will (*rise*) (*sink*).

144. Jet streams are found in the (*lower*) (*upper*) troposphere.

145. In the summer, the planetary wind at 10°N will be from the _____

ELECTROMAGNETIC SPECTRUM

- Radiant energy that can travel through empty space at the speed of light is known as **electromagnetic energy**.
- The **spectrum** is the different types of electromagnetic energies based on their specific wavelengths.

- As you move to the right on this chart, the wavelength increases.
- Energies with wavelengths shorter than visible light are dangerous to our health and to other organisms.
- **Visible light** is the only part of this spectrum that we can see.
- Visible light can be separated into a spectrum of colors (the rainbow); violet has a shorter wavelength and red has a longer wavelength.
- **Infrared radiation** is heat energy; also known as thermal energy.
- The wavelengths for x-rays overlap gamma rays and ultraviolet radiation.
- The wavelengths for microwaves overlap infrared and radio waves. This is why some microwaves are used for cooking; while longer wavelength microwaves are used in communication.

Questions

146. Compared to microwaves, the wavelength for ultraviolet is (*shorter*) (*longer*).

147. The electromagnetic energy with the shortest wavelength is _____.

148. Compared to blue light, yellow light has a (*shorter*) (*longer*) wavelength.

149. Which electromagnetic energy has the least range in wavelength? _____.

PAGE 15

CHARACTERISTICS OF STARS

- Stars are spherical masses of very hot gases and plasma (an electrically charged state of matter).
- On this diagram the characteristics of stars described are: luminosity, surface temperature, color, and life stage.
- The italic words are the names of stars.
- The words in bold capital letters name the various categories of stars that indicate their life stage.
- Stars go through a life cycle. In their early stage as a star they are **main sequence**. As they age they become a **giant** or **supergiant**. When stars reach old age (late stage) they become **dwarfs**.
- **Luminosity** refers to the apparent brightness of the star compared to the Sun. The Sun is the standard, so its luminosity is 1.
- The surface temperature of stars is recorded in Kelvin (K). On the graph axis the temperature decreases to the right.
- A star's temperature determines its color. Very hot stars are blue; colder stars are red.
- When the luminosity and temperature of stars are plotted on this graph most of the stars fall along the zone called **Main Sequence**. Our Sun is a main sequence star.
- **Supergiants** and **giants** are cool but very large in size causing them to be bright. They are the intermediate stage of a star's life cycle.
- **White dwarf** stars are hot but very small and therefore dim. They are stars at the end of their life cycle (late stage).

274

Questions

150. Complete this chart (*some parts of the chart are completed for you as an example*).

star name	40 Eridani B	Barnard's Star	Deneb	Spica
luminosity		0.005		
surface temperature (K)			8500 K	
color				
life stage	late stage			
classified as			Supergiant	

SOLAR SYSTEM DATA

- The solar system is heliocentric with the Sun towards the center.
- There are eight major planets orbiting the Sun.
- Earth's Moon is given as one example of the many planetary moons. A **moon** is a satellite of a planet and orbits that planet. Saturn has the most observed moons.
- For this chart, each body in the solar system is described by their physical features and orbit.
- **Mean distance** from Sun is recorded in million of kilometers. A value of 57.9 means 57.9 million kilometers or 57,900,000 kilometers.
- The **period of revolution** is the time it takes the planet to orbit the Sun or for the Moon to orbit Earth. This value can be in days (d) or years (y).
- A "year" on a planet is defined by its period of revolution.
- The **period of rotation** is the time it takes the object to make one spin on its axis. This value is recorded in days (d), hours (h), minutes (m) and/or seconds (s).
- A "day" on a planet is defined by its period of rotation.
- The size of the planet is described by its **equatorial diameter**. This is the distance from side to side through the center along the Equator zone.
- The mass of each object is compared to a standard: Earth. Therefore Earth is listed as a mass of 1. A mass of 17.15 means that the planet is 17.15 times more massive than Earth.
- **Eccentricity** describes the shape of the planet's elliptical orbit. The larger the eccentricity, the more elliptical the orbit is.
- Each object has a recorded density.
- The **terrestrial planets** are those relatively close to the Sun: Mercury, Venus, Earth, and Mars. These rocky planets have higher densities.
- The **Jovian planets** are found beyond the asteroid belt and include Jupiter, Saturn, Uranus, and Neptune. These are larger planets with low densities.

Questions

151. Compared to Jupiter, how many times further from the Sun is Saturn? _____
(*show solution, answer to the nearest tenth*)

152. How many times larger is Saturn than Earth? _____
(*show solution, answer to the nearest tenth*)

153. As distance from Sun increases, the period of revolution _____.

154. Compared to the terrestrial planets, the Jovian planets rotate (*slower*) (*faster*).

155. Name the planet that rotates slower than it revolves. _____

156. An unmanned space ship is 2,700,000,000 kilometers from the Sun. Name the planet it is closest to. _____

157. Name the planet with the least elliptical orbit. _____

158. Which planet's orbit has a similar shape to that of Earth's Moon? _____

159. Which planet is almost 100 times more massive than Earth? _____

160. Mercury has a density closest to _____.

PAGE 16

PROPERTIES OF COMMON MINERALS

- Minerals are identified by their physical characteristics.
- **Luster** is the ability of a mineral to reflect light. **Metallic luster** is shiny; **non-metallic luster** is dull and not very reflective.
- **Hardness** is the ability of a mineral to be scratched by a common material such as a nail or piece of glass or by another mineral. The lower the number the softer the mineral.
- When a mineral is broken it will either break evenly along a smooth surface (**cleavage**) or break unevenly (**fracture**).
- One mineral can appear in a number of different colors. A better way to identify a mineral is by its **streak** which is the color of the powdered mineral.
- A mineral may have a distinguishing feature that is unique to it and by which it can be identified.
- Minerals have many uses.
- Minerals are **compounds** composed of one or more elements which is written as a chemical formula. The chemical symbols for the elements are given at the bottom of the chart.

Questions

161. Name the mineral described:

  a. non-metallic, fracture, glassy luster. _____

  b. metallic luster, cleavage, an ore of lead. _____

  c. softer than calcite, fractures. _____

  d. will scratch quartz. _____

  e. cleavage in four directions. _____

  f. softest mineral with a metallic luster. _____

  g. hardness of 6.5, fracture, non-metallic, earthy red color. _____

162. Compared to calcite, pyrite is *(softer) (harder)*.

163. A compound of sodium and chlorine is _____

164. Two minerals that bubble with acid are _____ and _____.

165. Name the mineral also known as hornblende. _____

CHAPTER 14 REVIEW

1. What is the approximate straight-line distance from Mt. Marcy to Slide Mt.?

   (1) 120 km (2) 150 km (3) 205 km (4) 235 km

2. An earthquake occurred at 45°N, 75°W on September 5, 1994. Which location in New York State was closest to the epicenter of this earthquake?

   (1) Buffalo (2) Massena (3) Albany (4) New York City

3. In which atmospheric layer would a temperature of 95°C most likely occur?

   (1) troposphere (2) stratosphere (3) mesosphere (4) thermosphere

4. The ratio of oxygen to iron, by mass, in Earth's crust is approximately

   (1) 4:1 (2) 8:1 (3) 40:1 (4) 200:1

5. Which process releases the greatest amount of heat energy to the atmosphere?

   (1) melting (2) freezing (3) vaporization (4) condensation

6. The station model shows weather conditions at a location in Ohio at 9 a.m. on a day in June. Which statement is correct?

   (1) Wind direction is from the northwest at 15 knots.
   (2) The dew point is 74° F.
   (3) The air pressure is 900.2 and falling.
   (4) Visibility is 10 miles.

7. What is the estimated length of time of the Mesozoic Era?

   (1) 65 million years (3) 251 million years
   (2) 185 million years (4) 359 million years

8. Which event occurred at the start of the Cenozoic Era?

    (1) the extinction of many marine organisms
    (2) advance and retreat of the last continental ice
    (3) oceanic oxygen began to enter the atmosphere
    (4) the extinction of dinosaurs

9. Which statement describes the material found within Earth between depth of 2900 kilometers and 5200 kilometers?

    (1) It is in the liquid state.
    (2) It is sedimentary rock.
    (3) It has a higher temperature than Earth's inner core.
    (4) It has a lower pressure than the asthenosphere.

10. As one travels eastward from the Tug Hill Plateau into the Adirondack Mountains, the age of the bedrock changes from

    (1) Devonian to Silurian
    (2) Proterozoic to Cambrian
    (3) Ordovician to Silurian
    (4) Ordovician to Proterozoic

11. A seismic station records an arrival time difference of 5.5 minutes between the P-wave and the S-wave. How far is the seismic station from the epicenter of this earthquake?

    (1) 1.5×10^3 km
    (2) 2.0×10^3 km
    (3) 3.0×10^3 km
    (4) 4.0×10^3 km

12. Where in New York State are gneisses and marbles found?

    (1) Atlantic Coastal Plain
    (2) Hudson-Mohawk Lowland
    (3) Appalachian Upland
    (4) Adirondack Highlands

13. When the air temperature is 10.0° C and the wet-bulb temperature is 2.0° C, the dewpoint and relative humidity is approximately

    (1) 3.0°C and 74%
    (2) 6.0°C and 76%
    (3) -9.0°C and 19%
    (4) -14.0°C and 13%

14. A cold front moved eastward across New York State from Buffalo to Albany in 8 hours. What was the approximate rate at which this front moved?

    (1) 10 km/hr   (2) 25 km/hr   (3) 32 km/hr   (4) 51 km/hr

15. What type of landscape is located at 42° 30' N and 76° W?

    (1) coastal lowland   (2) mountain   (3) plateau   (4) plain

16. A temperature of 73° Fahrenheit is approximately equal to

    (1) 17° Celsius   (2) 23° Celsius   (3) 26° Celsius   (4) 162° Celsius

17. The prevailing winds at 45° S latitude are from the

    (1) southwest   (2) northwest   (3) southeast   (4) northeast

18. What is the largest rock particle that can be carried by a stream with velocity of 3 centimeters per second?

    (1) silt   (2) sand   (3) pebbles   (4) cobbles

19. Particles of which size would be found in shale?

    (1) 0.2 cm   (2) 0.02 cm   (3) 0.002 cm   (4) 0.0002 cm

20. A fine-grained igneous rock composed primarily of plagioclase, biotite, and amphibole is
    (1) basalt   (2) granite   (3) andesite   (4) rhyolite

21. Of the following index fossils, which is the oldest?
    (1) *Eucalyptocrinus*   (2) *Hexameroceras*   (3) *Platyceras*   (4) *Coelophysis*

22. Which rock formed as a result of biologic processes?
    (1) conglomerate   (2) limestone   (3) quartzite   (4) basalt

23. An iron-rich mineral with a metallic luster and a black streak is
    (1) olivine   (2) magnetite   (3) pyrite   (4) pyroxene

24. Compared to our Sun, this star is 100,000 times brighter but cooler?
    (1) *Rigel*   (3) *Barnard's star*
    (2) *Polaris*   (4) *Betelgeuse*

25. In which group are all Earth materials classified as minerals?
    (1) feldspar, granite, and olivine   (3) talc, olivine, and fluorite
    (2) granite, rhyolite, and pumice   (4) pebble, silt, and boulders

26. Which type(s) of rock can be the source of deposited sediments?
    (1) igneous and metamorphic rocks only
    (2) metamorphic and sedimentary rocks only
    (3) sedimentary rocks only
    (4) igneous, metamorphic, and sedimentary rocks

27. What is the name of the warm ocean current that flows along the east coast of South America?
    (1) Falkland Current   (3) Brazil Current
    (2) Gulf Stream   (4) Peru Current

28. What is the age of most of the surface bedrock found in New York State at a latitude of 43°N?
    (1) Middle Proterozoic   (3) Silurian and Ordovician
    (2) Triassic and Jurassic (4) Cambrian and Ordovician

29. Approximately how many years ago did a mass extinction of many land and marine organisms occur?
    (1) 542,000,000   (3) 65,500,000
    (2) 251,000,000   (4) 1,800,000

30. What great mountain building episode occurred in New York State during the Devonian Period?
    (1) Taconian orogeny   (3) Acadian orogeny
    (2) Grenville orogeny   (4) Alleghenian orogeny

31. What type of fossil could be found in the bedrock near Elmira, New York?
    (1) early mammal   (2) dinosaur   (3) flowering plant   (4) fish

32. Seafloor spreading is occurring at the boundary between
    (1) North American Plate and Pacific Plate
    (2) Nazca Plate and South American Plate
    (3) Pacific Plate and Philippine Plate
    (4) Indian-Australian Plate and Antarctic Plate

33. Which layer of Earth's interior most likely has a pressure of 2.5 million atmospheres?
    (1) asthenosphere   (2) stiffer mantle   (3) outer core   (4) inner core

34. In 6 minutes, a P-wave can travel a distance of:
    (1) 1600 km   (2) 3200 km   (3) 4400 km   (4) 9200 km

35. In which part of Earth are felsic rocks most likely to be found?
    (1) continental crust
    (2) oceanic crust
    (3) plastic mantle
    (4) rigid mantle

36. Olivine and pyroxene are commonly found in igneous rocks that are
    (1) felsic with low density
    (2) felsic with high density
    (3) mafic with high density
    (4) mafic with low density

37. During which geologic Period were the continents all part of one landmass with North America and South America joined to Africa?
    (1) Paleocene
    (2) Cretaceous
    (3) Jurassic
    (4) Carboniferous

38. What is the life stage for *Bernard's Star*?
    (1) protostar   (2) early stage   (3) intermediate stage   (4) late stage

39. The temperature in the mesosphere ranges from approximately
    (1) 0° F to -90° F
    (2) 0° C to -90° C
    (3) 15° F to -90° F
    (4) 15° C to -90° C

40. Compared to the wavelength of ultraviolet, the wavelength of infrared is
    (1) shorter   (2) longer   (3) the same

CHAPTER 15

EARTH SCIENCE SKILLS
Metric Measurements, Math Skills, Eccentricity, Locating Epicenters, Rock Types, Properties of Minerals, Mapping

METRIC MEASUREMENTS

In Earth Science length is a common measurement. The diameter of a sediment particle, the distance between the Sun and each planet, the equatorial diameter of a planet, and the distance to the epicenter all involve length measurements.

The meter is the standard unit of measurement. The centimeter is used for small distances; the kilometer is used for larger distances.

```
centimeter  =   10 millimeters
     meter  =  100 centimeters
 kilometer  = 1000 meters
```

Questions

1. Use the ESRT to find the value of the following (*include the unit of measurement*)

    a. mean distance from Sun to Uranus. _____

    b. distance that a S-wave will travel in 9 minutes. _____

    c. equatorial diameter of Mars. _____

    d. radius of Saturn. _____

    e. distance in miles from Elmira to Syracuse. _____

    f. distance in kilometers from Watertown to Albany. _____

    g. largest particle size that a stream moving 30 cm/s can move. _____

    h. largest particle size that a stream moving 0.2 cm/s can move. _____

    i. crystal size in basalt is _____

    j. crystal size in pegmatite is _____

MATH SKILLS

A. Rounding to the tenth.

Many times in science you will be asked to round to the tenth. Density, air pressure, and period of revolution are listed to the tenth place in the ESRT.

To round to the tenth place you need to only look at the value in the hundredth place. If it is 5 or greater then round up; if it is 0 to 4 then leave the number as is. *Examples*: 0.561 = 0.6; 0.537 = 0.5

B. Rounding to the hundredth.

To round to the hundredth place you only need to look at the value in the thousandth place. If it is 5 or more, round up; if it is 0 to 4 then leave the number as is. *Examples*: 6.7836 = 6.78; 6.7853 = 6.79

C. Rounding to the thousandth.

If you look on page 15 of the ESRT you will see in the "*Solar System Data*" chart that eccentricity is expressed to the thousandth place. When calculating eccentricity the final answer should follow this format: round to the thousandth place with NO unit of measurement.

To round to the thousandth, look at the ten thousandth place number. If it is 5 or more you round up; if it is 0 to 4 then leave the number as is.

Examples:
1.3286 ... the 6 is the ten thousandth place, so we should round up;
1.3286 = 1.329
1.3284 ... the 4 is the ten thousandth place, so we should not round up;
1.3284 = 1.328

Questions

2. a. Round to the tenth place: 23.906 = _____; 0.081 = _____; 6.049 = _____

    b. Round to the hundredth place: 8.932 = _____; 34.6053 = _____; 0.006 = _____

3. Round each number to the nearest thousandth.

    a. 0.2368 = _____   e. 1.98706 = _____

    b. 5.3276 = _____   f. 9.9886 = _____

    c. 0.9982 = _____   g. 1.9820 = _____

    d. 9.6555 = _____   h. 0.0019 = _____

4. Circle the smaller number in each pair of numbers.

    a. (0.00009) (0.0009)   e. (1.3244) (1.325)
    b. (0.2154) (0.216)   f. (2.0034) (2.0003)
    c. (1.0987) (1.09874)   g. (0.0023) (0.002288)
    d. (0.09) (0.009)   h. (0.0018) (0.0180)

5. Use the ESRT to place the planets in order of increasing eccentricity.

	planet name	eccentricity
least eccentric orbit		
most eccentric orbit		

6. Use a calculator to reduce each fraction to a number rounded to the tenth.

   a. $\dfrac{2.57}{32.86}$ = _____

   b. $\dfrac{4.5}{1.6}$ = _____

   c. $\dfrac{5.0}{4.3}$ = _____

   d. $\dfrac{0.3}{3.5}$ = _____

7. Use a calculator to perform each mathematical calculation. Express answer to the nearest tenth.

   a. 17.8×4.6 = _____
   b. $87.9 - 58.5$ = _____
   c. $13.5 + 342.8$ = _____
   d. $76.41 + 0.64 + 7.4$ = _____

   e. 0.67×3.4 = _____
   f. $0.56 \div 3.6$ = _____
   g. $82.4 \div 39.5$ = _____
   h. $159.4 - 83.8$ = _____

D. Adding and subtracting time of day.

Time of day is used as a variable in Earth Science when monitoring events such as weather, occurrence of an earthquake, and movements of celestial objects in the sky. The exact time of day can be expressed in hours, minutes, and seconds, such as 12:45:20 p.m. When adding or subtracting times remember that there are 60 seconds in a minute and 60 minutes in an hour. So when you borrow or add, every minute is 60 seconds and every hour is 60 minutes.

Examples:

lunar eclipse ended at… 11:40 p.m.
lunar eclipse began at … 8:50 p.m.
lunar eclipse lasted for …2 hours 50 minutes
	10:100
	    ~~11:40~~ p.m.
	− 8:50 p.m.
	     2:50
	  2 hrs 50 min

Earthquake occurred at …… 9:05:45 p.m.
Travel time for P-wave is… 6 min 20 sec
P-wave arrived at ……….. 9:12:05 p.m.
	9:05:45 p.m.
	+ 6:20
	   9:~~11:65~~
	   9:12:05 p.m.

Questions

8. Determine the time difference between the following times of day (*show solution*).

   a. 8:10 a.m. − 5:40 a.m. = _____ **c.** 6:30:26 p.m. − 4:43:56 p.m. = _____

   b. 3:40 p.m. − 2:20 p.m. = _____ **d.** 11:04:32 a.m. − 6:18:21 a.m. = _____

9. An earthquake occurred at 5:36:40 a.m. on the Pacific Ocean floor. Seismologists predict it will take 4 hours and 30 minutes for a tsunami to reach a coastline in Alaska. At what time will the tsunami arrive? _____
   (*show solution*)

10. Astronomers announce that a total solar eclipse will start at 12:30:20 p.m. It will last for 3 minutes and 45 seconds. At what time will the eclipse end? _____
    (*show solution*)

ECCENTRICITY

Eccentricity is the mathematical description of the shape of an elliptical orbit. It can be used to compare the shapes of orbits as to which is more circular or which is more elliptical. A circular orbit has an eccentricity of zero. A very eccentric (elliptical) orbit will have an eccentricity that is a larger value. Eccentricity can never be greater than 1.

Eccentricity is calculated using the equation:

$$\frac{\text{distance between the foci (d)}}{\text{length of the major axis (L)}}$$

Eccentricity is rounded to the nearest thousandth and there are no units.

For example, the eccentricity of the elliptical orbit to the right is calculated as $\frac{1.6 \text{ cm}}{6.2 \text{ cm}} = 0.258$

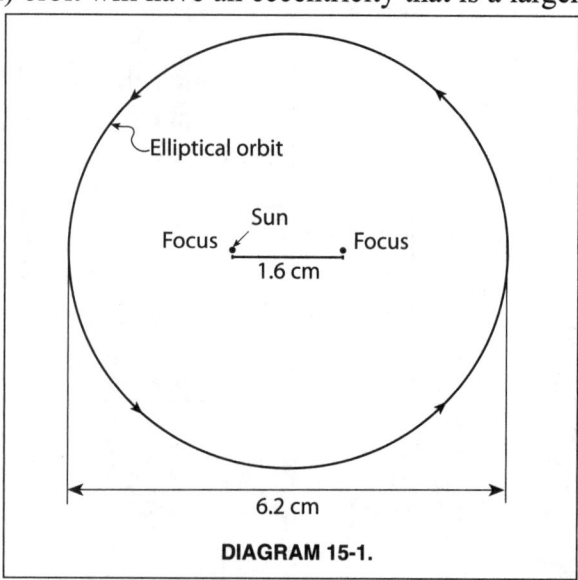

DIAGRAM 15-1.

284

Questions

11. Use the drawing below of an ellipse to do the following (*the foci are shown*):

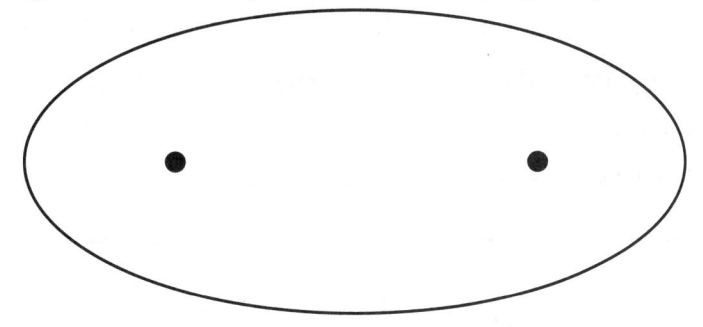

    a. Distance (d) between the foci = _____

    b. Length (L) of the major axis = _____

    c. Eccentricity (e) = _____ (*show solution*)

    d. Is this ellipse more or less eccentric than Venus's orbit? _____ Explain your answer based on the eccentricity of this ellipse and the eccentricity for Venus.
    _____
    _____

12. Use the drawing of the ellipse below to do the following (*the foci are shown*).

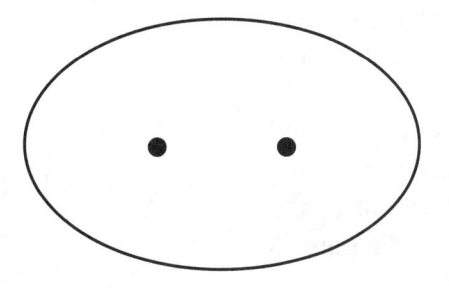

    a. Distance (d) between the foci = _____

    b. Length (L) of the major axis = _____

    c. Eccentricity (e) = _____ (*show solution*)

    d. Is this ellipse more or less eccentric than Mercury's orbit? _____ Explain your answer based on the eccentricity of this ellipse and the eccentricity for Mercury.
    _____
    _____

13. The diagram below shows the geocentric orbit of a satellite around Earth.
   a. Circle one of the foci, label this foci "*Earth*"
   b. Calculate eccentricity of this orbit. _____ (*show solution*)

   c. Write "**KE**" on the ellipse where the kinetic energy is the highest.
   d. Place an "**A**" on the ellipse where apogee occurs. Label this point "*apogee*."
   e. Why is this orbit classified as geocentric? _____
   _____

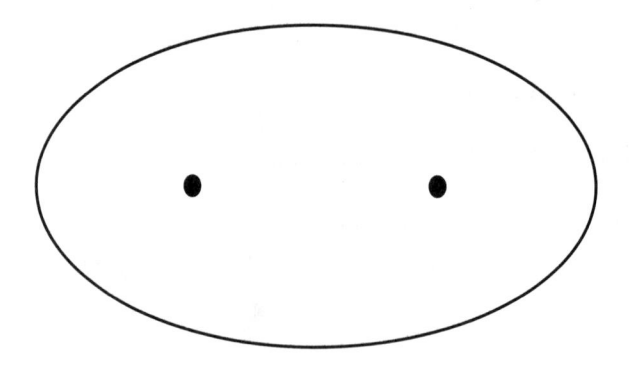

LOCATING EPICENTERS

One important task of geologists is to be able to locate the epicenter of an earthquake quickly and accurately. This enables rescue teams to aid and assist victims. If the epicenter is under an ocean, a tsunami warning will be issued.

To locate an epicenter, seismograms from three seismic stations that show P-wave and S-wave information from the same earthquake event are needed. For each seismogram the distance to the epicenter for the station can be determined. On a map, circles are drawn around each seismic station with the distance to the epicenter as the radius. The intersection of the three circles is the epicenter. On the diagram to the right the epicenter is at point **C**.

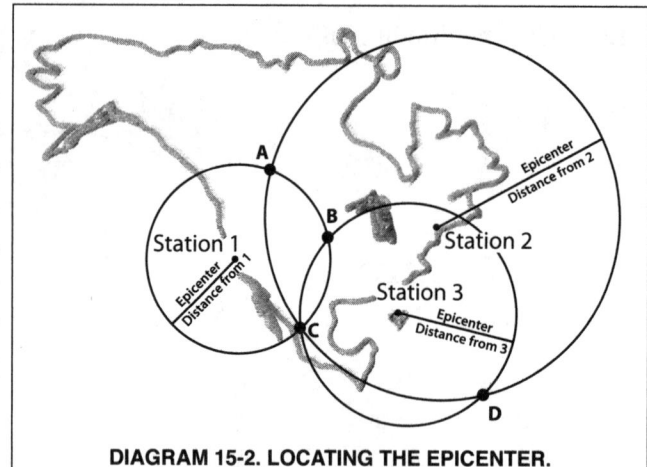

DIAGRAM 15-2. LOCATING THE EPICENTER.

When using a seismogram remember the following:
1. Read the seismogram carefully, the time may *NOT* start at 0.
2. Be sure you know the interval that each hour or minute is divided. It may be in intervals of 10, 20, or 30.
3. Use a straight edge when using the "*P-Waves and S-Wave Travel Time*" graph on page 11 of the ESRT. Check each time and distance at least twice.
4. When using the map scale, be precise! Measure carefully.
5. If your circles do not neatly intersect at one point, place your epicenter at the closest location to where they would appear to intersect.

Questions

14. The seismograms show the arrival of the P- and S-waves from the same earthquake at three different stations. Complete the data chart.

Station	P-wave arrival time	S-wave arrival time
A		
B		
C		

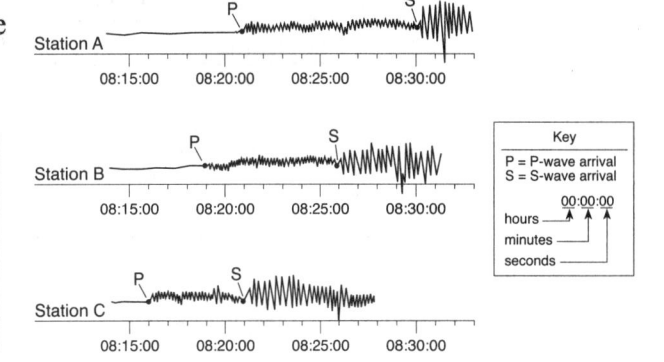

15. Determine the distance a **P**-wave and an **S**-wave will travel in each given time.
Express your answer to the nearest tenth in scientific notation.

Travel Time	P-wave travel distance	S-wave travel distance
3 min 40 sec		
6 min 10 sec		
2 min 30 sec		
5 min 00 sec		
12 min 20 sec		3.9×10^3 km

16. Determine the time it takes for a **P**-wave and an **S**-wave to travel each distance.
Express the answer in minutes and seconds.

Distance	P-wave travel time	S-wave travel time
3000 km		
1800 km		
8100 km		
2400 km	4 min 40 sec	
4200 km		

17. Determine the difference in arrival times of the **P**-wave and the **S**-wave for each distance.
Express answer in minutes and seconds.

Distance to Epicenter	Difference in arrival times
1600 km	
600 km	
8200 km	
4400 km	6 min
5800 km	

a. State the relationship between distance to epicenter and difference between arrival times between the seismic waves. _____

287

18. Determine the distance to an epicenter based on each time difference between the **P**-wave and the **S**-wave.

Time between P and S wave	Distance to the epicenter
6 min 40 sec	5200 km
5 min 10 sec	
3 min 30 sec	
9 min 0 sec	
2 min 50 sec	

19. Use the map scale on page 3 of the ESRT to do the following:

   a. Draw a circle around point **A** with a radius of 40 miles.

. A

   b. Draw a circle with a radius of 120 km around point **B**.

. B

20. Two seismic stations recorded the following seismograms from an earthquake.

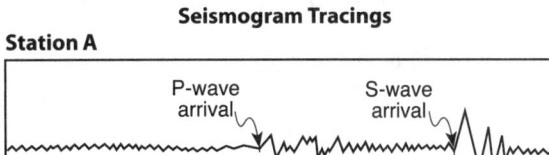

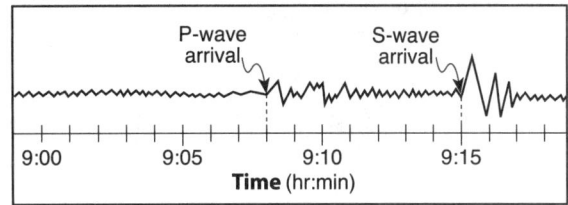

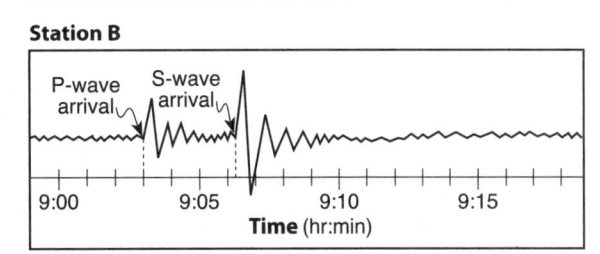

Seismogram Tracings

Station A

Station B

a. Complete the chart.

	station A	station B
arrival time of P-wave		
arrival time of S-wave		
difference in arrival times		
distance to epicenter		
travel time of P-wave		
travel time of S-wave		

b. Explain why the epicenter of the earthquake can not be determined from this data. _____

_____

c. A third seismic station submitted their seismogram but there was no information about the S-wave. Why did this occur? _____

_____

_____

d. Compare the speed of a P-wave in oceanic crust versus continental crust. Explain why this occurs. _____

_____

_____

_____

21. Station **Z** is 5000 km from the epicenter. Using a compass draw a circle with this radius around station **Z**. The epicenter is at point _____.

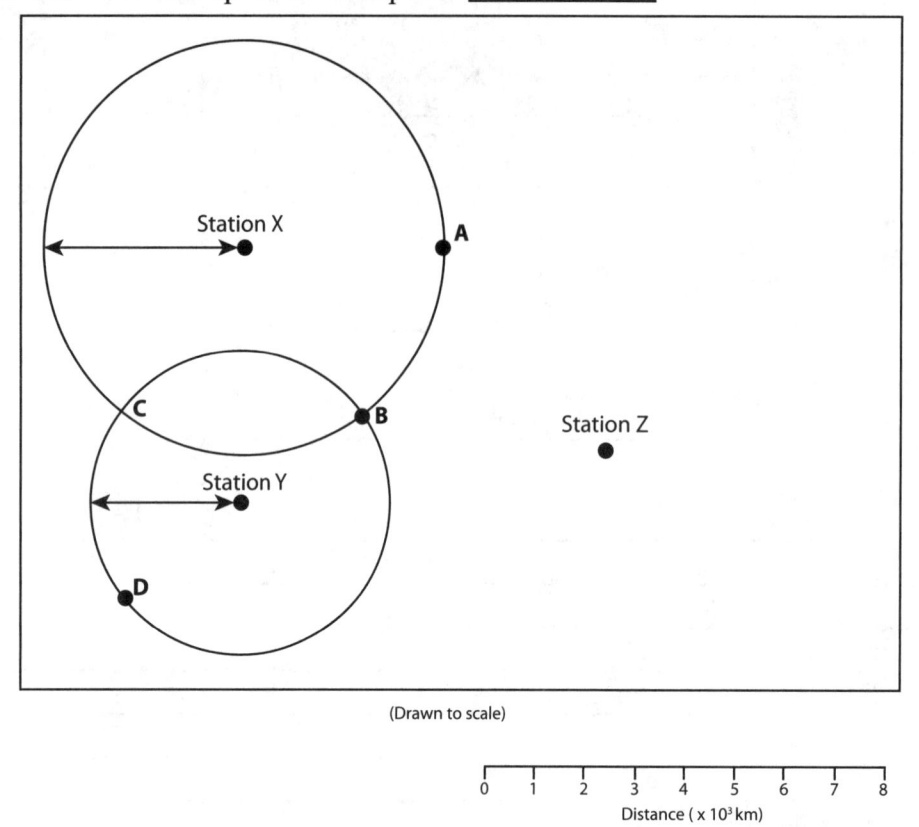

(Drawn to scale)

Distance (x 10³ km)

ROCK TYPES

Rocks are classified by their physical characteristics. These physical features are the result of the method of formation of the rock type. The chart below lists the physical features common to each of the three rock types.

Igneous Rocks (melting and solidification of molten rock material on surface or within Earth)	Sedimentary Rocks (compaction and cementation of land-derived sediments, bioclastic particles, or precipitating minerals)	Metamorphic Rocks (pre-existing rocks are exposed to intense heat and/or pressure without melting)
glassy texture may occur	fossils are common, such as imprints of shells	foliation or banding of the minerals
vesicular (empty gas pockets) may appear	may show layering of the sediments	mineral bands may be folded or distorted
often polyminerallic, minerals are intergrown/interlocking	sediments particles visible, ex. pebbles, sand grains	polyminerallic or monominerallic, minerals are intergrown
fine to coarse grain texture (mineral crystals are less than 1 mm to larger than 10 mm)	ripple marks and mud cracks sometimes appear	fine to coarse texture

Questions

22. The diagrams below represent five different rock samples and a description of each. Name the rock type (*igneous, sedimentary, metamorphic*) for each diagram.

A	B	C	D	E
Bands of alternating light and dark intergrown minerals	Easily splint layers of 0.0001-cm-diameter particles cemented together	Glassy black rock that breaks with a shell-shape fracture	Intergrown 0.5-cm-diameter crystals of various colors	Sand and pebbles cemented together

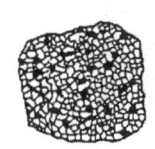

_____ _____ _____ _____ _____

23. For each description, name the rock type (*igneous, sedimentary, metamorphic*).

a. no mineral crystals visible, black color, glassy. _____

b. many trilobite fossils visible. _____

c. bird-like footprints. _____

d. polyminerallic, minerals are in distorted bands. _____

e. various sized particles cemented together. _____

f. evidence of water deposition by ripple marks. _____

g. evidence of gas pockets. _____

h. foliation of minerals. _____

i. pebbles embedded in clay. _____

j. polyminerallic, minerals are aligned. _____

k. polyminerallic, mineral cystrals are intergrown. _____

l. layering of sand-sized sediments. _____

24. The chart below shows drawings or photographs of rocks. Identify each as to rock type (*igneous, sedimentary, metamorphic*) and explain your reason.

	Rock type	Explanation
a.		
b.		
c.		
d.		
e.		
f.		
g.		
h.		
i.		

PROPERTIES OF MINERALS

Minerals are chemical compounds that make up rocks. Minerals form during processes that occur in Earth or at Earth's surface. Minerals are identified by their physical characteristics. The most common characteristics used are color, streak, luster, hardness, cleavage or fracture, density, and crystal shape.

Questions

25. The drawings below illustrate a physical property of a mineral. Name the physical property illustrated and describe the property.

Mineral Sample	Physical Property
a.	a. _____ define: _____ _____
b.	b. _____ define: _____ _____
c.	c. _____ define: _____ _____
d.	d. _____ define: _____ _____
e.	e. _____ define: _____ _____

26. Contrast metallic and non-metallic luster. _____
_____

293

27. Complete the chart to compare properties of some common minerals. Use the ESRT and the hardness scale below.

Hardness of Four Materials

Material	Hardness
human fingernail	2.5
copper penny	3.0
window glass	4.5
steel nail	6.5

mineral name	Olivine	Galena	Quartz	Calcite	Feldspar
luster					
hardness					
cleavage or fracture ?					
color					
other properties					
can a penny scratch it?					
can a steel nail scratch it ?					

MAPPING

Mapping is an important skill of geologists and meteorologists. They will often collect data in a given area called a field. After measurements are taken in the area, the data will be plotted on a map of the area. **Isolines** will be drawn that connect positions of the same value. This allows the scientist to see patterns in the data for the region. Diagram 15-3 shows two examples of field data and isolines.

Odor Field of a Barbeque

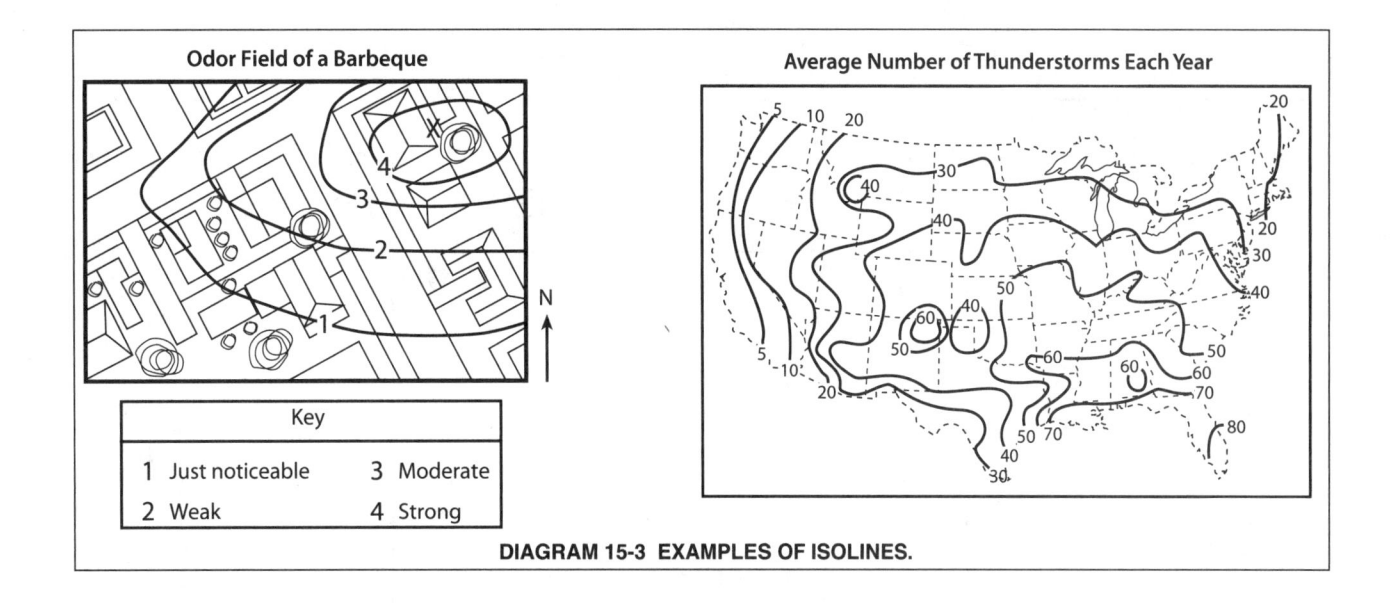

Average Number of Thunderstorms Each Year

Key
1 Just noticeable   3 Moderate
2 Weak   4 Strong

DIAGRAM 15-3 EXAMPLES OF ISOLINES.

Notice that on the maps above, the isolines follow these guidelines:
1. Isolines connect adjacent points of the same value with a smooth line.
2. Isolines do not touch or cross each other.
3. Isolines continue to the end of the map or form closed circles.
4. Isolines tend to be parallel to each other.
5. Isolines do not have sharp corners, they are smooth curved lines.

Isolines can be used to estimate the values of other positions which were not measured. A **profile** can be constructed which shows a side view of the map.

The **gradient** or change in the field value over a distance is calculated based on the map. Gradient can also be estimated by the degree of spacing between the isolines. Closely spaced isolines have steep gradients; isolines that are far apart have gentle gradients. Gradient is calculated using the equation: $\frac{change\ in\ field\ value}{distance}$.

$$gradient = \frac{change\ in\ field\ value}{distance}$$

$$= \frac{100\ ft}{4\ miles}$$

$$= 25.0\ ft\ \frac{ft}{mi}$$

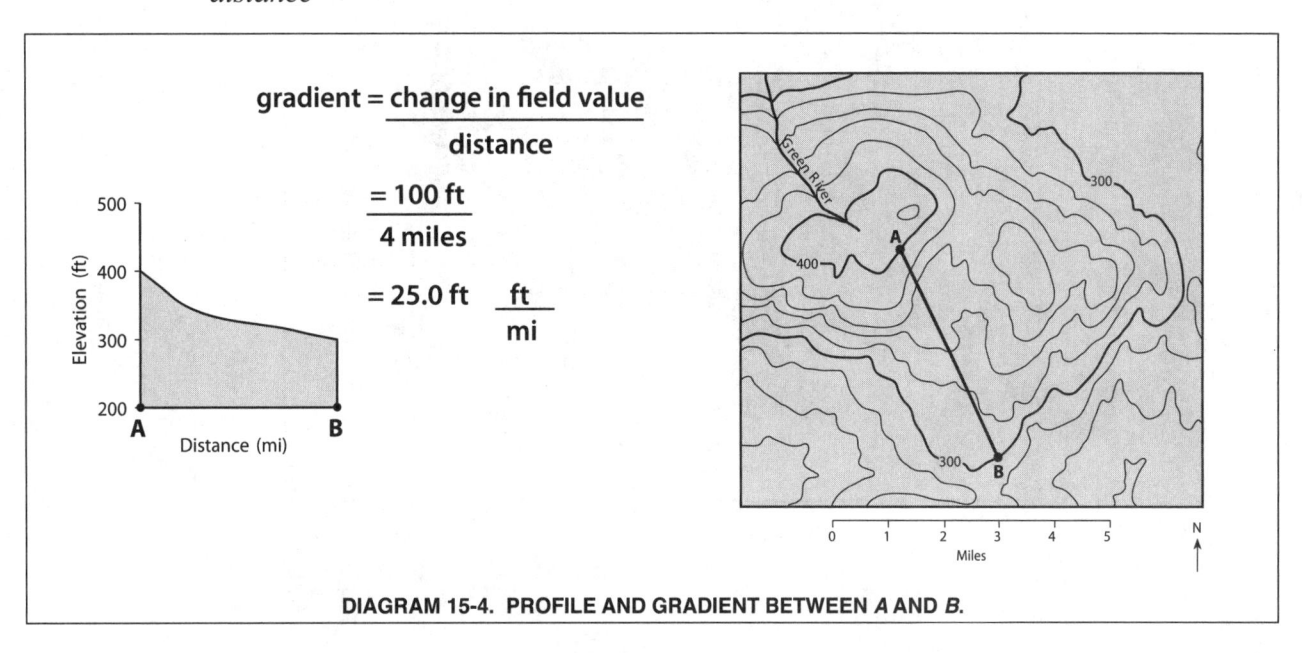

DIAGRAM 15-4. PROFILE AND GRADIENT BETWEEN *A* AND *B*.

Questions

28. The field map below shows air temperatures, in degrees Celsius, taken at the same elevation and same time in a classroom. Starting at 20°C, draw isotherms for every 2°C.

```
•A      •       •       •       •B
20      21      22      24      28

•       •       •       •       •
21      22      24      25      26

•       •       •       •       •
20      22      25      25      25

•       •       •       •       •
18      21      22      24      25

•       •       •       •       •
19      21      23      23      24
```

29. The map below shows noon temperatures, in degrees Fahrenheit, recorded by numerous weather stations in the eastern United States on a fall day. Draw isotherms for every 10° F starting at 30°F.

To construct a profile between two points, such as **A** and **B**, do the following:

1. label clearly the elevation of EVERY contour line;
2. use a piece of paper with a straight edge and line the straight edge of the paper along the bottom of the line **A-B**.
3. mark points **A** and **B** on the paper;
4. every time a contour line meets the straight edge, place a "tick" mark with the elevation of that line;
5. place the paper with the "tick" marks along the bottom of the profile grid;
6. for each "tick" mark, plot the elevation of that point on the grid;
7. connect the points with a smooth line; if there are two adjacent points with the same elevation do **not** connect them with a straight line, you must look at the map to determine if the elevation was lower or higher between those two points.

30. The map below shows elevations in feet for an island.

a. Calculate the gradient of Auroro Creek in feet per mile.

_____

(*show solution*)

b. Place an "X" on the map for a position that has an elevation of approximately 675 feet.

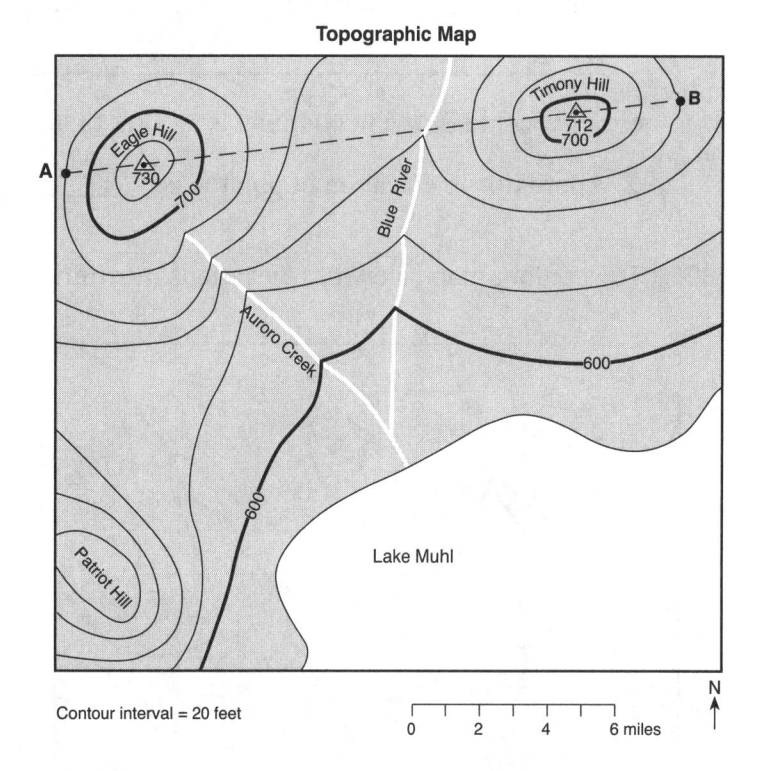

c. Construct a profile from **A** to **B**.

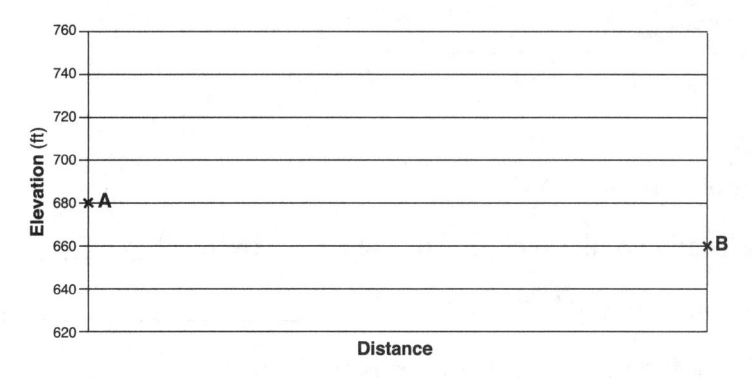

297

31. The contour map below has elevations measured in feet.

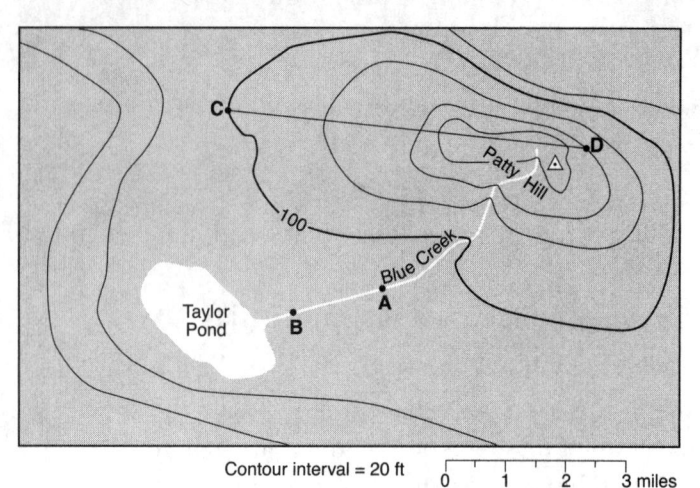

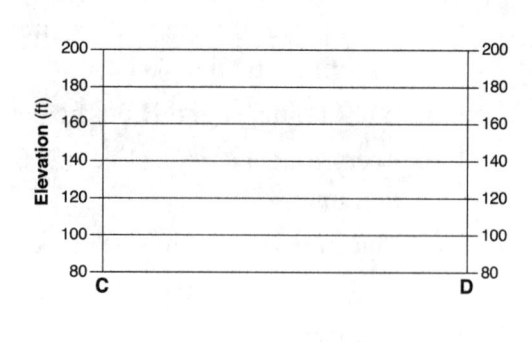

Contour interval = 20 ft

a. Construct a profile from **C** to **D**.
b. Blue Creek is flowing (*into*) (*out of*) Taylor Pond. How do you know? _____
_____
c. What could be the maximum elevation of Patty Hill? _____
d. Approximate elevation of point **A** is _____; **B** is _____

32. The contour map below has elevations measured in feet.

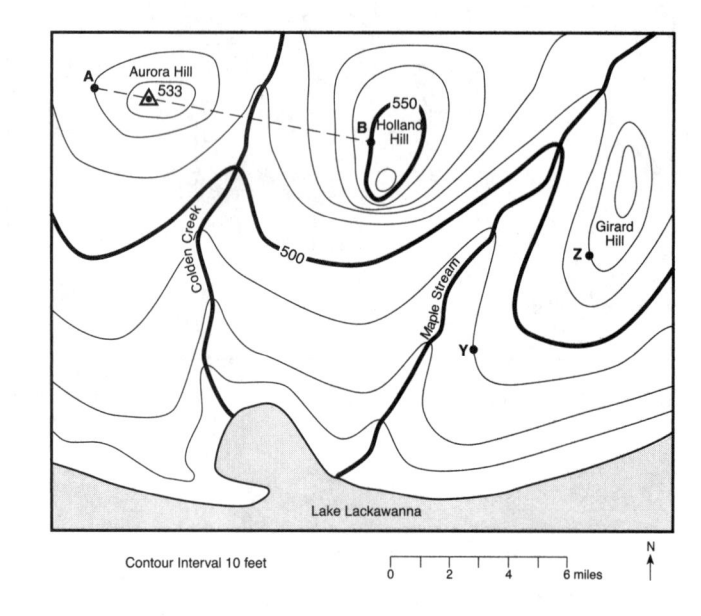

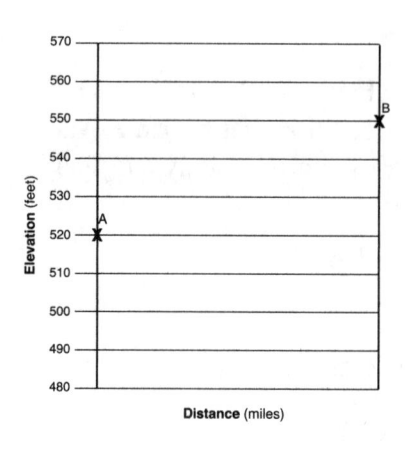

Contour Interval 10 feet

a. Construct a profile from **A** to **B**.
b. What is the elevation of the shoreline of Lake Lackawanna? _____
c. In what direction does Maple Stream flow? _____
d. What is the highest possible elevation for Girard Hill? _____
e. Place an "**X**" in an area that has a steep slope.

298

33. The map below shows elevations, measured in feet, for a region. Contour lines for **800** feet to **900** feet have been drawn.

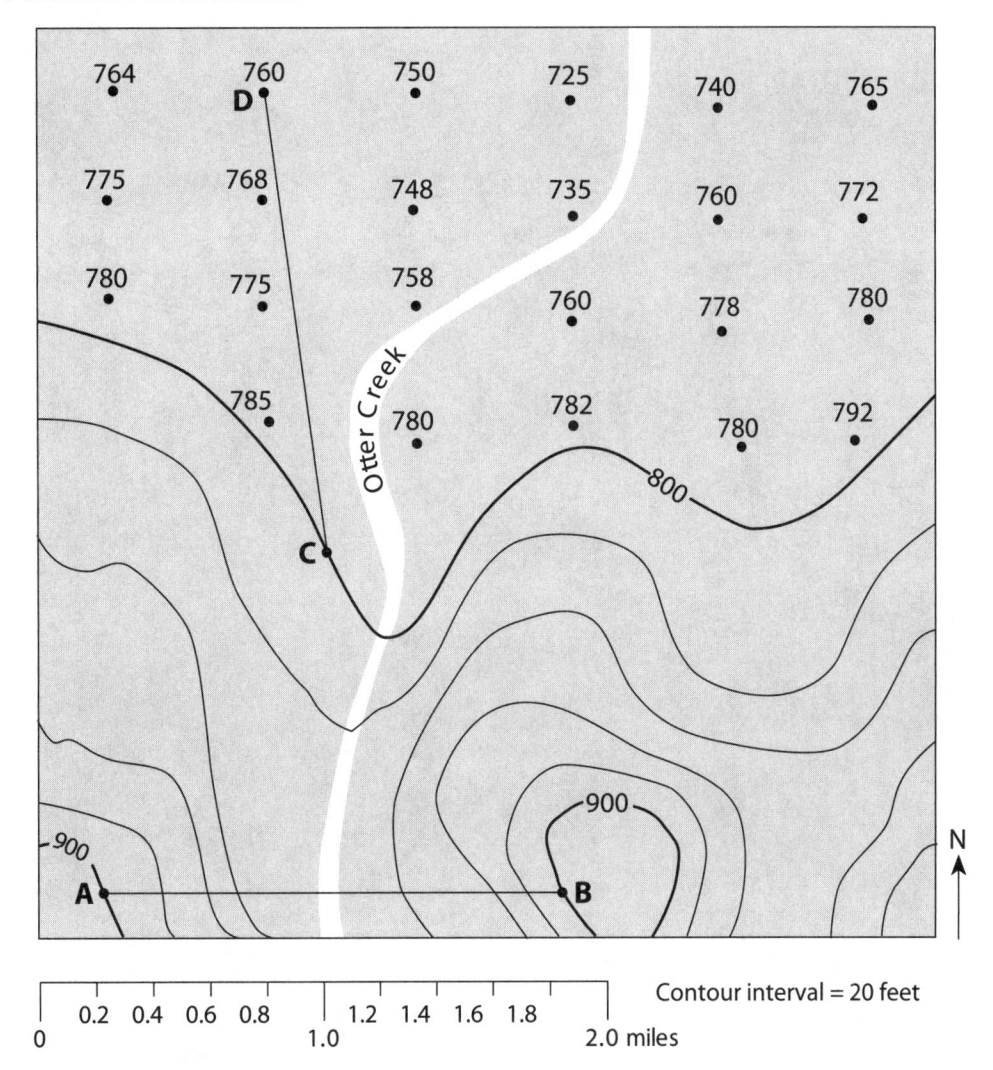

a. Draw the contour lines for **780**, **760**, and **740** feet.

b. Calculate the gradient from **C** to **D**. _____
(show solution)

299

The University of the State of New York • THE STATE EDUCATION DEPARTMENT • Albany, New York 12234 • www.nysed.gov

Reference Tables for Physical Setting/EARTH SCIENCE

Radioactive Decay Data

RADIOACTIVE ISOTOPE	DISINTEGRATION	HALF-LIFE (years)
Carbon-14	$^{14}C \rightarrow {}^{14}N$	5.7×10^3
Potassium-40	$^{40}K \rightarrow {}^{40}Ar$ $^{40}K \rightarrow {}^{40}Ca$	1.3×10^9
Uranium-238	$^{238}U \rightarrow {}^{206}Pb$	4.5×10^9
Rubidium-87	$^{87}Rb \rightarrow {}^{87}Sr$	4.9×10^{10}

Equations

$$\text{Eccentricity} = \frac{\text{distance between foci}}{\text{length of major axis}}$$

$$\text{Gradient} = \frac{\text{change in field value}}{\text{distance}}$$

$$\text{Rate of change} = \frac{\text{change in value}}{\text{time}}$$

$$\text{Density} = \frac{\text{mass}}{\text{volume}}$$

Specific Heats of Common Materials

MATERIAL	SPECIFIC HEAT (Joules/gram • °C)
Liquid water	4.18
Solid water (ice)	2.11
Water vapor	2.00
Dry air	1.01
Basalt	0.84
Granite	0.79
Iron	0.45
Copper	0.38
Lead	0.13

Properties of Water

Heat energy gained during melting	334 J/g
Heat energy released during freezing	334 J/g
Heat energy gained during vaporization	2260 J/g
Heat energy released during condensation	2260 J/g
Density at 3.98°C	1.0 g/mL

Average Chemical Composition of Earth's Crust, Hydrosphere, and Troposphere

ELEMENT (symbol)	CRUST Percent by mass	CRUST Percent by volume	HYDROSPHERE Percent by volume	TROPOSPHERE Percent by volume
Oxygen (O)	46.10	94.04	33.0	21.0
Silicon (Si)	28.20	0.88		
Aluminum (Al)	8.23	0.48		
Iron (Fe)	5.63	0.49		
Calcium (Ca)	4.15	1.18		
Sodium (Na)	2.36	1.11		
Magnesium (Mg)	2.33	0.33		
Potassium (K)	2.09	1.42		
Nitrogen (N)				78.0
Hydrogen (H)			66.0	
Other	0.91	0.07	1.0	1.0

2011 EDITION

This edition of the Earth Science Reference Tables should be used in the classroom beginning in the 2011–12 school year. The first examination for which these tables will be used is the January 2012 Regents Examination in Physical Setting/Earth Science.

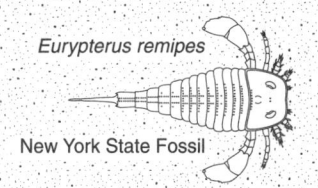

Eurypterus remipes

New York State Fossil

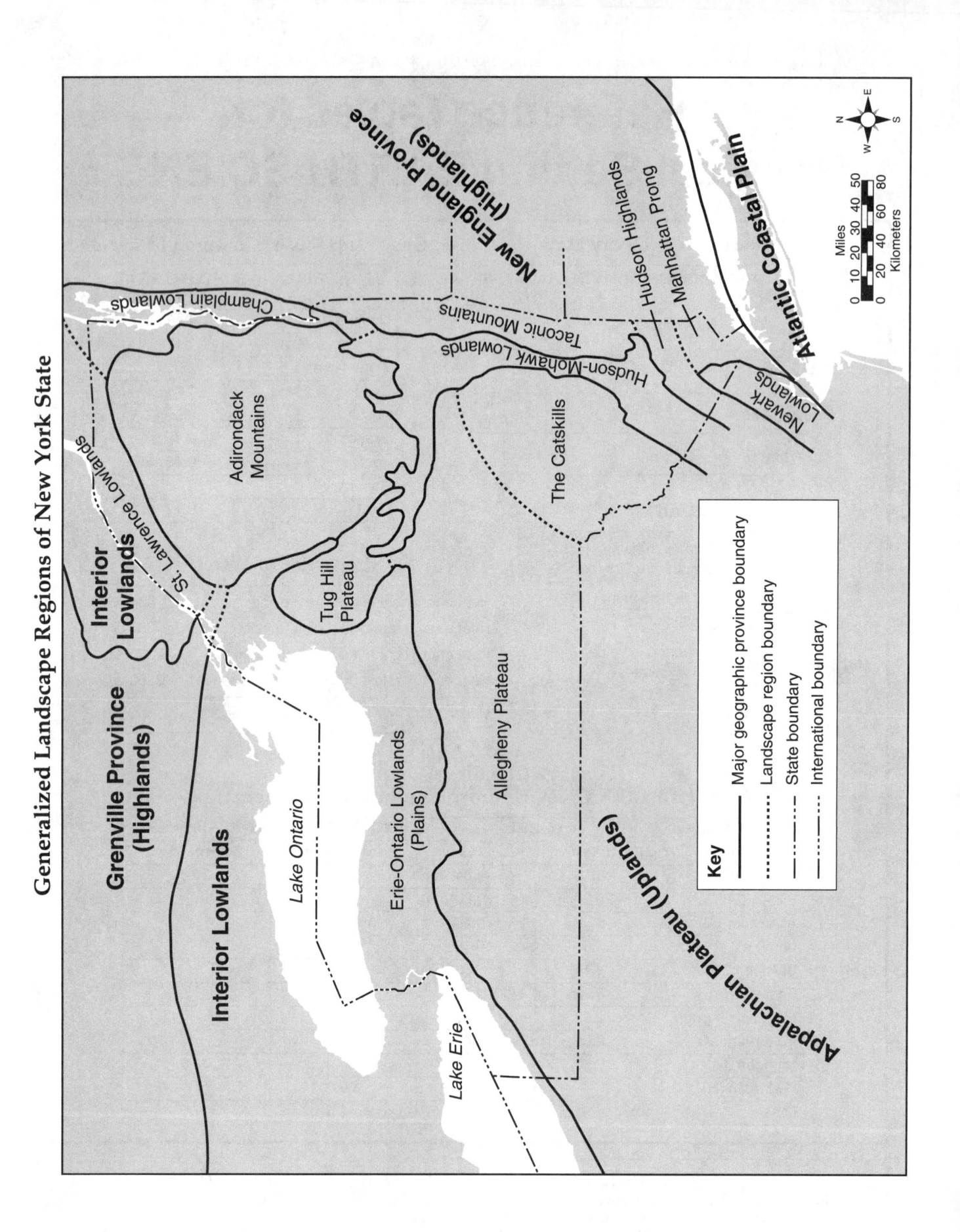

Generalized Landscape Regions of New York State

Generalized Bedrock Geology of New York State

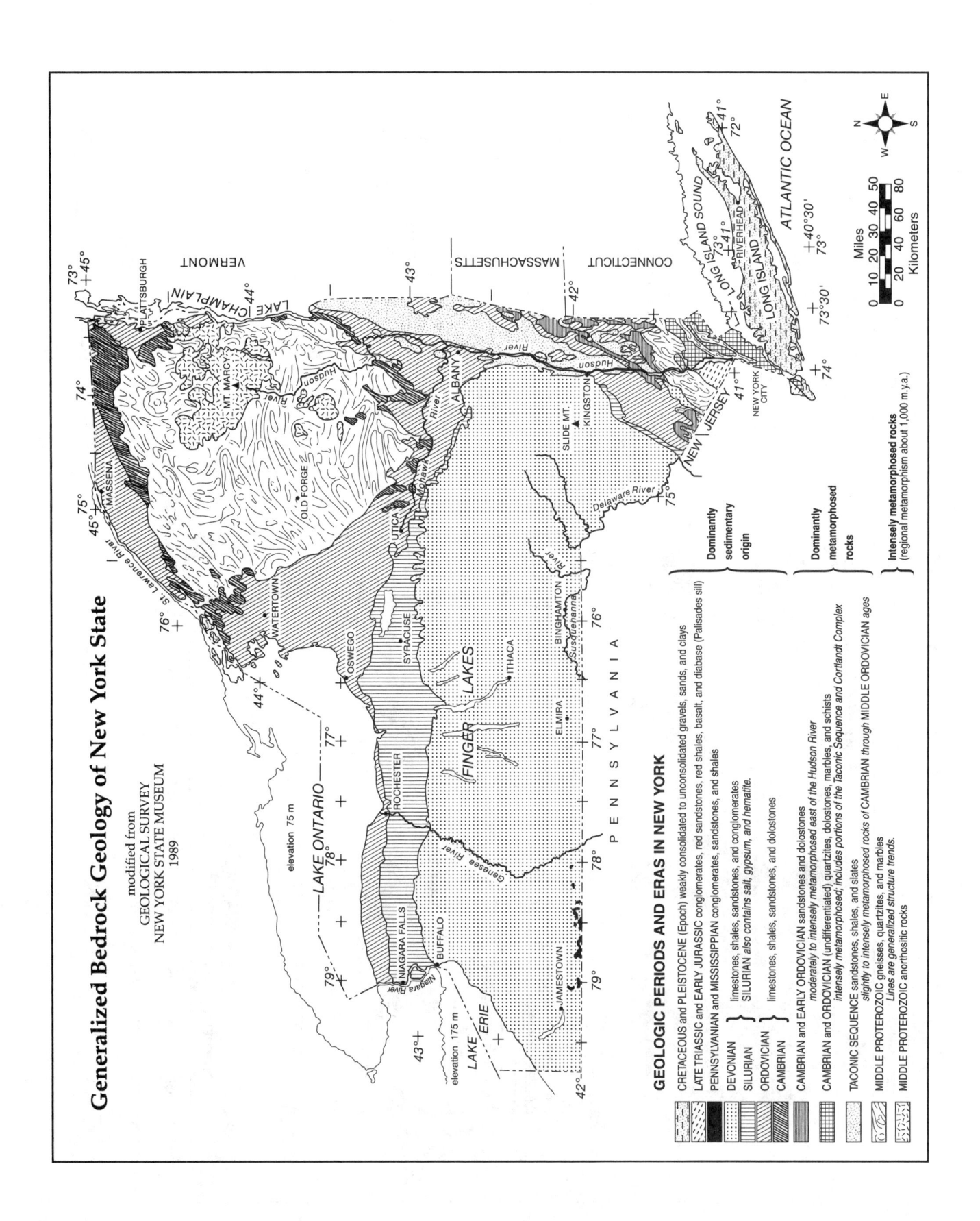

modified from
GEOLOGICAL SURVEY
NEW YORK STATE MUSEUM
1989

GEOLOGIC PERIODS AND ERAS IN NEW YORK

- CRETACEOUS and PLEISTOCENE (Epoch) weakly consolidated to unconsolidated gravels, sands, and clays
- LATE TRIASSIC and EARLY JURASSIC conglomerates, red sandstones, red shales, basalt, and diabase (Palisades sill)
- PENNSYLVANIAN and MISSISSIPPIAN conglomerates, sandstones, and shales
- DEVONIAN limestones, shales, sandstones, and conglomerates
- SILURIAN *also contains salt, gypsum, and hematite.*
- ORDOVICIAN limestones, shales, sandstones, and dolostones
- CAMBRIAN
- CAMBRIAN and EARLY ORDOVICIAN sandstones and dolostones *moderately to intensely metamorphosed east of the Hudson River*
- CAMBRIAN and ORDOVICIAN (undifferentiated) quartzites, dolostones, marbles, and schists *intensely metamorphosed; includes portions of the Taconic Sequence and Cortlandt Complex*
- TACONIC SEQUENCE sandstones, shales, and slates *slightly to intensely metamorphosed rocks of CAMBRIAN through MIDDLE ORDOVICIAN ages*
- MIDDLE PROTEROZOIC gneisses, quartzites, and marbles
  Lines are generalized structure trends.
- MIDDLE PROTEROZOIC anorthositic rocks

Dominantly sedimentary origin

Dominantly metamorphosed rocks

Intensely metamorphosed rocks (regional metamorphism about 1,000 m.y.a.)

Surface Ocean Currents

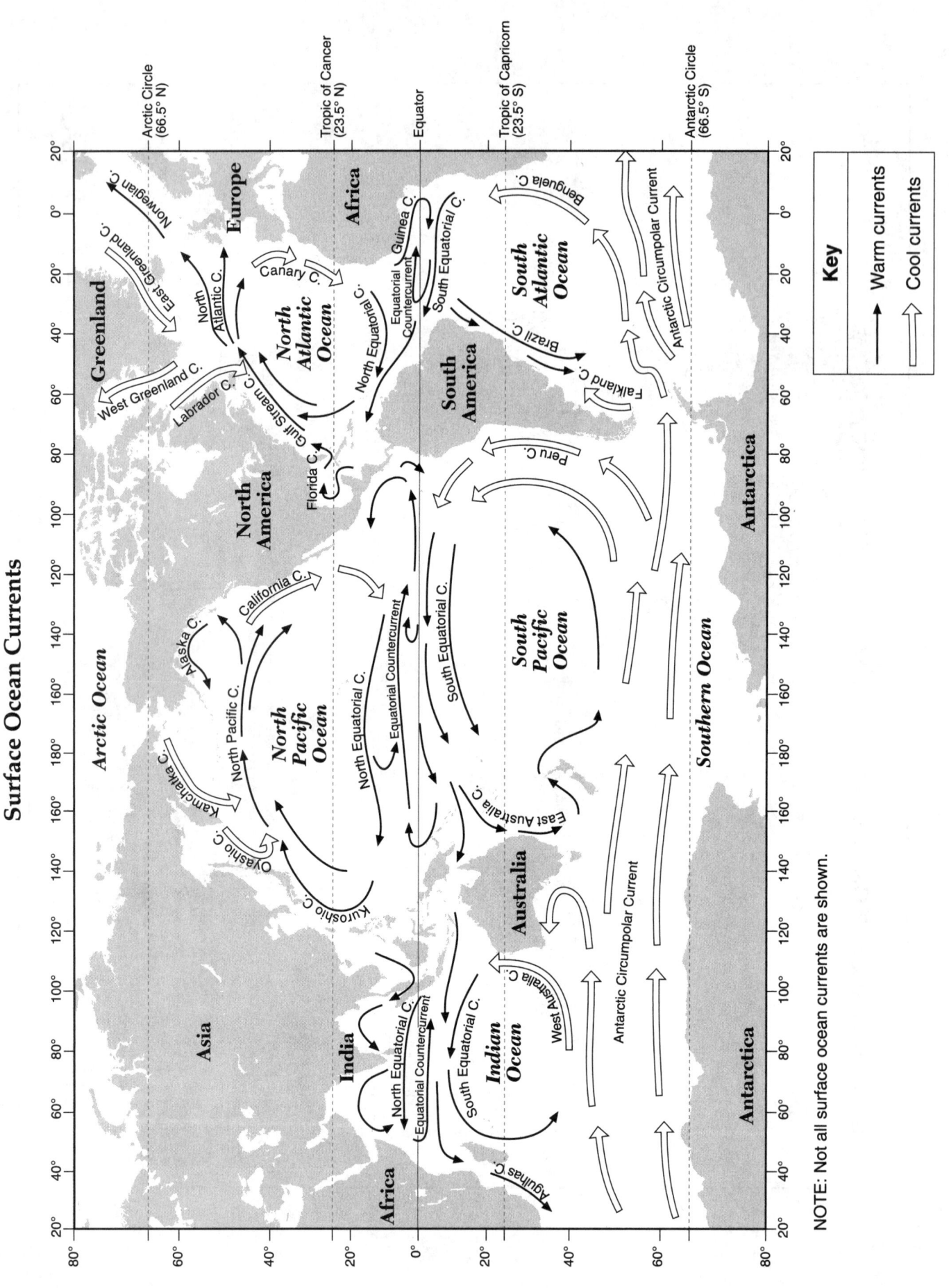

NOTE: Not all surface ocean currents are shown.

Physical Setting/Earth Science Reference Tables — 2011 Edition

Tectonic Plates

Key

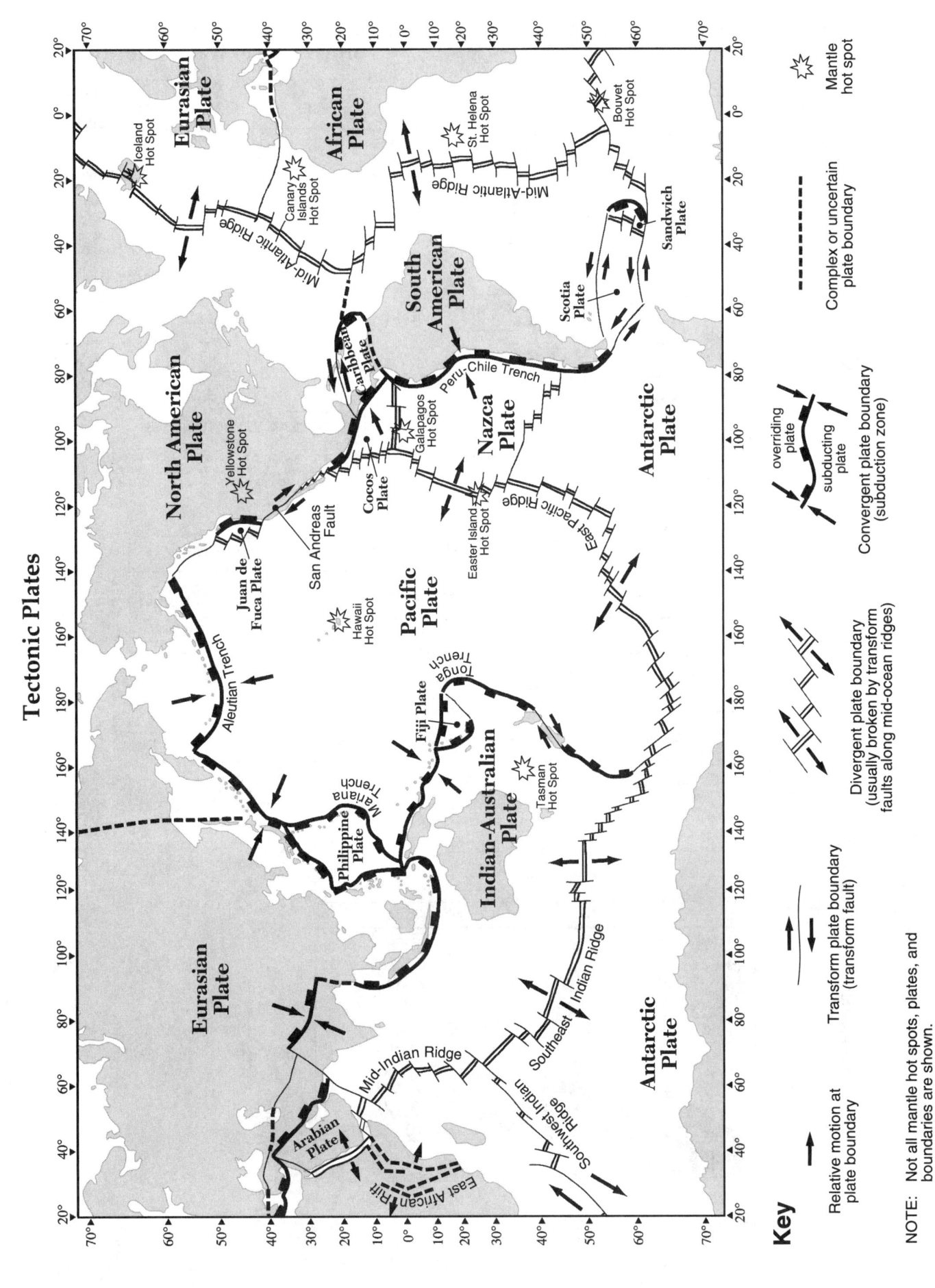

- Mantle hot spot
- Complex or uncertain plate boundary
- Convergent plate boundary (subduction zone) — overriding plate / subducting plate
- Divergent plate boundary (usually broken by transform faults along mid-ocean ridges)
- Transform plate boundary (transform fault)
- Relative motion at plate boundary

NOTE: Not all mantle hot spots, plates, and boundaries are shown.

Physical Setting/Earth Science Reference Tables — 2011 Edition

305

Rock Cycle in Earth's Crust

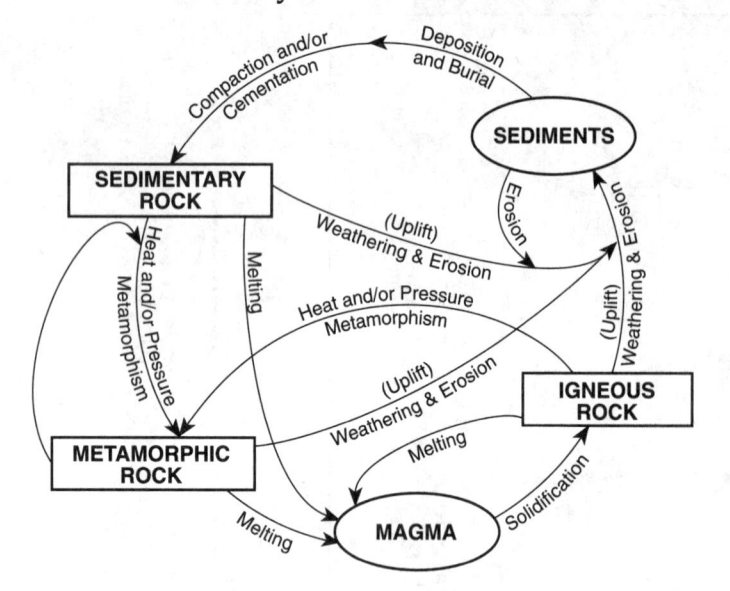

Relationship of Transported Particle Size to Water Velocity

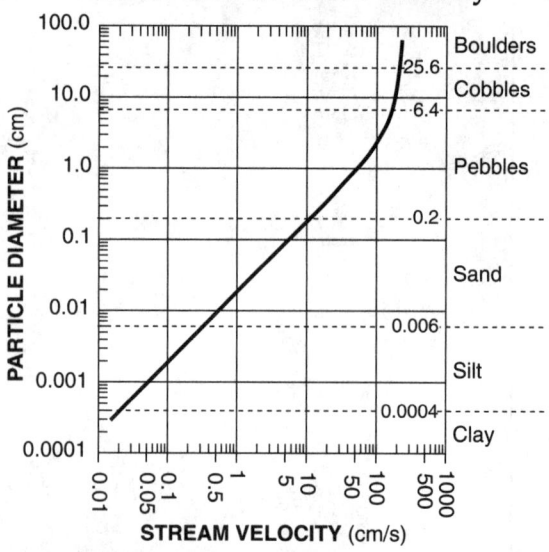

This generalized graph shows the water velocity needed to maintain, but not start, movement. Variations occur due to differences in particle density and shape.

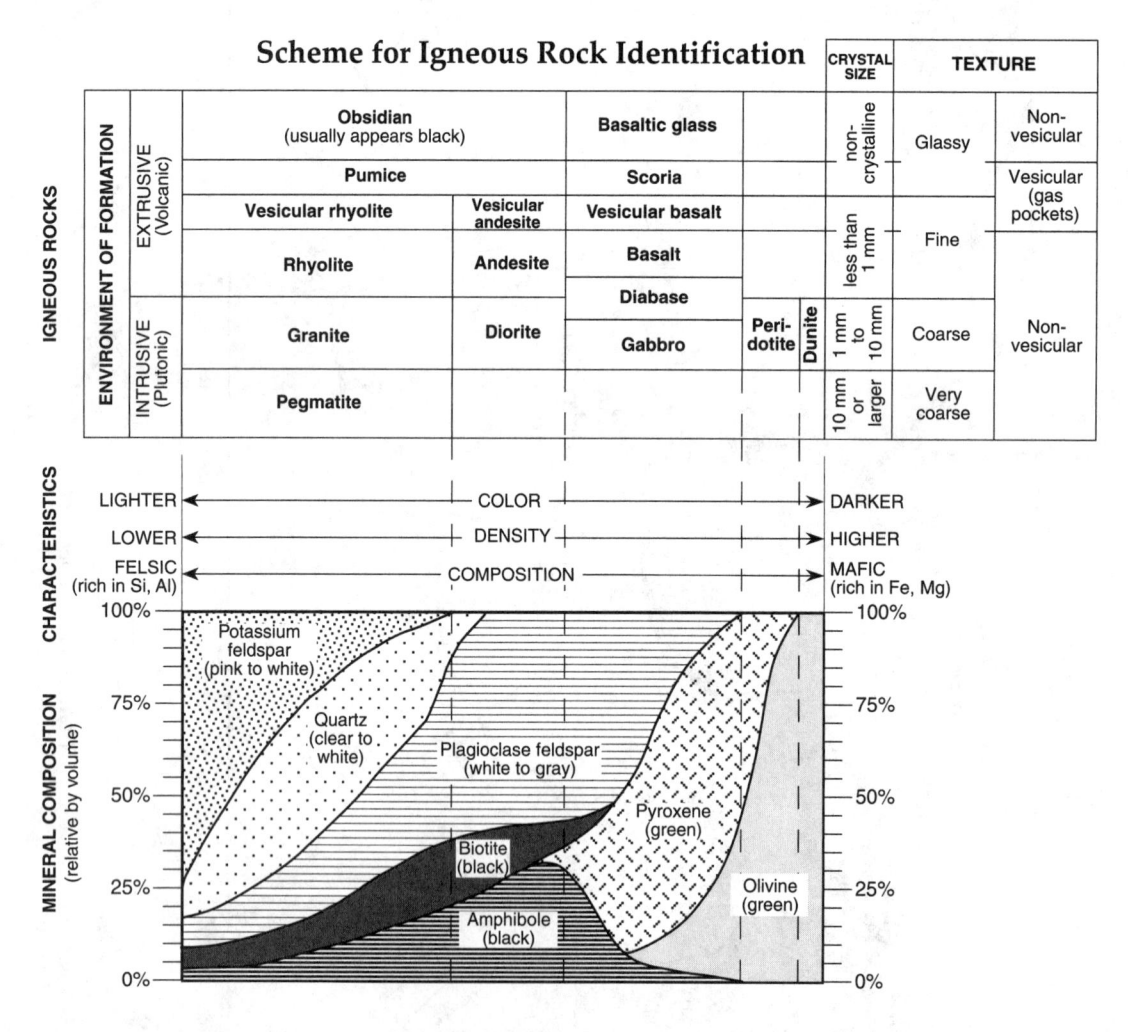

Scheme for Igneous Rock Identification

ENVIRONMENT OF FORMATION						CRYSTAL SIZE	TEXTURE
EXTRUSIVE (Volcanic)	Obsidian (usually appears black)		Basaltic glass			non-crystalline	Glassy / Non-vesicular
	Pumice		Scoria				Vesicular (gas pockets)
	Vesicular rhyolite	Vesicular andesite	Vesicular basalt			less than 1 mm	Fine
	Rhyolite	Andesite	Basalt				
INTRUSIVE (Plutonic)	Granite	Diorite	Diabase	Peridotite	Dunite	1 mm to 10 mm	Coarse / Non-vesicular
			Gabbro				
	Pegmatite					10 mm or larger	Very coarse

CHARACTERISTICS

LIGHTER ←—— COLOR ——→ DARKER
LOWER ←—— DENSITY ——→ HIGHER
FELSIC (rich in Si, Al) ←—— COMPOSITION ——→ MAFIC (rich in Fe, Mg)

Minerals: Potassium feldspar (pink to white), Quartz (clear to white), Plagioclase feldspar (white to gray), Biotite (black), Amphibole (black), Pyroxene (green), Olivine (green)

Physical Setting/Earth Science Reference Tables — 2011 Edition

Scheme for Sedimentary Rock Identification

INORGANIC LAND-DERIVED SEDIMENTARY ROCKS

TEXTURE	GRAIN SIZE	COMPOSITION	COMMENTS	ROCK NAME	MAP SYMBOL
Clastic (fragmental)	Pebbles, cobbles, and/or boulders embedded in sand, silt, and/or clay	Mostly quartz, feldspar, and clay minerals; may contain fragments of other rocks and minerals	Rounded fragments	Conglomerate	
			Angular fragments	Breccia	
	Sand (0.006 to 0.2 cm)		Fine to coarse	Sandstone	
	Silt (0.0004 to 0.006 cm)		Very fine grain	Siltstone	
	Clay (less than 0.0004 cm)		Compact; may split easily	Shale	

CHEMICALLY AND/OR ORGANICALLY FORMED SEDIMENTARY ROCKS

TEXTURE	GRAIN SIZE	COMPOSITION	COMMENTS	ROCK NAME	MAP SYMBOL
Crystalline	Fine to coarse crystals	Halite	Crystals from chemical precipitates and evaporites	Rock salt	
		Gypsum		Rock gypsum	
		Dolomite		Dolostone	
Crystalline or bioclastic	Microscopic to very coarse	Calcite	Precipitates of biologic origin or cemented shell fragments	Limestone	
Bioclastic		Carbon	Compacted plant remains	Bituminous coal	

Scheme for Metamorphic Rock Identification

TEXTURE	GRAIN SIZE	COMPOSITION	TYPE OF METAMORPHISM	COMMENTS	ROCK NAME	MAP SYMBOL
FOLIATED (MINERAL ALIGNMENT)	Fine	MICA, QUARTZ, FELDSPAR, AMPHIBOLE, GARNET, PYROXENE	Regional (Heat and pressure increases)	Low-grade metamorphism of shale	Slate	
	Fine to medium			Foliation surfaces shiny from microscopic mica crystals	Phyllite	
				Platy mica crystals visible from metamorphism of clay or feldspars	Schist	
FOLIATED (BANDING)	Medium to coarse			High-grade metamorphism; mineral types segregated into bands	Gneiss	
NONFOLIATED	Fine	Carbon	Regional	Metamorphism of bituminous coal	Anthracite coal	
	Fine	Various minerals	Contact (heat)	Various rocks changed by heat from nearby magma/lava	Hornfels	
	Fine to coarse	Quartz	Regional or contact	Metamorphism of quartz sandstone	Quartzite	
		Calcite and/or dolomite		Metamorphism of limestone or dolostone	Marble	
	Coarse	Various minerals		Pebbles may be distorted or stretched	Metaconglomerate	

Physical Setting/Earth Science Reference Tables — 2011 Edition

GEOLOGIC HISTORY

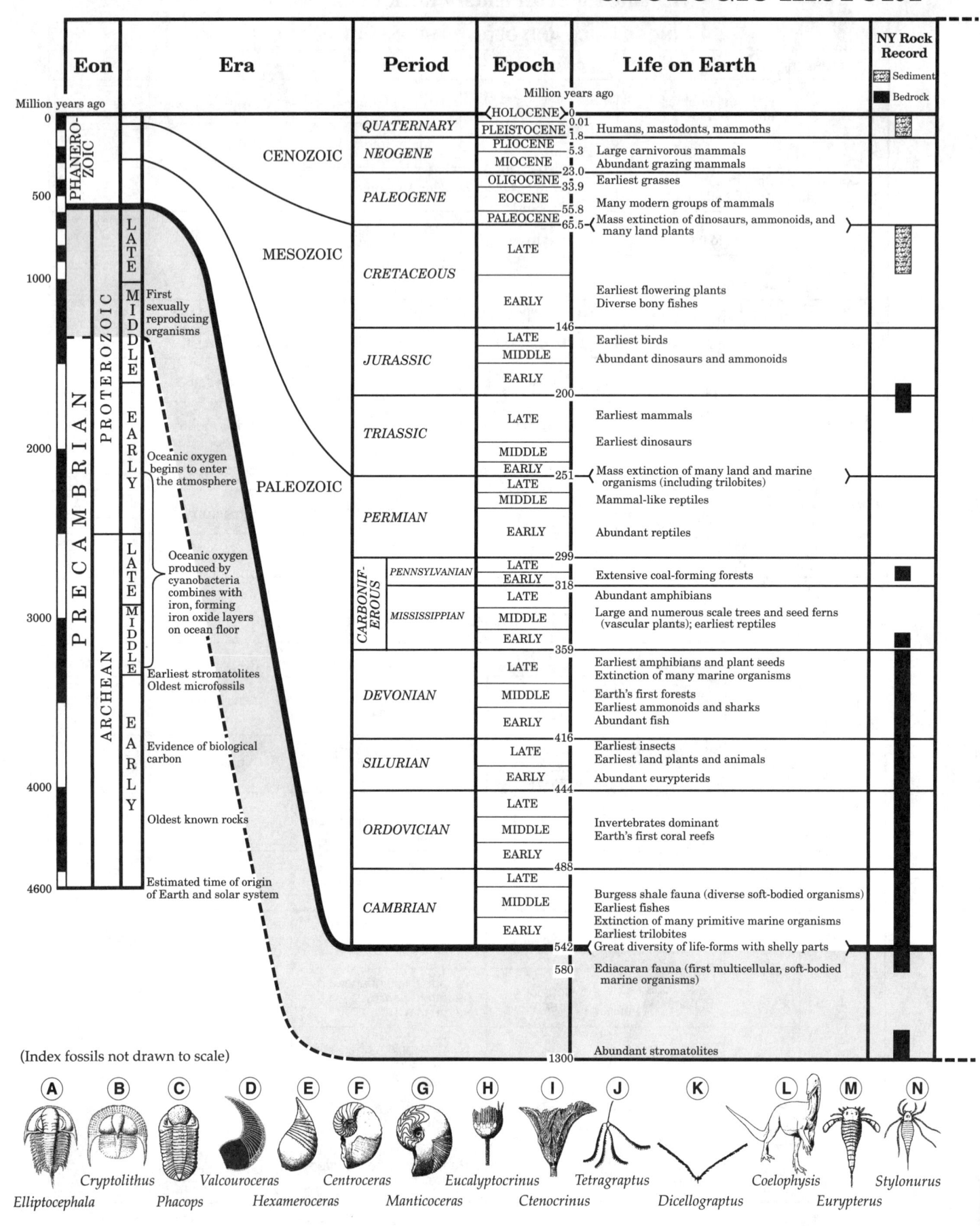

Eon	Era	Period	Epoch	Life on Earth	NY Rock Record
					Sediment / Bedrock

Million years ago (Eon axis) / Million years ago (Epoch axis)

Phanerozoic Eon:

- **Cenozoic Era**
  - **Quaternary**: Holocene (0–0.01), Pleistocene (0.01–1.8) — Humans, mastodonts, mammoths
  - **Neogene**: Pliocene (1.8–5.3) — Large carnivorous mammals; Miocene (5.3–23.0) — Abundant grazing mammals
  - **Paleogene**: Oligocene (23.0–33.9) — Earliest grasses; Eocene (33.9–55.8) — Many modern groups of mammals; Paleocene (55.8–65.5) — Mass extinction of dinosaurs, ammonoids, and many land plants

- **Mesozoic Era**
  - **Cretaceous**: Late; Early — Earliest flowering plants, Diverse bony fishes (–146)
  - **Jurassic**: Late — Earliest birds; Middle — Abundant dinosaurs and ammonoids; Early (–200)
  - **Triassic**: Late — Earliest mammals; Middle — Earliest dinosaurs; Early (–251) — Mass extinction of many land and marine organisms (including trilobites)

- **Paleozoic Era**
  - **Permian**: Late; Middle — Mammal-like reptiles; Early — Abundant reptiles (–299)
  - **Carboniferous – Pennsylvanian**: Late; Early (–318) — Extensive coal-forming forests
  - **Carboniferous – Mississippian**: Late — Abundant amphibians; Middle — Large and numerous scale trees and seed ferns (vascular plants); earliest reptiles; Early (–359)
  - **Devonian**: Late — Earliest amphibians and plant seeds, Extinction of many marine organisms; Middle — Earth's first forests, Earliest ammonoids and sharks; Early — Abundant fish (–416)
  - **Silurian**: Late — Earliest insects, Earliest land plants and animals; Early — Abundant eurypterids (–444)
  - **Ordovician**: Late; Middle — Invertebrates dominant, Earth's first coral reefs; Early (–488)
  - **Cambrian**: Late; Middle — Burgess shale fauna (diverse soft-bodied organisms), Earliest fishes; Early — Extinction of many primitive marine organisms, Earliest trilobites (–542) — Great diversity of life-forms with shelly parts

Precambrian:

- **Proterozoic**: Late / Middle / Early
  - First sexually reproducing organisms
  - Oceanic oxygen begins to enter the atmosphere
  - Oceanic oxygen produced by cyanobacteria combines with iron, forming iron oxide layers on ocean floor
  - 580 — Ediacaran fauna (first multicellular, soft-bodied marine organisms)
  - 1300 — Abundant stromatolites

- **Archean**: Late / Middle / Early
  - Earliest stromatolites, Oldest microfossils
  - Evidence of biological carbon
  - Oldest known rocks

- 4600 — Estimated time of origin of Earth and solar system

(Index fossils not drawn to scale)

Index fossils:
- A — Elliptocephala
- B — Cryptolithus
- C — Phacops
- D — Valcouroceras
- E — Hexameroceras
- F — Centroceras
- G — Manticoceras
- H — Eucalyptocrinus
- I — Ctenocrinus
- J — Tetragraptus
- K — Dicellograptus
- L — Coelophysis
- M — Eurypterus
- N — Stylonurus

Physical Setting/Earth Science Reference Tables — 2011 Edition

OF NEW YORK STATE

Geologic History of New York State

Time Distribution of Fossils
(including important fossils of New York)

The center of each lettered circle indicates the approximate time of existence of a specific index fossil (e.g. Fossil Ⓐ lived at the end of the Early Cambrian).

Important Geologic Events in New York

- Advance and retreat of last continental ice
- Sands and clays underlying Long Island and Staten Island deposited on margin of Atlantic Ocean
- Dome-like uplift of Adirondack region begins
- Initial opening of Atlantic Ocean North America and Africa separate
- Intrusion of Palisades sill
- Pangaea begins to break up
- **Alleghenian orogeny** caused by collision of North America and Africa along transform margin, forming Pangaea
- Catskill delta forms
- Erosion of Acadian Mountains
- **Acadian orogeny** caused by collision of North America and Avalon and closing of remaining part of Iapetus Ocean
- Salt and gypsum deposited in evaporite basins
- Erosion of Taconic Mountains; Queenston delta forms
- **Taconian orogeny** caused by closing of western part of Iapetus Ocean and collision between North America and volcanic island arc
- Widespread deposition over most of New York along edge of Iapetus Ocean
- Rifting and initial opening of Iapetus Ocean
- Erosion of Grenville Mountains
- **Grenville orogeny:** metamorphism of bedrock now exposed in the Adirondacks and Hudson Highlands

Inferred Positions of Earth's Landmasses

- 59 million years ago
- 119 million years ago
- 232 million years ago
- 359 million years ago
- 458 million years ago

Ⓞ Mastodont, Beluga Whale
Ⓟ Cooksonia
Ⓠ Naples Tree, Aneurophyton
Ⓡ Bothriolepis
Ⓢ Condor
Ⓣ Lichenaria
Ⓤ Cystiphyllum, Pleurodictyum
Ⓥ Pleurodictyum
Ⓦ Maclurites, Platyceras
Ⓧ Platyceras
Ⓨ Eospirifer
Ⓩ Mucrospirifer

Physical Setting/Earth Science Reference Tables — 2011 Edition

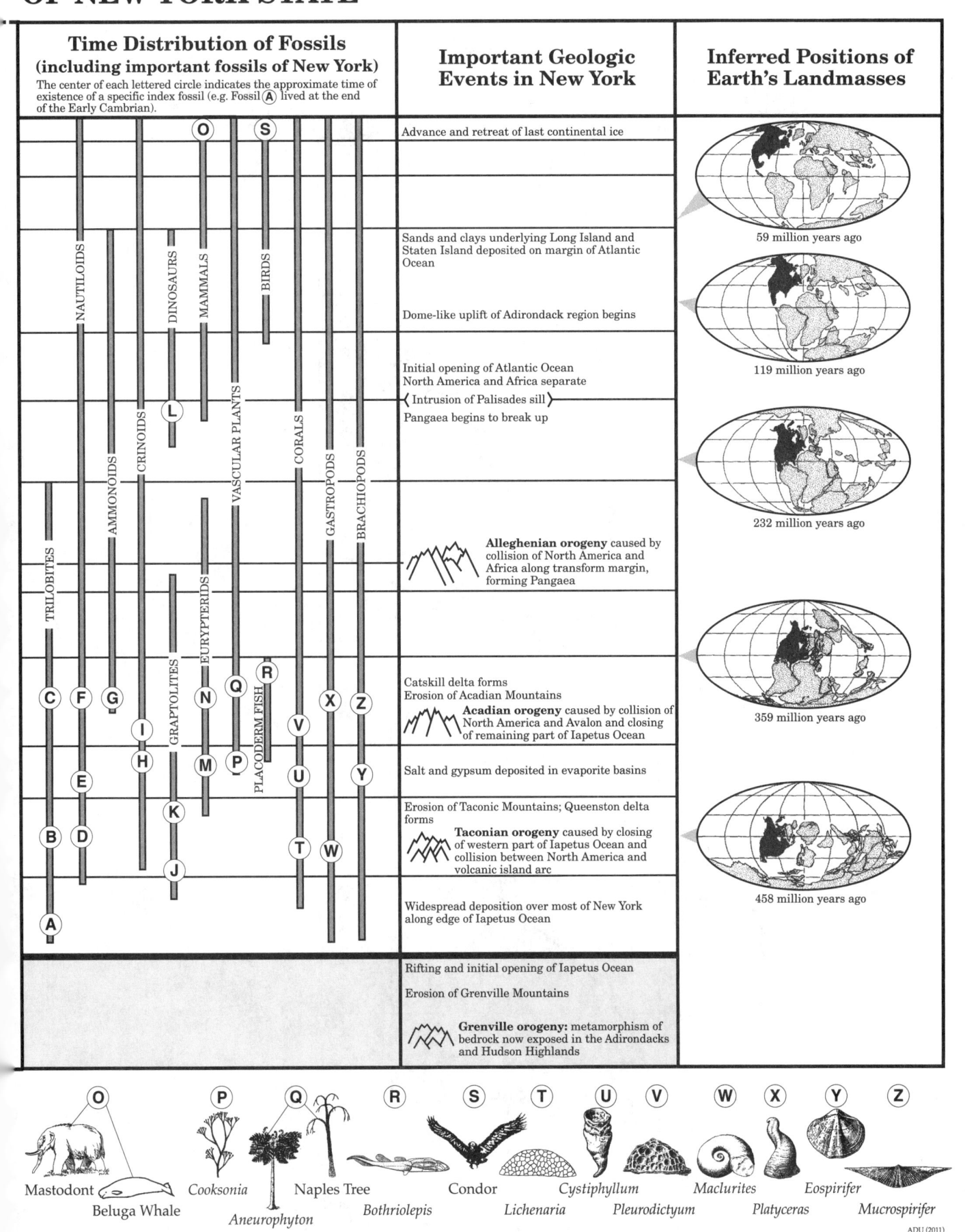

309

Inferred Properties of Earth's Interior

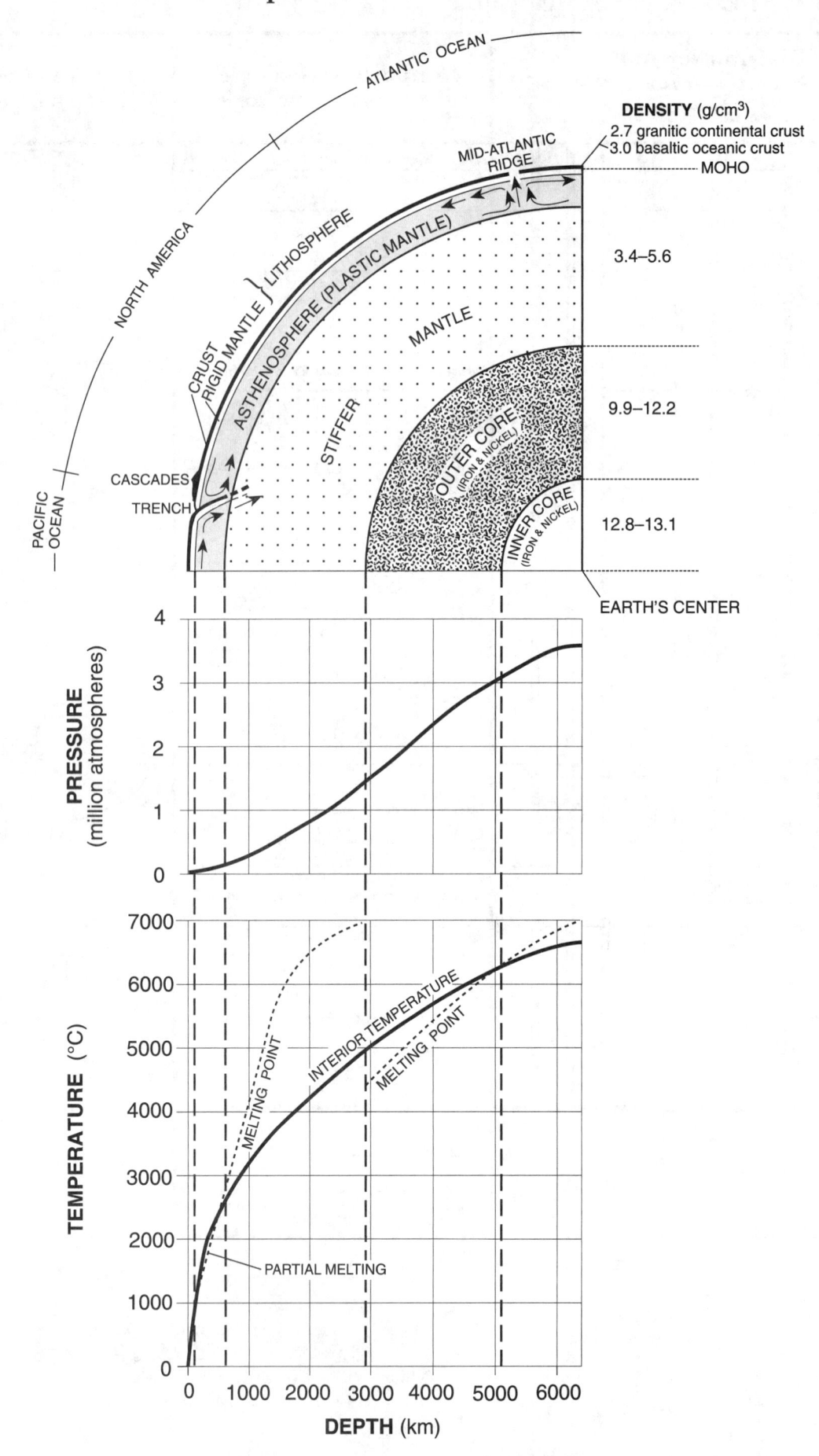

DENSITY (g/cm³)
- 2.7 granitic continental crust
- 3.0 basaltic oceanic crust
- MOHO
- 3.4–5.6
- 9.9–12.2
- 12.8–13.1

Physical Setting/Earth Science Reference Tables — 2011 Edition

310

Earthquake P-Wave and S-Wave Travel Time

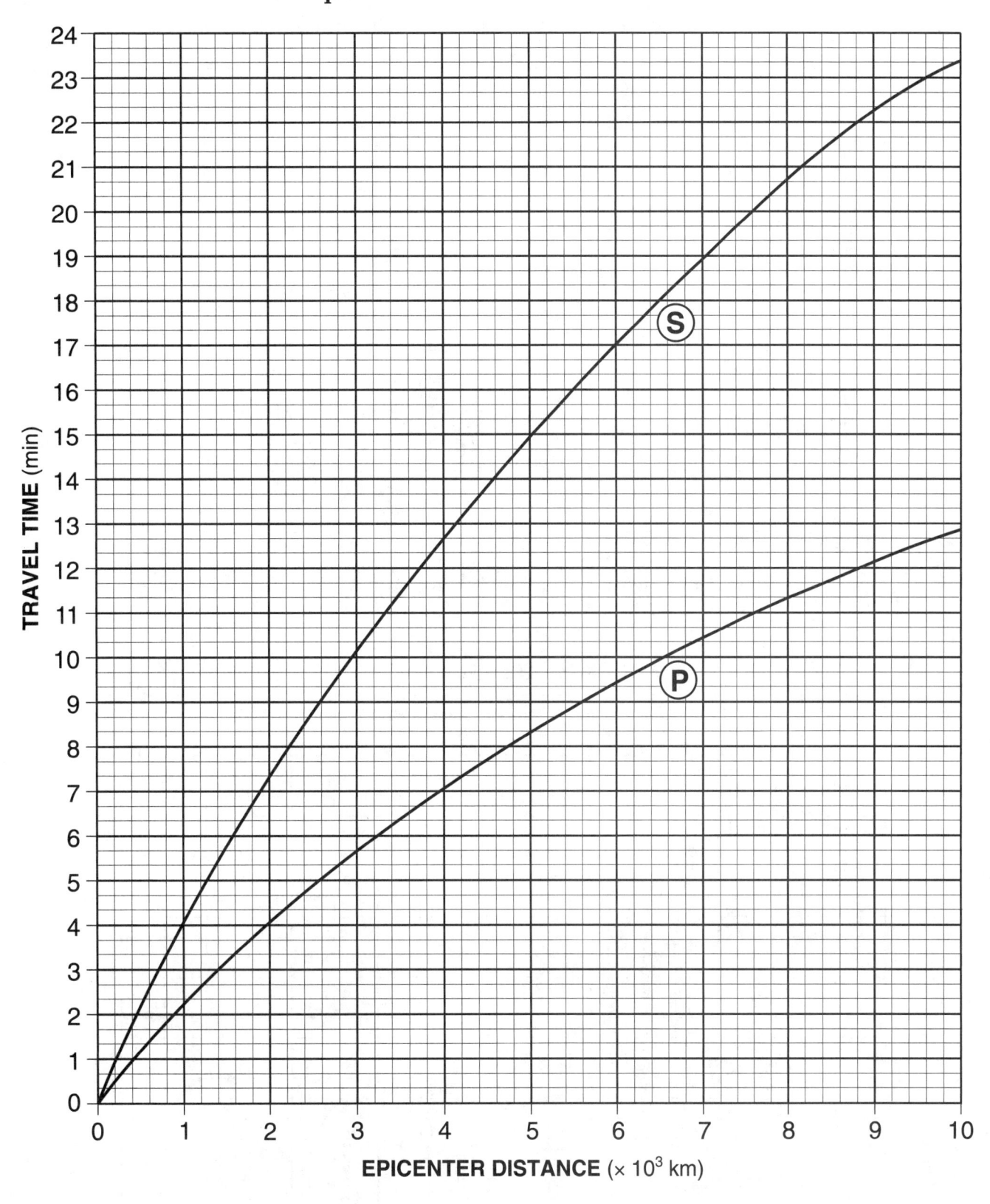

Physical Setting/Earth Science Reference Tables — 2011 Edition

Dewpoint (°C)

Dry-Bulb Temperature (°C)	\multicolumn{16}{c}{Difference Between Wet-Bulb and Dry-Bulb Temperatures (C°)}
	0	1	2	3	4	5	6	7	8	9	10	11	12	13	14	15
−20	−20	−33														
−18	−18	−28														
−16	−16	−24														
−14	−14	−21	−36													
−12	−12	−18	−28													
−10	−10	−14	−22													
−8	−8	−12	−18	−29												
−6	−6	−10	−14	−22												
−4	−4	−7	−12	−17	−29											
−2	−2	−5	−8	−13	−20											
0	0	−3	−6	−9	−15	−24										
2	2	−1	−3	−6	−11	−17										
4	4	1	−1	−4	−7	−11	−19									
6	6	4	1	−1	−4	−7	−13	−21								
8	8	6	3	1	−2	−5	−9	−14								
10	10	8	6	4	1	−2	−5	−9	−14	−28						
12	12	10	8	6	4	1	−2	−5	−9	−16						
14	14	12	11	9	6	4	1	−2	−5	−10	−17					
16	16	14	13	11	9	7	4	1	−1	−6	−10	−17				
18	18	16	15	13	11	9	7	4	2	−2	−5	−10	−19			
20	20	19	17	15	14	12	10	7	4	2	−2	−5	−10	−19		
22	22	21	19	17	16	14	12	10	8	5	3	−1	−5	−10	−19	
24	24	23	21	20	18	16	14	12	10	8	6	2	−1	−5	−10	−18
26	26	25	23	22	20	18	17	15	13	11	9	6	3	0	−4	−9
28	28	27	25	24	22	21	19	17	16	14	11	9	7	4	1	−3
30	30	29	27	26	24	23	21	19	18	16	14	12	10	8	5	1

Relative Humidity (%)

Dry-Bulb Temperature (°C)	\multicolumn{16}{c}{Difference Between Wet-Bulb and Dry-Bulb Temperatures (C°)}
	0	1	2	3	4	5	6	7	8	9	10	11	12	13	14	15
−20	100	28														
−18	100	40														
−16	100	48														
−14	100	55	11													
−12	100	61	23													
−10	100	66	33													
−8	100	71	41	13												
−6	100	73	48	20												
−4	100	77	54	32	11											
−2	100	79	58	37	20	1										
0	100	81	63	45	28	11										
2	100	83	67	51	36	20	6									
4	100	85	70	56	42	27	14									
6	100	86	72	59	46	35	22	10								
8	100	87	74	62	51	39	28	17	6							
10	100	88	76	65	54	43	33	24	13	4						
12	100	88	78	67	57	48	38	28	19	10	2					
14	100	89	79	69	60	50	41	33	25	16	8	1				
16	100	90	80	71	62	54	45	37	29	21	14	7	1			
18	100	91	81	72	64	56	48	40	33	26	19	12	6			
20	100	91	82	74	66	58	51	44	36	30	23	17	11	5		
22	100	92	83	75	68	60	53	46	40	33	27	21	15	10	4	
24	100	92	84	76	69	62	55	49	42	36	30	25	20	14	9	4
26	100	92	85	77	70	64	57	51	45	39	34	28	23	18	13	9
28	100	93	86	78	71	65	59	53	47	42	36	31	26	21	17	12
30	100	93	86	79	72	66	61	55	49	44	39	34	29	25	20	16

Physical Setting/Earth Science Reference Tables — 2011 Edition

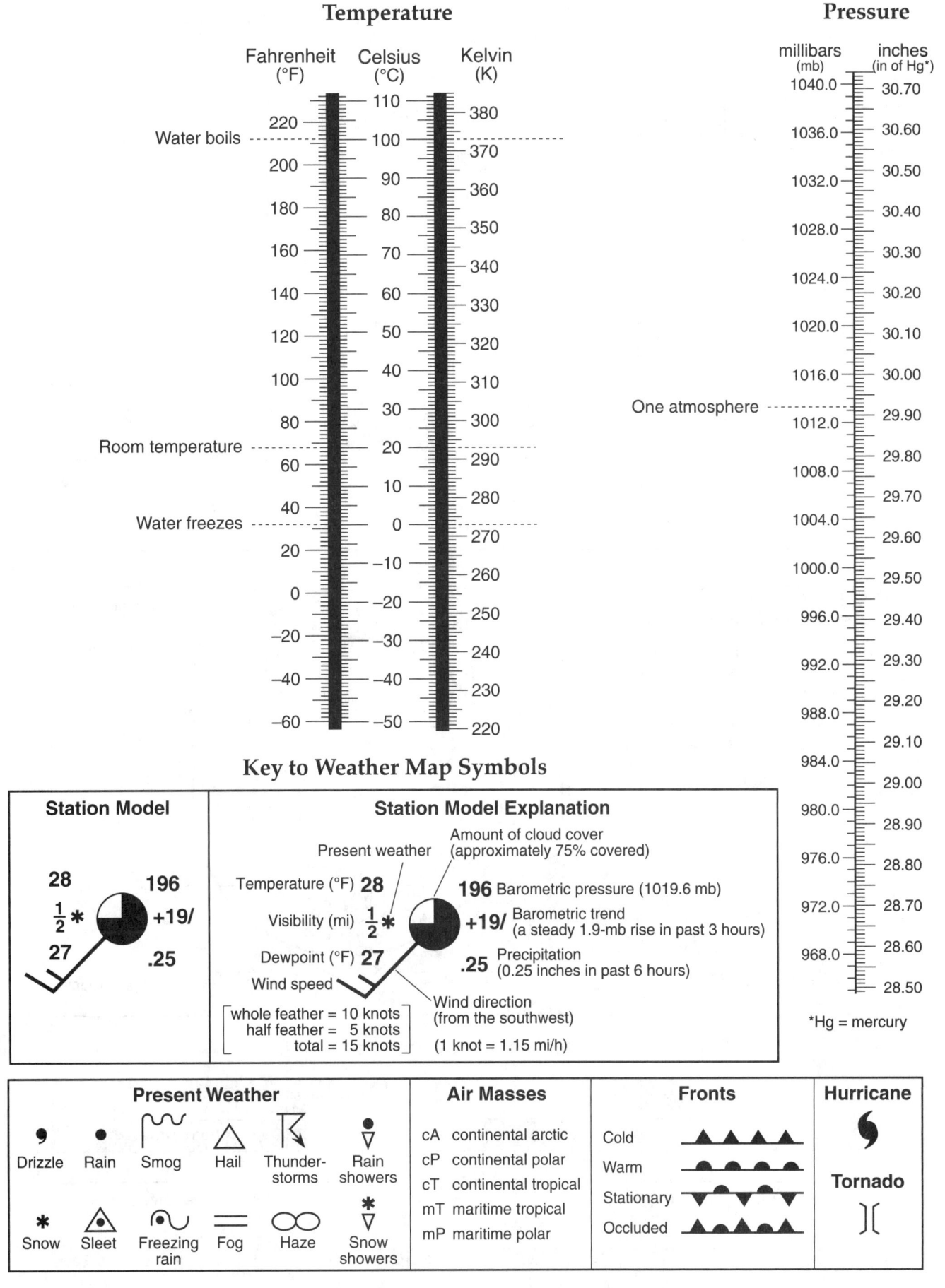

Temperature

Fahrenheit (°F)	Celsius (°C)	Kelvin (K)
220	110	380
200	100 — Water boils	370
180	90	360
160	80	350
140	70	340
120	60	330
100	50	320
80	40	310
60	30	300
40	20 — Room temperature	290
20	10	280
0	0 — Water freezes	270
−20	−10	260
−40	−20	250
−60	−30	240
	−40	230
	−50	220

Pressure

millibars (mb)	inches (in of Hg*)
1040.0	30.70
1036.0	30.60
1032.0	30.50
1028.0	30.40
1024.0	30.30
1020.0	30.20
1016.0	30.10
1012.0 — One atmosphere	30.00
1008.0	29.90
1004.0	29.80
1000.0	29.70
996.0	29.60
992.0	29.50
988.0	29.40
984.0	29.30
980.0	29.20
976.0	29.10
972.0	29.00
968.0	28.90
	28.80
	28.70
	28.60
	28.50

*Hg = mercury

Key to Weather Map Symbols

Station Model

```
   28      196
   ½ *    +19/
   27      .25
```

Station Model Explanation

- Temperature (°F): 28
- Present weather
- Amount of cloud cover (approximately 75% covered)
- Barometric pressure (1019.6 mb): 196
- Visibility (mi): ½
- Barometric trend (a steady 1.9-mb rise in past 3 hours): +19/
- Dewpoint (°F): 27
- Precipitation (0.25 inches in past 6 hours): .25
- Wind speed: whole feather = 10 knots, half feather = 5 knots, total = 15 knots
- Wind direction (from the southwest)
- (1 knot = 1.15 mi/h)

Present Weather

- Drizzle
- Rain
- Smog
- Hail
- Thunderstorms
- Rain showers
- Snow
- Sleet
- Freezing rain
- Fog
- Haze
- Snow showers

Air Masses

- cA continental arctic
- cP continental polar
- cT continental tropical
- mT maritime tropical
- mP maritime polar

Fronts

- Cold
- Warm
- Stationary
- Occluded

Hurricane

Tornado

Physical Setting/Earth Science Reference Tables — 2011 Edition

Selected Properties of Earth's Atmosphere

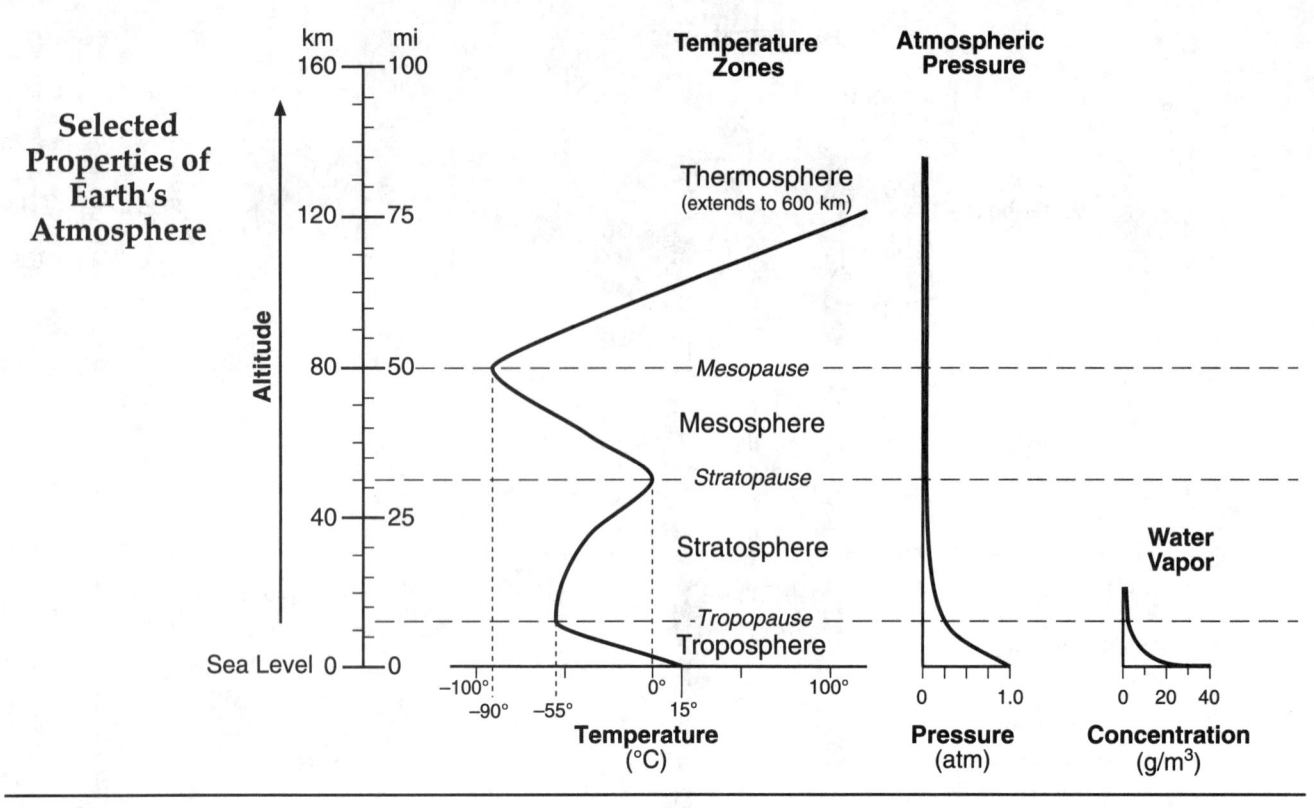

Temperature Zones (from sea level upward):
- Troposphere
- Tropopause
- Stratosphere
- Stratopause
- Mesosphere
- Mesopause
- Thermosphere (extends to 600 km)

Altitude axis: 0 to 160 km (0 to 100 mi)

Temperature (°C) reference points: −90°, −55°, 0°, 15°, 100°

Atmospheric Pressure: 0 to 1.0 atm

Water Vapor Concentration (g/m³): 0, 20, 40

Planetary Wind and Moisture Belts in the Troposphere

The drawing on the right shows the locations of the belts near the time of an equinox. The locations shift somewhat with the changing latitude of the Sun's vertical ray. In the Northern Hemisphere, the belts shift northward in the summer and southward in the winter.

(Not drawn to scale)

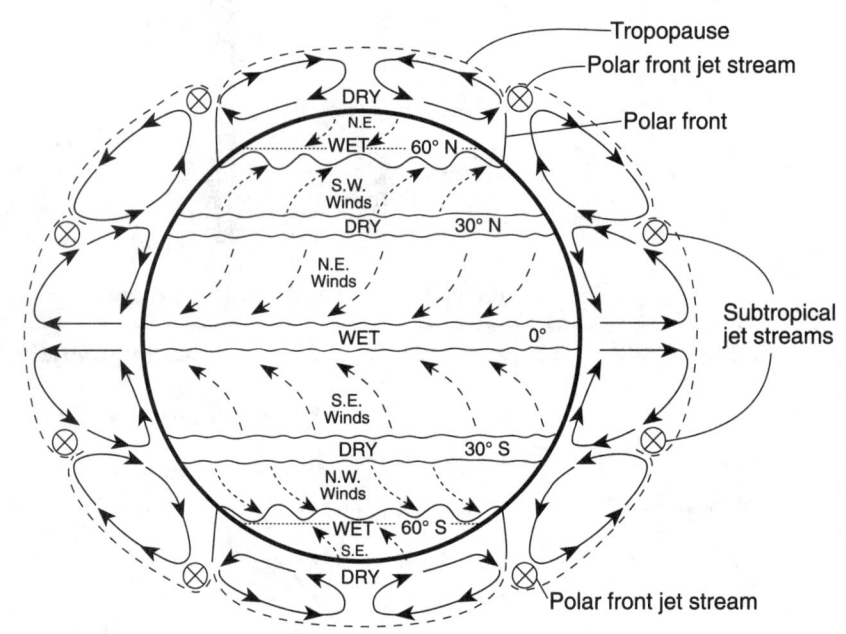

Labels on diagram:
- Tropopause
- Polar front jet stream
- Polar front
- DRY / N.E. / WET — 60° N
- S.W. Winds
- DRY — 30° N
- N.E. Winds
- WET — 0°
- S.E. Winds
- DRY — 30° S
- N.W. Winds
- WET — 60° S
- S.E.
- DRY
- Polar front jet stream
- Subtropical jet streams

Electromagnetic Spectrum

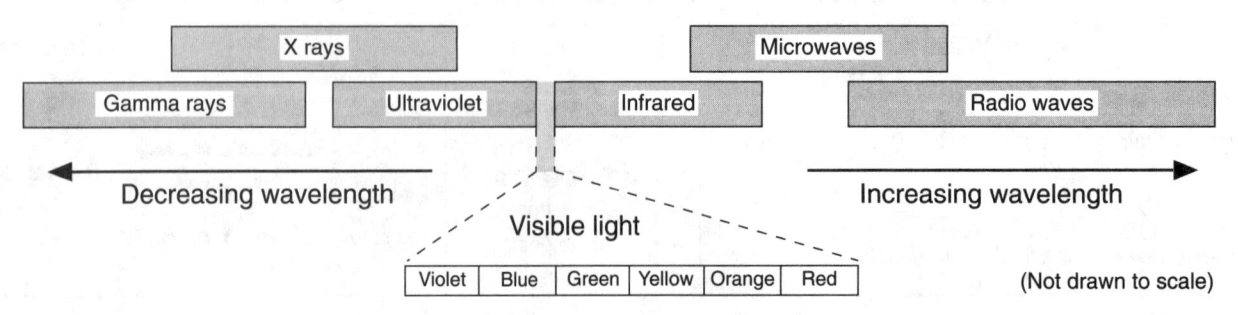

Decreasing wavelength ← | → Increasing wavelength

Bands: Gamma rays, X rays, Ultraviolet, Visible light, Infrared, Microwaves, Radio waves

Visible light: Violet | Blue | Green | Yellow | Orange | Red

(Not drawn to scale)

Physical Setting/Earth Science Reference Tables — 2011 Edition

Characteristics of Stars

(Name in italics refers to star represented by a ⊕.)
(Stages indicate the general sequence of star development.)

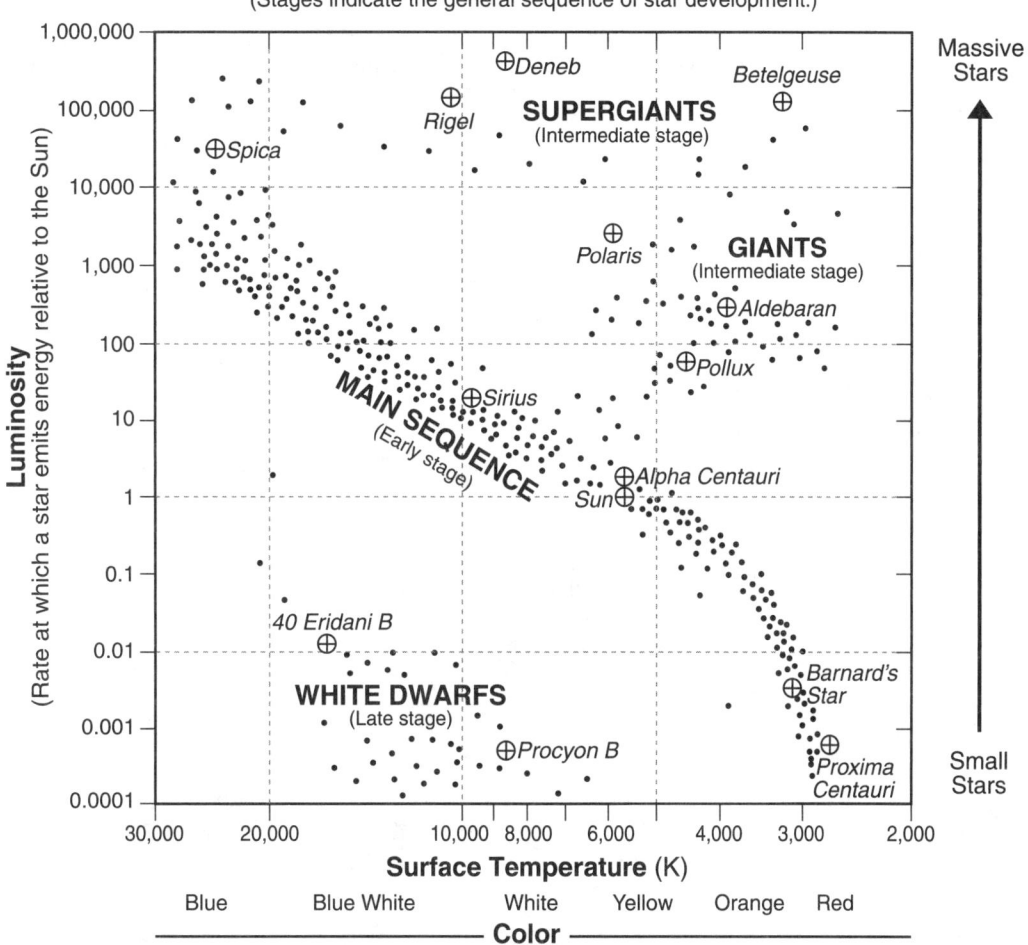

Solar System Data

Celestial Object	Mean Distance from Sun (million km)	Period of Revolution (d=days) (y=years)	Period of Rotation at Equator	Eccentricity of Orbit	Equatorial Diameter (km)	Mass (Earth = 1)	Density (g/cm^3)
SUN	—	—	27 d	—	1,392,000	333,000.00	1.4
MERCURY	57.9	88 d	59 d	0.206	4,879	0.06	5.4
VENUS	108.2	224.7 d	243 d	0.007	12,104	0.82	5.2
EARTH	149.6	365.26 d	23 h 56 min 4 s	0.017	12,756	1.00	5.5
MARS	227.9	687 d	24 h 37 min 23 s	0.093	6,794	0.11	3.9
JUPITER	778.4	11.9 y	9 h 50 min 30 s	0.048	142,984	317.83	1.3
SATURN	1,426.7	29.5 y	10 h 14 min	0.054	120,536	95.16	0.7
URANUS	2,871.0	84.0 y	17 h 14 min	0.047	51,118	14.54	1.3
NEPTUNE	4,498.3	164.8 y	16 h	0.009	49,528	17.15	1.8
EARTH'S MOON	149.6 (0.386 from Earth)	27.3 d	27.3 d	0.055	3,476	0.01	3.3

Physical Setting/Earth Science Reference Tables — 2011 Edition

Properties of Common Minerals

LUSTER	HARD-NESS	CLEAVAGE	FRACTURE	COMMON COLORS	DISTINGUISHING CHARACTERISTICS	USE(S)	COMPOSITION*	MINERAL NAME
Metallic luster	1–2	✔		silver to gray	black streak, greasy feel	pencil lead, lubricants	C	Graphite
Metallic luster	2.5	✔		metallic silver	gray-black streak, cubic cleavage, density = 7.6 g/cm³	ore of lead, batteries	PbS	Galena
Metallic luster	5.5–6.5		✔	black to silver	black streak, magnetic	ore of iron, steel	Fe_3O_4	Magnetite
Metallic luster	6.5		✔	brassy yellow	green-black streak, (fool's gold)	ore of sulfur	FeS_2	Pyrite
Either	5.5–6.5 or 1		✔	metallic silver or earthy red	red-brown streak	ore of iron, jewelry	Fe_2O_3	Hematite
Nonmetallic luster	1	✔		white to green	greasy feel	ceramics, paper	$Mg_3Si_4O_{10}(OH)_2$	Talc
Nonmetallic luster	2		✔	yellow to amber	white-yellow streak	sulfuric acid	S	Sulfur
Nonmetallic luster	2	✔		white to pink or gray	easily scratched by fingernail	plaster of paris, drywall	$CaSO_4 \cdot 2H_2O$	Selenite gypsum
Nonmetallic luster	2–2.5	✔		colorless to yellow	flexible in thin sheets	paint, roofing	$KAl_3Si_3O_{10}(OH)_2$	Muscovite mica
Nonmetallic luster	2.5	✔		colorless to white	cubic cleavage, salty taste	food additive, melts ice	NaCl	Halite
Nonmetallic luster	2.5–3	✔		black to dark brown	flexible in thin sheets	construction materials	$K(Mg,Fe)_3$ $AlSi_3O_{10}(OH)_2$	Biotite mica
Nonmetallic luster	3	✔		colorless or variable	bubbles with acid, rhombohedral cleavage	cement, lime	$CaCO_3$	Calcite
Nonmetallic luster	3.5	✔		colorless or variable	bubbles with acid when powdered	building stones	$CaMg(CO_3)_2$	Dolomite
Nonmetallic luster	4	✔		colorless or variable	cleaves in 4 directions	hydrofluoric acid	CaF_2	Fluorite
Nonmetallic luster	5–6	✔		black to dark green	cleaves in 2 directions at 90°	mineral collections, jewelry	$(Ca,Na)(Mg,Fe,Al)(Si,Al)_2O_6$	Pyroxene (commonly augite)
Nonmetallic luster	5.5	✔		black to dark green	cleaves at 56° and 124°	mineral collections, jewelry	$CaNa(Mg,Fe)_4(Al,Fe,Ti)_3Si_6O_{22}(O,OH)_2$	Amphibole (commonly hornblende)
Nonmetallic luster	6	✔		white to pink	cleaves in 2 directions at 90°	ceramics, glass	$KAlSi_3O_8$	Potassium feldspar (commonly orthoclase)
Nonmetallic luster	6	✔		white to gray	cleaves in 2 directions, striations visible	ceramics, glass	$(Na,Ca)AlSi_3O_8$	Plagioclase feldspar
Nonmetallic luster	6.5		✔	green to gray or brown	commonly light green and granular	furnace bricks, jewelry	$(Fe,Mg)_2SiO_4$	Olivine
Nonmetallic luster	7		✔	colorless or variable	glassy luster, may form hexagonal crystals	glass, jewelry, electronics	SiO_2	Quartz
Nonmetallic luster	6.5–7.5		✔	dark red to green	often seen as red glassy grains in NYS metamorphic rocks	jewelry (NYS gem), abrasives	$Fe_3Al_2Si_3O_{12}$	Garnet

*Chemical symbols: Al = aluminum, Cl = chlorine, H = hydrogen, Na = sodium, S = sulfur, C = carbon, F = fluorine, K = potassium, O = oxygen, Si = silicon, Ca = calcium, Fe = iron, Mg = magnesium, Pb = lead, Ti = titanium

✔ = dominant form of breakage

Physical Setting/Earth Science Reference Tables — 2011 Edition

GLOSSARY

abrasion: wearing away by rubbing or scraping
absorb: to take in
abundant: to be present in a large amount
acid precipitation: any other form of precipitation (ex: rain, snow) that is unusually acidic, having a low pH
adhere: to attach or stick to
adiabatic temperature change: change in air temperature due to expansion or compression of air molecules with no addition or removal of heat energy
aerosols: solid or liquid particles suspended in the air, such as soot, dust, pollen, water droplets
agent of erosion: any moving system that can transport sediment
air (barometric) pressure: force of the atmosphere pushing downward
air mass: large body of air in the troposphere that has similar weather conditions throughout
altitude: height above sea level; angular height above the horizon
anemometer: instrument which measures wind speed
angle of insolation: angle at which the Sun's rays strike the observer's horizon
anticyclone: system of high pressure air with winds that move clockwise and out from the center in the Northern Hemisphere; characterized by cool, dry, clear weather
aphelion: the position in a planet's or comet's orbit when it is furthest away from the Sun
apparent: visible or evident
arid climate: climate where average precipitation is less than the average potential evapotranspiration
ascend: to move upward
asteroid: irregularly-shaped rock which orbits the Sun
asteroid belt: region between the orbits of Mars and Jupiter in which most asteroids are found
asthenosphere: gel-like, plastic layer of the mantle that the lithospheric plates move upon
atmosphere: thin layer of gases surrounding Earth
atmospheric variables: factors which describe the conditions in the atmosphere for a location
aurora: luminous bands of colorful light in the night sky caused by the interaction of charged particles from the Sun and the gases in the atmosphere; also known as northern or southern lights since they are most visible in polar regions
barometer: instrument which measures air pressure in millibars or inches of mercury (Hg)

barometric (air) pressure: force of the atmosphere pushing downward
barometric trend: the increase or decrease of air pressure in a given amount of time
barrier island: long, narrow island of sand parallel to the mainland built up by the action of waves and currents; protects the mainland from ocean waves and currents
bedrock: solid rock beneath the soil
benchmark: position where the exact elevation is known
Big Bang Theory: the idea that a sudden expansion of a singular dense mass was the beginning of time, matter, energy, and the expanding universe
bioclastic: sediments that have an organic origin, such as plant debris and shell fragments
black hole: object in space so dense that light cannot escape; this is the late stage in a massive star's life cycle
blizzard: winter storms with heavy snowfall and winds over 35 mph
capillarity: attractive force between water molecules and the surface of soil particles; causes water to be held in the zone of aeration; can cause water to move upward through the soil
capillary action: ability of water to rise upward through the pore spaces in the zone of aeration
capillary water: water held in the zone of aeration that adheres to the surfaces of soil particles
celestial: of the universe such as stars, planets, comets, moons
celestial sphere: represents an observer's view of the universe
cementation: process by which sediments are joined together by minerals or clay
chemical precipitate: dissolved minerals in water that settle out to form a sedimentary rock
chemical weathering: action on a rock that results in a change in its chemical composition and appearance
chronometer: instrument that indicates the time on the Prime Meridian (0° longitude)
circumpolar: describes stars which appear to make circular star trails around *Polaris* as Earth rotates
classification system: the organization of observations into categories based on similar properties
clastic: refers to land-derived sediments; fragments of rocks
cleavage: when a mineral breaks smoothly and evenly along flat surfaces

317

climate: expected precipitation and temperature for a region over a long period of time
cloud: microscopic water droplets and/or ice crystals suspended in the air
cloud base: the bottom of a cloud
cloud cover: amount of clouds in the sky; can range from overcast, partly cloudy, to scattered clouds
coarse texture: in igneous rocks, refers to mineral crystals larger than 1mm, caused by the slow cooling of magma inside Earth
colloid: very tiny pieces of sediment that can stay suspended in water; includes clay and silt
comet: a mass of ice and rock which orbits the Sun in a very elliptical path
compaction: to be closely and firmly pushed together; to make more dense
component: to be part of something
composed of: made of
compound: composed of two or more elements chemically combined
compress: to press together; to make more compact, denser, and smaller
compressional (P) wave: primary, seismic vibration generated by an earthquake; travels the fastest through any medium; first to arrive at the seismograph
condensation: change of a gas (vapor) to a liquid by the removal of heat energy
condensation nuclei: solid particles that water vapor condenses on to form a raindrop or snowflake
conduction: energy is transferred within a solid from molecule to molecule as they interact
constellation: group of stars that appear in a pattern in our sky
contact (thermal) metamorphism: the changing of the texture and mineral composition of pre-existing rock caused by direct contact with hot, intruding magma
continental (inland) climate: semi-arid to arid climate which has large range in seasonal temperatures due to its inland location
continental air mass: air mass which forms over a landmass resulting in dry air
continental crust: rock layers which are the surface land of Earth
continental drift: the idea that the continents move and change position over time
contour interval: change in elevation from one contour line to the next contour line
contour line: isoline which connects points of the same elevation

contour map: map which shows the elevation and shape of the landscape
contract: to shrink in size
convection: energy is transferred within a liquid or a gas due to density differences
convection currents: flow of energy in a liquid or a gas caused by uneven heating and density differences
converge: to come together
convergent plate boundary: place in lithosphere where two "plates" are coming together
Coriolis effect: apparent deflection or curving of moving objects, air, and ocean currents on Earth's surface due to Earth rotation
correlate: to match up as being the same age
cosmic background radiation: energy created from the formation of the universe by the big bang expansion
crust: outer, solid layer of Earth
cyclic: occurring at regular intervals
cyclone: system of low pressure air with winds that move counterclockwise and toward the center in the Northern Hemisphere; characterized by warm, moist, stormy weather
deforestation: the clearing of a forest which includes removal of trees and vegetation
delta: deposit of sediment at the mouth of a river formed when the river enters a body of water and slows down
density: concentration of mass in a given volume of a substance
denudation: the removal of soil, surface sediment, and any vegetative covering
deplete: to become less by gradual usage
deposition: process by which transported rock debris and sediment are released from the agent of erosion; change in phase from a gas directly to a solid
descend: to go downward
dewpoint: temperature at which water vapor condenses to a liquid or becomes a solid
direct Sun rays: Sun rays which hit a surface at a 90 degree angle; perpendicular rays; vertical rays
disintegration: for radioactive elements, to change or "decay" to a more stable element
distorted strata: rock layers which are folded, faulted, or tilted
diverge: to move apart from each other
divergent plate boundary: place in lithosphere where two plates are moving apart; occurs at ridges and rift valleys
dominant: having the greatest effect

doppler shift: change in wave frequency as an energy source moves towards or away from an observer; for light, the spectrum shifts to the red for objects moving away and towards blue for objects moving towards the observer

drumlin: glacial landform in the form of an elongated hill composed of unsorted till, the drumlin points in the direction of the glacier's movement

duration of insolation: length of daylight hours

dwarf planet: in the solar system, solid spheres, smaller than Mercury, whose orbits cross the paths of other orbiting objects; most are beyond Neptune

earthquake: natural shaking of Earth's lithosphere

eccentricity: numerical description of the shape of an ellipse; describes the degree to which the ellipse varies from a circle

eclipse: temporary darkening of the Sun or the full Moon due to the alignment of Earth, Sun, and Moon

El Niño: abnormal warming of surface ocean waters in the eastern tropical Pacific, off the west coast of South America; this causes an interruption of the normal ocean circulation in the Pacific Ocean and affects weather patterns

electromagnetic spectrum: types of energies that travel as transverse waves at the speed of light and can travel through empty space

elevation: height above sea level, measured in feet or meters

ellipse: oval-shaped; describes the shape of an orbit relative to two focus points

eon: largest division of geologic time

epicenter: location on Earth's surface above where an earthquake occurred

epoch: smallest division of geologic time; smaller part of a geologic period

Equator: reference line for latitude, 0° latitude; divides Earth into Northern and Southern Hemispheres

equatorial diameter: distance from side to side through the center of a planet along its equator

equilibrium: a state of balance; a stable situation where opposing forces cancel each other out so that changes do not occur

era: smaller part of an eon

erosion: process by which weathered rock fragments and sediment are transported to a different location by a moving system

erratic: boulder or large rock transported and deposited to a new location by a glacier

escarpment: steep-sloped, long cliff that occurs from faulting or erosion

evaporation: change of a liquid to a gas (vapor) by the addition of heat energy

evaporite: sedimentary rock formed when dissolved minerals in water are deposited when the water evaporates

evapotranspiration: the combined processes of evaporation and transpiration which add moisture to the atmosphere

extrusion: igneous rock formed by the cooling of lava on Earth's surface

fault: crack in Earth's crust

felsic: igneous rock containing a high percentage of aluminum (Al) and silicon (Si)

fine texture: in igneous rocks, refers to mineral crystals less than 1 mm in size, caused by fast cooling of lava near or on Earth's surface

focus: one of the two fixed points in an ellipse; the place in the crust where an earthquake occurred

fog: ground level cloud

foliated: the alignment or banding of minerals in a metamorphic rock

fossil: evidence of past life forms; the remains, molds, or casts of plants and animals that are found in sedimentary rock

fossil fuels: coal, oil, and natural gas formed from plant or microscopic animal remains millions of years ago; used as an energy source

Foucault pendulum: a freely swinging pendulum that consists of a heavy weight hung on a long, thin wire which swings in a constant direction; it appears to change its direction of swing as Earth rotates beneath it

fracture: tendency of a mineral to break unevenly

fragment: a piece broken off from a larger rock

front: boundary between two different air masses

frost action: the expansion of freezing water in cracks of rocks which causes further physical weathering

Fujita scale: rating system for the severity of tornadoes based on wind speed

galaxy: huge system of billions of stars and matter in the universe

geocentric: describes a system where objects orbit Earth

geologic event: occurrence during Earth's history that affected the solid Earth such as volcanic eruption, earthquake, faulting, deposition, erosion

glacier: mass of ice which slowly moves downhill

glassy texture: refers to an igneous rock which has no mineral crystals, caused by rapid cooling of lava on Earth's surface

global warming: a sustained increase in the average temperature of Earth's atmosphere

gradient: degree of change in a measured value over a distance; slope of the landscape

graph: diagram which illustrates the relationship between two variables

greenhouse effect: warming of the lower atmosphere as the surface of a planet absorbs sunlight and re-radiates its heat energy which is then absorbed by certain atmospheric gases

greenhouse gas: a gas in the atmosphere which absorbs and retains heat energy; includes water vapor, carbon dioxide, methane

Greenwich Mean Time (GMT): the time on the Prime Meridian (0° longitude); recorded on a chronometer

groundwater: water in the pore spaces of saturated soil; subsurface water

hachure marks: short lines placed on contour lines which indicate a decrease in elevation into a depression or hole; on a cross-section of rock layers indicates contact metamorphism

half-life: time required for one-half of an unstable radioactive element to change (disintegrate) to a stable element

hardness: a mineral's resistance to being scratched

headwaters: the small streams that feed into a river; the source of a river's water

heliocentric: describes a system where objects orbit the Sun

high pressure (H) system: a cool, dry air mass where air movement is clockwise and away from the center in the Northern Hemisphere; also known as an anticyclone

highland: an elevated region, such as a mountain or plateau

horizon: where Earth and sky meet as seen by an observer

horizontal: from side to side; parallel to the horizon; the x-axis on a graph

hot spot: place in the crust where a mass or plume of hot magma exists

humid climate: climate where the average precipitation is greater than the average potential evapotranspiration

humidity: the amount of water vapor in the air

humus: partially decomposed organic (plant and animal) matter which enriches the topsoil

hurricane: low pressure system that forms over warm ocean water; characterized by strong sustained winds and heavy precipitation

hydrologic (water) cycle: the continual movement of water between the hydrosphere, lithosphere, and atmosphere

hydrosphere: the waters of Earth; includes oceans, lakes, rivers, snowfields, ice caps, groundwater

Ice Age: an extended period of time characterized by cooler temperatures and continental glaciers

igneous rock: rock which forms from the cooling and solidification of molten magma or lava

impact event: the collision of a large meteoroid, asteroid, or comet with a body such as a planet or moon

impermeable: not allowing fluids to pass through

index fossil: fossil of a plant or animal which lived in a large geographic range for a short period of geologic time

inference: conclusion, opinion, or explanation of what was observed

infiltrate: to seep or soak into; to pass through

infrared radiation: heat or thermal energy; long-wave radiation compared to visible light

inner core: innermost layer of Earth; inferred to be a very hot solid of nickel and iron under high pressure

inorganic: not from a living organism

insolation: incoming solar radiation; includes visible light, ultraviolet radiation, and infrared radiation from the Sun

intensity of insolation: the strength and amount of solar radiation received

interface: the boundary between adjoining regions, objects, substances, or phases

intergrown crystals: mineral crystals that are interlocked with no spaces or material between them

intrusion: igneous rock formed by the cooling of magma inside Earth

isobar: line which connects points of the same air pressure on a weather map

isoline: line that connects points of the same value

isotherm: line which connect points of the same air temperature

isotope: atom of the same element that has the same number of protons but different number of neutrons

jet stream: high altitude air movement from west to east which encircles Earth moving air masses and weather systems

jovian planets: planets beyond the asteroid belt characterized by large size, low density, rings, many moons; also known as gas giants

kinetic energy (KE): the energy of a moving object

landfill: the disposal method for waste material and garbage where the waste is buried between layers of soil

landscape: shape of the land due to leveling and uplifting forces; the result of interactions of the atmosphere and hydrosphere on Earth's surface

landslide: the rapid downward sliding of a mass of soil, loose sediment, and/or rock

latitude: angular distance north and south of the Equator; ranges from 0° to 90° North and South

lava: hot, liquid rock material on the surface of Earth

leeward: on the side opposite from the way the wind blows; sheltered from the wind

leveling forces: those actions which wear away the Earth's surfaces causing elevation and hillslopes to decrease

light year (ly): the distance that light travels in one year; equal to about 9.5 trillion kilometers (5.9 trillion miles)

lithosphere: outermost solid rock layer of Earth above the atshenophere; includes the crust and rigid mantle

lithospheric plates: separate pieces of the lithosphere that move and shift above the asthenosphere; continents and landmasses are "passengers" on these plates

longitude: angular distance east and west of the Prime Meridian; ranges from 0° to 180° East and West

longshore current: flow of seawater parallel to the shoreline

low pressure (L) system: a warm, moist air mass where air movement is counterclockwise and toward the center in the Northern Hemisphere; also known as a cyclone

lowland: a region of low elevation, such as a plain

luminosity: rate at which a star gives off energy compared to the Sun

lunar eclipse: occurs when Earth is located directly between the Sun and Moon so that Earth's shadow darkens the full Moon

luster: the way a mineral reflects light

mafic: igneous rocks containing a high percentage of iron (Fe) and magnesium (Mg)

magma: liquid rock material found inside Earth

magnetic pole: the location on Earth toward which a compass needle will point

main sequence stars: stars in their early stage of development whose luminosity is directly related to temperature; most stars are classified as main sequence

mantle: interior layer of Earth between the crust and the cores; includes the upper rigid mantle, "plastic" asthenosphere, and stiffer mantle

mantle convection current: flow of heat energy in the asthenosphere that is responsible for the movement of the lithospheric plates

map scale: used to measure the actual distance between points on a map in miles or kilometers

marine: related to the ocean

marine (coastal) climate: humid to semi-humid climate with mild seasonal temperature changes due to its location near a large body of water

maritime air mass: moist air mass that formed over water

mass: amount of matter in an object, measured in grams

mass movement: slow or rapid movement of sediment downhill due to force of gravity

maximum: the greatest quantity

meander: curves or bends in a mature river

measurement: an amount or size as determined by an instrument

Mercalli scale: rating system for the intensity of an earthquake based upon people's perceptions and the damage that occurred

mesopause: boundary between the mesosphere and the thermosphere

mesosphere: layer of the atmosphere between the stratosphere and the thermosphere where temperature decreases with altitude

metamorphic rock: rock which forms when a pre-existing rock is changed by intense heat and/or pressure

metamorphism: the changing of the texture and mineral composition of a pre-existing rock exposed to extreme heat and/or pressure; this occurs without melting

meteor: a meteoroid which enters Earth's atmosphere and disintegrates due to friction with atmospheric gases; often misnamed "shooting star"

meteorite: a meteoroid which impacts another solid object; often forms a crater

meteoroid: rock fragment in space

meteorology: scientific study of weather and atmospheric conditions

mid-latitude: latitudes between 30° and 60° North and South

Milky Way Galaxy: large system of billions of stars of which the Sun and its solar system is part of

millibar (mb): unit of measurement for barometric air pressure

mineral: solid, inorganic compound of one or more elements; smallest component of a rock

minimum: the least quantity

Moh's scale of hardness: the ten reference minerals that are the standard of hardness against which other minerals are compared to

Moho: interface or boundary between the crust and the mantle where a rapid increase in the velocity of seismic waves occurs

molten: melted; refers to liquid rock material
moon: solid object which orbits a planet
moraine: hills of unsorted sediment (till) deposited by a glacier along its side or front
mountain: landscape of high elevations, steep slopes, and rock layers which are distorted (folded, faulted, tilted)
natural resources: parts of Earth used by humans, plants, and animals
nebula: huge "cloud" of dust and gases in the universe; contains materials which can form stars
non-clastic: refers to sediments which formed as a chemical precipitates and evaporites or from fragments of plant and animal remains
non-foliated: crystalline metamorphic rock whose minerals are not aligned or banded
non-renewable resource: natural resources that cannot be replaced in our lifetime; they are in limited supply; they are often used faster than they can be replaced
North Star: also known as *Polaris*; located almost directly above the axis of rotation and the geographic North Pole
nuclear fusion: process by which stars produce energy; the reaction in which two light nuclei (such as hydrogen) combine to form heavier nuclei (such as helium) and energy
oblate spheroid: not a perfect sphere because the equatorial diameter is larger than the polar diameter
observation: description or measurement of an object or an event
ocean trench: long narrow, steep-sided depression in Earth's oceanic crust; usually along a subduction zone where two plates are converging as one plate overrides the subducting plate
oceanic crust: solid rock layer which makes up the seafloor
orbit: to revolve around another object
orbital velocity: speed at which an object moves around another object
organic: originating from a plant or an animal
orogeny: uplifting event that forms mountains
outer core: layer of Earth's interior between the mantle and inner core; inferred to be hot, liquid iron and nickel
outgassing: release of gases from Earth's interior into the atmosphere; examples include volcanic eruptions, geysers
oxbow: forms when a meander of a river is cut off from the flow of water; this isolated meander can become a lake

ozone (O_3): compound of oxygen which absorbs dangerous ultraviolet radiation from the Sun; most of atmospheric ozone is found as a thin layer in the stratosphere
P-wave: primary, compressional seismic vibration generated by an earthquake; travels the fastest through any medium; first to arrive at seismograph
parcel: a part of a larger amount
perihelion: the position in a planet's or comet's orbit when it is closest to the Sun
period: smaller divisions of geologic eras
period of revolution: time it takes for an object to complete one orbit
period of rotation: time it takes for an object to complete one turn on its axis
permeability: property of soil that allows water to enter and flow through it
permeable: allowing fluids to pass through
perpendicular rays of Sun: Sun rays which hit a surface at a 90 degree angle; direct rays; vertical rays
phases of the Moon: the illuminated portion of the Moon seen from Earth; caused by Moon's revolution around Earth
phenomena: an observable fact or event
physical weathering: processes that breaks apart surface bedrock into smaller pieces with no change in chemical composition
plain (lowland): landscape with low elevation, flat relief, and horizontal rock or sediment layers
planet: large spherical body that revolves around a star
planetary (global) winds: worldwide air movements in specific latitude zones caused by uneven heating of the atmosphere
plate: any of the rigid pieces of Earth's lithosphere which moves and shifts
plate tectonics: the movements of and interactions between the lithospheric plates as they shift and move over the asthenosphere; plate tectonics is caused by mantle convection currents
plateau: landscape with high elevation and horizontal rock layers
polar: locations at high latitudes beyond the Arctic and Antarctic Circles
polar reversal: a change in a planet's magnetic field so that the positions of magnetic north pole and magnetic south pole are switched
***Polaris*:** the North Star; located almost directly above the axis of rotation and geographic North Pole
pollutant: any substance(solid, liquid, gas) or form of energy that is harmful to life

polyminerallic: made of more than one mineral
porosity: percentage of open pore space in a volume of soil; the empty spaces between the soil particles
porous: containing openings through which water can flow or be stored
potential energy (PE) stored energy that is the result of object's position
Precambrian: the first and longest geologic eon; represents about 85% of Earth's history; during this time Earth's environment was evolving to become suitable for life
precession: change in the tilt of the axis of a rotating body
precipitation: the falling of water droplets, snow, and/or ice from clouds; when dissolved ions and minerals settle out of water
prediction: an inference about what will happen in the future
pressure gradient: the rate of change in air pressure between two locations
prevailing (planetary) wind: general movement of winds in specific latitude zones, such as the westerlies in the mid-latitude and the easterlies in the tropics
Prime Meridian: reference line for longitude, 0° longitude; extends from North Pole to South Pole through Greenwich, England
principle of original horizontality: states that sedimentary rocks and extrusive igneous rocks form as horizontal layers
principle of superposition: states that older rocks are found below younger rocks
profile: side view of a lansdscape, illustrates the slope and elevations of the land
protostar: the beginning of a star before nuclear fusion begins
psychrometer: instrument used to measure relative humidity and dewpoint temperature
radiation: energy transfer from one location to another without any material in between; can occur in empty space
radioactive disintegration ("decay"): occurs when the nucleus of an unstable element changes to a more stable element
radioactive isotope (element): an element with an unstable nucleus; it will emit radiation as it changes (disintegrates) to a stable element
rain gauge: instrument which measures rainfall in inches
random: having no specific pattern
rate of change: change in a measured value in a period of time

recrystallize: occurs during metamorphism when heat and/or pressure cause the minerals in pre-existing rock to recombine and change; this occurs without melting
red giant: star which is cool but large in size causing it to be luminous; the intermediate stage in a star's life cycle
reflection: energy which is returned unchanged after it strikes a material
refraction: the bending of a wave of energy as it moves into a new material
regional metamorphism: metamorphism of pre-existing rock caused by intense heat and pressure affecting a large area
relative age: the geologic age of a fossil, rock, or geologic feature or event compared to other fossils, rocks, or features; not given in terms of years
relative dating: the sequencing of geologic events and rock layers from oldest to youngest by comparison of position and geologic features
relative humidity: ratio of water vapor in the air to the amount of water vapor the air can hold, expressed as a percent
relief: the variation in elevations in an area
renewable resource: natural resources which can be replaced by nature in our lifetime
reradiate: to give off energy that was previously absorbed
residual soil: soil formed from the bedrock below it; soil which formed where it is found
resistant: a rock which is not easily weathered and eroded
revolve: to move around another object; to orbit
Richter scale: rating system for earthquakes based on the amount of energy released
ridge: boundary between diverging plates where new crust is formed due to upwelling of magma, often in the ocean crust
rift valley: forms along the center of the ridge where two plates are diverging
rigid mantle: upper portion of the mantle above the asthenosphere; the rigid mantle and the crust are the lithosphere
rock: naturally formed solid, made of minerals or bioclastic sediments
rock cycle: diagram which shows how each type of rock can be changed into another type of rock
rotate: to turn on an axis; to spin
runoff: water which flows on Earth's surface; often from precipitation or melting ice and snow
S-wave: secondary, shear wave generated by an earthquake; can only travel through solids; arrives at the seismograph after the P-wave

Safir-Simpson scale: system that rates the severity of hurricanes based on wind speed and air pressure

sand dune: hill of sorted, sand-sized sediment which was carried and deposited by the wind

satellite: solid body which orbits another body, can be natural or man-made

saturated: filled to capacity

scattered: occurs when waves of energy are sent in many directions

seafloor spreading: upwelling at ridges forms new oceanic crust which pushes old oceanic crust away from the ridge in both directions

seasonal Sun path: daily changes in the apparent path of the Sun in an observer's sky throughout the year; this is a cyclic pattern

sediment: weathered pieces of rock; fragments of rock

sedimentary rock: rock which forms from the compaction and cementation of sediment and/or organic material

seismic wave: vibrations caused by shifting of Earth's crust during an earthquake; includes the P- and S-waves

seismogram: the record of seismic vibrations printed by a seismograph

seismograph: instrument which detects and records seismic vibrations from an earthquake

sequence: to arrange in a specified order

severe weather: dangerous meteorological event with the potential to cause injury and damage to property

sextant: instrument used to measure the altitude of a star above the horizon, measured as an angle

shadow zone: zone on Earth's surface where no seismic waves are received from an earthquake

shallow marine organism: plant or animal that lives in seawater near the shoreline

shear (S) wave: secondary wave generated by an earthquake; can only travel through solids; arrives at the seismograph after the P-wave

smog: combination of smoke, aerosols, and fog in stagnant air

soil: loose surface material formed from weathered rock and the remains of decayed plants and animals

solar eclipse: Sun is blocked from our view when the Moon is located directly between Earth and Sun

solar noon: time of day when the Sun reaches its highest altitude in the sky

solar system: all the solid objects that orbit the Sun

solar wind: stream of high-speed, ionized particles ejected from the Sun's corona

solidification: to become a solid; change from liquid to solid

sorted particles: sediment particles that are all the same shape and size

source region: location over which an air mass forms and acquires the characteristics of that location

specific heat: amount of heat energy needed to raise the temperature of one gram of a substance by one degree Celsius; measured in joules

star: spherical mass of gases, primarily hydrogen, that produces energy by nuclear fusion

star trails: arc-shaped or circular paths that stars appear to make in our sky as Earth rotates; at night, star trails can be photographed by leaving the camera shutter open for a period of time

stiffer mantle: the largest portion of Earth's interior beneath the asthenosphere

storm surge: sudden rise in sea level as a low pressure system impacts a coastline

storm track: path that a storm follows

storm warning: issued by the National Weather Service when a severe storm will impact a region

storm watch: issued by the National Weather Service when there is a possibility of a severe storm impacting a region

strata: refers to layers of rock

stratopause: the boundary between the stratosphere and the mesosphere

stratosphere: layer of the atmosphere between the troposphere and the mesosphere; in this zone temperature rises with altitude due to absorption of ultraviolet radiation by the ozone layer

streak: color of the powdered mineral

stream channel: the sides and bottom that confine a stream, river, or other waterway

stream discharge: volume or amount of water moving in a stream

stream drainage pattern: the natural design that streams form as they flow over a land surface; the pattern depends on slope and resistance of the bedrock

striations: elongated, parallel scratches in bedrock caused by the rocks dragged along the bottom of a moving glacier

subduction: in plate tectonics, the denser plate will slide under the less dense overriding plate

sublimation: process which changes a solid to a gas without going through a liquid phase

subsoil: soil layer below the topsoil

sunpath: apparent arc that the Sun traces in our sky each day as Earth rotates

sunspot: dark, cool areas on the surface of the Sun

supernova: the explosion of a star, usually occurs in the more massive stars; material from a star explosion usually forms a nebula
talus: deposit of unsorted, angular-shaped sediment found at the base of a hill; usually a result of erosion by gravity
terrestrial: related to Earth
terrestrial planets: the dense, rocky planets with few moons and no rings that are between the Sun and the asteroid belt
terrestrial radiation: heat energy given off from Earth's surface
texture: refers to a surface's roughness or smoothness; in rocks, refers to the size and alignment of mineral crystals or sediments
theory of plate tectonics: theory that Earth's lithosphere is divided into pieces called "plates" which move and shift due to convection currents in the asthenosphere
thermometer: instrument which measures temperature in Celsius, Fahrenheit, or Kelvin
thermosphere: upper layer of the atmosphere before outer space begins
thunderstorm: a storm accompanied by lightning, thunder, and sometimes hail; usually forms along a cold front
tide: the cyclic rising and falling of the ocean waters along the shoreline; caused primarily by the Moon's gravitational pull on Earth
till: unsorted sediment deposited by a glacier
tilt (inclination) of axis: the angle between an object's rotational axis and the line perpendicular to its orbit
tilted strata: rock layers which are not horizontal; rock layers that are slanted at an angle
time zone: a region that has the same time throughout
topographic map: contour map that illustrates elevation, slopes, and natural and man-made features of a region
topography: shape and features of Earth's surface
topsoil: dark, humus-rich, loose sediment found near the surface
tornado: rapidly rotating funnel of air extending down from a thundercloud with extremely low pressure
transform plate boundary: boundary between two plates that are sliding past each other in opposite directions
transition zone: zone of contact between hot intruding magma and the surrounding bedrock; contact metamorphism of the rock will occur at this interface

transpiration: process by which plants release water vapor into the atmosphere
transported soil: soil which after it formed was moved to a new location by an agent of erosion
travel time: the time it takes for a P-wave or an S-wave to move a given distance; expressed in minutes and seconds
tributary: a stream that flows into a larger stream
tropics: the regions of Earth between the latitudes 23°30′ North and 23°30′ South
tropopause: the boundary between the troposphere and the stratosphere
troposphere: lowest layer of the atmosphere near Earth's surface that contains most of the atmospheric gases and moisture, this is where weather occurs
tsunami: ocean wave caused by an earthquake which occurred in the oceanic crust; often misnamed "tidal wave"
ultraviolet radiation: invisible short-wave radiation from the Sun; this dangerous energy is absorbed by ozone in the stratosphere
uncomformity: a buried erosional surfaces which results in time gaps in the geologic history of a region; caused by uplift and erosion followed by subsidence and deposition
uniform: to be the same
universe: consists of empty space, matter, and energy
unsaturated: not filled to capacity
unsorted particles: sediment particles that are different shapes and sizes
uplifting forces: those actions which raise or elevate the land
vapor: gas phase of a substance
variable: any factor in an experiment which affects the results of the experiment
velocity: rate at which an object changes its position; depends on speed and direction of motion
vertical: up and down; perpendicular to the horizon; the y-axis on a graph
vertical rays of Sun: Sun rays which hit a surface at a 90 degree angle; direct rays; perpendicular rays
vesicular: a texture of empty gas pockets found in extrusive igneous rocks
visibility: the distance that one can see clearly
visible light: the only part of the electromagnetic spectrum that we can see
volcano: weak spots in the crust where hot, less dense molten magma rise to the surface
volume: amount of space an object occupies; can be measured in cubic centimeters or milliliters

water (hydrologic) cycle: the continual movement of water between the hydrosphere, lithosphere, and atmosphere

water table: boundary between the zone of aeration and the zone of saturation; the top of the groundwater zone

water vapor: gaseous form of water

watershed: the region drained by a river and all its tributaries

wavelength: the distance between the crests of waves

weather: the current conditions of the atmosphere for a location; described by the weather variables

weather station model: an illustration which shows the values for the atmospheric variables for a specific location for a particular time

weathering: process by which bedrock is broken into smaller pieces by physical and chemical processes

white dwarf: star which is very hot but small in size causing it to be dim; the late stage in a low mass star's life cycle

wind: horizontal air movements which move parallel to the ground from high pressure to low pressure

wind vane: instrument which indicates the direction of the wind

windward: on the side facing the prevailing wind

zenith: the point in the sky directly above the observer; position in the sky with a 90° altitude above the horizon

zone of aeration: upper part of the soil and loose sediment where the pore spaces are mostly filled with air

zone of saturation: lower part of soil and loose sediment where pore spaces are filled with water

Index

A
absorbed 65
acid precipitation 82
adiabatically 87
aerosols 79
agent of erosion 151
air mass 107, 271
air temperature 104, 270
altitude 272
anemometer 92
angle of insolation 66
anticyclone 93
aphelion 29, 44
arctic air masses 271
arid (dry) climate 131
asteroids 26
asthenosphere 209
atmosphere 41, 79
atmospheric variables 87
auroras 16, 80

B
barometer 92
barometric air pressure 92, 104
barometric trend 104, 271
benchmark 175
Big Bang theory 13
bioclastic 197, 263
black hole 15
blizzards 114

C
capillarity 128
capillary water 128
chemical weathering 148
chronometer 48
circumpolar 44
classification systems 1
clastic 197, 263
cleavage 276
climate 131
cloud cover 270
clouds 90
coarse texture 195
colloids 147
comets 26
compounds 276
condensation nuclei 90
contact metamorphism 264
contact (thermal) metamorphism 198
continental air masses 271
continental crust 209
continental drift 212
continental (inland) climates 133
contour interval 175
contour map 175
convection currents 87, 213
convergent plate boundary 260
Coriolis Effect 43
correlate 243
crust 209, 258, 267
crystalline 263

D
deforestation 125
delta 157
density 2, 256
deposition 147, 160
dew point 89, 104, 270
disintegrates 255
divergent plate boundary 260
doppler red shift 13
drumlins 153
duration of insolation 67
dwarf planets 26
dwarfs 274

E
earthquakes 220
eccentricity 29, 256, 275, 284
eclipses 32
electromagnetic energy 273
electromagnetic spectrum 13
elevation 175
ellipses 29
environment of formation 262
eons 234, 265
epicenter 220, 268
epochs 234
equator 47
equatorial diameter 275
eras 234
erosion 147, 151
erratics 153
escarpments 173
evaporation 89, 125
evapotranspiration 89
extrusive (volcanic) 262

F
felsic 195, 262
fine texture 195
focus 29, 220
foliated 264
foliated metamorphic 198
fossils 234
Foucault pendulum 43
fracture 276
front 108
frost action 147

G
geocentric 32, 45
geologic events 240
giant(s) 15, 274
glaciers 153
glassy texture 195
global warming 81
GMT 48
gradient 155, 175, 256, 295
graphs 5
greenhouse effect 81
groundwater 128

H
hachure marks 175
half-life 237, 255
hardness 276
heating lag 67
heliocentric 25, 45
hot spots 218, 261
humid climate 131
humus 149
hurricanes 115
hydrosphere 41, 258

I
igneous rocks 195
index fossil(s) 243, 266
inference 1
infiltrate 125
Infrared radiation 274
inner core 209, 267
insolation 65
intensity of insolation 65
intrusive (plutonic) 262
isobars 92
isolines 294
isotherms 88
isotopes 237

J
jet streams 93, 273
jovian planets 26, 275

K
kettle hole 153

L
landscape 171
latitude 47
lava 195
leeward side 134
leveling forces 172
lithosphere 41, 209, 212, 260, 267
longitude 48
longshore current 158
luminosity 15, 274
lunar eclipse 32
luster 276

M
mafic 195, 262
magma 195, 261
main sequence 15, 274
mantle 213, 214
mantle convection currents 267
map scale 175
marine (coastal) climates 133
maritime air masses 271
mass 2
mass movement 151
meander 156
mean distance 275
measurement 1
Mercalli scale 223
mesopause 80, 272
mesosphere 80
metallic luster 276
metamorphic rock 198
meteor 26

meteorites 26
meteoroid 26
meteorology 103
metric (SI) system 1
Milky Way galaxy 16
minerals 191, 293
moho 209
moon(s) 26, 275
moraines 153
mountains 258

N
natural resources 189
nebula 15, 25
non-clastic 197
non crystalline 262
non-foliated 198, 264
non-metallic luster 276
non-renewable 189
nuclear fusion 15

O
oblate spheroid 41
observation 1
ocean currents 93
oceanic crust 209
one atmosphere 270
orbit 25
orogeny 234, 266
outer core 209, 267
outwash plain 153
ozone 80, 81

P
perihelion 29, 44
period of revolution 275
period of rotation 275
periods 234
permeability 127
phases of the Moon 32
photosynthesis 79
physical weathering 147
plains 258
planetary (global) winds 93
planets 26
plate(s) 212, 260
plateaus 258
polar air masses 271
polar climates 132
Polaris 44
polar reversals 213
polyminerallic 195
porosity 127
precession 44
precipitation 90, 104, 271
prediction 1
present weather 104, 270
pressure 270
primary waves 221
prime meridian 48
principle of original horizontality 211, 240
principle of superposition 211, 240
profile 175, 295
protostar 15
psychrometer 90
P-waves 221

R
radioactive dating 237
radioactive elements 237
radioactive isotope 237, 255
rate of change 5, 256
red supergiants 15
reflected 65
refracted 65
regional metamorphism 198, 264
relative dating 240
relative humidity 89
renewable 189
residual soil 149
revolve(s) 25, 44
Richter scale 223
ridges 213
rift valleys 213
rigid mantle 209
rock cycle 193
rocks 193
rotate(s) 25, 43
runoff 125

S
Safir-Simpson scale 115
sand bars 157
sand dunes 152
scattered 65
scientific notation 1
seafloor spreading 213
secondary waves 221
sediment(s) 147, 261
sedimentary rocks 196
seismic waves 220
seismogram 220
seismograph 220
severe weather 113
sextant 48
shadow zone 221
smog 81
soil 149
solar eclipse 32
solar system 25
solar wind 16
source region 107
specific heat 83, 133, 256
spectrum 273
stars 15
station model 270
stiffer mantle 209
storm warning 113
storm watch 113
stratopause 80, 272
stratosphere 80
streak 276
stream discharge 125, 155
striations 153
Sun 16
sunspots 16
supergiant(s) 274
supernova 15
S-waves 221

T
talus 151
temperate mid-latitude climates 132
terrestrial planets 26, 275

texture 195
theory of plate tectonics 212
thermometer 88
thermosphere 80
thunderstorms 114
tides 33
till 153
topographic maps 175
topsoil 149
tornado 113
transform plate boundary 260
transition zone 198
transpiration 89, 125
transported soil 149
travel time 268
trenches 213
tropical air masses 271
tropical climates 132
tropopause 80, 272
troposphere 41, 80, 258
tsunamis 223

U
unconformities 242
unconformity 266
universe 13
uplifting forces 172

V
vesicular 262
vesicular texture 195
visibility 104, 270
visible light 274
volcanoes 218
volume 2

W
water cycle 125
watershed 157
water table 128
weather 103
weathering 147
weather station model 103
white dwarf 15, 274
wind direction 104, 270
wind speed 104, 270
wind vane 92
windward side 134

Z
zone of aeration 128
zone of saturation 128

The University of the State of New York

REGENTS HIGH SCHOOL EXAMINATION

PHYSICAL SETTING
EARTH SCIENCE
PRACTICE EXAMS

The possession or use of any communications device is strictly prohibited when taking this examination. If you have or use any communications device, no matter how briefly, your examination will be invalidated and no score will be calculated for you.

Use your knowledge of Earth science to answer all questions in this examination. Before you begin this examination, you must be provided with the *2011 Edition Reference Tables for Physical Setting/Earth Science*. You will need these reference tables to answer some of the questions.

You are to answer all questions in all parts of this examination. You may use scrap paper to work out the answers to the questions, but be sure to record your answers on your answer sheet and in your answer booklet. A separate answer sheet for Part A and Part B–1 has been provided to you. Follow the instructions from the proctor for completing the student information on your answer sheet. Record your answers to the Part A and Part B–1 multiple-choice questions on this separate answer sheet. Record your answers for the questions in Part B–2 and Part C in your separate answer booklet. Be sure to fill in the heading on the front of your answer booklet.

All answers in your answer booklet should be written in pen, except for graphs and drawings, which should be done in pencil.

When you have completed the examination, you must sign the declaration printed on your separate answer sheet, indicating that you had no unlawful knowledge of the questions or answers prior to the examination and that you have neither given nor received assistance in answering any of the questions during the examination. Your answer sheet and answer booklet cannot be accepted if you fail to sign this declaration.

Notice . . .

A four-function or scientific calculator and a copy of the *2011 Edition Reference Tables for Physical Setting/Earth Science* must be available for you to use while taking this examination.

PHYSICAL SETTING
EARTH SCIENCE
PRACTICE EXAM #1
Part A

Answer all questions in this part.

Directions (1–35): For *each* statement or question, choose the word or expression that, of those given, best completes the statement or answers the question. Some questions may require the use of the *2011 Edition Reference Tables for Physical Setting/Earth Science*. Record your answers on your separate answer sheet.

1 Evidence that the universe is expanding is best supported by the observation that the wavelengths of light from distant galaxies are shifted toward the
 (1) red end of the spectrum because they are shortened
 (2) red end of the spectrum because they are lengthened
 (3) blue end of the spectrum because they are shortened
 (4) blue end of the spectrum because they are lengthened

2 Scientists infer that the Big Bang occurred approximately
 (1) 4.6 billion years ago
 (2) 7 billion years ago
 (3) 9 billion years ago
 (4) 13.8 billion years ago

3 Which process produces the largest amount of energy given off by stars?
 (1) nuclear fusion of lighter elements into heavier elements
 (2) nuclear fusion of heavier elements into lighter elements
 (3) radioactive decay of lighter elements into heavier elements
 (4) radioactive decay of heavier elements into lighter elements

4 Which two factors caused the interior layering of Earth and other planets in our solar system during their formation?
 (1) cosmic background radiation and density differences
 (2) cosmic background radiation and specific heat
 (3) gravity and density differences
 (4) gravity and specific heat

5 The map below shows the location of the Chicxulub impact crater, which was formed in the Gulf of Mexico approximately 65.5 million years ago. With an estimated 108-mile diameter, this crater, which is now buried beneath surface crustal rocks, is one of the largest craters on Earth.

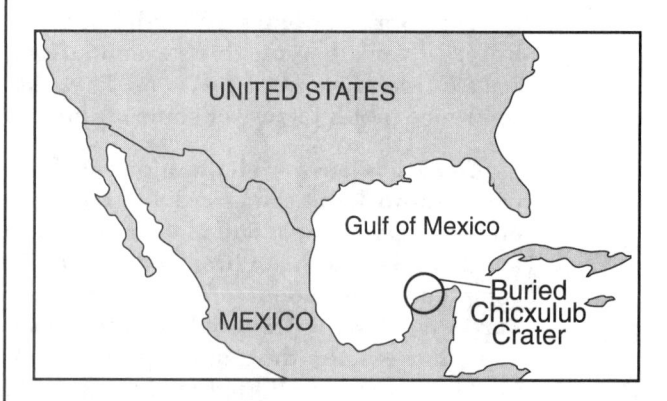

The asteroid impact that formed this large crater is theorized to have caused
 (1) a drop in sea level from the sea water draining into the large crater
 (2) warmer than normal worldwide ocean temperatures from the hot asteroid
 (3) mass extinctions of many species on Earth from extreme climate changes
 (4) an increase in worldwide greenhouse gases from vaporizing crustal rocks

6 The altitude of *Polaris* measured by an observer at the Tropic of Cancer is
 (1) 15°   (3) 66.5°
 (2) 23.5°   (4) 90°

7 During which month does the Sun rise north of due east in New York State?
 (1) February   (3) October
 (2) July   (4) December

8 The graph below shows the change in tide heights of the Hudson River at Newburgh, New York.

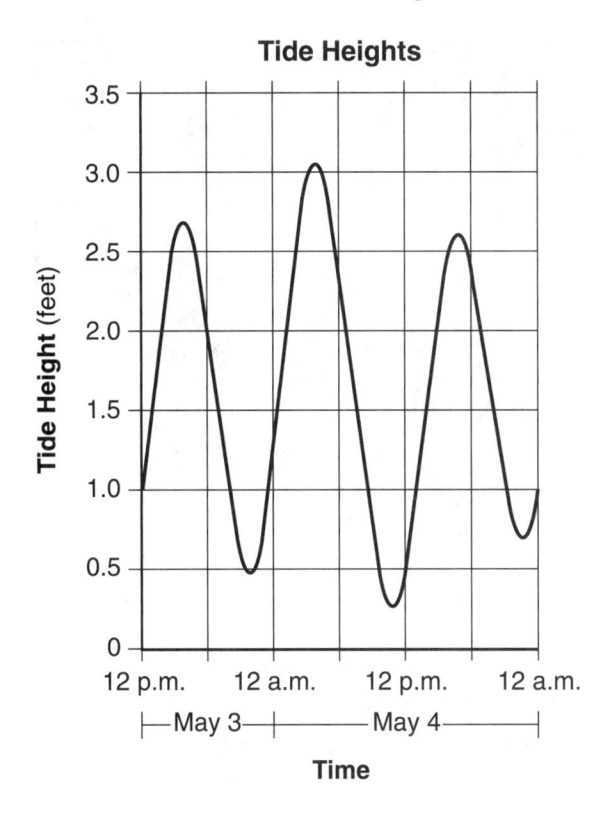

Tide Heights

According to the graph, the time difference between high tide and the next low tide is approximately

(1) 2 hours (3) 6 hours
(2) 3 hours (4) 12 hours

9 During a rainstorm, water is flowing down the side of a hill composed of solid bedrock. What will be the effect on the relative amounts of runoff and infiltration when the water reaches an area of unsaturated soil with a gentler slope?

(1) Runoff will decrease as infiltration decreases.
(2) Runoff will decrease as infiltration increases.
(3) Runoff will increase as infiltration decreases.
(4) Runoff will increase as infiltration increases.

10 What is the dewpoint when the dry bulb temperature is 20°C and the relative humidity is 17%?

(1) –5°C (3) 11°C
(2) –2°C (4) 15°C

11 The diagram below represents the apparent changes in the direction of swing of a Foucault pendulum.

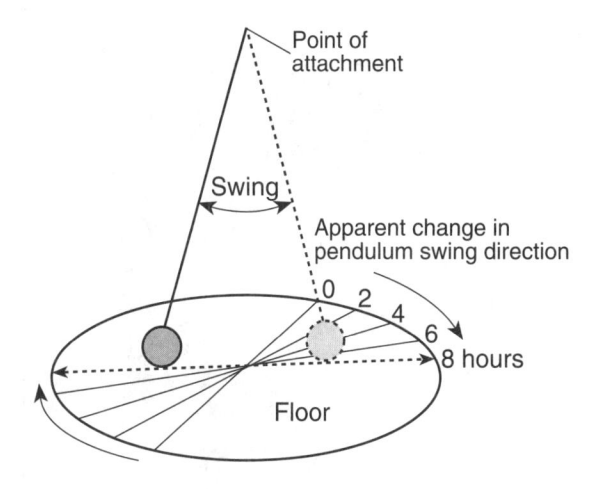

This apparent change in direction of swing provides evidence that Earth

(1) has a spherical shape
(2) is tilted on its axis
(3) orbits around the Sun
(4) turns on its axis

12 Which change in the heat energy content of water occurs when water changes phase from a liquid to a solid?

(1) gain of 334 Joules of heat energy per gram
(2) release of 334 Joules of heat energy per gram
(3) gain of 2260 Joules of heat energy per gram
(4) release of 2260 Joules of heat energy per gram

13 What is the primary source of energy for Earth's weather systems?

(1) incoming solar radiation
(2) subtropical jet streams
(3) precipitation from clouds
(4) heat from Earth's interior

14 The diagram below represents a view of Earth from above the North Pole. Points *A* and *B* represent locations on Earth's surface.

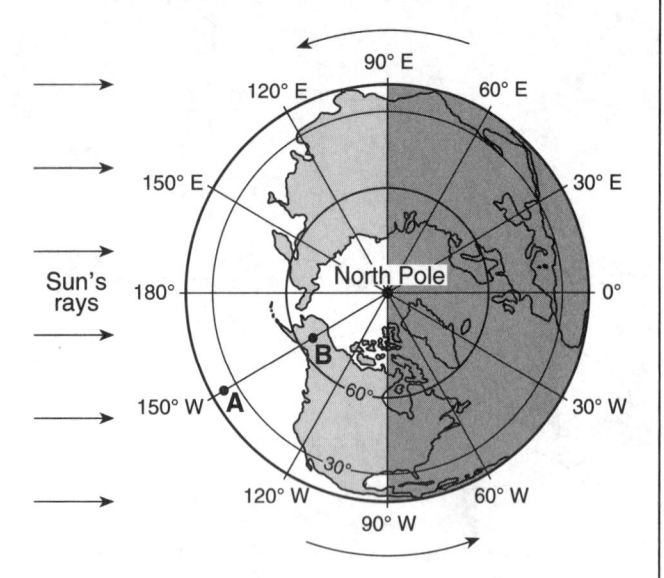

Locations *A* and *B* have the same

(1) latitude and local time
(2) latitude and elevation
(3) longitude and local time
(4) longitude and elevation

15 The diagram below shows an instrument used in weather forecasting.

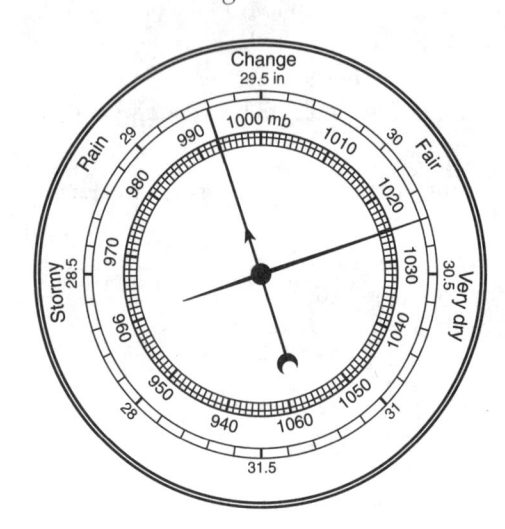

This instrument measures atmospheric

(1) wind speed
(2) wind direction
(3) pressure
(4) temperature

16 The map of North America below shows the position of the polar front jet stream on January 7, 2014, and the location of Atlanta, Georgia.

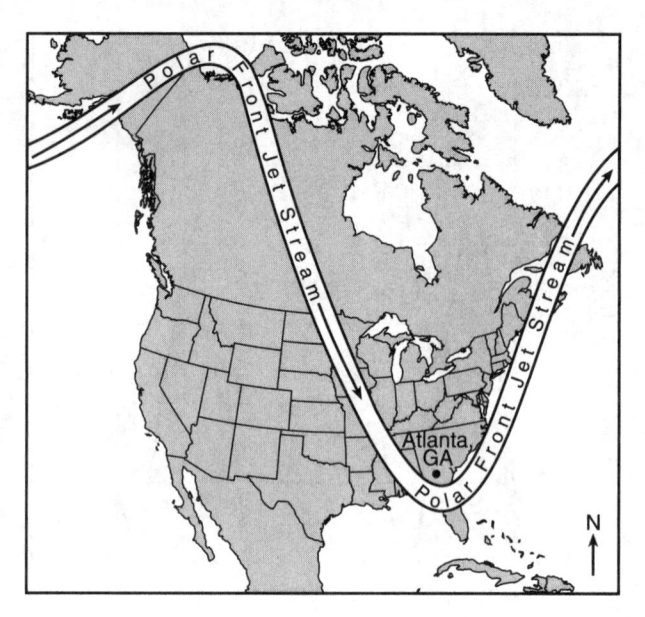

Which type of air mass was most likely located over Atlanta, Georgia?

(1) mT
(2) mP
(3) cT
(4) cP

17 As altitude increases in the troposphere and stratosphere, the air temperature

(1) decreases in the troposphere and increases in the stratosphere
(2) decreases in both the troposphere and stratosphere
(3) increases in the troposphere and decreases in the stratosphere
(4) increases in both the troposphere and stratosphere

18 Which factor causes the surface of Lake Ontario to cool at a slower rate than the surface of the land along the shore of the lake?

(1) Evaporating water releases more heat into the lake than into the land.
(2) Lake water has a higher specific heat than land.
(3) Water vapor cools the lake as it condenses.
(4) Sunlight passes through the top layers of the lake water.

19 The map below shows location *X* in northern India.

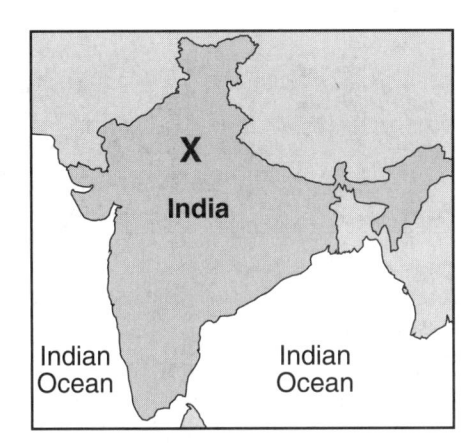

Summer monsoon rains normally occur in India when

(1) high pressure exists near location *X*, pulling moisture in from the Indian Ocean
(2) high pressure exists near location *X*, pushing moisture out to the Indian Ocean
(3) low pressure exists near location *X*, pulling moisture in from the Indian Ocean
(4) low pressure exists near location *X*, pushing moisture out to the Indian Ocean

20 Which gas absorbs some of the harmful insolation in Earth's upper atmosphere before that insolation reaches Earth's surface?

(1) nitrogen
(2) ozone
(3) oxygen
(4) hydrogen

21 Which ocean current brings warm water to the southeastern coast of Africa?

(1) Agulhas Current
(2) Benguela Current
(3) West Australian Current
(4) Equatorial Countercurrent

22 The intensity of insolation at solar noon from November 1 to February 1 in New York State will

(1) decrease, only
(2) increase, only
(3) decrease, then increase
(4) increase, then decrease

23 Most scientists infer that a major factor in the increased rate of melting of Earth's glaciers is

(1) a decrease in the output of energy from the Sun
(2) a decrease in Earth's atmospheric transparency
(3) an increase in Earth's orbital distance from the Sun
(4) an increase in carbon dioxide in Earth's atmosphere

24 What is the approximate percentage of geologic time that humans have existed on Earth since its origin?

(1) less than 1%
(2) 1.8%
(3) 11.8%
(4) more than 25%

25 The photograph below shows the East African Rift Valley in Africa. Which tectonic movement of Earth's crust is most likely responsible for this feature?

(1) convergence of continental crust
(2) convergence of oceanic crust
(3) divergence of continental crust
(4) divergence of oceanic crust

26 Radioactive dating of fossils and rocks is possible because radioactive isotopes

(1) are found in all fossils and rocks
(2) are easily collected and measured
(3) disintegrate into organic substances
(4) disintegrate at a predictable rate

27 The photograph below shows the bedrock structure of a limestone outcrop.

Which process is responsible for the deformation of this bedrock?

(1) folding   (3) mass movement
(2) weathering   (4) volcanic activity

28 Which particles will be transported by a stream moving at a velocity of 5 cm/s?

(1) pebbles, sand, silt, and clay, only
(2) sand, silt, and clay, only
(3) silt and clay, only
(4) clay, only

29 The surface bedrock of New York State that is most likely to contain the mineral garnet can be found in an area 30 miles

(1) north of Binghamton
(2) south of Mt. Marcy
(3) east of Oswego
(4) west of Utica

30 The geologic cross section below represents surface landscape features that developed in an arid climate.

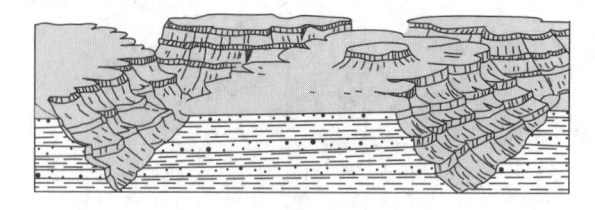

A change in climate to one that is more humid would cause the

(1) shape of the hills to become more rounded
(2) elevation of the area to become higher
(3) porosity of the bedrock to increase
(4) rate of chemical weathering to decrease

31 The cross section below represents an outcrop of sedimentary rock layers exposed on Earth's surface. Rock layers A, B, C, and D are labeled.

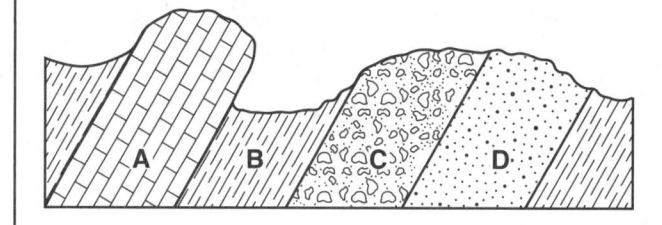

Which rock layer shows the greatest resistance to weathering and erosion?

(1) A   (3) C
(2) B   (4) D

32 The diagrams below represent the compass direction and altitude of the Sun's rays at noon for a location on Earth on four different dates.

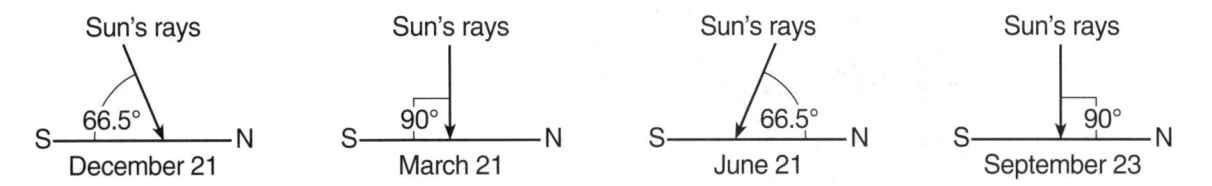

```
    Sun's rays              Sun's rays              Sun's rays              Sun's rays
        ↓                       ↓                       ↓                       ↓
   S─66.5°─N              S─90°─N                S─66.5°─N              S─90°─N
   December 21            March 21                June 21                September 23
```

What is the latitude of this location?

(1) 0°
(2) 23.5° N
(3) 23.5° S
(4) 90° N

33 The map below shows the present-day positions of the continents. Points A through D represent locations on Earth's surface. The location of New York State on the North American continent is indicated.

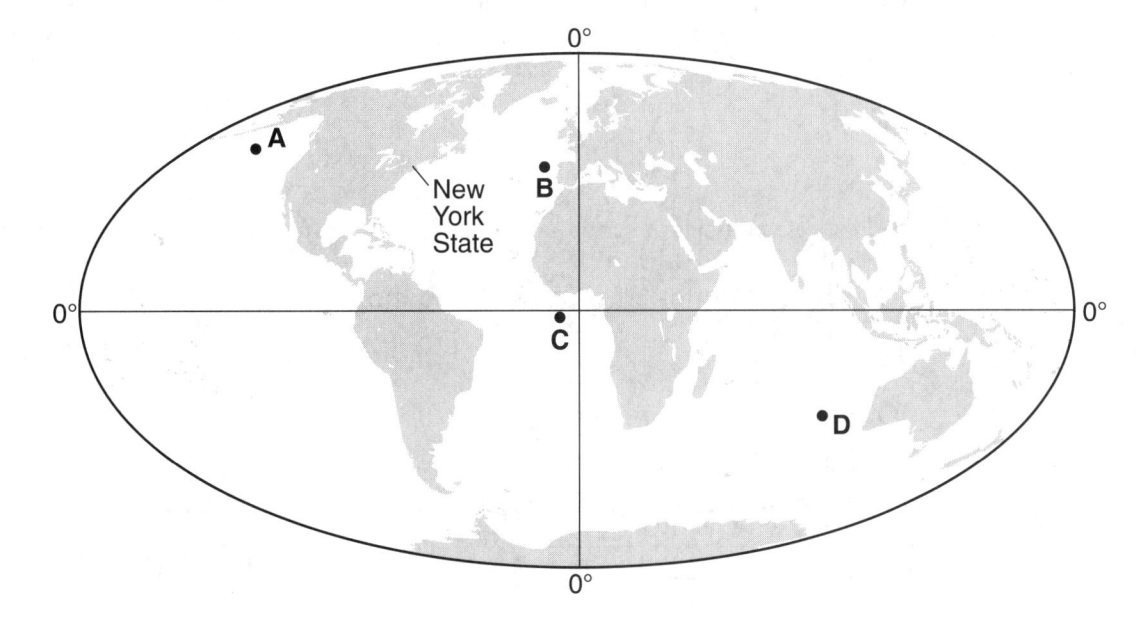

Which letter best represents the inferred position of the New York State region on Earth at the end of the Devonian Period?

(1) A
(2) B
(3) C
(4) D

Practice Exam 1   335

34 The image below shows a spear point embedded in part of a mastodon's rib bone, found near Seattle, Washington.

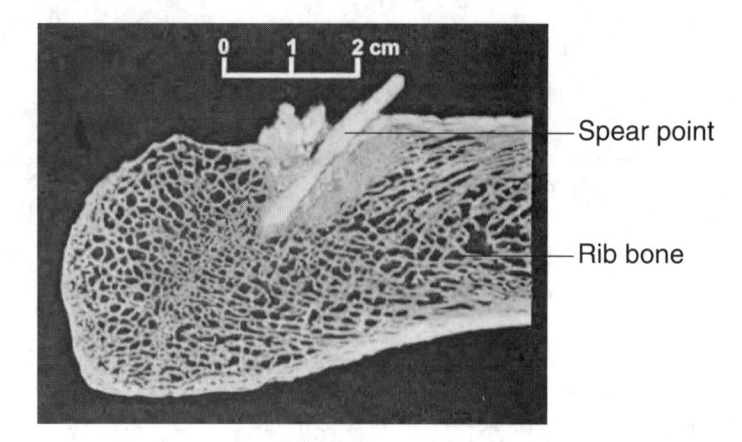

Scientists infer that early North American humans hunted the mastodon. Carbon-14 dating of the rib bone indicates that 2.4 half-lives have passed since the mastodon was killed. Approximately how many years ago did the mastodon die?

(1) 5700
(2) 11,400
(3) 13,700
(4) 17,100

35 The cross section below represents two types of sorted-sand depositional features found at a coastal location.

Cross Section of a Coastal Location

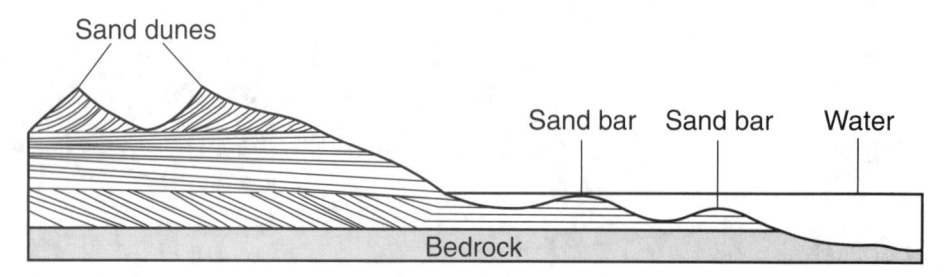

Which table correctly pairs these depositional features with the agents of erosion that formed them?

Depositional Feature	Agent of Erosion
sand dune	mass movement
sand bar	wind

(1)

Depositional Feature	Agent of Erosion
sand dune	glaciers
sand bar	waves

(2)

Depositional Feature	Agent of Erosion
sand dune	mass movement
sand bar	glaciers

(3)

Depositional Feature	Agent of Erosion
sand dune	wind
sand bar	waves

(4)

Part B–1

Answer all questions in this part.

Directions (36–50): For *each* statement or question, choose the word or expression that, of those given, best completes the statement or answers the question. Some questions may require the use of the *2011 Edition Reference Tables for Physical Setting/Earth Science*. Record your answers on your separate answer sheet.

Base your answers to questions 36 through 38 on the diagram below and on your knowledge of Earth science. The diagram represents Earth's position in its orbit on the first day of each of the four seasons, one of which is labeled *A*. The North Pole is labeled *N*. Earth's closest distance to the Sun and Earth's farthest distance from the Sun are labeled in kilometers.

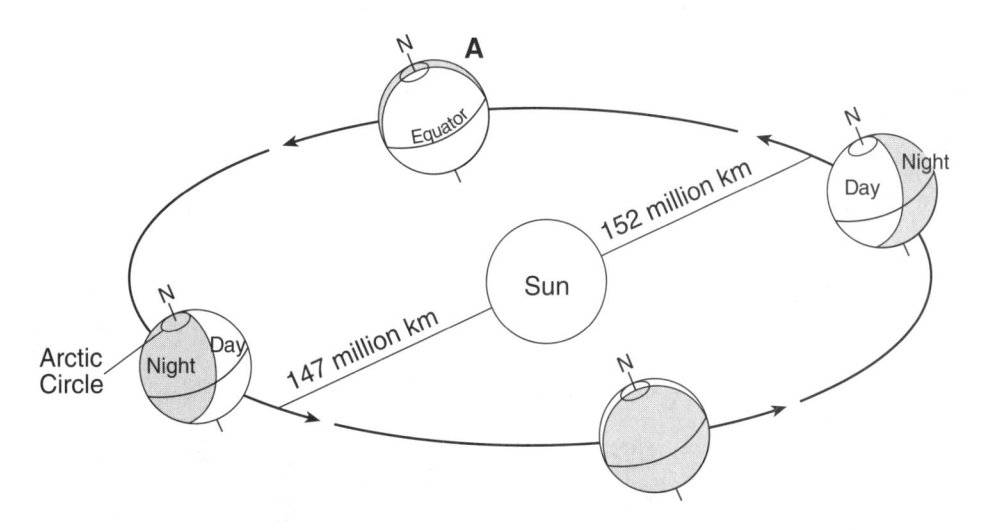

(Not drawn to scale)

36 How many hours (h) of daylight are received at the Arctic Circle when Earth is at position *A*?
 (1) 0 h   (3) 18 h
 (2) 12 h   (4) 24 h

37 When Earth is closest to the Sun, which season is occurring in the Northern Hemisphere?
 (1) spring   (3) fall
 (2) summer   (4) winter

38 What would most likely happen to New York State's summer and winter temperatures if the tilt of Earth's axis increased from 23.5° to 30°?
 (1) Both the summers and winters would become cooler.
 (2) Both the summers and winters would become warmer.
 (3) The summers would become cooler and the winters would become warmer.
 (4) The summers would become warmer and the winters would become cooler.

Base your answers to questions 39 and 40 on the diagrams below and on your knowledge of Earth science. The diagrams represent four columns, labeled A, B, C, and D, that are partially filled with equal volumes of dry, sorted sediments. A fine wire mesh screen covers the bottom of each column to prevent the sediment from falling out. The lower part of each column has been placed in a beaker of water.

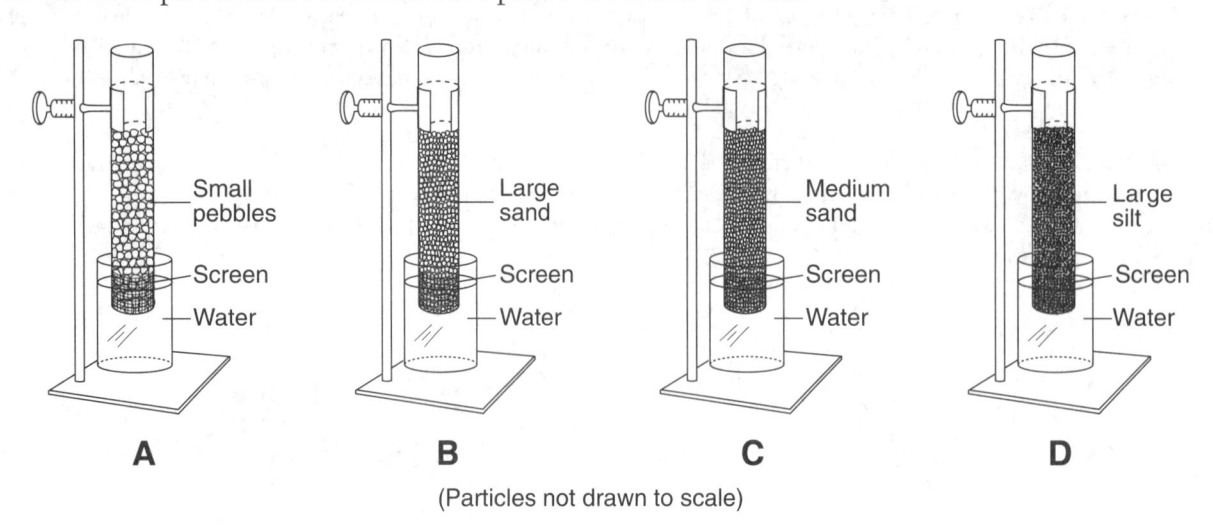

(Particles not drawn to scale)

39 Capillarity will cause water to rise highest in column
(1) A
(2) B
(3) C
(4) D

40 Equal volumes of sediments from all four columns are mixed and poured into a column of water. Which diagram best represents how the sediments will most likely settle?

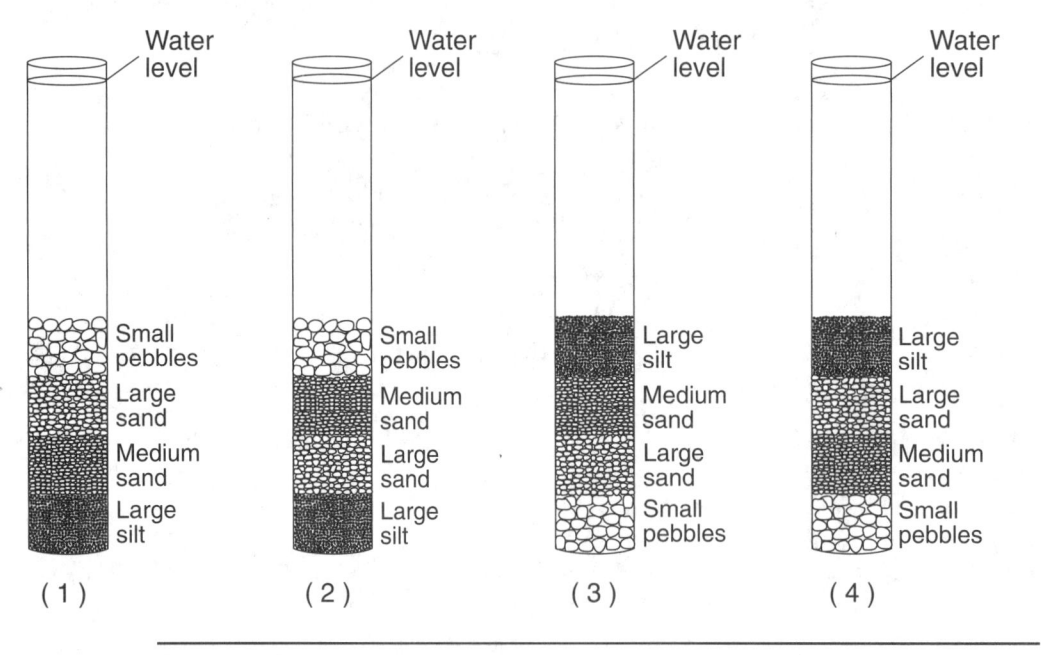

Base your answers to questions 41 and 42 on the cross sections below and on your knowledge of Earth science. The cross sections represent two rock outcrops, labeled I and II, located 10 miles apart. Letters A through E and numbers 1 through 8 identify rock units. The rock units have *not* been overturned.

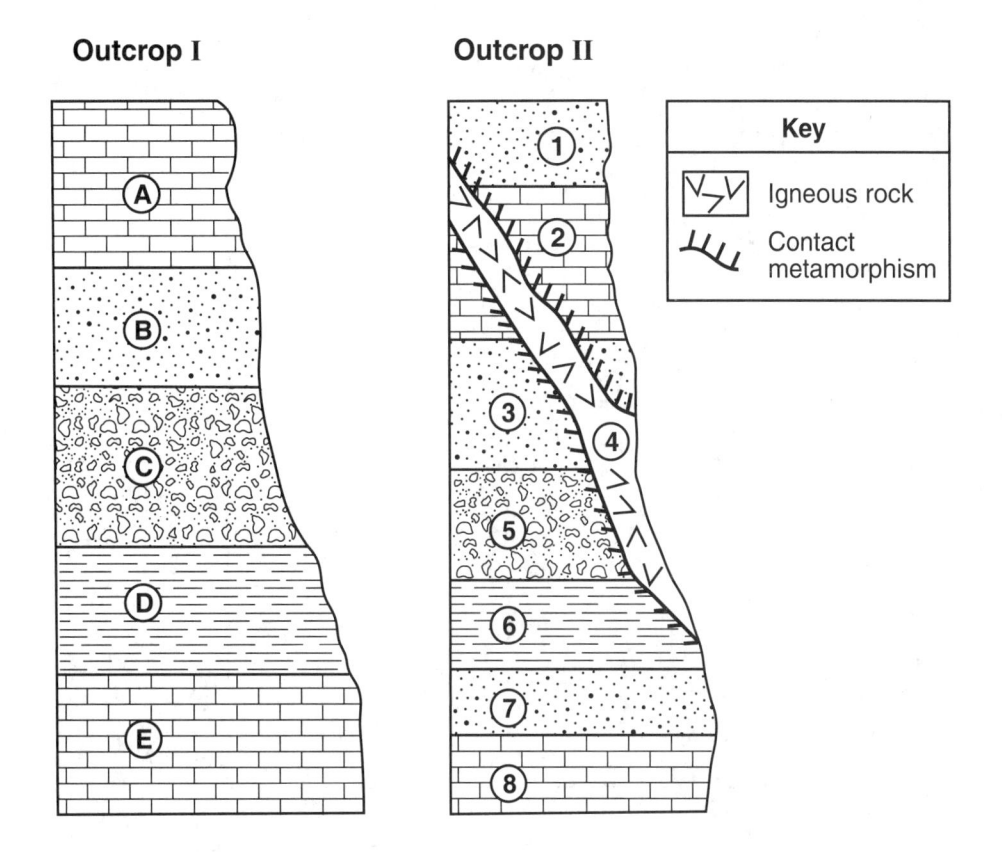

41 In outcrop I, which geologic principle is best represented by the rock units?
   (1) crosscutting relationships
   (2) correlation
   (3) original horizontality
   (4) inclusion

42 The rock record in outcrop II suggests that an unconformity probably exists in outcrop I between rock units
   (1) A and B
   (2) B and C
   (3) C and D
   (4) D and E

Base your answers to questions 43 and 44 on the map and table below and on your knowledge of Earth science. The map shows the zones of observed effects reported after a 1944 earthquake that occurred near Massena, New York. The isolines on the map are boundaries between zones of observed effects described in the Modified Mercalli Scale table. Four cities are labeled on the map.

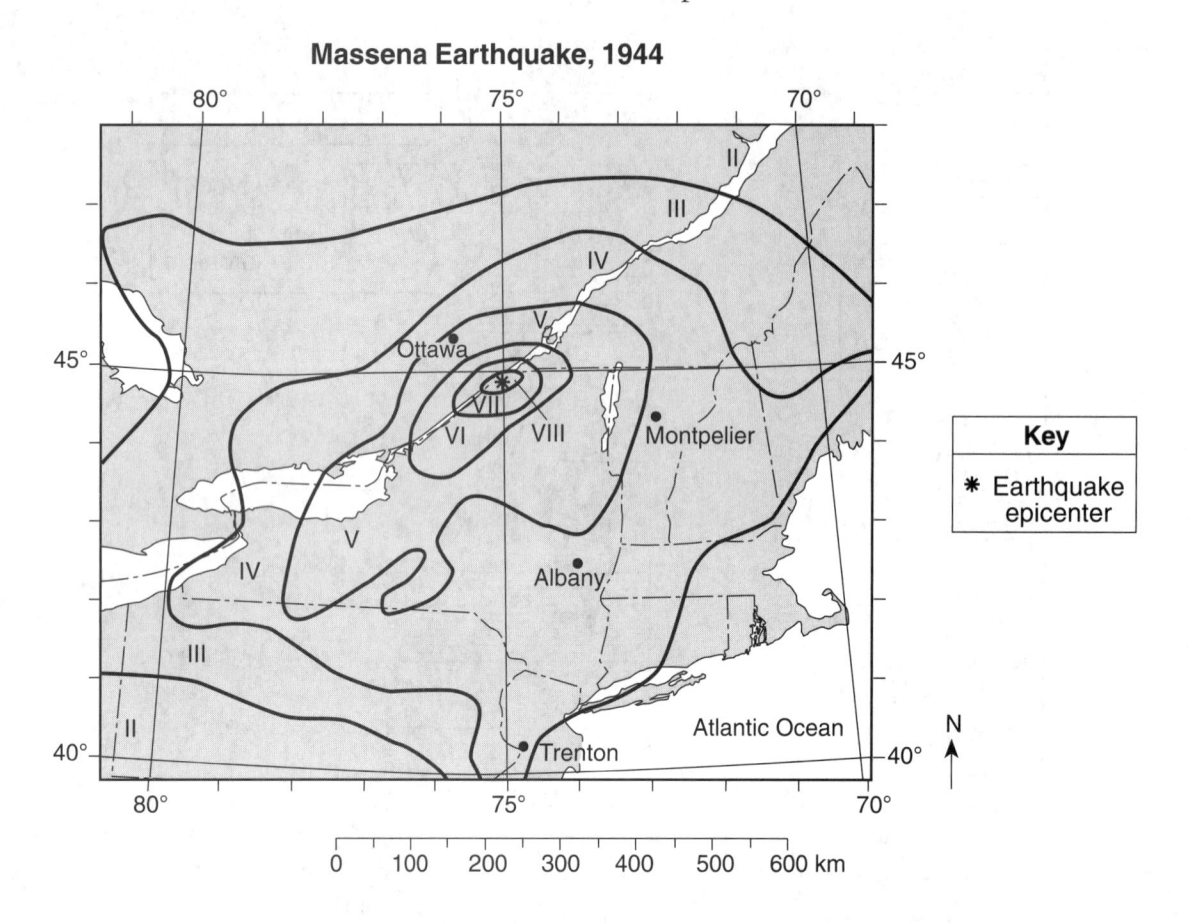

Massena Earthquake, 1944

Key
* Earthquake epicenter

Modified Mercalli Scale

Intensity Value	Description of Observed Effects
I	Usually detected only by instruments
II	Felt by a few persons at rest, especially on upper floors
III	Hanging objects swing; vibration like passing of truck; noticeable indoors
IV	Felt indoors by many, outdoors by few; sensation like heavy truck striking building; parked automobiles sway
V	Felt by nearly everyone; sleepers awakened; liquids disturbed; unstable objects overturned; some dishes and windows broken
VI	Felt by all; many frightened and run outdoors; some heavy furniture moved; glassware broken; books off shelves; damage slight
VII	Difficult to stand; noticed in moving automobiles; damage to some masonry; weak chimneys broken at roofline
VIII	Partial collapse of masonry; chimneys, factory stacks, columns fall; heavy furniture overturned; frame houses moved on foundations

340   Practice Exam 1

43 How long did it take for the first *P*-wave to travel from the epicenter of this earthquake to a seismic station in Trenton, New Jersey?

(1) 1 minute 10 seconds
(2) 2 minutes 10 seconds
(3) 3 minutes 20 seconds
(4) 4 minutes 20 seconds

44 Based on the Modified Mercalli Scale, the darker shading on which map shows the area where the Massena earthquake was felt by nearly everyone?

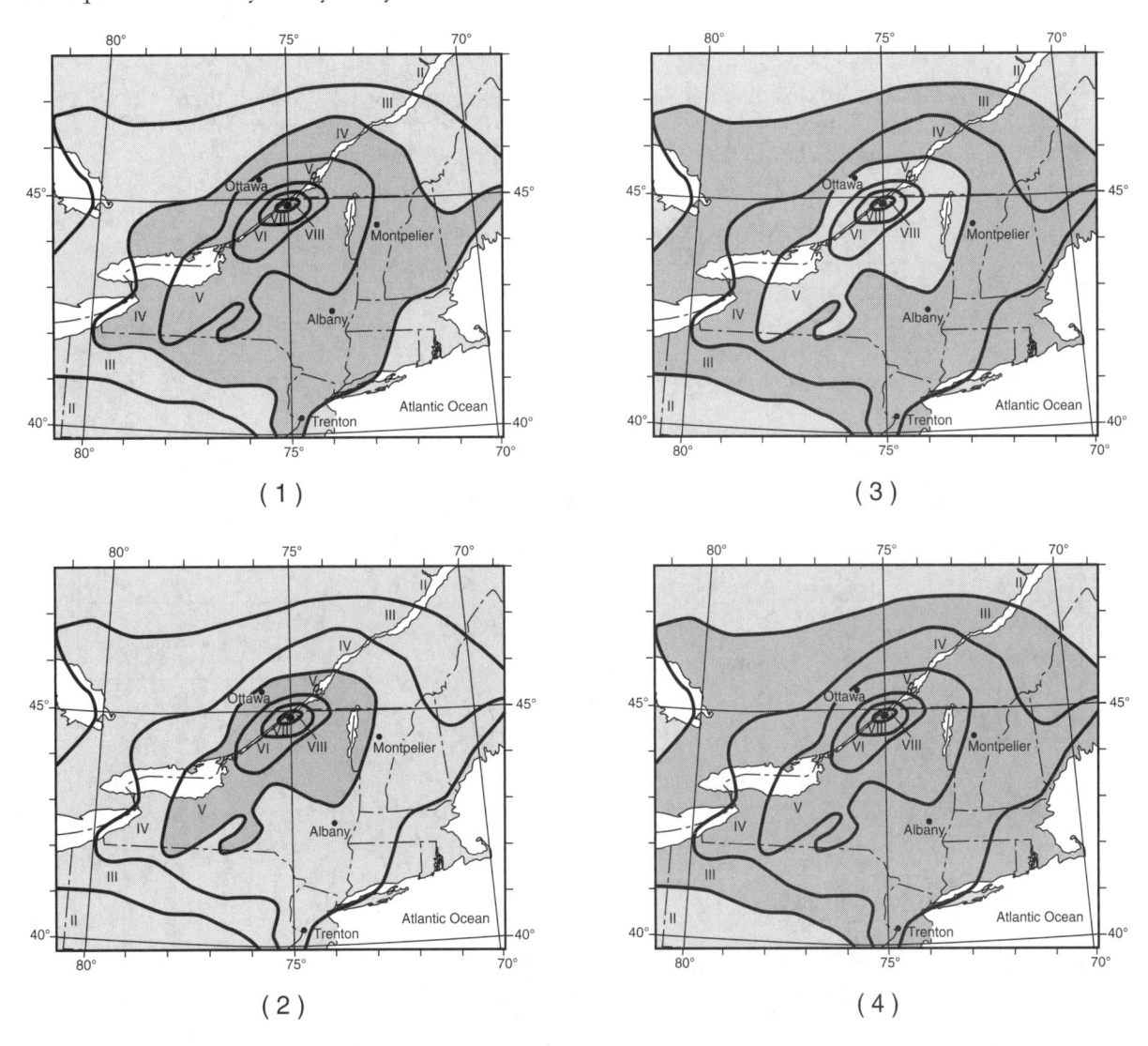

Base your answers to questions 45 through 48 on the passage and map below and on your knowledge of Earth science. The map shows glacial features found in Mendon Ponds Park.

Mendon Ponds Park

Mendon Ponds Park, in New York State, is listed in the National Registry of National Landmarks due to its outstanding glacial landscape features. Glacial ice that covered most of New York State retreated northward at the end of the last ice age. As this glacial ice melted, great amounts of sediments were deposited at the glacier's southern edge. Four glacial features dominate the park's landscape. Kettles are bowl-shaped depressions formed when buried blocks of glacial ice melt. If the depressions fill with water, they are called kettle lakes. The Mendon Park ponds are all kettle lakes. Eskers are ridges of sorted sediments deposited within streams flowing beneath the melting glacier. Kames are small hills of unsorted sediment deposited at the base of waterfalls formed by streams flowing over the edge of a melting glacier.

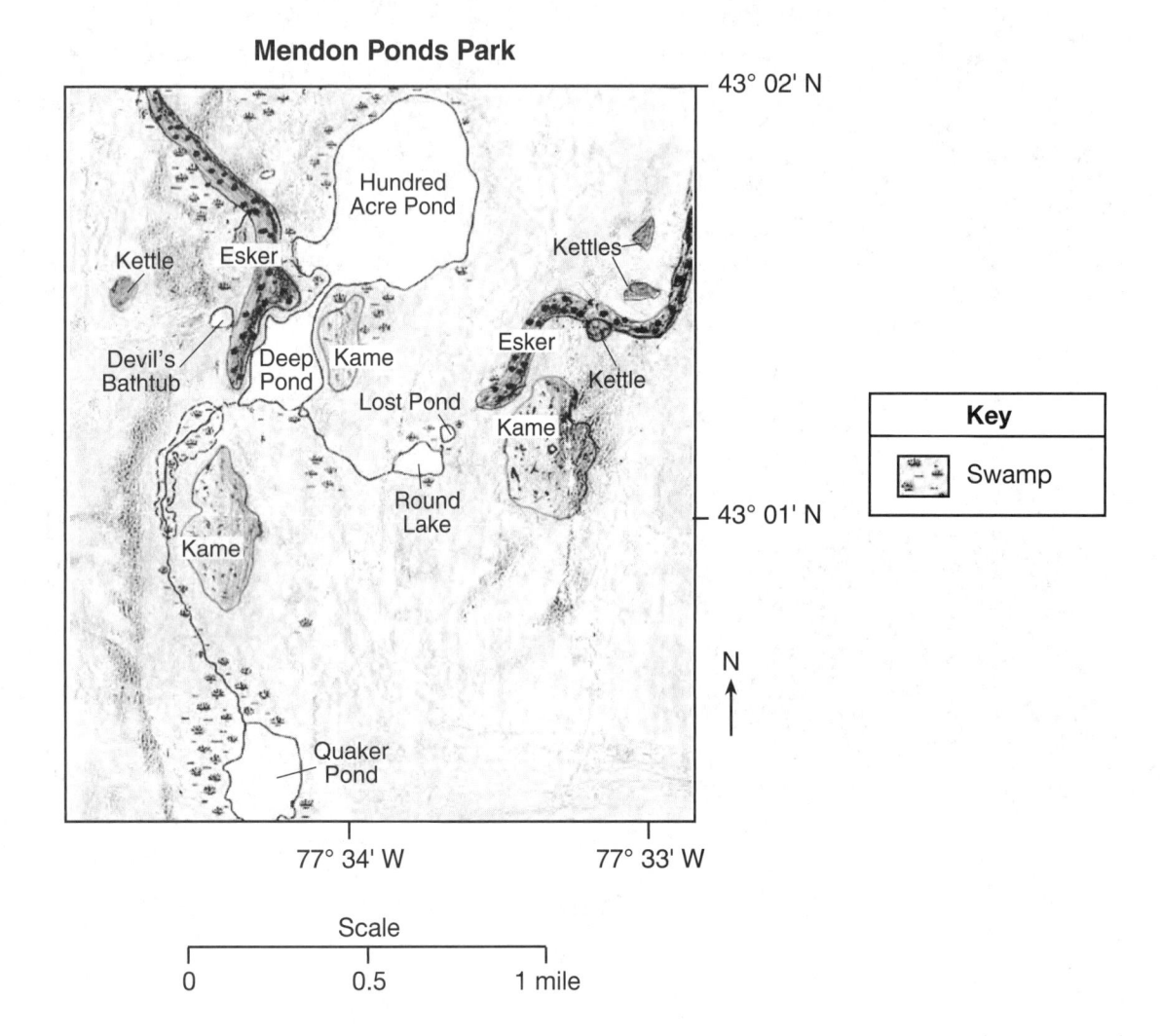

342   Practice Exam 1

45 The last continental ice sheet retreated northward across New York State during which geologic epoch?
   (1) Pleistocene
   (2) Pliocene
   (3) Eocene
   (4) Paleocene

46 The cross sections below represent how a present-day glacial landscape feature was formed in Mendon Ponds Park and its appearance at present.

11,000 Years Ago — Outwash sediment, Buried ice block, Melting glacier

Present Day — Glacial landscape feature, Water

Which glacial landscape feature is indicated in the present-day cross section?
   (1) esker
   (2) kame
   (3) finger lake
   (4) kettle lake

47 Based on the map, in which New York State landscape region is Mendon Ponds Park located?
   (1) Allegheny Plateau
   (2) Tug Hill Plateau
   (3) Erie-Ontario Lowlands
   (4) Hudson-Mohawk Lowlands

48 Which landscape feature is also formed directly by glacial deposition?
   (1) drumlin
   (2) delta
   (3) barrier island
   (4) escarpment

Base your answers to questions 49 and 50 on the graph below and on your knowledge of Earth science. The graph shows the average global air temperature changes that have occurred since the late 1800s. Five volcanoes that experienced major eruptions during this time period are indicated.

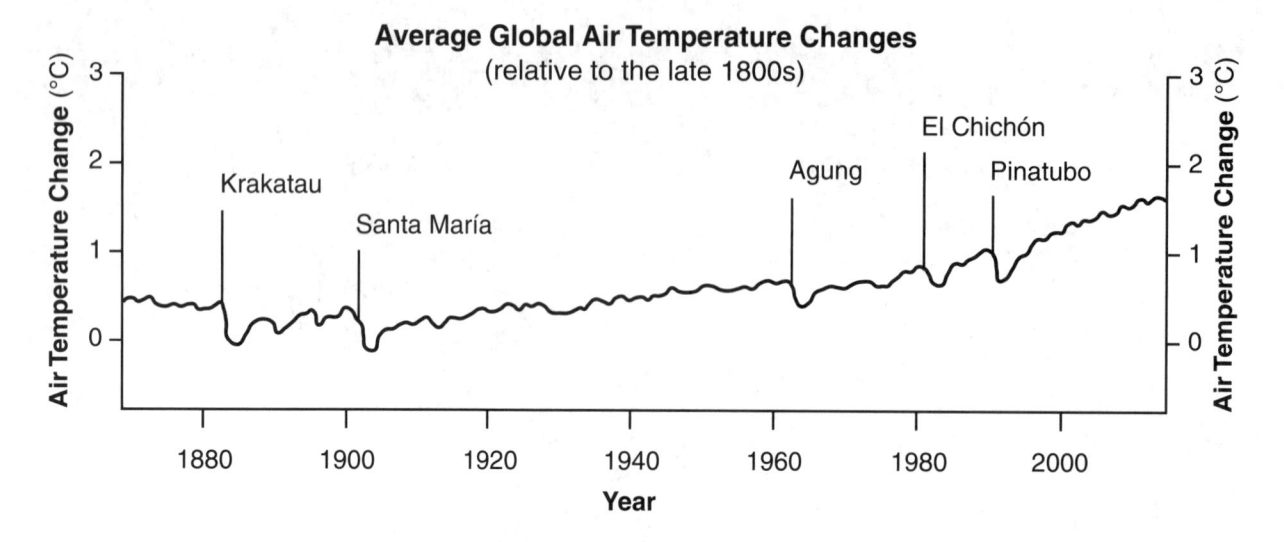

Average Global Air Temperature Changes
(relative to the late 1800s)

49 In the years immediately after each volcanic eruption occurred, average global air temperatures
  (1) decreased because volcanic gases and dust blocked insolation
  (2) decreased because molten rock released heat
  (3) increased because volcanic gases and dust blocked insolation
  (4) increased because molten rock released heat

50 Which conclusion can be made from the data shown in the graph?
  (1) Volcanic eruptions occur in a cyclic and predictable pattern.
  (2) Volcanic eruptions have generally increased in strength since the late 1800s.
  (3) Global air temperatures are warmer today than they were in the late 1800s.
  (4) Global air temperatures have had fewer changes since 1950.

Part B–2

Answer all questions in this part.

Directions (51–65): Record your answers in the spaces provided in your answer booklet. Some questions may require the use of the *2011 Edition Reference Tables for Physical Setting/Earth Science*.

Base your answers to questions 51 through 53 on the diagram below and on your knowledge of Earth science. The diagram represents a planetary system, discovered in 2013, with seven exoplanets (planets that orbit a star other than our Sun) labeled *b* through *h* orbiting a star. The exoplanet orbits are represented with solid lines. For comparison, the orbits of three planets of our solar system are shown with dashed lines. The sizes of the star, exoplanets, and planets are not drawn to scale.

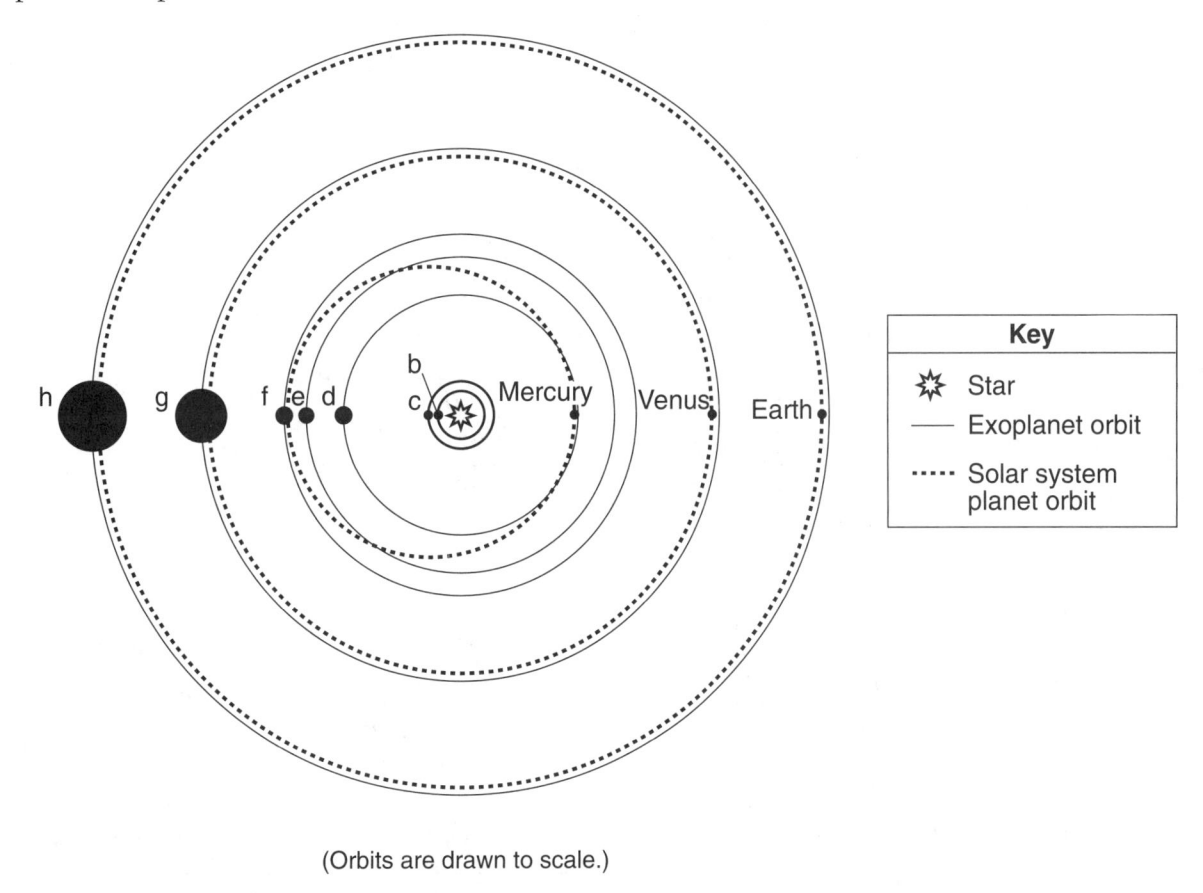

(Orbits are drawn to scale.)

51 Identify the name of the planet represented in the diagram that has the most eccentric orbit. [1]

52 *In your answer booklet*, circle the type of planet (terrestrial or Jovian) to indicate the classification of the three solar system planets shown in the diagram. Describe *one* characteristic of this type of planet that distinguishes it from the other type of planet. [1]

53 Identify the letter of the exoplanet with the shortest period of revolution and explain why that exoplanet has the shortest period of revolution. [1]

Practice Exam 1   345

Base your answers to questions 54 through 56 on the star chart in your answer booklet and on your knowledge of Earth science. The star chart shows the approximate locations of the Big Dipper, Little Dipper, and Cassiopeia visible in the night sky from Syracuse, New York, at a particular time of night. The dots represent individual stars. During the night, these stars appear to move counterclockwise around the star in the center of the chart. Straight lines are at 15-degree intervals. The stars *Caph*, *Kochab*, and *Merak* are labeled.

54 On the star chart *in your answer booklet*, circle the dot that represents the star *Polaris*. [1]

55 On the star chart *in your answer booklet*, place an **X** to indicate the location of the star *Merak* after five hours have passed. [1]

56 Identify the Earth motion that causes the apparent counterclockwise movement of these stars. [1]
_____

Base your answers to questions 57 through 59 on the table below and on your knowledge of Earth science. The table shows how many million years each group of organisms existed on Earth before they became extinct.

Existence on Earth

Group of Organisms	Duration of Existence (million years)
Ammonoids	340
Eurypterids	200
Graptolites	195
Placoderm fish	70
Trilobites	270

57 Identify the group of organisms listed on the data table that was the first group to exist on Earth. [1]

58 Identify the name of *one* specific index fossil from the eurypterid group that is found in New York State bedrock. [1]

59 Identify the type of environment on Earth where all of these groups of organisms appear to have lived. [1]
_____

Base your answers to questions 60 through 62 on the cross sections below and on your knowledge of Earth science. The cross sections represent three different stages in the development of Denali (Mt. McKinley) and the growth of the North American Plate in Alaska near the boundary with the Pacific Plate. Arrows represent the direction of plate movement.

Formation of Denali (Mt. McKinley)

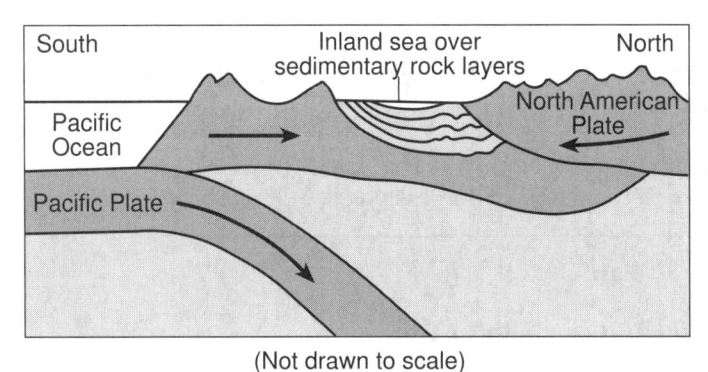

Stage 1
100 Million Years Ago:
Sedimentary rocks that would later form Denali's (Mt. McKinley's) north peak began as sediments deposited under an inland sea.

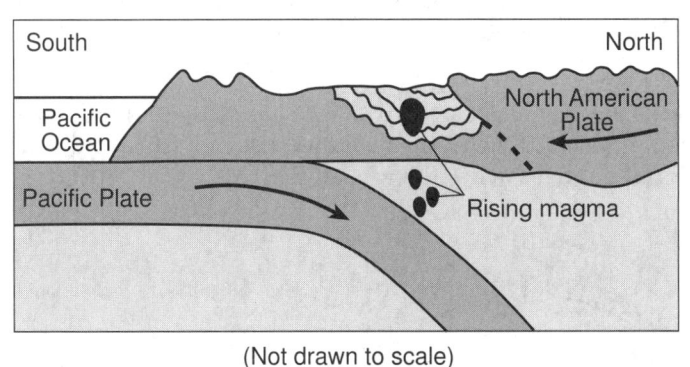

Stage 2
56 Million Years Ago:
Magma rose into the sedimentary rocks. This would later form the granite rock making up Denali's (Mt. McKinley's) south peak. Tectonic forces continued to push up the land surface.

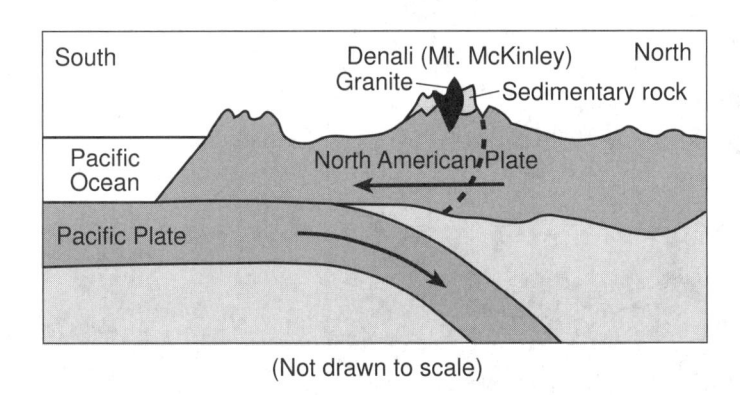

Stage 3
Today:
Tectonic forces continue to cause uplift in the region.

60 Identify the type of plate boundary represented in the cross sections. [1]

61 *In your answer booklet*, circle either volcanic or plutonic to identify the environment of formation of the granite found on Denali (Mt. McKinley). Describe the cooling rate of the magma that produced this granite. [1]

62 State the average density of the continental crust of the North American Plate and the average density of the oceanic crust of the Pacific Plate. [1]

Base your answers to questions 63 through 65 on the map in your answer booklet and on your knowledge of Earth science. The map shows the total amount of snowfall, measured in inches, from a lake-effect snow storm that affected western New York from November 17 through November 21, 2014. The 20-inch and 40-inch snowfall isolines have been drawn. Niagara Falls and Cowlesville are labeled on the map.

63 On the map *in your answer booklet*, draw the 60-inch snowfall isoline. Extend the isoline to the edge of Lake Erie. [1]

64 Cowlesville, New York, received a total of 88 inches of snow in 85 hours. Calculate the average rate of snowfall in inches per hour (in/h) for Cowlesville. [1]

65 Describe *two* actions that people could take to prepare for a forecasted lake-effect snowstorm. [1]

Part C

Answer all questions in this part.

Directions (66–85): Record your answers in the spaces provided in your answer booklet. Some questions may require the use of the *2011 Edition Reference Tables for Physical Setting/Earth Science*.

Base your answers to questions 66 through 68 on the diagram below and on your knowledge of Earth science. The diagram represents Earth as viewed from above the North Pole. The nighttime side of Earth and the Moon have been shaded. The Moon is represented in eight positions in its orbit around Earth.

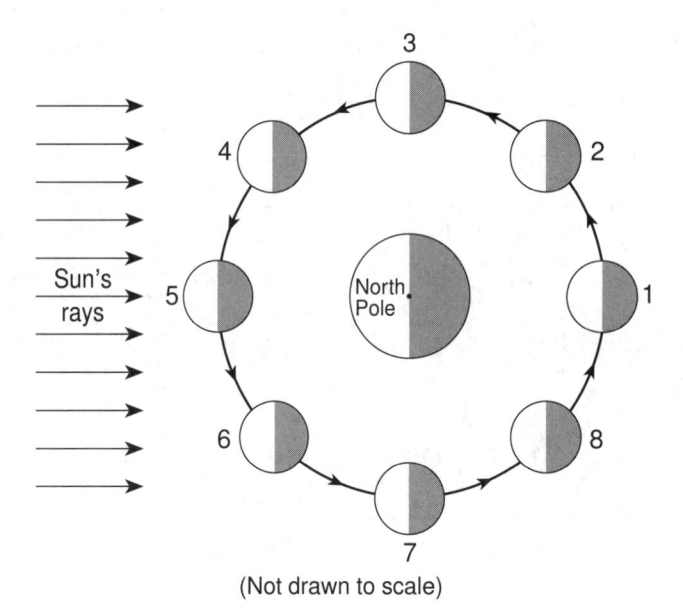

(Not drawn to scale)

66 Identify by number the Moon's position where a solar eclipse might be observed from Earth. [1]

67 The photographs *in your answer booklet* show the changing appearance of the Moon as viewed from New York State during three consecutive Moon phases. In the space below each photograph, identify the number of the Moon position that matches each of these phases. [1]

68 Explain how the Moon's rotation and revolution cause the same side of the Moon to always face Earth. [1]

Base your answers to questions 69 through 71 on the map in your answer booklet and on your knowledge of Earth science. The weather map shows isobars, recorded in millibars (mb).

69 On the map *in your answer booklet*, place an **L** to indicate the location of the center of a low-pressure system and place an **H** to indicate the location of the center of a high-pressure system. [1]

70 A weather station recorded the barometric pressure on a weather station model as shown below.

026

On the map *in your answer booklet*, place an **X** to represent a possible location for this weather station. [1]

71 The table below lists some weather conditions for another location on this map.

Temperature (°F)	Dewpoint (°F)	Precipitation (inches in past 6 hours)	Present Weather
76	74	0.85	Rain showers

On the weather station model *in your answer booklet*, using the proper format, record the weather conditions listed in the table. [1]

Base your answers to questions 72 through 75 on the topographic map in your answer booklet and on your knowledge of Earth science. The map is centered on the peak of New York State's Slide Mountain at 42° North. Points *A*, *B*, and *X* represent locations on the map. Line *AB* is a reference line on the map. Elevations are shown in feet.

72 On the map *in your answer booklet*, draw a line showing the most likely path of a stream that begins at point *X* and flows to the edge of the map. [1]

73 Determine *one* possible elevation of point *X*. [1]

74 On the grid *in your answer booklet*, construct a topographic profile along line *AB* by plotting the elevation of each contour line that crosses line *AB*. Points *A* and *B* have already been plotted. Connect *all ten* plots with a line, starting at *A* and ending at *B*, to complete the profile. [1]

75 Describe *one* piece of evidence shown on the map that indicates that the northeastern side of Slide Mountain has the steepest slope. [1]

Base your answers to questions 76 through 79 on the passage and data table below, on the graph in your answer booklet, and on your knowledge of Earth science. The data table shows the average percentages of sodium and calcium, and the average densities of samples from each of the six varieties of plagioclase feldspar. The graph in your answer booklet shows the range of sodium and calcium percentages for each of the six varieties of plagioclase feldspar.

Plagioclase Feldspars

The plagioclase feldspars are a family of six silicate minerals that are difficult to tell apart. They have the same crystal structure, cleavage, and hardness, and can be similar in color; however, they do differ slightly in chemical composition and density. The general chemical composition for plagioclase is $(Na,Ca)AlSi_3O_8$. The percentages of sodium (Na) and calcium (Ca) vary relative to each other, causing the differences in density. The mineral albite is sodium-rich, with little or no calcium, while anorthite is calcium-rich, with little or no sodium. The plagioclase feldspars with higher sodium content are more likely to be found in felsic igneous rocks, while the plagioclase feldspars with higher calcium content are more likely to be found in mafic igneous rocks.

Data Table

Variety of Plagioclase Feldspar	Average Percentage of Sodium (%)	Average Percentage of Calcium (%)	Average Density (g/cm³)
albite	100	0	2.63
oligoclase	80	20	2.65
andesine	60	40	2.67
labradorite	40	60	2.69
bytownite	20	80	2.71
anorthite	0	100	2.73

76 Complete the line graph *in your answer booklet*, by plotting the average density for the average percentages of sodium and calcium of each sample shown on the data table. The data for albite and oligoclase have been plotted for you. Connect *all six* plots with a line. [1]

77 A sample of plagioclase feldspar was found to have a ratio of 35% sodium to 65% calcium. Based on the graph, state the name of this variety of plagioclase feldspar. [1]

78 A sample of plagioclase feldspar has a mass of 534 grams and a volume of 200 cubic centimeters. State the name of this variety of plagioclase feldspar. [1]

79 State the name of *one* variety of plagioclase feldspar that is more likely to be found in the igneous rock pegmatite. [1]

Base your answers to questions 80 through 83 on the passage and the map of South America below and on your knowledge of Earth science.

Two South American Deserts

South America is an excellent example of the influence that plate tectonic features have on climates. The Andes mountain range, formed by plate tectonics, is on the western edge of South America. When prevailing winds come from the southeast, which usually occurs between 0° and 30° S latitudes, rainfall is increased on the eastern side of the mountain range. The Atacama Desert lies in the rain shadow (dry area) to the west of the mountains. Farther south, the reverse pattern is found, due to different prevailing winds blowing between 30° S and 60° S latitudes. The Patagonian Desert lies on the eastern side of the Andes, between the Andes and the South Atlantic Ocean.

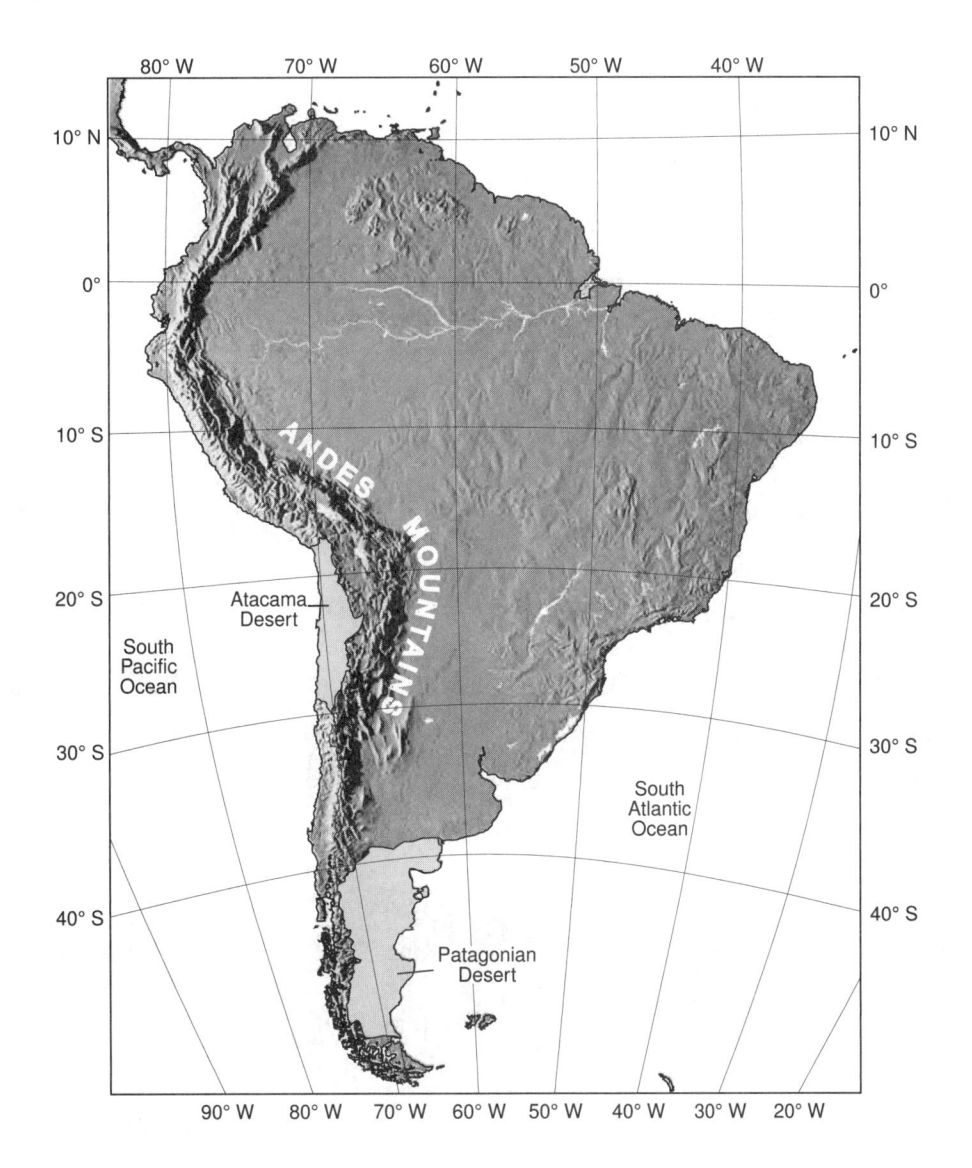

80 Name *one* tectonic plate that is interacting with the South American Plate to uplift the Andes Mountains. [1]

81 On the map *in your answer booklet*, draw *one* arrow in the box located on the Andes Mountains to indicate the surface planetary wind direction that helped produce the Atacama Desert. [1]

82 Glaciers are found on some of the mountains in the Andes near the equator. Identify *one* climate factor that causes the cold temperatures on these mountains. [1]

83 Andesite makes up much of the volcanic rock of the Andes Mountains. Name *three* minerals that are commonly found in a single andesite rock. [1]

Base your answers to questions 84 and 85 on the map and block diagrams below and on your knowledge of Earth science. The map shows a stream and its tributaries. Enlarged block diagrams, labeled *A*, *B*, and *C*, indicate the relative widths of floodplains in the rectangular areas along the stream. Points *W*, *X*, *Y*, and *Z* are locations on the stream banks.

Stream and Its Tributaries

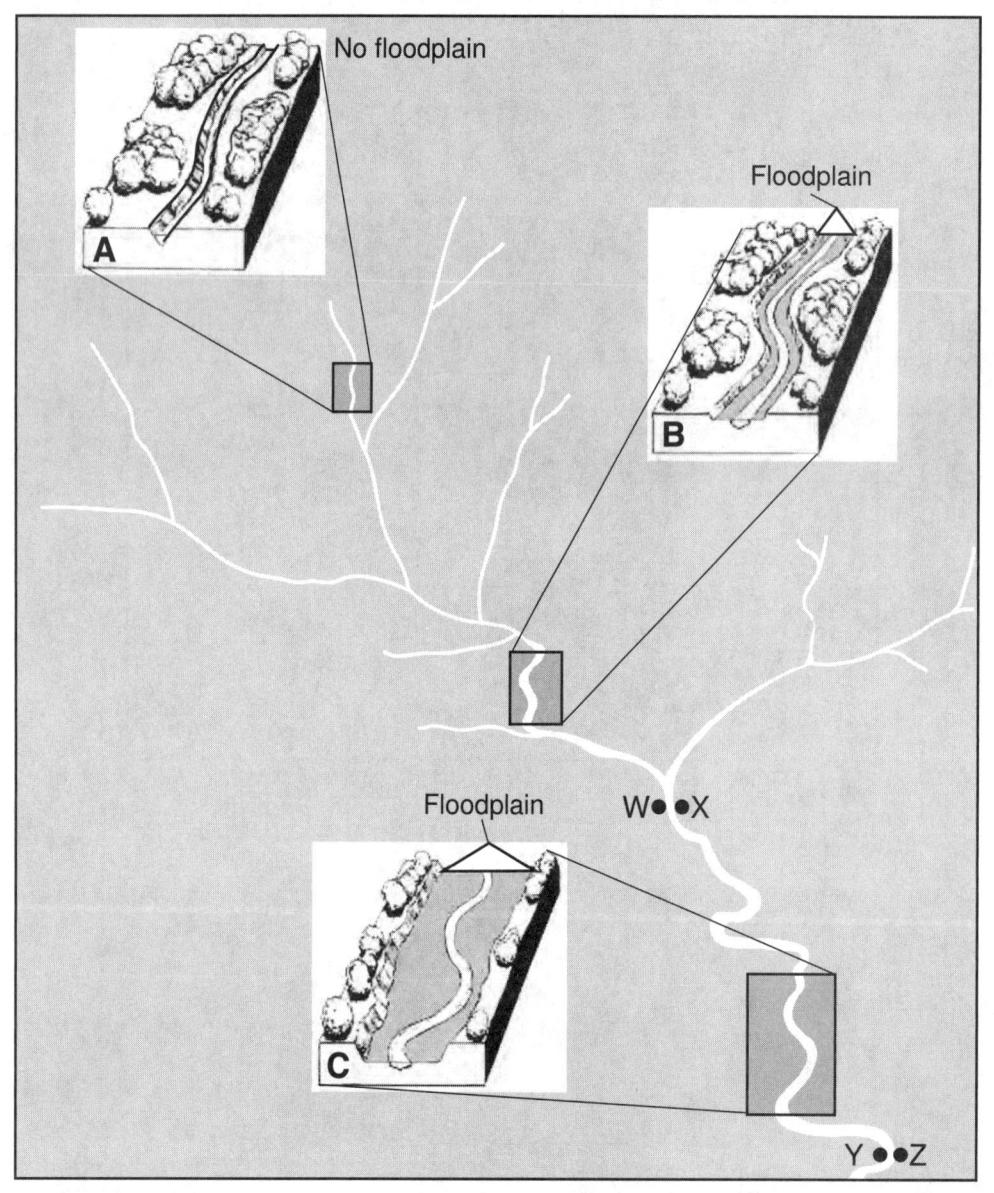

84 The slope of the stream in area *A* is steeper than the slope of the stream in area *C*. Describe *one* piece of evidence shown by the block diagrams that supports this statement. [1]

85 The cross section below represents the shape of the stream channel between *W* and *X*.

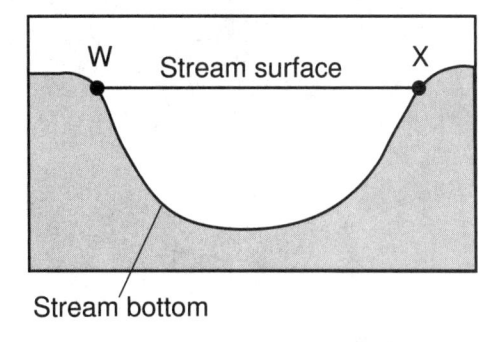

W Stream surface X
Stream bottom

On the cross section *in your answer booklet*, draw the shape of the stream bottom between *Y* and *Z*. [1]

The University of the State of New York
REGENTS HIGH SCHOOL EXAMINATION

PHYSICAL SETTING
EARTH SCIENCE

Practice Exam #1

ANSWER BOOKLET

Student .... Sex: ☐ Male ☐ Female

Teacher ....

School .... Grade ....

Record your answers for Part B–2 and Part C in this booklet.

Part B–2

51 _____

52 Circle one: terrestrial planet   Jovian planet

   Characteristic of this type of planet: _____

53 Exoplanet: _____

   Explanation: _____

   _____

Practice Exam 1   353

54-55

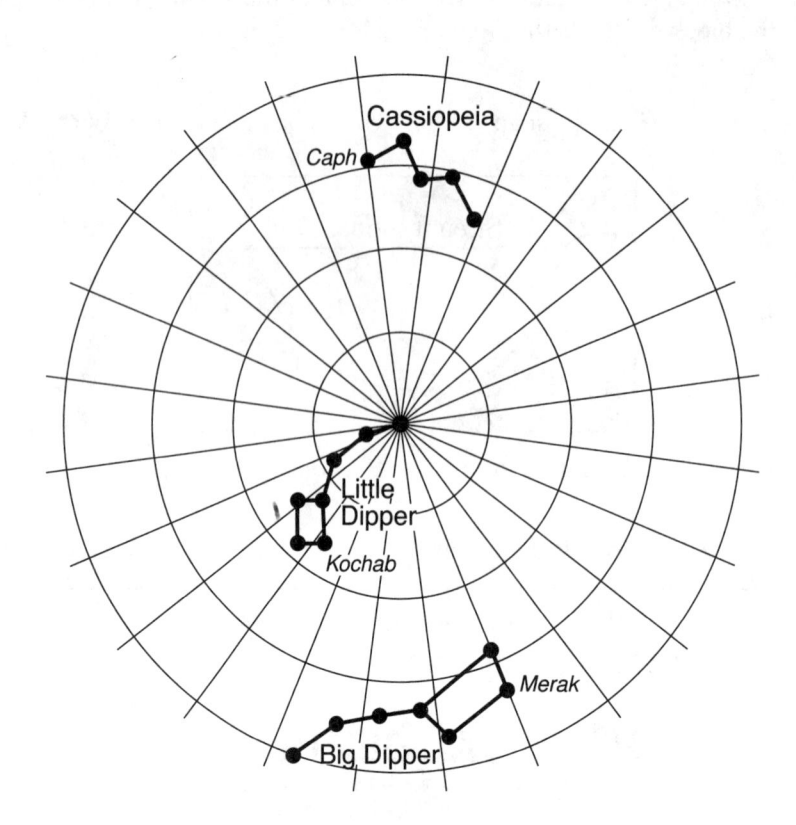

56 _____

57 _____

58 _____

59 _____

60 _____

61 Circle one: volcanic   plutonic

   Cooling rate: _____

62 North American Plate continental crust: _____ **g/cm³**

   Pacific Plate oceanic crust: _____ **g/cm³**

354   Practice Exam 1

63

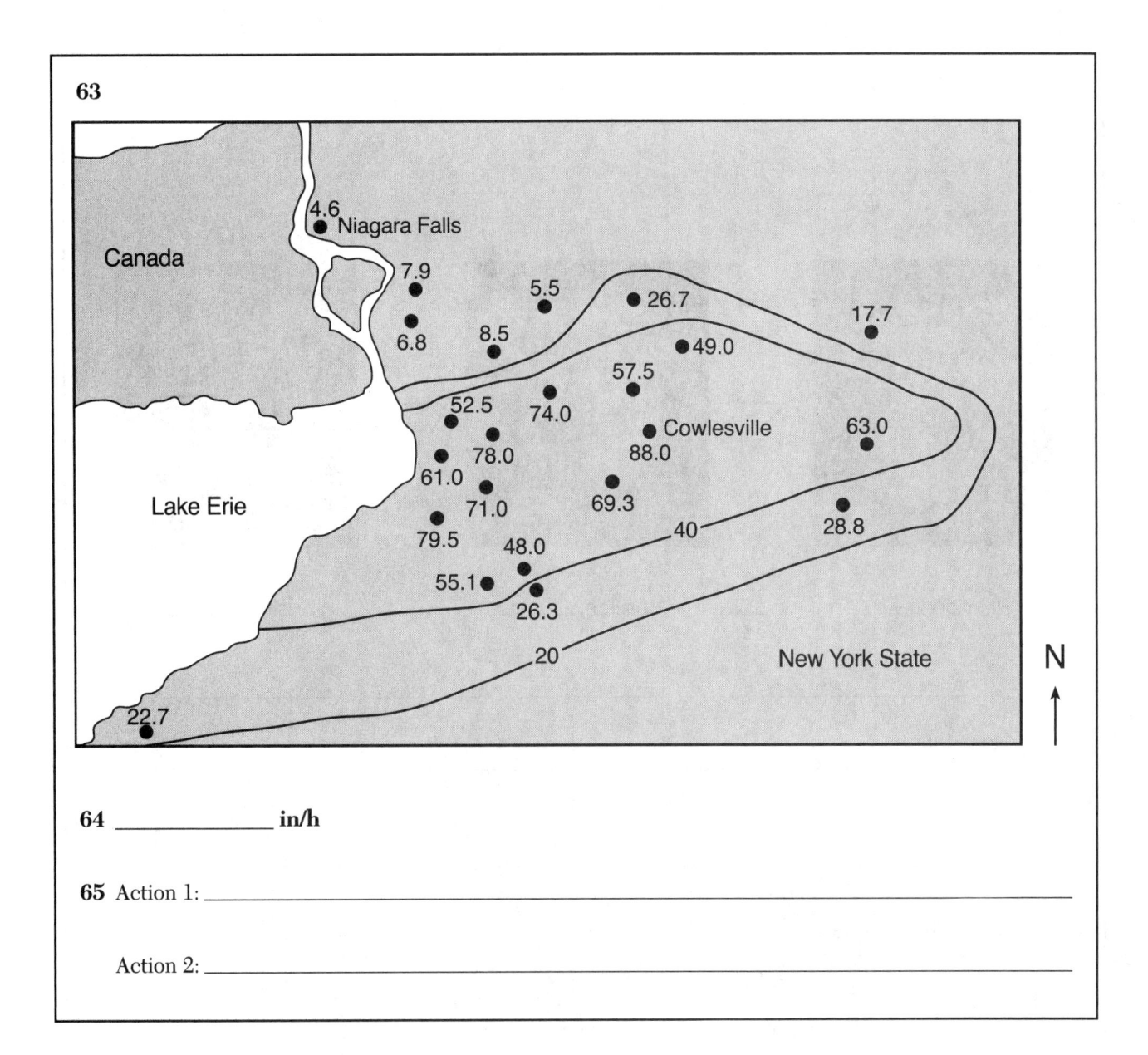

64 _____ in/h

65 Action 1: _____

Action 2: _____

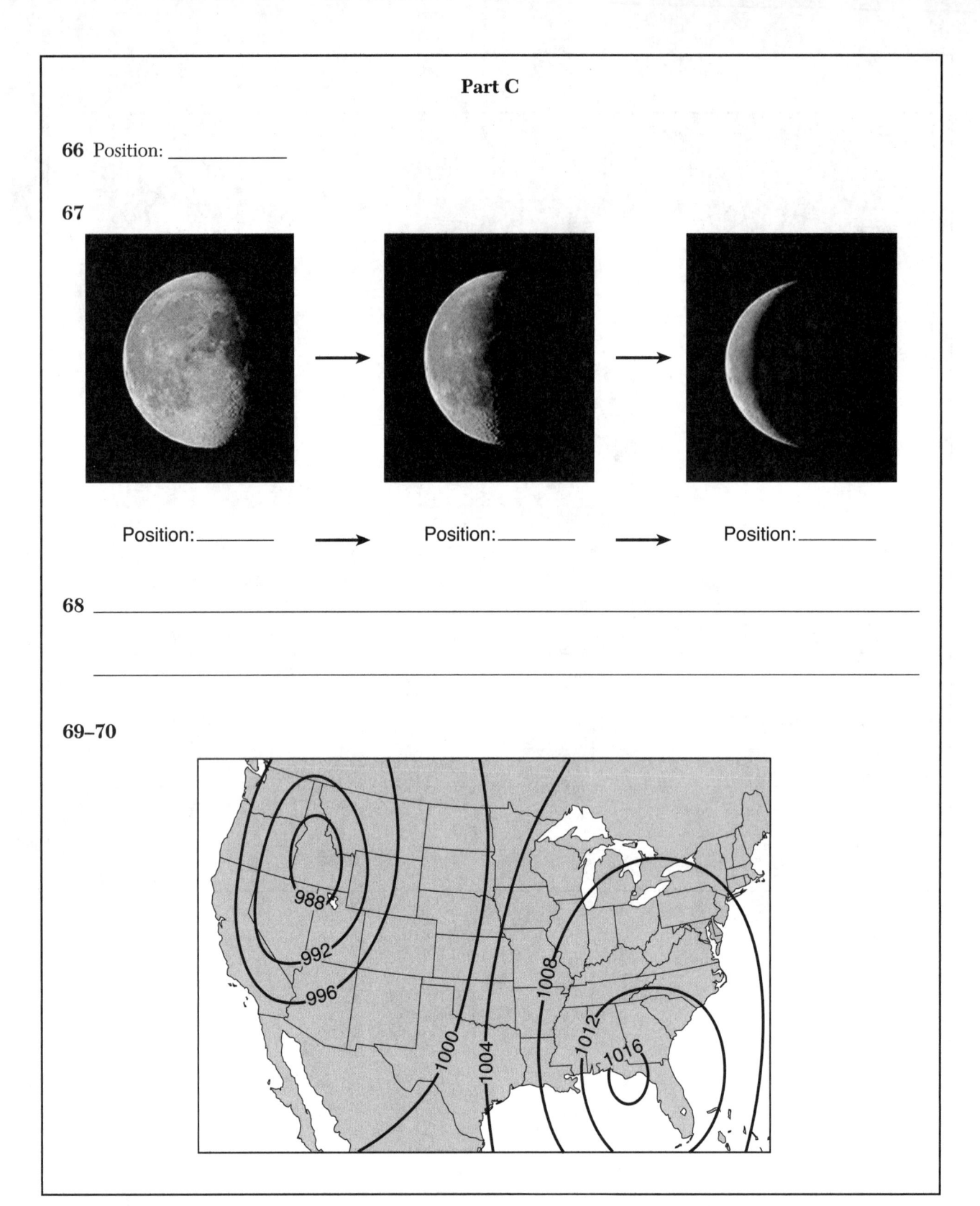

Part C

66 Position: _____

67

Position: _____ → Position: _____ → Position: _____

68 _____

_____

69–70

71

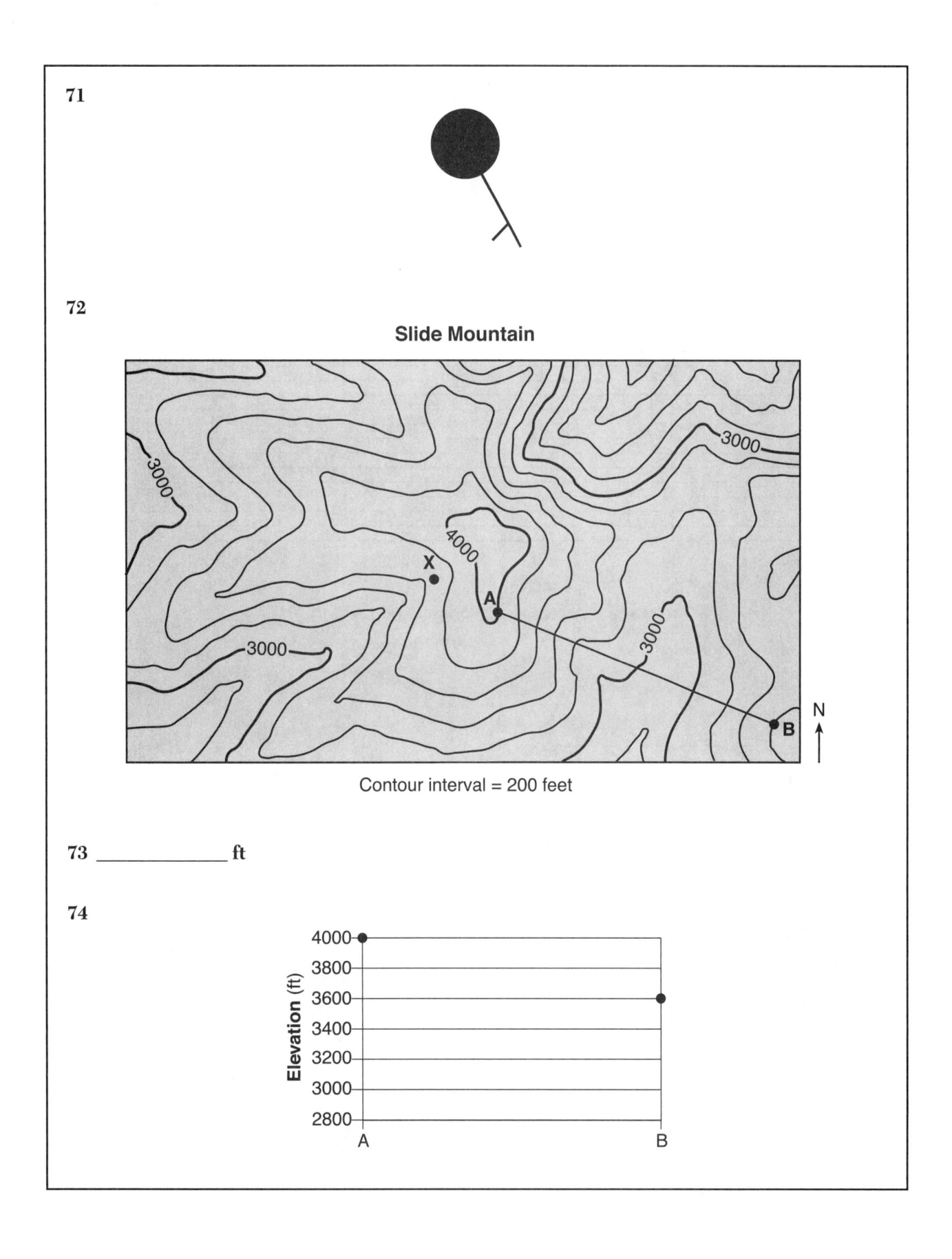

72

Slide Mountain

Contour interval = 200 feet

73 _____ ft

74

75 _____

76

Six Varieties of Plagioclase Feldspar

| Albite | Oligoclase | Andesine | Labradorite | Bytownite | Anorthite |

77 _____

78 _____

79 _____

80 _____ Plate

81

82 _____

83 Mineral 1: _____

Mineral 2: _____

Mineral 3: _____

84 Evidence: _____

85

Y • Stream surface • Z

PHYSICAL SETTING
EARTH SCIENCE
PRACTICE EXAM #2

Part A

Answer all questions in this part.

Directions (1–35): For *each* statement or question, choose the word or expression that, of those given, best completes the statement or answers the question. Some questions may require the use of the *2011 Edition Reference Tables for Physical Setting/Earth Science*. Record your answers on your separate answer sheet.

1 The photographs below show two types of solar eclipses. Letters *A* and *B* represent two celestial objects.

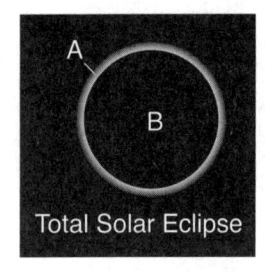

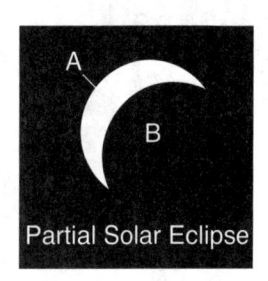

Total Solar Eclipse | Partial Solar Eclipse

Which two celestial objects are represented by letters *A* and *B*?

(1) *A*-Moon; *B*-Sun
(2) *A*-Moon; *B*-Earth
(3) *A*-Sun; *B*-Moon
(4) *A*-Sun; *B*-Earth

2 Compared to the terrestrial planets, the Jovian planets

(1) are less massive
(2) are more dense
(3) have greater orbital velocities
(4) have shorter periods of rotation

3 Which event occurred more than 10 billion years ago?

(1) Big Bang
(2) origin of life on Earth
(3) Pangaea begins to break up
(4) origin of Earth and its Moon

4 In 1851, French physicist Léon Foucault used a swinging pendulum to demonstrate that Earth

(1) is rotating
(2) is revolving
(3) has a curved surface
(4) has a gravitational pull

5 Approximately how many degrees does Earth travel in its orbit in one month?

(1) 1°
(2) 15°
(3) 30°
(4) 360°

6 What is the relative humidity when the dry-bulb temperature is 16°C and the wet-bulb temperature is 10°C?

(1) 6%
(2) 14%
(3) 33%
(4) 45%

7 Boarding up windows would be one emergency action most likely taken to prepare for which natural disaster?

(1) earthquake
(2) hurricane
(3) flood
(4) tsunami

8 Which diagram best represents the general position and direction of flow of the polar front jet stream in the Northern Hemisphere during the winter months?

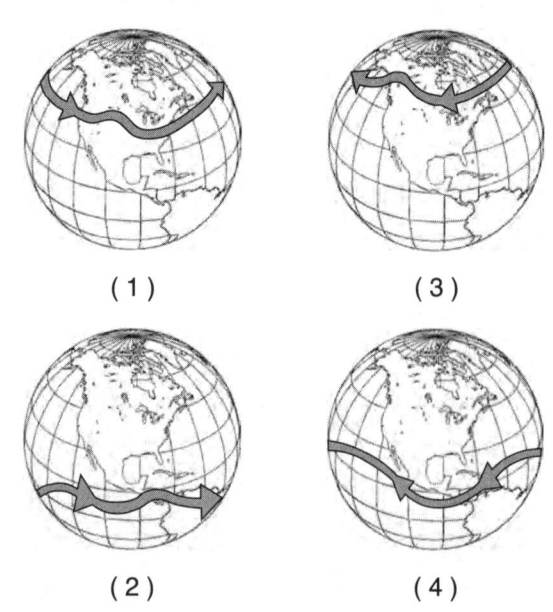

(1) (3)
(2) (4)

360 Practice Exam 2

9 The diagram below represents four positions of the Moon, labeled A through D, as it orbits Earth.

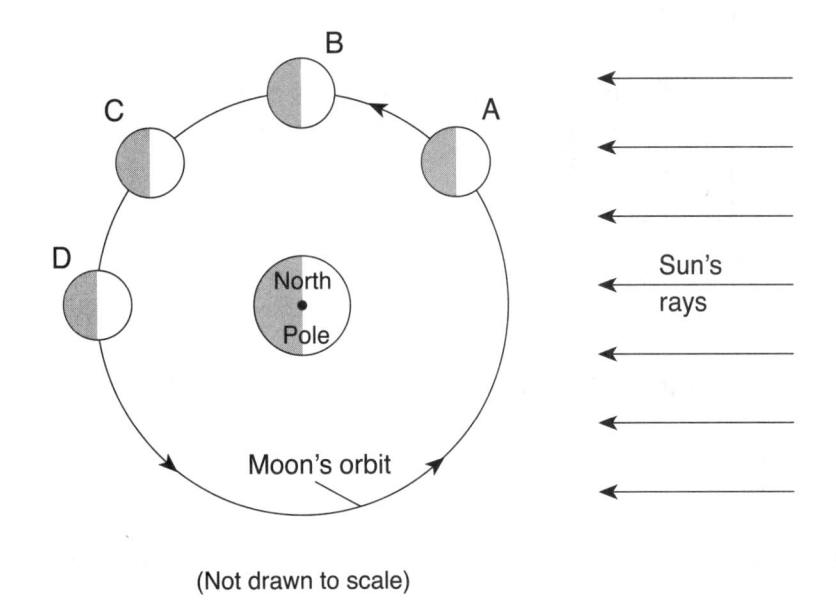

(Not drawn to scale)

Which diagram best represents the sequence of Moon phases, as seen by an observer in New York State, when the Moon travels from position A to position D in its orbit around Earth?

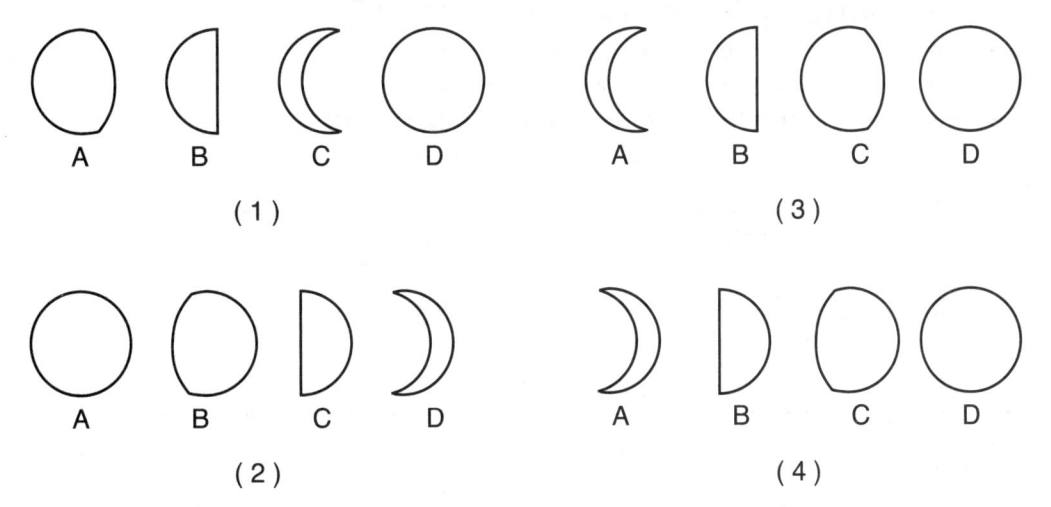

(1) A: full, B: half (left-lit), C: crescent (left), D: new
(2) A: new, B: full, C: half (right-lit), D: crescent (right)
(3) A: crescent (left), B: half (left-lit), C: full, D: full
(4) A: crescent (right), B: half (right-lit), C: new, D: full

10 The diagrams below represent spectral lines of hydrogen gas observed in a laboratory and the spectral lines of hydrogen gas observed in the light from a distant star.

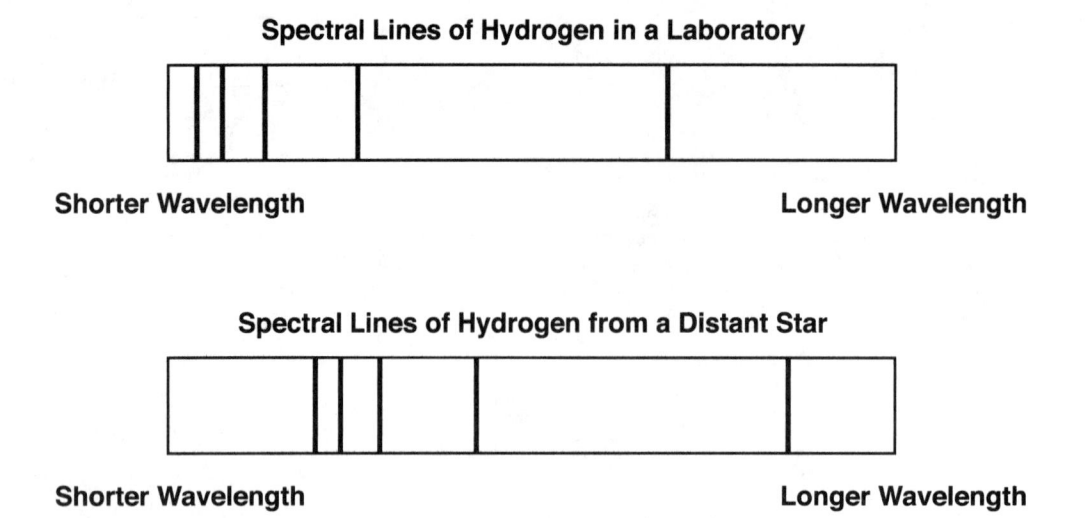

Spectral Lines of Hydrogen in a Laboratory

Shorter Wavelength — Longer Wavelength

Spectral Lines of Hydrogen from a Distant Star

Shorter Wavelength — Longer Wavelength

Compared to the spectral lines observed in the laboratory, the spectral lines observed in the light from the distant star have shifted toward the

(1) red end of the spectrum, indicating the star's movement toward Earth
(2) red end of the spectrum, indicating the star's movement away from Earth
(3) blue end of the spectrum, indicating the star's movement toward Earth
(4) blue end of the spectrum, indicating the star's movement away from Earth

11 The diagram below represents a cross-sectional view of the plane of Earth's orbit around the Sun. A line drawn perpendicular to the plane of Earth's orbit is shown on the diagram.

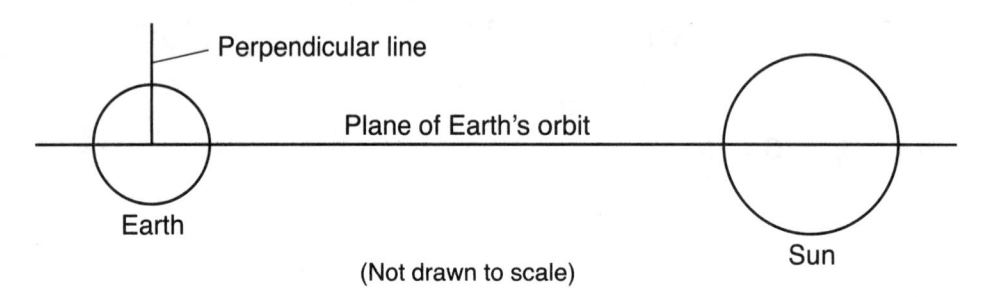

How many degrees is Earth's rotational axis tilted with respect to the perpendicular line shown in the diagram?

(1) 15°
(2) 23.5°
(3) 90°
(4) 180°

12 The larger white dots in the diagrams below represent stars in the constellations Scorpius and Orion. Information indicating when these constellations are visible from New York State is provided below the diagrams.

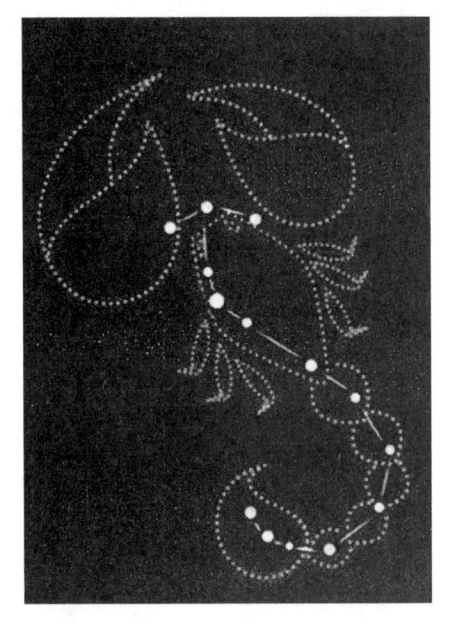

Scorpius
Visible in the New York State nighttime sky during July; not visible at all in January

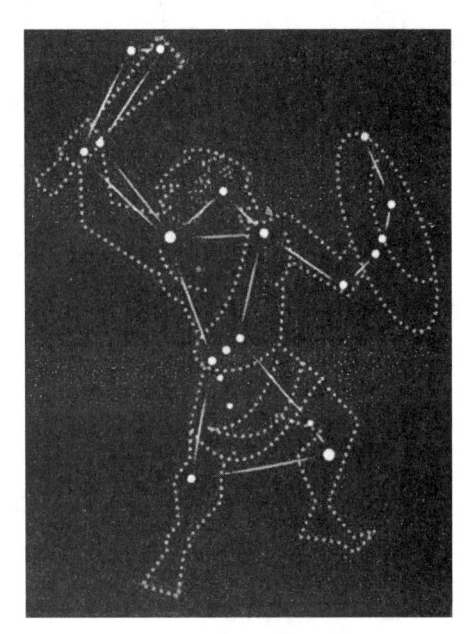

Orion
Visible in the New York State nighttime sky during January; not visible at all in July

Which statement best explains why these two constellations are visible in the night sky in the months identified?

(1) Earth spins on its axis at a constant rate during a 24-hour period.
(2) Earth spins on its axis at a variable rate during the year.
(3) The nighttime side of Earth is facing different parts of our galaxy as Earth orbits the Sun.
(4) The nighttime side of Earth is facing different parts of our galaxy as the stars orbit Earth.

13 Which table correctly shows the interior temperature, melting point, and state (phase) of matter of the materials located 4000 kilometers below Earth's surface?

Interior Temperature (°C)	Melting Point (°C)	State of Matter
5700	5400	solid

(1)

Interior Temperature (°C)	Melting Point (°C)	State of Matter
5400	5700	solid

(3)

Interior Temperature (°C)	Melting Point (°C)	State of Matter
5700	5400	liquid

(2)

Interior Temperature (°C)	Melting Point (°C)	State of Matter
5400	5700	liquid

(4)

14 Which gas is a greenhouse gas that has increased in Earth's atmosphere partly as a result of deforestation over the last 100 years?

(1) ozone
(2) oxygen
(3) nitrogen
(4) carbon dioxide

15 Which ocean current brings warm water to the southeastern tip of Africa?

(1) Brazil Current
(2) Agulhas Current
(3) Guinea Current
(4) Benguela Current

16 Which pie graph is shaded to best represent the approximate percentage of time that humans have existed during Earth's entire history?

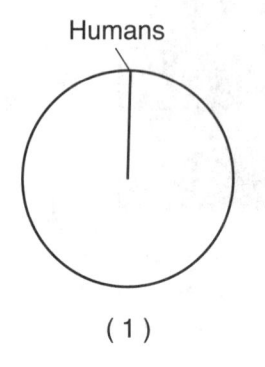

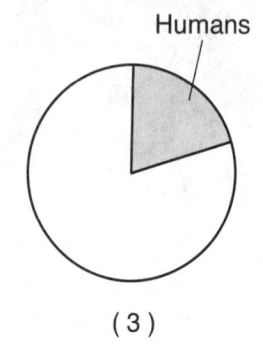

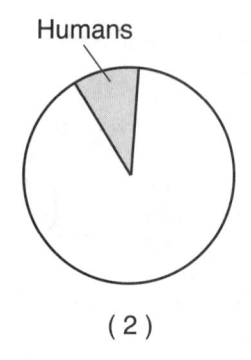

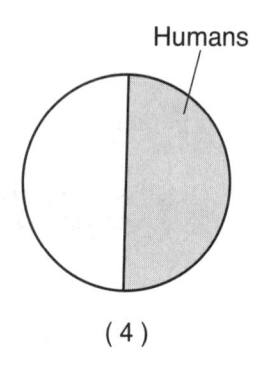

17 Volcanic ash can be used as a time marker to correlate rock layers because the ash

(1) is deposited rapidly over a large area
(2) represents a buried erosional surface
(3) forms intrusive igneous rock
(4) cuts across rock layers

18 The cross section below represents a mountain range. Points A and B represent locations on Earth's surface.

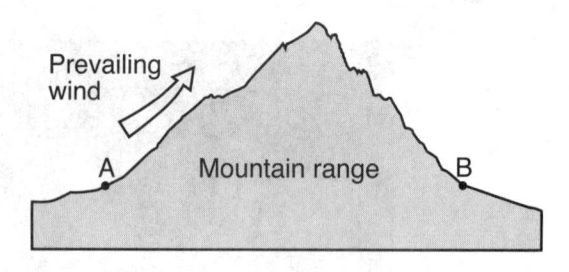

Compared to the climate of location A, the climate of location B is most likely

(1) cooler and wetter
(2) cooler and drier
(3) warmer and wetter
(4) warmer and drier

19 The photograph below shows conglomerate composed of pebbles cemented together with calcite.

Compared to the ages of the calcite cement and the conglomerate, the relative age of the pebbles is

(1) younger than both the calcite cement and the conglomerate
(2) younger than the calcite cement, but the same age as the conglomerate
(3) older than both the calcite cement and the conglomerate
(4) older than the calcite cement, but the same age as the conglomerate

20 The cross section below represents some parts of Earth's water cycle. Letters A, B, C, and D represent processes that occur during the cycle.

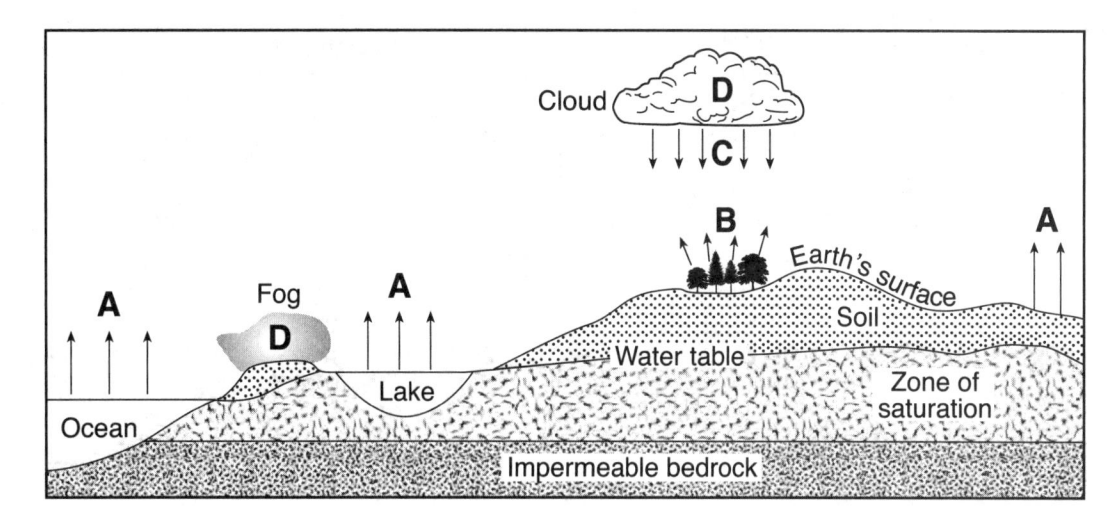

Which table correctly matches each letter with the process it represents?

Letter	Process
A	Condensation
B	Transpiration
C	Precipitation
D	Evaporation

(1)

Letter	Process
A	Condensation
B	Evaporation
C	Precipitation
D	Transpiration

(3)

Letter	Process
A	Evaporation
B	Precipitation
C	Transpiration
D	Condensation

(2)

Letter	Process
A	Evaporation
B	Transpiration
C	Precipitation
D	Condensation

(4)

21 Which table best shows the relationship between latitude and general climate conditions on Earth?

Latitude	Climate Conditions
90°N	Arid
60°N	Arid
30°N	Humid
0°	Humid
30°S	Humid
60°S	Arid
90°S	Arid

(1)

Latitude	Climate Conditions
90°N	Arid
60°N	Humid
30°N	Arid
0°	Humid
30°S	Arid
60°S	Humid
90°S	Arid

(2)

Latitude	Climate Conditions
90°N	Humid
60°N	Arid
30°N	Humid
0°	Humid
30°S	Humid
60°S	Arid
90°S	Humid

(3)

Latitude	Climate Conditions
90°N	Humid
60°N	Arid
30°N	Humid
0°	Arid
30°S	Humid
60°S	Arid
90°S	Humid

(4)

22 The photograph below shows different-sized rounded sediment.

Which table shows the most likely process and agent of erosion responsible for this rounded sediment?

Process	Agent of Erosion
sandblasting	running water

(1)

Process	Agent of Erosion
land slide	mass movement

(3)

Process	Agent of Erosion
abrasion	wave action

(2)

Process	Agent of Erosion
deposition	wind

(4)

Practice Exam 2

23 The photograph below shows an outcrop with two basaltic intrusions, labeled A and B, in a rock unit, labeled C.

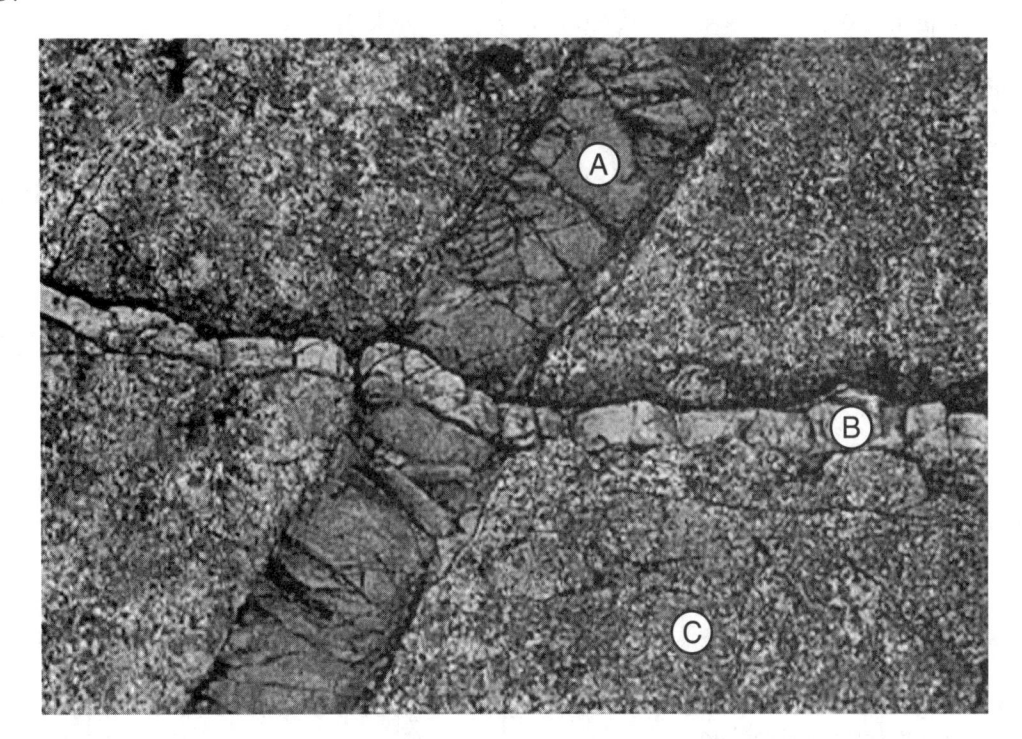

What is the relative age of these three rock units from oldest to youngest?
(1) B → A → C
(2) B → C → A
(3) C → A → B
(4) C → B → A

24 The world map below shows Earth's major tectonic plate boundaries. Letters A through D represent four surface locations.

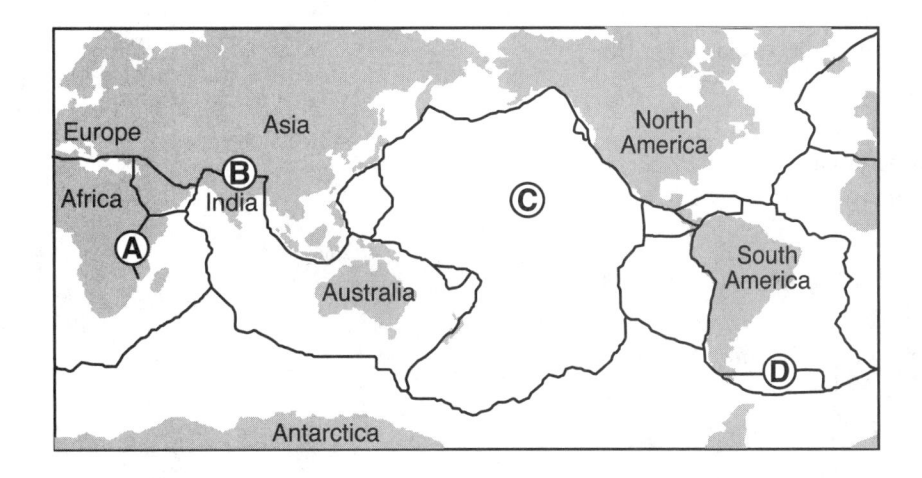

Which location is on a major rift valley?
(1) A
(2) B
(3) C
(4) D

Practice Exam 2 367

25 The first *P*-wave of an earthquake took 11 minutes to travel to a seismic station from the epicenter of the earthquake. What is the seismic station's distance to the epicenter of the earthquake and how long did it take for the first *S*-wave to travel that distance?

(1) Distance to epicenter: 3350 km
    S-wave travel time: 4 min 50 sec
(2) Distance to epicenter: 3350 km
    S-wave travel time: 6 min 10 sec
(3) Distance to epicenter: 7600 km
    S-wave travel time: 9 min
(4) Distance to epicenter: 7600 km
    S-wave travel time: 20 min

26 The Catskills are commonly called mountains, but are actually part of the Allegheny Plateau. The Catskills are classified as a plateau because of their

(1) low elevation
(2) bedrock structure
(3) bedrock age
(4) high degree of metamorphism

27 The minimum stream velocity necessary to transport a sediment particle that is 0.1 centimeter in diameter is closest to

(1) 0.1 cm/s   (3) 5.5 cm/s
(2) 0.002 cm/s   (4) 10.0 cm/s

28 Which rock is classified as an evaporite?

(1) clastic shale
(2) foliated phyllite
(3) nonfoliated marble
(4) crystalline rock salt

29 Which pair of elements makes up most of Earth's crust by volume?

(1) nitrogen and potassium
(2) oxygen and silicon
(3) hydrogen and oxygen
(4) potassium and oxygen

30 The cross section below represents zones of soil labeled *A*, *B*, and *C*. Letter *D* represents underlying bedrock.

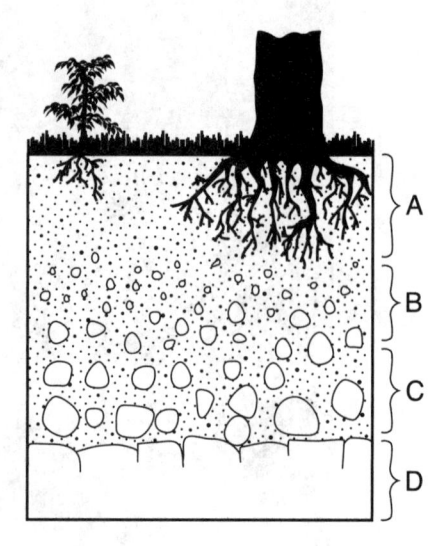

Which letter identifies the zone having the most organic and weathered material?

(1) *A*   (3) *C*
(2) *B*   (4) *D*

31 Which type of surface bedrock is most commonly found in the Utica, New York area?

(1) sedimentary, with limestone, shale, sandstone, and dolostone
(2) sedimentary, with limestone, shale, sandstone, and conglomerate
(3) metamorphic, with quartzite, dolostone, marble, and schist
(4) metamorphic, with gneiss, quartzite, marble, and slate

32 The diagram below represents a geologic landscape.

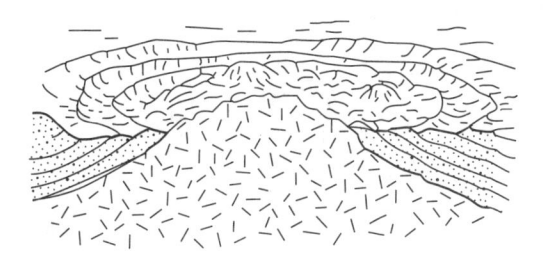

Which type of stream drainage pattern formed on this landscape?

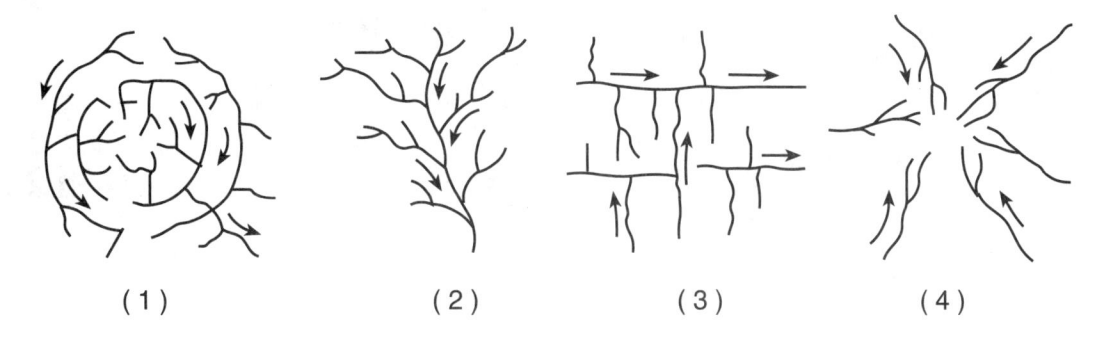

(1)   (2)   (3)   (4)

33 The north polar view maps below show the average area covered by Arctic Sea ice in September of 1980, 2000, and 2011.

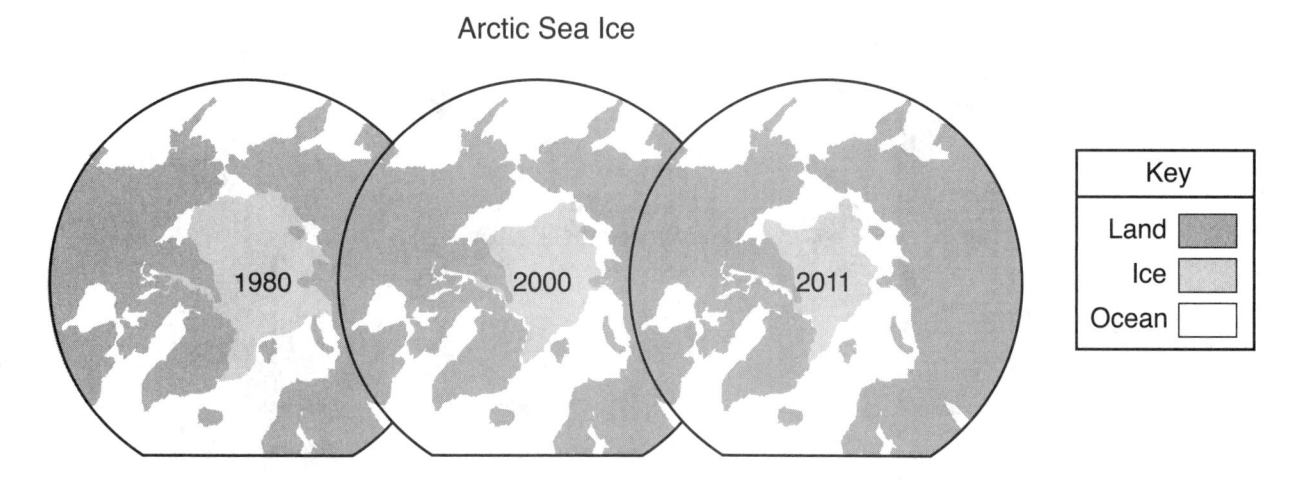

Arctic Sea Ice

Key
Land
Ice
Ocean

The maps best support the inference that Earth's climate is
(1) cooling, because the average area covered by Arctic Sea ice is decreasing
(2) cooling, because the average area covered by Arctic Sea ice is increasing
(3) warming, because the average area covered by Arctic Sea ice is decreasing
(4) warming, because the average area covered by Arctic Sea ice is increasing

34 The photographs below shows two depositional features labeled A and B.

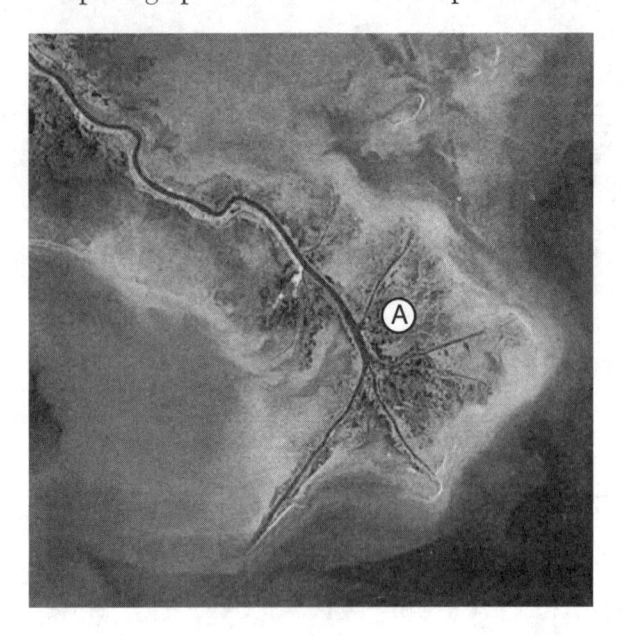

Which terms correctly identify depositional features A and B?
(1) A-delta; B-barrier island
(2) A-sand bar; B-island arc
(3) A-barrier island; B-delta
(4) A-island arc; B-sand bar

35 Diagrams A and B represent magnified views of the arrangement of mineral crystals in a rock before and after being subjected to geologic processes.

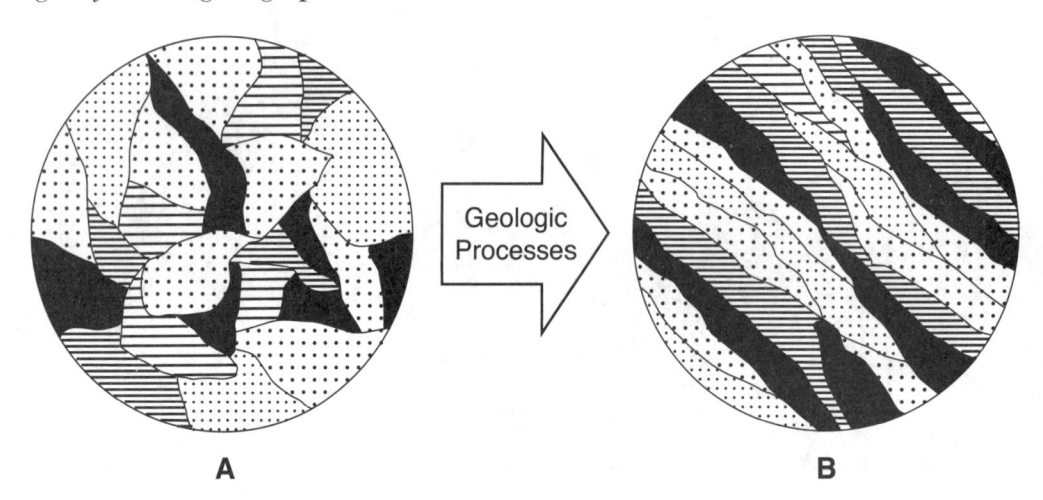

Which geologic processes are most likely responsible for the banding and alignment of mineral crystals represented in diagram B?
(1) melting and solidification
(2) heating and increasing pressure
(3) compaction and cementation
(4) weathering and erosion

Part B–1

Answer all questions in this part.

Directions (36–50): For *each* statement or question, choose the word or expression that, of those given, best completes the statement or answers the question. Some questions may require the use of the *2011 Edition Reference Tables for Physical Setting/Earth Science*. Record your answers on your separate answer sheet.

Base your answers to questions 36 through 39 on the graph below and on your knowledge of Earth science. The graph shows the observed water levels, in feet (ft), for a tide gauge located at Montauk, New York, on the easternmost end of Long Island, from January 24, 2008 to noon on January 25, 2008.

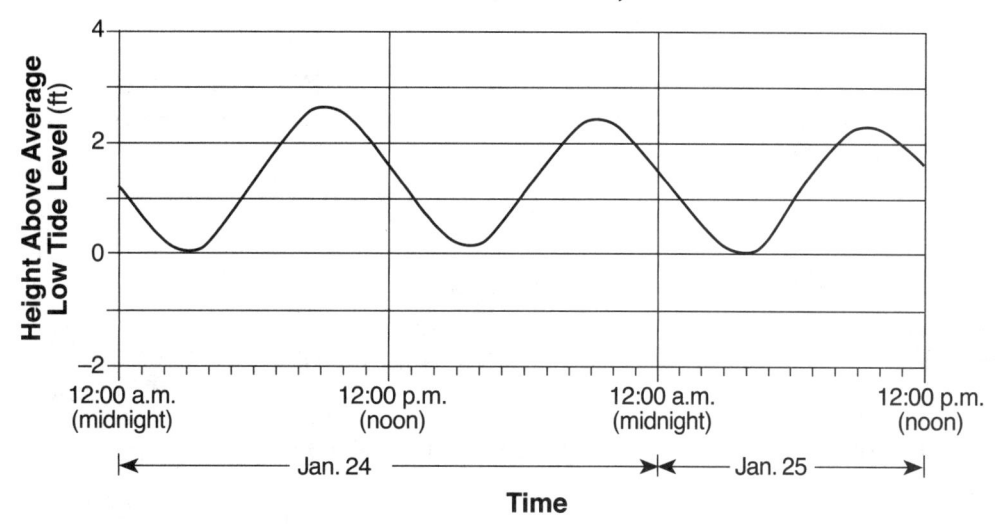

Water Levels at Montauk, New York

36 What was the height of the water above average low tide level at noon on January 24?
 (1) 1.2 ft
 (2) 1.6 ft
 (3) 2.2 ft
 (4) 2.6 ft

37 These changing water levels at Montauk can best be described as
 (1) cyclic and predictable
 (2) cyclic and not predictable
 (3) noncyclic and predictable
 (4) noncyclic and not predictable

38 What causes the water-level variation pattern shown by the graph?
 (1) changes in wind velocity produced by coastal storms
 (2) changes in magnetic orientation of the North American Plate
 (3) Earth's revolution and the distance from the equator
 (4) Earth's rotation and the gravitational pull of the Moon

39 What is the approximate latitude and longitude of the tide gauge?
 (1) 40°30′ N 72°00′ W
 (2) 40°30′ N 74°00′ W
 (3) 41°00′ N 72°00′ W
 (4) 41°00′ N 74°00′ W

Base your answers to questions 40 through 42 on the diagram below and on your knowledge of Earth science. The diagram represents some of the inferred stages in the life cycle of stars according to their original mass.

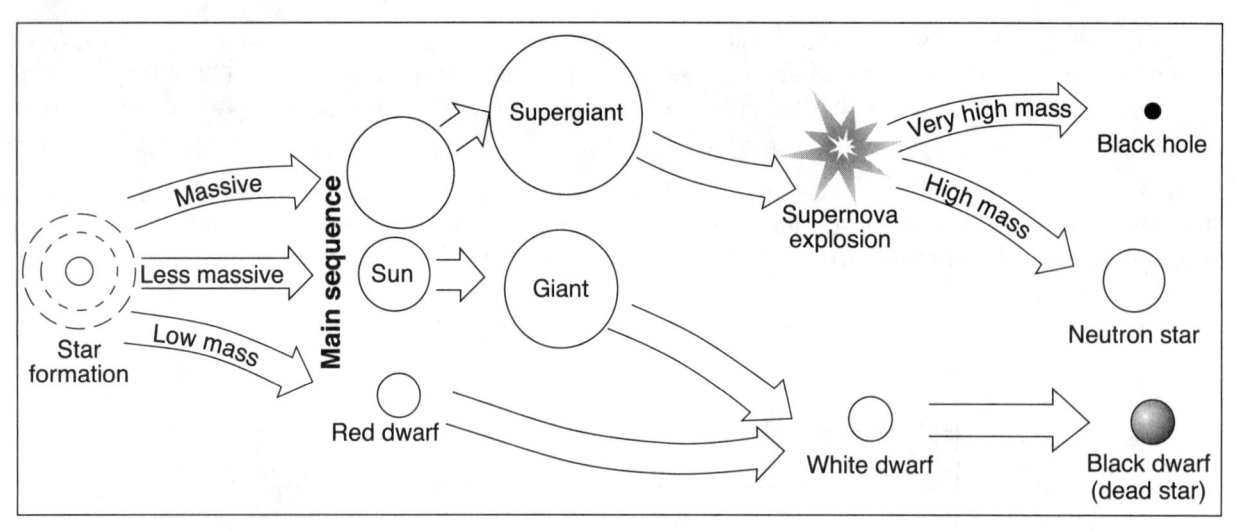

(Not drawn to scale)

40 The final stage in the life cycle of the most massive stars is a
(1) black hole
(2) black dwarf
(3) supergiant
(4) white dwarf

41 Which star may once have been similar to our Sun in mass and luminosity?
(1) *Deneb*
(2) *Spica*
(3) *Procyon B*
(4) *Proxima Centauri*

42 Energy is produced in the cores of main sequence stars when
(1) lighter elements undergo fusion into heavier elements
(2) heavier elements undergo fusion into lighter elements
(3) cosmic background radiation is absorbed
(4) cosmic background radiation is released

Base your answers to questions 43 and 44 on the graph below and on your knowledge of Earth science. The graph shows the number of radioactive Isotope X atoms present as a sample of the isotope undergoes radioactive decay.

Radioactive Decay of Isotope X

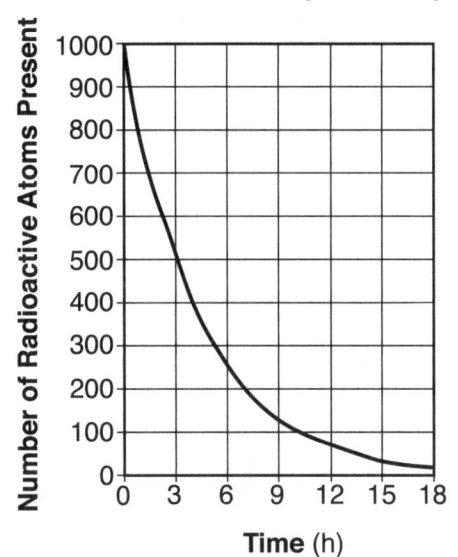

43 Based on the graph, the half-life of this radioactive isotope is
   (1) 6 h
   (2) 9 h
   (3) 3 h
   (4) 12 h

44 Based on the graph, what is the approximate number of radioactive atoms of Isotope X that are present when 8 hours of decay has occured?
   (1) 90
   (2) 115
   (3) 155
   (4) 200

Base your answers to questions 45 through 47 on the diagram below and on your knowledge of Earth science. The arrows in the diagram show air movement in a thunderstorm cloud. Point A represents a location in the atmosphere.

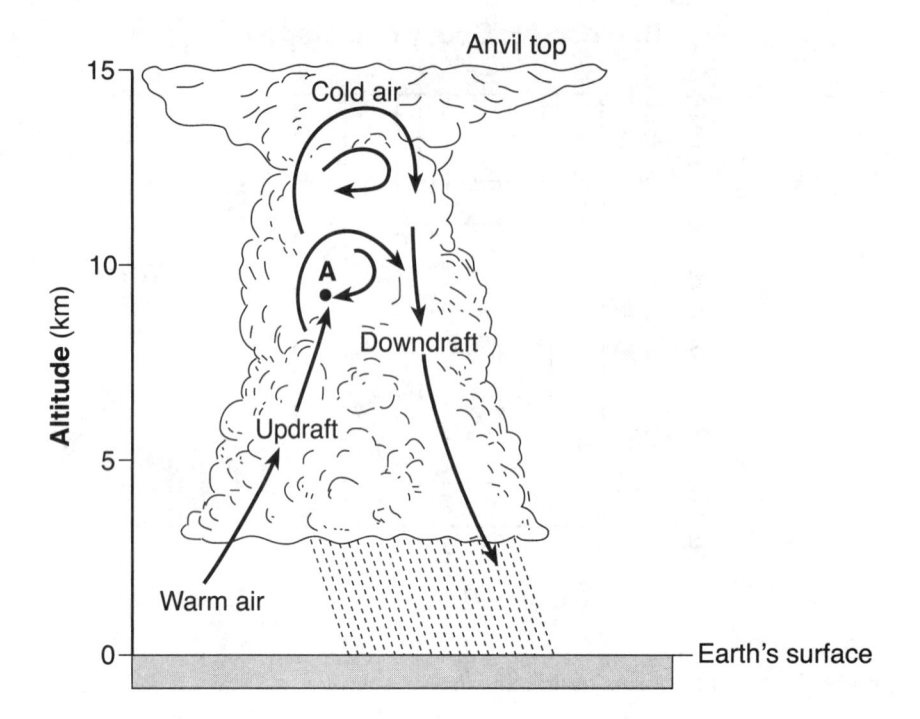

45 In which temperature zone of the atmosphere is point A located?
(1) thermosphere
(2) mesosphere
(3) stratosphere
(4) troposphere

46 The updrafts and downdrafts represented within this cloud are primarily caused by differences in
(1) altitude above sea level
(2) air density
(3) relative humidity
(4) specific heat

47 Which weather symbol would be placed on a station model to represent this weather event?

(1) (2) (3) (4)

Base your answers to questions 48 through 50 on the topographic map below and on your knowledge of Earth science. On the map, points A, B, C, and D represent surface locations. The dashed line between points C and D represents a hiking trail. Elevations are in feet (ft).

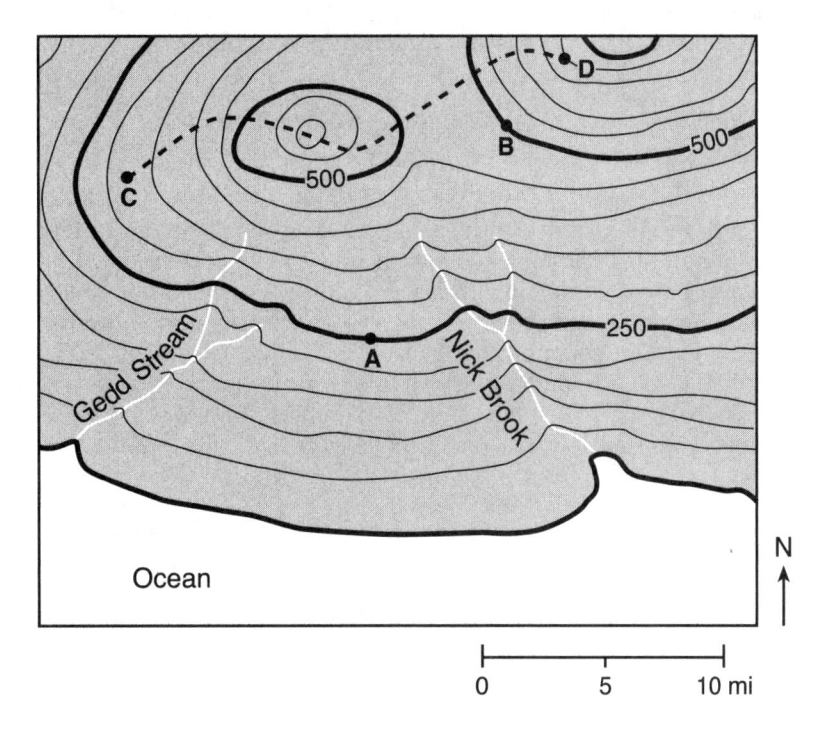

48 What is the contour interval on this map?
(1) 25 ft
(2) 50 ft
(3) 150 ft
(4) 250 ft

49 The gradient between location A and location B is approximately
(1) 0.04 ft/mi
(2) 25 ft/mi
(3) 40 ft/mi
(4) 50 ft/mi

50 A person walks along the trail from location C to location D. The person will be walking
(1) downhill then uphill, only
(2) downhill, then uphill, then downhill again
(3) uphill then downhill, only
(4) uphill, then downhill, then uphill again

Practice Exam 2   375

Part B–2

Answer all questions in this part.

Directions (51–65): Record your answers in the spaces provided in your answer booklet. Some questions may require the use of the *2011 Edition Reference Tables for Physical Setting/Earth Science*.

Base your answers to questions 51 through 54 on the passage below and on your knowledge of Earth science.

The Mica Family

The familiar term "mica" is not the name of a specific mineral, but rather the name for a family of more than 30 minerals that share the same properties. All members of the mica family have high melting points and are similar in density, luster, hardness, streak, type of breakage, and crystal shape. As a result, telling the micas apart can be difficult. However, some common members of the family can be identified by color. For example, biotite is black to dark brown while muscovite can be light shades of several colors, or even colorless. When less common members of the mica family have any of these colors, or have similar colors, chemical tests are needed to tell them apart.

51 Identify the *two* chemical elements present in biotite mica that are *not* present in muscovite mica. [1]

52 Identify the luster, hardness, and dominant form of breakage for members of the mica family. [1]

53 State the name of the igneous rock in which crystals of biotite mica are larger than 10 millimeters in diameter. [1]

54 Large crystals of mica, sometimes weighing several hundred tons, have been found in igneous rock in Canada. Identify the environment of formation and the relative rate of cooling of the magma that formed the igneous rock containing these large crystals. [1]

Base your answers to questions 55 through 58 on the diagram in your answer booklet and on your knowledge of Earth science. The diagram represents the Sun's apparent daily path for the first day of three seasons at 43° North latitude. The solid lines represent daytime paths as seen by an observer at this latitude. The dashed lines represent the nighttime paths that can *not* be seen by the observer.

55 On the diagram *in your answer booklet*, draw an **X** to represent the solar noon position of the Sun as seen by the observer on April 21. [1]

56 Identify the rate of the Sun's apparent movement, in degrees per hour, along its path on December 21. [1]

57 Identify the compass direction toward which the observer's shadow would point at solar noon on March 21. [1]

58 List the three dates shown on the diagram from the least number of nighttime hours to the greatest number of nighttime hours. [1]

Base your answers to questions 59 through 62 on the information below, on the map in your answer booklet, and on your knowledge of Earth science. The map shows a portion of the tectonic plates map from the *2011 Edition Reference Tables for Physical Setting/Earth Science*. Letters A and B represent locations on the ocean floor.

    The area between North America and South America is a tectonically active region of Earth. This region contains all of the types of tectonic plate boundaries, and it has frequent earthquake and volcanic activity. The tectonic plates on either side of the East Pacific Ridge move at an average rate of 7.5 cm/year.

59 On the map *in your answer booklet*, draw *one* arrow in each of the two boxes to show the relative motion of the Caribbean Plate and the North American Plate. [1]

60 On the set of axes *in your answer booklet*, draw a line to represent the relative age of the ocean floor bedrock from location A to location B. [1]

61 Identify the name of the hot spot shown on the map, and identify the name of the tectonic plate under which the center of this hot spot is located. [1]

62 Identify the type of mafic igneous bedrock that is most likely to make up the oceanic crust at location A, and state the average density of this oceanic crust. [1]

Base your answers to questions 63 through 65 on the diagram and data table below and on your knowledge of Earth science. The diagram represents laboratory materials used for an investigation of the effects of particle diameter on permeability and porosity (percentage of pore space). Four separate plastic tubes were filled to the same level with different particles.

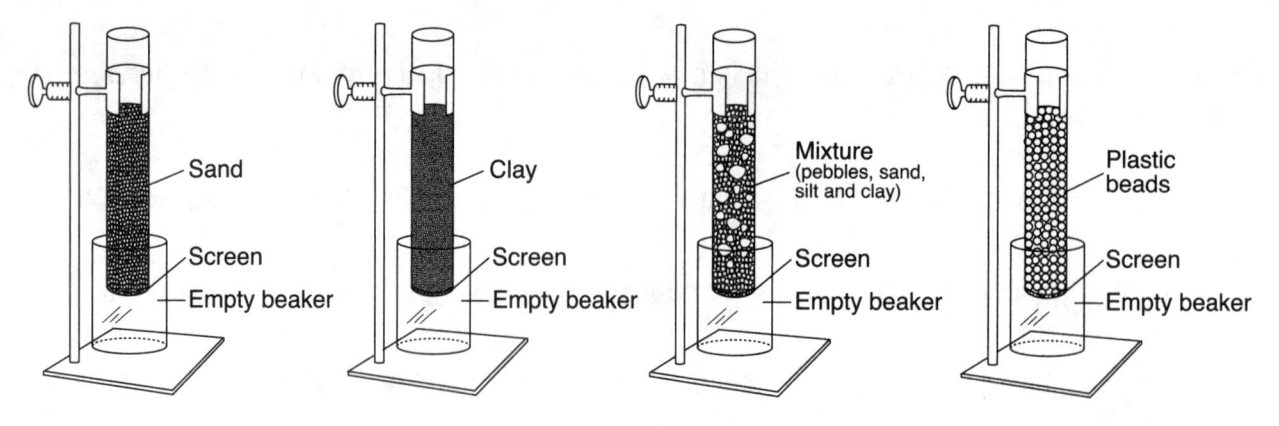

(Not drawn to scale)

Particle Type	Particle Diameter (cm)	Time for Water to Infiltrate (s)	Porosity (%)
Sand	0.1	7	42.0
Clay	0.0003	322	40.0
Mixture	from 0.0003 to 0.8	15	34.0
Plastic beads	0.4	4	44.0

63 Explain why the particle sizes fit together more closely in the mixture, resulting in the lowest porosity of all these particle types. [1]

64 The height of the column of sand is 28 centimeters. Calculate the rate of infiltration, in centimeters per second, for the water that flowed through the column of sand. [1]

65 Based on the particle diameter of the plastic beads, identify the type of sediment represented by these beads. [1]

Part C

Answer all questions in this part.

Directions (66–85): Record your answers in the spaces provided in your answer booklet. Some questions may require the use of the *2011 Edition Reference Tables for Physical Setting/Earth Science*.

Base your answers to questions 66 through 68 on the weather map below and on your knowledge of Earth science. The map shows the location of a wintertime low-pressure system over Lake Ontario with two fronts extending into New York State. Isobar values are recorded in millibars. Partial weather station data are shown for several locations.

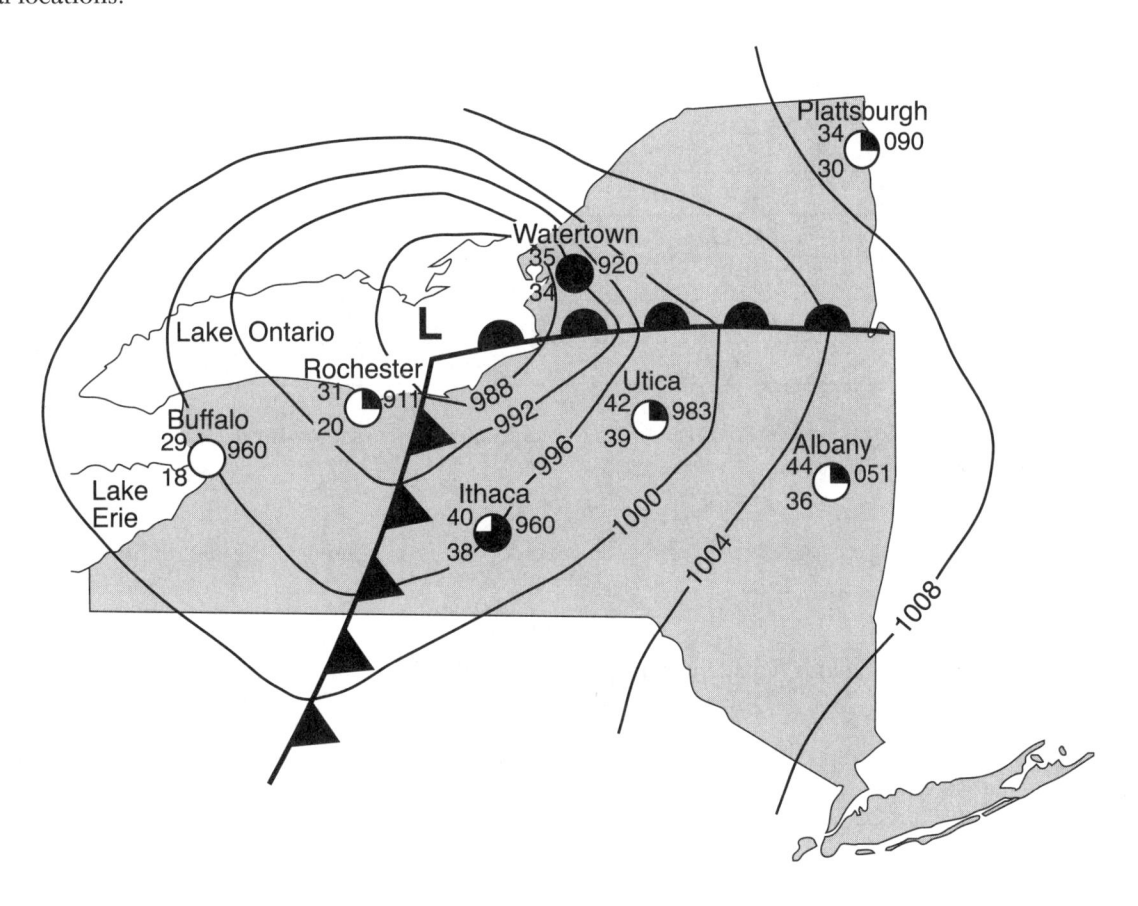

66 Describe the evidence shown on the map that indicates that the highest wind speeds occurred near Watertown, New York. [1]

67 Complete the table *in your answer booklet* by recording the weather data shown on the station model for Albany, New York. [1]

68 State the compass direction toward which the center of this low-pressure system moved over the next two days if the low followed a normal storm track. [1]

Practice Exam 2

Base your answers to questions 69 through 72 on the information and data table below and on your knowledge of Earth science. The data table shows the average body volume, including the shell, of a brachiopod at certain times in geologic history. The geologic ages are shown in million years ago (mya). The average body volumes including the shell are shown in milliliters (mL).

Cope's Rule

Cope's Rule states that the average size of animals preserved in the fossil record tends to increase as each group evolves from a previous group. This rule was first proposed in the 1800s by Edward Drinker Cope, a famous fossil hunter of that time. Recent research, involving well over 10,000 fossil groups spanning the time since the start of the Cambrian Period until today, has shown that Cope's Rule is accurate for most animal groups. Brachiopod data support Cope's Rule.

Brachiopod Data Table

Geologic Age (mya)	Average Body Volume Including the Shell (mL)
480	0.1
460	0.2
430	0.6
410	1.0
380	1.1

69 On the grid *in your answer booklet*, plot the average brachiopod body volume for each of the geologic ages listed in the data table. Connect *all five* plots with a line. [1]

70 Identify, by name, *two* geologic periods when the brachiopods represented in the data table were living. [1]

71 State the names of the *two* brachiopod index fossils found in New York State bedrock. [1]

72 The earliest horses appeared in the Eocene epoch and were about the size of a large dog of today. Explain how the evolution of horses supports Cope's Rule. [1]

Base your answers to questions 73 through 75 on the snowfall map in your answer booklet and on your knowledge of Earth science. The snowfall map shows some average yearly snowfall values, measured in inches, recorded for a portion of New York State. Some average yearly snowfall isolines have been drawn. Line XY is a reference line on the map. The cities of Watertown and Oswego are shown on the map.

73 On the map *in your answer booklet*, draw the 240-inch average yearly snowfall isoline. [1]

74 On the grid *in your answer booklet*, construct a profile of the average annual snowfall along line XY by plotting the value of each isoline that crosses line XY. Connect *all six* plots with a line to complete the profile. [1]

75 The diagram *in your answer booklet* represents an observer standing next to the side of a building. Using the scale shown, draw an **X** on the side of the building to represent the height of the greatest amount of average yearly snowfall that is indicated on the map. [1]

Base your answers to questions 76 through 78 on the diagram below and on your knowledge of Earth science. The diagram represents a cutaway view of a flat-plate solar collector used to heat water at a New York State location.

Solar Collector

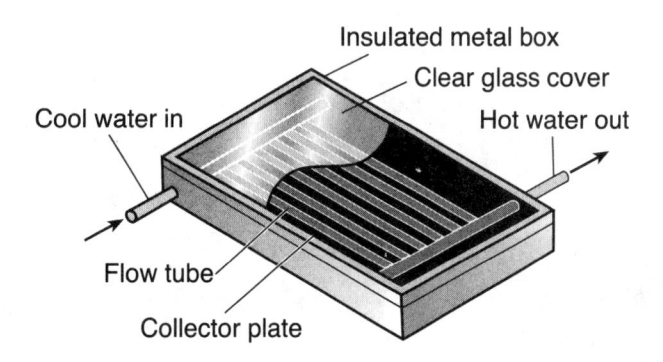

76 Identify the energy transfer process by which light travels through space from the Sun to the solar collector. [1]

77 Explain why the flow tubes and collector plate inside the solar collector are black in color. [1]

78 The glass cover on this solar collector allows visible light to enter the collector. Identify the type of electromagnetic energy emitted by the flow tubes and collector plate that is trapped inside the collector by the glass cover. Also, circle the relative wavelength of this trapped electromagnetic energy compared to wavelengths of visible light. [1]

Base your answers to questions 79 through 82 on the passage and diagram below, on the data table on the next page, and on your knowledge of Earth science. The diagram compares the inner planets of our solar system to the planetary system surrounding the star *Kepler-62*, which is located in our galaxy. The data table shows some data for the planets in the *Kepler-62* system.

Kepler-62 Planetary System

Five planets orbit a seven-billion-year-old star, *Kepler-62*, which has a surface temperature of approximately 4900 Kelvin. Two of these planets are located within the habitable zone, which is the region around a star where life may exist due to the possible presence of water in the liquid phase. The shaded areas in the orbital diagrams below indicate the habitable zone of each system.

***Kepler-62* Planetary System**

62f   62e   62d   62c   62b

Key
☐	Non habitable zone
▨	Habitable zone

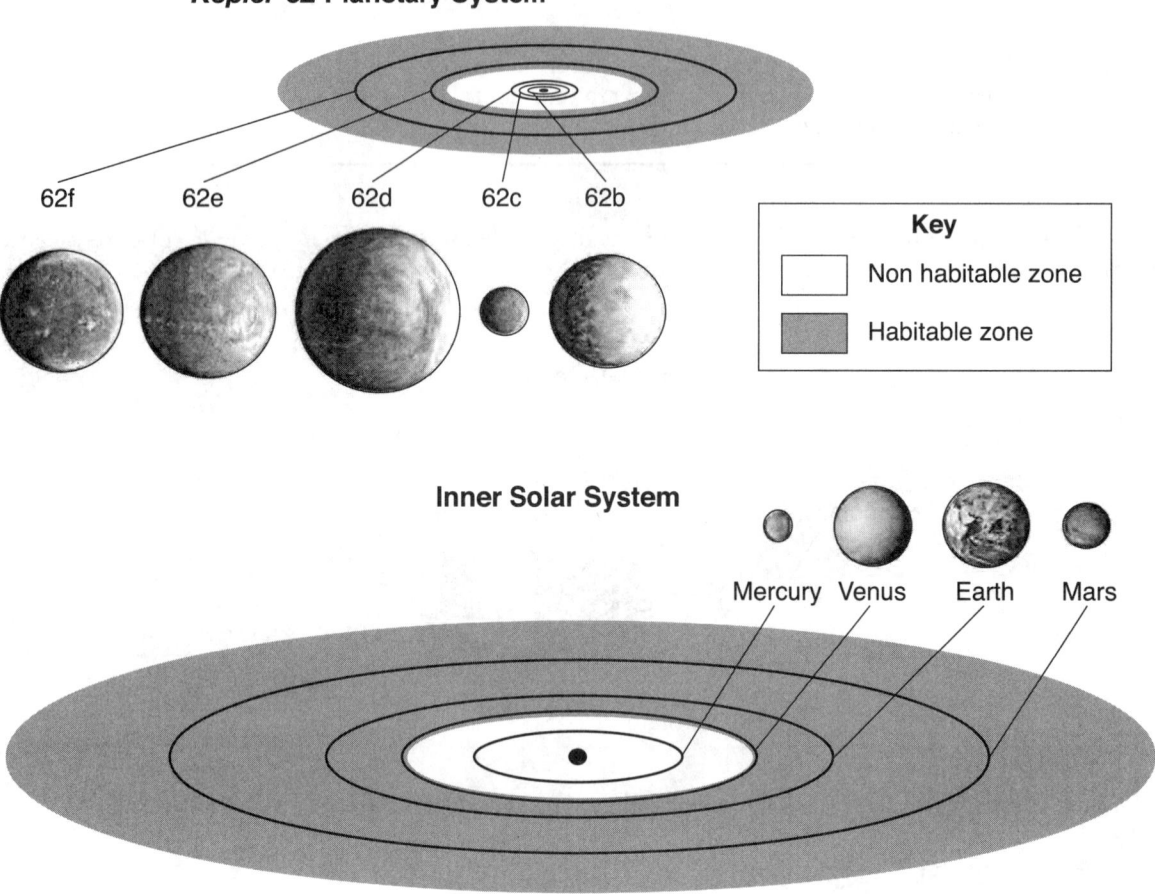

Inner Solar System

Mercury   Venus   Earth   Mars

(Planets and orbits are drawn to scale)

382 Practice Exam 2

Data Table

Name of Planet	Distance from Kepler-62 (million kilometers)	Equatorial Diameter (compared to Earth's diameter)
62b	8.23	1.31
62c	13.76	0.54
62d	17.95	1.95
62e	63.88	1.6
62f	107.41	1.4

79 Identify the name of the galaxy where the *Kepler-62* planetary system is located. [1]

80 Identify the name of the planet in our solar system that has an equatorial diameter most similar in size to the equatorial diameter of planet Kepler-62c. [1]

81 Identify the name of the planet in the *Kepler-62* planetary system that has the shortest period of revolution, and explain why this planet has the shortest period of revolution. [1]

82 Identify the names of the *two* planets in the *Kepler-62* planetary system that may have liquid water on their surfaces, and explain why these planets may have liquid water on their surfaces. [1]

Base your answers to questions 83 through 85 on the block diagram below and on your knowledge of Earth science. The diagram represents glacial features formed by a continental glacier and its melt water.

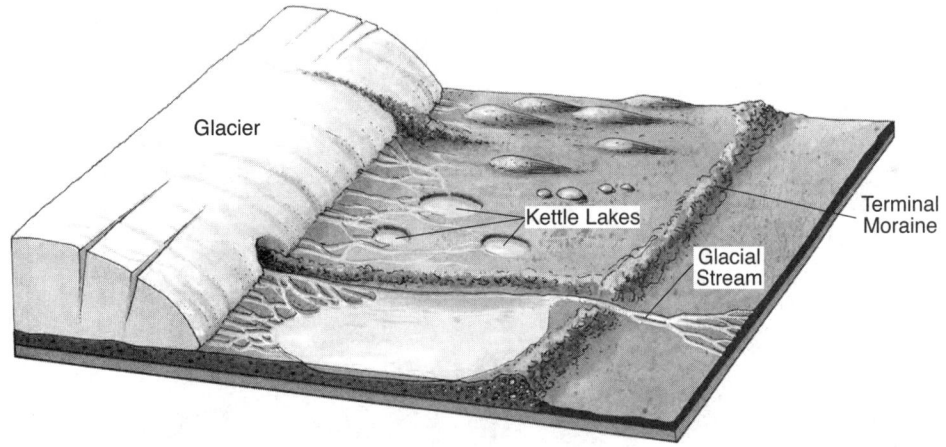

83 Describe the arrangement of the sediments found within the terminal moraine, which marks the farthest advance of the glacier. [1]

84 The cross sections below, labeled *A*, *B*, *C*, and *D*, represent four stages in the development of a kettle lake. The stages are *not* shown in the correct order.

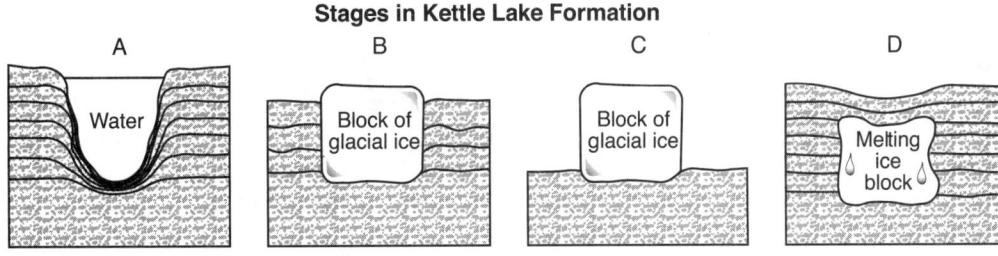

Stages in Kettle Lake Formation

In your answer booklet, place the letters in the correct order to indicate the sequence of development of a kettle lake from earliest stage to latest stage. [1]

85 Terminal moraines found on Long Island were deposited during the advance and retreat of glacial ice during the last ice age. Identify, by name, the geologic epoch during which these moraines were deposited. [1]

The University of the State of New York

REGENTS HIGH SCHOOL EXAMINATION

PHYSICAL SETTING
EARTH SCIENCE

Practice Exam #2

ANSWER BOOKLET

Student ....

Teacher ....

School .... Grade ....

Record your answers for Part B–2 and Part C in this booklet.

Part B–2

51 _____ and _____

52 Luster: _____

   Hardness: _____

   Dominant form of breakage: _____

53 _____

54 Environment of formation: _____

   Relative rate of cooling: _____

55

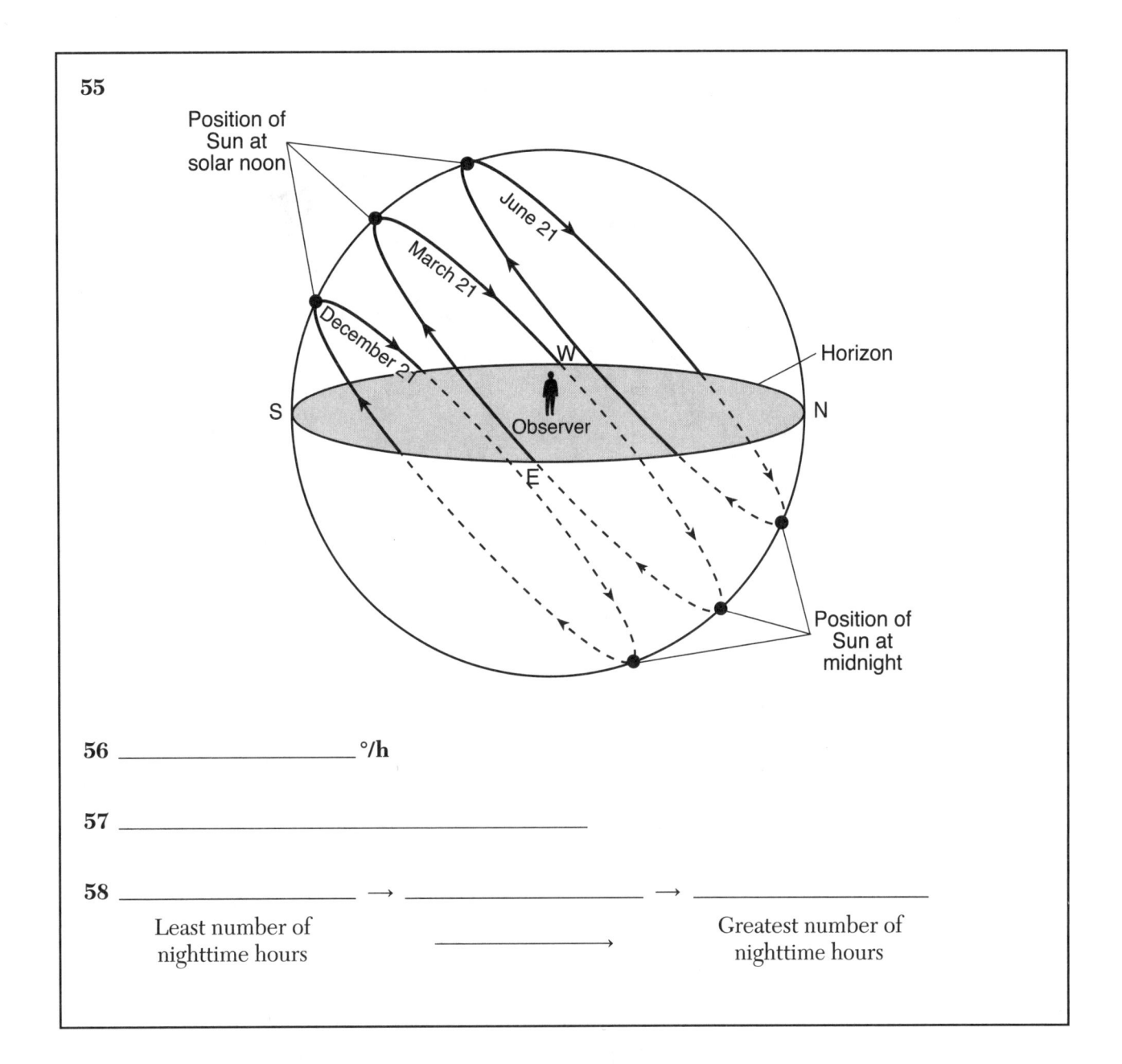

Position of Sun at solar noon

June 21
March 21
December 21

Horizon

S
W
Observer
E
N

Position of Sun at midnight

56 _____ °/h

57 _____

58 _____ → _____ → _____
    Least number of   Greatest number of
    nighttime hours   nighttime hours

59

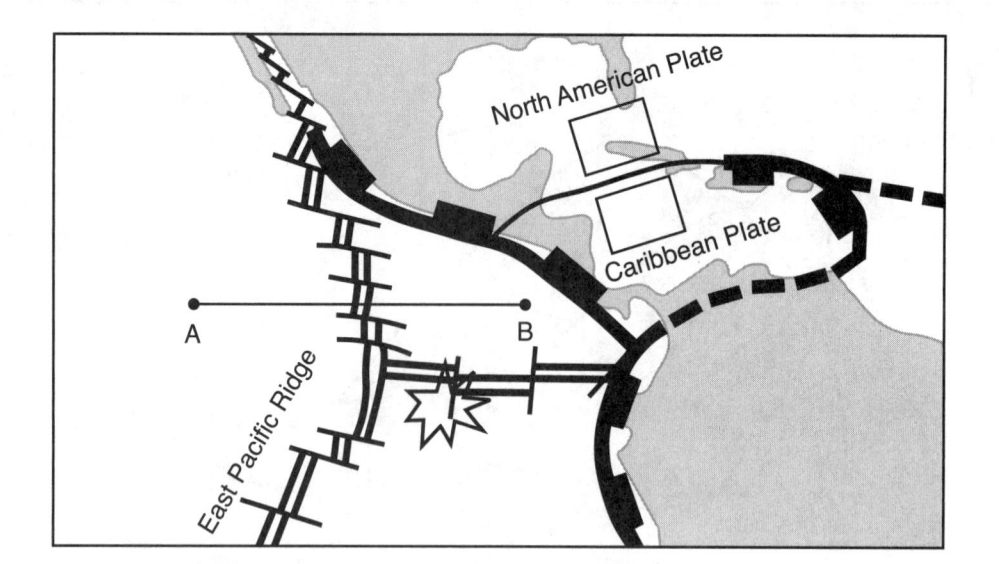

60

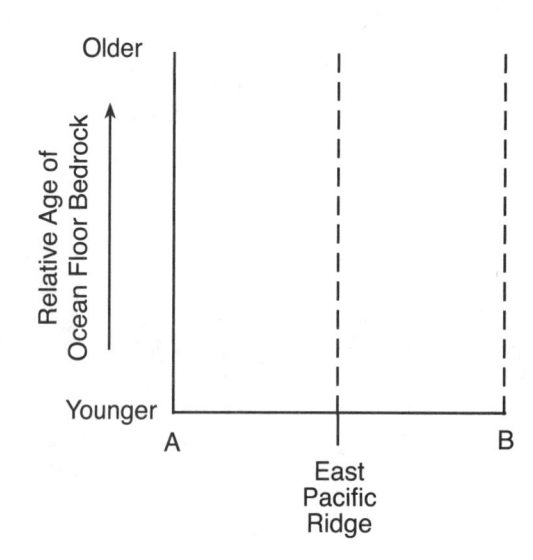

61 _____ **Hot Spot**

_____ **Plate**

62 Type of bedrock: _____

Density: _____ **g/cm³**

63 _____

_____

64 _____ cm/s

65 _____

Part C

66 _____

_____

67

Albany, New York

Weather Variable	Weather Data
Dewpoint	°F
Cloud cover	%
Actual barometric pressure	mb

68 _____

69

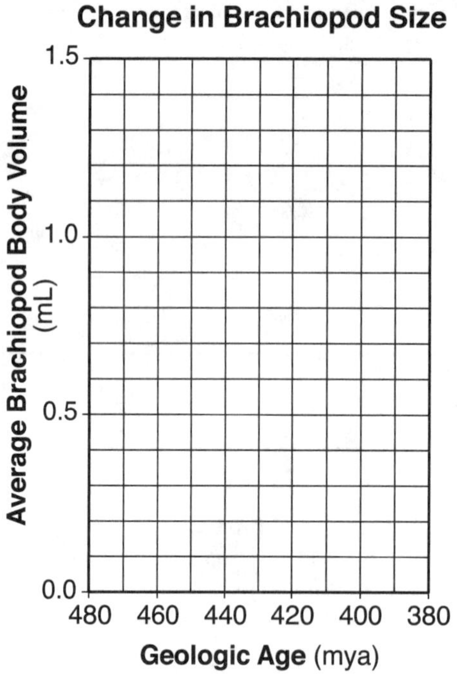

Change in Brachiopod Size
Average Brachiopod Body Volume (mL)
1.5
1.0
0.5
0.0
480
460
440
420
400
380
Geologic Age (mya)

70 ______________________ **Period** and ______________________ **Period**

71 ______________________ and ______________________

72 ____

____

____

73

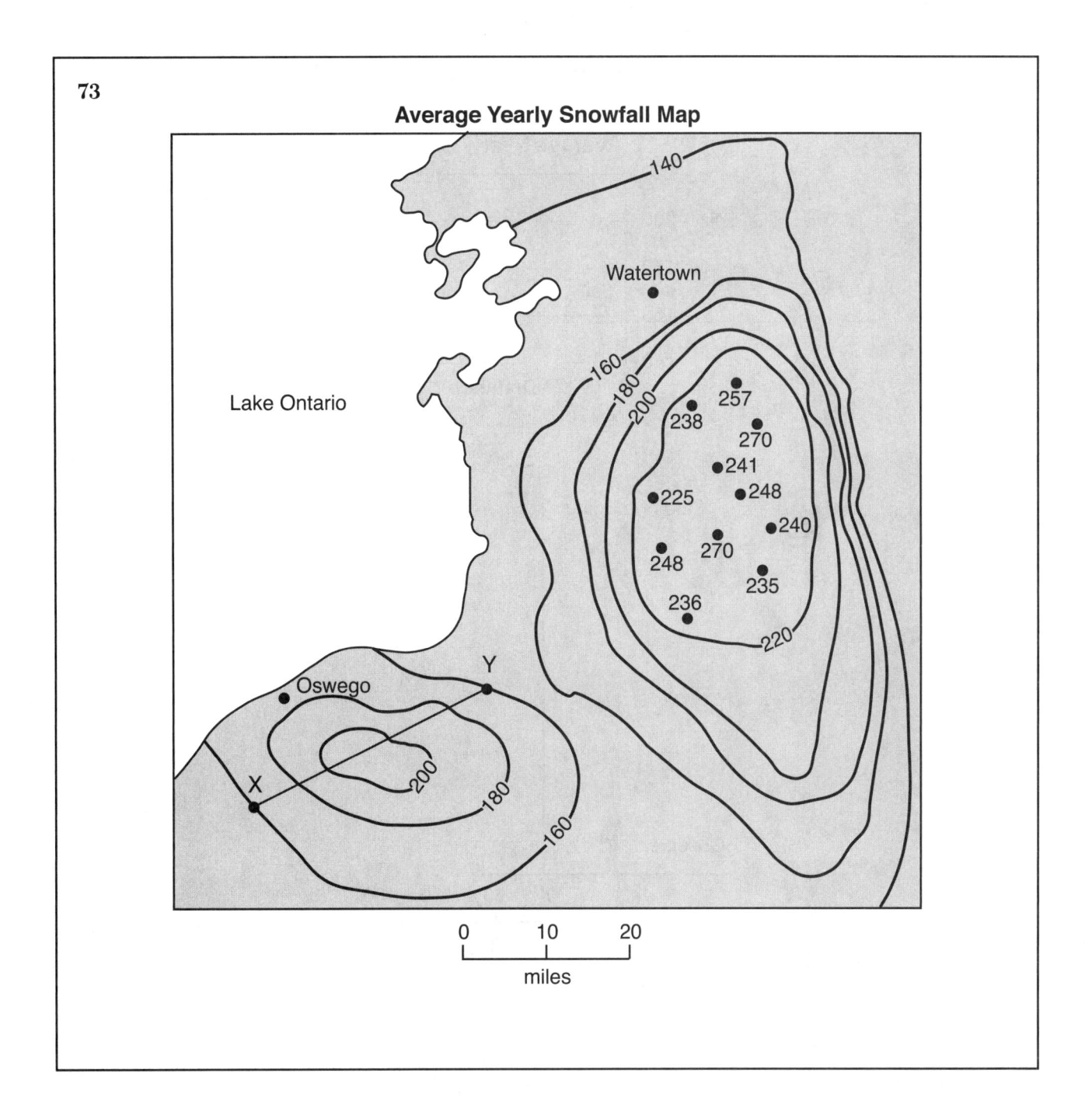

Average Yearly Snowfall Map

74

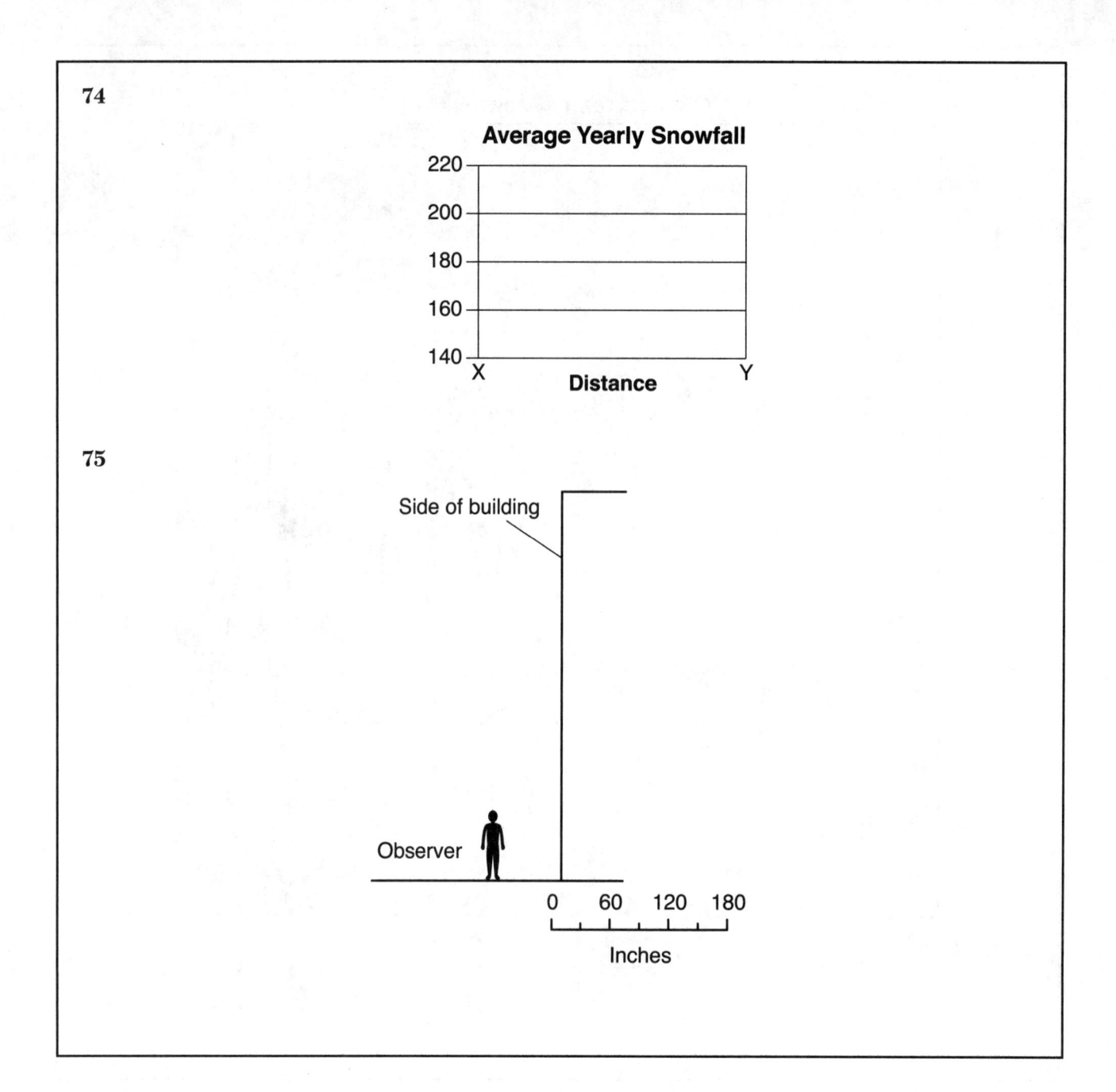

Average Yearly Snowfall

```
220
200
180
160
140
   X        Distance        Y
```

75

Side of building

Observer

0   60   120   180
Inches

76 _____

77 _____

_____

78 Emitted electromagnetic energy: _____

   Relative wavelength (circle one):   shorter   longer   the same

79 _____

80 _____

81 Name of planet: _____

   Explanation: _____

   _____

82 _____ and _____

   Explanation: _____

   _____

83 _____

   _____

84 _____ → _____ → _____ → _____
   Earliest stage   Latest stage

85 _____ **epoch**

Practice Exam 2 391